Volatiles and Food Security

Devendra K. Choudhary • Anil K. Sharma
Prachi Agarwal • Ajit Varma
Narendra Tuteja
Editors

Volatiles and Food Security

Role of Volatiles in Agro-ecosystems

Springer

Editors
Devendra K. Choudhary
Amity Institute of Microbial Technology (AIMT)
Amity University
Noida, Uttar Pradesh, India

Anil K. Sharma
GB Pant University of Agriculture & Technology
Department of Biological Sciences, CBSH
US Nagar, Uttarakhand, India

Prachi Agarwal
Ministry of Science and Technology
Biotechnology Industry Research Assistance Council
New Delhi, India

Ajit Varma
Amity Institute of Microbial Technology (AIMT)
Amity University
Noida, Uttar Pradesh, India

Narendra Tuteja
Amity Institute of Microbial Technology (AIMT)
Amity University
Noida, Uttar Pradesh, India

ISBN 978-981-10-5552-2 ISBN 978-981-10-5553-9 (eBook)
DOI 10.1007/978-981-10-5553-9

Library of Congress Control Number: 2017951767

Printed on acid-free paper

This Springer imprint is published by Springer Nature
The registered company is Springer Nature Singapore Pte Ltd.
The registered company address is: 152 Beach Road, #21-01/04 Gateway East, Singapore 189721, Singapore

Preface

One major challenge for the twenty-first century will be the production of sufficient food – the United Nations Population Fund estimates that the global human population may well reach ten billion by 2050 (www.unfpa.org). This means increasing agricultural productivity of food crops, as plants form the basis of every food chain. If global food production is to keep pace with an increasingly urbanized and growing population, while formulating new food production strategies for developing countries, the great challenge for modern societies is to boost plant productivity in an environmentally sustainable manner. The task of providing food security to our country's burgeoning population is becoming increasingly difficult. This challenge must and needs to be met in the face of the changing consumption patterns, impacts of the climate change and degradation of the finite land and water resources. Management of land resources, in general, and potentially culturable lands, in particular, encompasses crop production methods that will keep pace with a country's food needs, sustaining environment, blunting impacts of climate change, preserving and enhancing natural resources and supporting the livelihood of farmers and rural population in the country. Thus, there is a pressing need for enlarging area under arable lands, by the way of reclaiming degraded lands for sustainable intensification of agriculture, in which crop yields can be increased without compromising and yielding to adverse environmental impacts and without reducing area under forests. The science of crop management and agricultural practices suited to lands exposed to different stresses at present demands a specific orientation for meeting challenges of food insecurity. In this scheme of agricultural development, effective utilization, rejuvenation and management of degraded and wastelands by public and private investments become imperative. In addition to the type and the extent of degradation the lands have undergone or are undergoing, appropriate management strategies need to be designed and implemented in a defined time frame to bring these lands to 'productive health'.

Over the last decade, bacterial volatile emissions under stress and their possible practical applications on plant growth have received increasing attention. We determined bacterial VOC-mediated antifungal activity using NA + PDA agar plates. As shown in the Fig. 1 bacterial culture, SJ-5 was found to be antifungal in nature as it strongly inhibited the growth of fungus mycelia 2 days after inoculation. The same result was obtained in the volatile plate assay, and volatiles produced by SJ-5 inhibited the growth of mycelium.

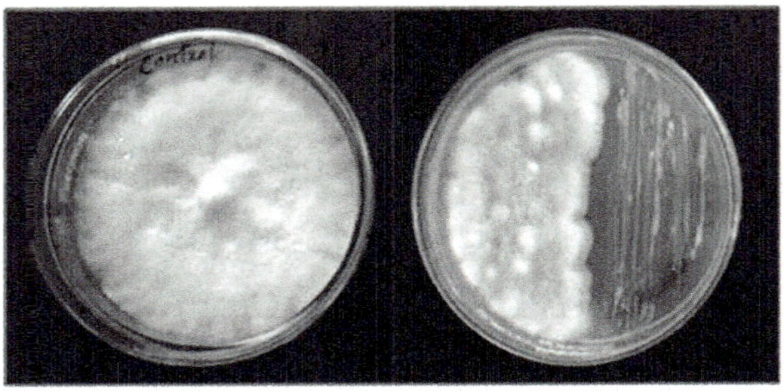

Fig. 1 Volatile activity of *Bacillus* sp. SJ-5 against *Fusarium oxysporum*

Fig. 2 Effect of bacteria-mediated volatiles on mung bean seed germination under 100 mM NaCl. Left plate, without strain, whereas right plate, with bacterial strain AU. Right plate showed significant radical growth in the presence of bacterial VOCs

In continuation of VOC production, *Pseudomonas simiae* AU (MTCC no. 12057) was grown on nutrient agar media for 24 h before plant experiment and scraped into sterile, distilled water. The liquid suspension culture was diluted with water to yield 10^9 colony-forming units/ml. Bacterial culture were inoculated into one compartment of a petri plate containing half Murashige and Skoog medium amended with 0.8 % agar and 1.5 % sucrose. *Vigna radiata* (mung bean) seeds were surface sterilized with 0.1 % $HgCl_2$ for 1 min and 70% ethanol for 3 min and washed three times after both treatments with Milli-Q water (Millipore, Germany) and transferred to the other compartment of the petri plate. Plates were sealed with Parafilm and arranged in a completely randomized design in a plant growth chamber for 7 days at 27 °C. There were two experimental groups with three replicates (control +100 mM NaCl, without bacterial inoculation; AU inoculated +100 mM NaCl, with bacterial culture inoculation). Seedlings were harvested after 7 days of incubation period and total plant fresh weight, length and protein content were measured (Fig. 2). The sphere of microbial VOCs for rhizobacteria could be within the soil or above ground, and the possibility existed that VOCs are produced at sufficient levels for aerial tissues to perceive and respond to bacterial volatiles. It is interesting to analyse whether VOCs are by-products of various plant processes or they are actively produced and used as a sophisticated 'language' by plants to pursue communication with other organisms. Plants are capable of disseminating information to their environment by employing VOCs, and plants have the capacity to change

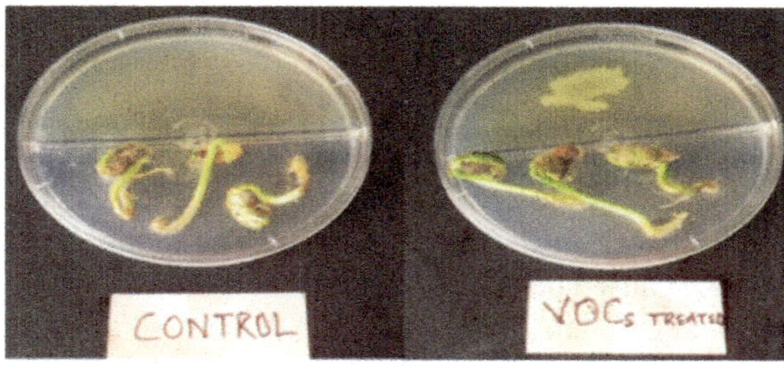

Fig. 3 Effect of *Pseudomonas simiae*-mediated volatile treatment on soybean seed germination in bipartite petri plates (I-plates). Left plate devoid of strain showed stunted growth, whereas right plate with strain exhibited significant growth

the growth condition employing reactive VOCs. No doubt plants have evolved with the capacity to release and detect VOCs in their environment; the emission of plant odours transmits signals to other organisms and members of its own species.

Case Study Authors and team have performed experiments on bacterial VOC-mediated IST in soybean.

Plant root-associated rhizobacteria elicit plant immunity referred to as IST against multiple abiotic stresses. Among multibacterial determinants, bacterial VOCs that induce IST and promote growth are reported in this study. This research calls attention on the role played by *P. simiae* AU VOCs in salt tolerance in soybean plants directly by the reduction of Na^+ ions in the root and shoot and the induction of proline content in the root and indirectly on the expression of VSPs, GGH and RuBisCO large chain protein. Emissions of AU VOCs significantly increased shoot and root growth. This was attributed to the higher IAA content in the root and lower uptake of Na^+ ions, and the increase in chlorophyll content was related to the up-regulation of chloroplast-specific protein RuBisCO. Furthermore, the up-regulation of GGH protein in salt stress indicates its role in plant metabolism and development through folate homeostasis. Additionally, overexpression of VSPs induced by salt stress suggests its role in salt tolerance by regulating Na^+ homeostasis, maintaining phosphorus content through acid phosphatase activity. Hence, the selected PGPR *P. simiae* have great potential for improving crop yield. Firstly, seed germination was checked on a bipartite petri plate containing half MS incorporated with 100 mM NaCl in one partition and King's B inoculated with strain AU in another partition. All seeds were germinated but could not grow more because of space limitations (Fig. 3).

Over the last decade, bacterial volatile emissions and their possible practical applications on plant growth have received increasing attention. The PGPR strain *P. simiae* AU was previously found to promote seedling growth of *Glycine max* through its plant growth promotion activities like IAA, siderophore production, phosphate solubilization and ACC deaminase activity under salt stress condition. Herein, the study reported for the first time that *P. simiae* produces a volatile blend that can enhance soybean seedling growth and elicit IST against 100 mM NaCl stress condition. To overcome the difficulty encountered with plants of larger size than *Arabidopsis* for studying the effects of bacterial VOCs, in a petri plate, a new VOC assay system, based on a magenta box, was developed. Using this test system, we found that IST elicited by VOCs released from strain AU is mediated by the induction of proteins in plant tissues. After 10 days of seedling growth, the shoot and root length were significantly higher in soybean seedlings treated with strain AU VOCs. It is reported that NaCl caused the depletion and precipitation of available phosphorus that caused restricted growth and impaired the delivery of available phosphate in plant cells. In this study, VOC-treated seedlings showed higher phosphate content in salt stress than non-treated seedlings. It may be caused by the acid phosphatase activity of VSPs that release soluble phosphate from their insoluble compounds inside the cell.

A greater understanding of how plants and microbes live together and benefit each other can therefore provide new strategies to improve plant productivity while helping to protect the environment and maintain global biodiversity. To date, the application of chemicals to enhance plant growth or induce resistance in plants is limited due to some negative effects of chemical treatment and difficulty in determining the optimal concentrations to benefit the plant. For alternative means to solve these problems, biological applications have been extensively studied. Collectively, our description on VOCs eliciting growth promotion suggested that these compounds could be an environmentally sound means to grow and protect plants under greenhouse or field conditions better. From the whole plant perspective, it still remains to be determined whether growth promotion by microbial VOCs occurs in soil or soil-less media. To understand the nature of VOCs and gene expression profiling of plant genes, studies of these compounds can be conducted. It is possible that VOCs produced by microbes while colonizing roots are generated at sufficient concentrations to trigger plant responses. In conclusion, positive or negative effects of VOCs on plant productivity will be dependent upon their specific microbial strains, plant genotype and the presence/absence of abiotic/biotic stresses.

Economic Implications of the Proposed Work

The proposal will have the following long-term socio-economic impact:

- The innovative microbial strategies developed will be employed to plants grown under pathogen-conducive stress in the soil.
- The environment will also benefit from reduced fertilizer and pesticide use.
- The stakeholder will be able to exploit the inoculants developed.
- A dissemination of the proposed book will ensure that the whole agricultural community can benefit from the project.

Hence, in this book, editors compiled researches carried out on volatiles produced by microbes and plants along with their biotechnological implications for sustainable agriculture.

Chapter 1 summarizes the role of plant VOCs produced in various tissues against stresses regarding herbivores, plant viruses, pathogens, temperature, humidity, light ozone, food usability, etc., and their implication for physiological processes such as plant development, seed formation and germination, pollination and fruit ripening.

Chapter 2 briefly describes the VOCs released by bacteria in the air that interact with their surrounding environment. Soil bacterial volatiles are known to contribute to plant interactions, and several studies also identified their influence on plant stress tolerance. This chapter describes the characterization of different bacterial VOCs and their roles in enhancing plant abiotic stress tolerance, a new research area, with potential agriculture applications.

Chapter 3 emphasizes the insight of the phytoextraction and phytovolatilization mechanisms that are involved in the decontamination of the soil. Phytoremediation is a green emerging technology used to remove pollutants from environment components.

Chapter 4 describes techniques like basic chromatography and mass spectrometry to understand the chemical structure and function of plant and microbial VOCs. In addition to this, modern OMICS methods give opportunity to a deep insight of microbial diversity and strengthen the concept of volatile compound function by providing real-time pictures of their expression and signalling. Besides, the incorporation of computational tools with molecular biology techniques incredibly creates a reservoir of knowledge-based database of volatile compound structure, function, diversity, signalling and even prediction through statistical tools.

Chapter 5 highlights the diversity of VOCs present in the plant rhizosphere. The rhizovolatiles discussed here include those produced by plants as well as by microorganisms inhabiting the rhizosphere. The chapter focuses on the role of these volatiles in the establishment of a successful association between plants and other organisms and their beneficial effects on plant growth and development. This will value-add to the present understanding of the chemical cues defining the complexity and dynamism of rhizosphere functioning.

Chapter 6 highlights the current knowledge on the expression patterns and functions of some leguminous plant proteome in response to VOCs released during biotic and abiotic stresses. The biogenesis of VOCs and their functional role in plant-plant signalling and environmental and biological stress responses are highlighted. Experimental evidences revealed that a plant symbiont produces VOCs that induce resistance to phytopathological species and PGPR.

Chapter 7 describes the role of plant VOCs for direct and indirect defence against various abiotic stresses (like temperature, water stress, ozone, salt stress and heavy metals) and biotic stresses (herbivores and pathogen) with above ground and below ground impact.

Chapter 8 briefly describes the importance, chemistry and role of microbial VOCs in defence in general.

Chapter 9 highlights the chemical measures necessary for controlling plant pathogens and their negative impact on human health and/or the environment. It also defines the different sources of essential oils and their antimicrobial activity with particular emphasis on the antifungal properties exhibited against some serious pathogenic fungi and postharvest disease.

Chapter 10 briefly discusses the role of VOCs in microbial-microbial and microbial-plant interactions. The effect of VOCs as inducers for enhancing crop productivity is reviewed. Problems associated with field applications are also highlighted.

Chapter 11 summarizes volatile-mediated interactions in a microbial community mimicking the natural conditions of a heterogeneous soil environment along the rhizosphere as well as the biological and ecological significance of VOC-mediated resistance.

Chapter 12 describes the role of MVOCs wherein they can be exploited as eco-friendly, cost-effective, disease-resistant and sustainable strategies for agricultural practices and aims to provide a comprehensive discussion on below and above ground interactions of microbial volatile diversity and their role against pathogenic fungi.

Chapter 13 elaborately describes the intricate relationship of plant, herbivore and carnivore in this tritrophic ecosystem to determine whether evolution plays any role and to calculate the overall ecological cost to govern this complex machinery of life.

Chapter 14 focuses on the diversity of MVOCs and further discusses their potential in exploiting these bioactive molecules in sustainable eco-friendly agriculture for improving plant growth, production and protection.

Chapter 15 elaborately describes the functionality of herbivore-induced plant volatiles (HIPVs) in communicating with the parasitoids to prey upon attacking herbivores. HIPVs not only help plants to interact with the natural enemies of herbivores but also warn the neighbouring plants of the imminent danger. Thus, HIPVs provide a reliable mechanism for natural control of insect pests.

Chapter 16 describes PGPR-emitted VOCs that can directly and/or indirectly mediate increases in plant biomass, disease resistance and abiotic stress tolerance. Bacterial VOCs promote plant growth by eliciting different hormone signalling pathways. In particular, the volatile components 2,3-butanediol and acetoin were released exclusively from two bacterial strains that trigger the greatest level of growth promotion and induced resistance against fungal pathogens. This chapter focuses on recent research studies and the role of bacterial volatiles in plant growth promotion and protection against pathogens.

Chapter 17 presents an overview of current insights of fungal VOCs on growth and development and the defence system of plants. Numerous fungal VOCs contribute to dynamic processes, leading to myriad interactions between plants, antagonists and mutualistic symbionts. For better understanding of the role of fungal VOCs at field level, more studies will offer further constructive scientific evidences on cost-effective, eco-friendly, and ecologically produced fungal VOCs for crop welfare.

Chapter 18 describes the role of MVOCs as an alternative strategy in lieu of chemicals to protect plants from pathogens which provides a setting for better crop welfare. MVOCs can modulate the physiology of plants and microorganisms and thus can be exploited as eco-friendly, cost-effective, and sustainable strategies for agricultural practices.

Chapter 19 envisages the role of MVOCs with special emphasis on plant defence and health, alluding to the potential of these compounds to control various processes that result in plants' elevated health and defence.

Finally, we would like to express our gratitude to the contributors upon their consent to be a part of this book.

Noida, Uttar Pradesh, India — Devendra K. Choudhary
Pantnagar, India — Anil K. Sharma
New Delhi, India — Prachi Agarwal
Noida, Uttar Pradesh, India — Narendra Tuteja
Noida, Uttar Pradesh, India — Ajit Varma

Contents

About the Editors

Devendra K. Choudhary Dr. Choudhary has 15 years of experience in microbial ecology and is currently working as an associate professor at Amity University, Noida. Dr. Choudhary has worked on GOI-sponsored major projects as principal investigator. As an active researcher, Dr. Choudhary has published more than 50 research and review articles along with several book chapters for reputed journals and edited books. Dr. Choudhary has been recognized as member of the National Academy of Sciences (MNASc).

Anil K. Sharma Dr. Sharma has completed his Ph.D. (1988) in the area of mycorrhiza and rhizosphere biology with special emphasis on basic and applied aspects in bioinoculants, composting and microalgae research. Upon completion of his Ph.D., he went abroad for postdoctoral research work as research associate at the Department of Biological Sciences, Grambling State University, Louisiana, USA (November 1992–May 1994); visiting scientist at the Department of Botany, University of Basel, Basel, Switzerland (August 2003–November 2003); and visiting scientist at the Department of Environment Sciences, University of Helsinki, Lahti, Finland (September 2013). Dr. Sharma has worked on more than ten international and GOI-sponsored major projects as principal investigator. He has published more than 100 research review papers in reputed journals with impact factor.

Prachi Agarwal Dr. Agarwal presently works as 'programme manager' with Biotechnology Industry Research Assistance Council, a Govt. of India Enterprise, Ministry of Science and Technology, New Delhi (since June 2012). Dr. Agarwal has pursued Ph.D. in energy and environment (2004–2009), from the Indian Institute of Technology Delhi, and M.Sc. in environmental science (2002–2004), from the University of Agriculture and Technology, Pantnagar. Dr. Agarwal has published more than 20 research and review articles along with several book chapters for reputed journals and edited books.

Narendra Tuteja Dr. Narendra Tuteja, presently acts as director at the Amity Institute of Microbial Technology (AIMT), Amity University Uttar Pradesh (AUUP), Noida. Before joining AIMT, he spent years as group leader and senior scientist at the International Centre for Genetic Engineering and Biotechnology

(ICGEB), New Delhi. Based on his contribution to science and society, he has been elected as fellow for various science academies in India/abroad: fellow of 'Third Word Academy of Sciences', Trieste, Italy (FTWAS); fellow of 'National Academy of Agricultural Sciences', New Delhi (FNAAS); fellow of 'National Environmental Science Academy', New Delhi (FNESA); fellow of 'Indian Academy of Sciences', Bangalore (FASc.), fellow of 'Indian National Science Academy', and fellow of 'National Academy of Sciences', Allahabad (FNASc.).

Ajit Varma Dr. Varma currently is distinguished scientist and professor of eminence at the Amity Institute of Microbial Technology (AIMT), Amity University Uttar Pradesh (AUUP), Noida. Dr. Varma has successfully completed major projects as PI sponsored by DBT, DST, DRDO and ICAR. He has supervised more than 69 Ph.D. students and published over 374 research articles for national and international journals of repute, as well as several major review articles and chapters in books. He has published 83 books in the area of microbial technology and has been the series editor for Springer-Verlag's series on soil biology and has edited 50 volumes on *Soil Biology*. Dr. Varma has been a member of the National Academy of Agricultural Sciences (FNAAS) International Society of Symbiosis, Boston, USA; and Indian Science Congress Association, Executive Council, Amity University Uttar Pradesh.

Plant Volatile Compounds in Growth

1

Demet Altındal and Nüket Altındal

Abstract

Plants synthesise volatile organic compounds (VOCs) in various tissues against stresses regarding herbivores, plant viruses, pathogens, temperature, humidity, light ozone, food usability, etc., and for physiologic processes such as plant development, seed formation and germination, pollination and fruit ripening. These compounds are synthesised in all parts of plants, especially flowers, fruits, roots, xylems and cells, and just as they may be effective in the tissues they are produced, they may be transferred to other parts of the plants and show their effect there.

Plants communicate with living things around them by emitting numerous different volatile compounds. They develop morphological and physiological defence mechanisms by repulsing or attracting their enemies with these compounds. Plants store these compounds produced for defence and release them in the form of volatile gases when needed. Plant volatile compounds include isoprene, terpene, fatty acid derivatives, alcohols, esters, volatile oils, plant development regulators (abscisic acid, auxin, cytokinin, etc.), phenolic compounds and secondary metabolites.

Keywords

Plant • Volatile compounds

D. Altındal (✉)
Fethiye Ali Sıtkı Mefharet Koçman Vocational School, Department of Crop and Animal Production, Organic Farming Programme, Muğla Sıtkı Koçman University, Muğla, Turkey
e-mail: daltindal@gmail.com

N. Altındal
Sivaslı Vocational School, Department of Crop and Animal Production, Programme of Medical and Aromatic Plants, Uşak University, Uşak, Turkey

D.K. Choudhary et al. (eds.), *Volatiles and Food Security*,
DOI 10.1007/978-981-10-5553-9_1

Headings and Heading Numbering

1. Plants produce volatile organic compounds (VOCs).
2. Plants can communicate by releasing volatile compounds.
3. Plant volatile compounds act in physiological processes such as growing, development, seed formation and germination.
4. Plants release volatile compounds they produce for the morphological and physiological defence mechanisms they develop.

1.1 Introduction

Plants, which are sources of food for bacteria, fungi, viruses, insects, nematodes, humans and animals, have developed defence mechanisms against different stress factors, and they regulate these mechanisms by releasing volatile organic compounds (VOCs). Plant volatile compounds may be classified as hydrocarbons, alcohols, aldehydes, ketones, ethers and esters based on their chemical structures. In botany, plant volatiles are usually classified into volatile terpenes, volatile phenylpropanoid/benzenoid and volatile fatty acid derivatives based on their varying biosynthetic functions (Dong et al. 2016).

Volatiles are evaporating compounds with small molecules produced by plants (flowers, fruits, vegetables, herbs, etc.). There are more than 2000 known volatiles in plants to date. Each volatile compound has a different smell. β-Ionone volatile compound is produced in floral, woody, sweet, fruity berry and greens, while dimethyl disulphide sulphurous substance is produced by vegetables, onions, garlic and leeks, and myrcene substance is produced by peppery, spicy berry and plants (Peter 2012). Natural aroma or smell consists of hundreds of volatile compounds.

During their growth and development, plants are exposed to biotic and abiotic stress factors, and thus they develop defence systems. Among these stress factors, herbivores, plant viruses and pathogens are biotic stresses, while temperature, humidity, ozone and food usability are abiotic stresses. Volatile organic compounds (volatiles) are synthesised in different plant tissues. In order for plant volatiles to be released into the atmosphere, secretory cells in sweet basil can release phenylpropanoids, or epidermal cells in mint can release volatiles such as p-methane, diterpenes and monoterpenes (Kant et al. 2009). The places plastid and cytosol volatiles are synthesised are plant cells. These volatiles are secondary metabolites whose place of synthesis is determined biochemically, and they are classified as terpenoids, phenylpropanoids/benzenoids, fatty acid derivatives and amino acid derivatives based on their source of production. Volatile compounds in plants are transferred from primary and secondary roots to leaves, flowers, fruits and the body by way of transpiration. These compounds are released from the living secretion tissue with cells that have abundant cytoplasms and big nuclei.

Table 1.1 Functions of plant volatile compounds in some plants

Process/stress/ interactions	Plant	Volatile compounds	References
Plant growth	*Arabidopsis thaliana*	Brassinosteroid	Zhang et al. (2015)
	Arabidopsis	Abscisic acid, gibberellins	Li et al. (2016)
			Procko et al. (2014)
	Arabidopsis, *Brassica rapa*	Auxin	Altındal and Altındal (2013)
		Sage oil and thyme oil	
	Sainfoin		
Ripening	Maize	Ethylene	Louis et al. (2015)
	Tomato	Ethylene	Kim et al. (2015)
Herbivores	Tomato	Alkaloids, phenolics	Reisenman and Riffell (2015)
	Maize	Terpene	Fiers vd. (2013)
	Cowpea	Thyme oil	Altındal and Altındal (2011)
Pathogens	Tobacco	Flavonoids, phenolics	Büchel vd. (2015)
	Barley	Methyl salicylate	Shulaev vd. (1997)
Pollination	Entomophilous plants	Terpenes and benzenoid	Farre-Armengol vd. (2015)
	Monkey flower	Monoterpenes	Byers et al. (2014)

Plant volatiles which can be stored in the plant cell in liquid form are organic molecules that can evaporate in contact with air and form compounds with carbon atoms. VOCs are exposed to air when mint leaves or pine needles are crushed and they become usable for perfumery. Again, when grass is mown (green leaves are cut), a nice smell is released.

1.2 Types of Plant Volatile Compounds

Plant volatile compounds include different organic classes such as isoprene, terpene, fatty acid derivatives, alcohols, alkanes, esters and acids. These compounds are found in various plants and they serve various physiological purposes (Table 1.1).

1.2.1 Terpenes

They are a broad and diverse class of hydrocarbons which are produced by plants, especially Coniferales. Terpene types are hemiterpenes, monoterpenes, sesquiterpenes, diterpenes, sesterterpenes, triterpenes, tetraterpenes and polyterpenes. Some particular terpenes have well-known functions in plant growth and development; therefore, they are considered primary metabolites.

1.2.2 Alkaloids

They are naturally produced chemical compounds which have amine structures. Alkaloids are seen mostly in Sclanaceae and Papaveraceae plant families and found less in Rosaceae, Graminaceae and Labiatae families. The highest amounts of alkaloids are found in plants' roots and leaves, while the lowest amounts are found in skins, seeds and stems. Almost all alkaloids have a bitter, hot taste and they are odourless. Some important alkaloids in plants are piperidine, pyridine and tetrahydropyridine alkaloids.

1.2.3 Glycosides

They are formed by the combination of glycosidic hydroxyl (−OH) in sugar with another substance containing OH such as cellulose, polysaccharide, phenol and extraction of water. These are essential compounds. Glycosides are found extensively in plants. Glycosides with their bitter and hot taste are probably effective for protection against ruminants, and they act in regulation of various biochemical processes. While glycosides exist in plants' leaves too, they are usually found in fruit shells and roots.

1.2.4 Essential Oils

Essential oils in plants are essential compounds for plants as they provide transfer of all chemicals to cells and increase the plants' resistance to diseases and pests by strengthening their defence mechanisms. While they form the structure of hormones in plant cells, they play roles in transfer of information among cells and plants' defence mechanisms.

1.2.5 Plant Volatile Oils

Volatile oils are oil-like mixtures that are obtained from plants and herbal drugs by various methods which are in liquid form in room temperature, crystallisable, pungent and towable by water vapour. The smell of several aromatic, fragrant and scented plants comes from their content of active ingredients (volatile oils). Volatile oils exist in plants' rhizome, body and shells, leaves, fruits and flowers. Many plant volatile oils and monoterpenoid compounds are rapidly degradable in nature, and they do not accumulate in the body or the environment. So, they are used to fight against insects (Tunaz et al. 2009).

Limonene, 1,8-cineole, camphor, linalool, citronellol, citronellal and anethole are chemical compounds with known insect repellent activities, and they are found in volatile oils obtained from plants. Table 1.2 shows volatile oils as insect repellents and their active ingredients.

Table 1.2 Volatile oils as insect repellents and their active ingredients (Öz 2013)

Volatile oil	Active ingredients
Eucalyptus oil	Globulol, epiglobulol, b-pinene, a-pinene,1,8-cineole, limonene, terpinen-4-ol, aromadendrene, piperitone and a-phellandrene
Cedar oil	Cedrol, thujopsene (a ketone), widdrol, cedrene, copaene, thujaplicin, methyl thujate and thujic acid
Fennel oil	Myrcene, a-pinene, trans-anethole, fenchone, anisic aldehyde, limonene, 1,8-cineole and methyl chavicol
Clove oil	Alcohols, phenols (eugenol, acetyl-eugenol), a small amount of esters, ketones and sesquiterpenes (a and b caryophyllenes)
Laurel oil	Myrcene, methyl chavicol, a-terpineol, neral, geranyl acetate, eugenol, chavicol, limonene, linalool, a-pinene and b-pinene
Thyme oil	Camphene, thujone, a-terpinene, b-pinene, p-cymene, borneol, linalool, carvacrol, thymol, b-caryophyllene and a-pinene
Lavender oil	Limonene, 1,8-cineole, cis-ocimene, 3-octanone, camphor, linalyl acetate, terpinen-4-ol, a-pinene, lavandulyl acetate, trans-ocimene, caryophyllene and linalool
Mint oil	Menthol, menthone, menthofuran, limonene, b-pinene, a-pinene, methyl acetate, germacrene-d, trans-sabinene hydrate, isomenthone, 1,8-cineole and pulegone
Basil oil	Limonene, linalool, camphene, a-pinene, b-pinene, myrcene, cis-ocimene, geraniol, camphor, methyl chavicol, eugenol, y-terpineol, camphor, methyl cinnamate and citronellol
Rosemary oil	Borneol, 1,8-cineole, camphor, bornyl acetate, a-pinene, b-pinene, camphene and limonene

1.2.6 Secondary Metabolites

Secondary metabolites produced by plants and assumed to be waste outputs are products of highly complicated mechanisms developed for defence, protection, adaptation and proliferation. All plant metabolisms have phenolic compounds in different quantities and qualities which are considered to act in plants' protection against some pests.

1.2.7 Phenolic Compounds

All plants create phenolic compounds as secondary metabolites in their metabolisms, roles of which are not sufficiently known. This is why all foods of plant origin contain phenolic compounds in different quantities and qualities. Phenolic compounds in plants are phenolic acids (or phenol carbonic acids), flavonoids and compounds with small molecules that are mostly volatile.

1.2.8 Ethylene

It is a colourless alkene gas and an unsaturated hydrocarbon. The reaction capability of ethylene is very high. Plants produce ethylene as a reaction to stresses such as drought, floods, mechanical pressure and infections. Ethylene acts in processes of fruit ripening, conversion of polysaccharide into sugar, defoliation, seed germination and budding.

1.3 Volatile Compounds in Plant Development

Substances that regulate plant development are volatile organic compounds produced by the plant; they regulate growth, development and other physiological processes, and while they can be effective in the parts they are produced, they can be transferred to other parts of the plant and utilised there. Plants produce these fundamental substances needed for their growth, development and transformation. These substances that are produced in the plants' body and regulate growth and development (physiological processes) are called hormones or phytohormones (plant hormones). There are several phytohormones. They are classified in five main groups as gibberellin (GA), abscisic acid (ABA), auxin, cytokinin and ethylene based on their similarities and effects.

1.3.1 Volatile Compounds in Seed Germination and Vegetative Development

Abscisic acid (ABA) and gibberellins (GA) are well-known phytohormones that act in seed germination and regulation. These two hormones affect germination in opposite ways. While ABA inhibits germination, GA catalyses this biological process (Li et al. 2016).

Gibberellins (GA) play an important role in eradicating seed and bud dormancy, as well as controlling and stimulating seed germination. They are found abundantly in developing seeds. Their amount is lower in more developed seeds of especially dicotyledon plants. GA provides that the enzymes taking part in this process are stimulated and play a role in the conversion of polysaccharide into sugar by moving from the embryo to endosperm and stimulating the α-amylase enzyme to provide energy (Zadeh et al. 2015). GA stimulates the growth potential of the embryo and weakens the structures surrounding the embryo. Endo-β-mannanase produced in the endosperm based on GA may help germination by providing degradation of endosperm cell walls (Ogawa et al. 2003). Naturally occurring ABA plays a role in promoting dormancy in seed germination, buds and seeds. Abscisic acid (ABA) prevents germination in seeds by sustaining dormancy and ensures that germination does not take place in arid conditions. The juglone substance that can be found in the roots and leaves of the walnut (*Juglans* sp.) plant prevents germination of some plant seeds.

Naturally occurring ABA plays a preventive role in not only seed germination but also plant growth, especially root development. It is effective in developing an adaptation mechanism against stress. The auxin hormone which produces cotyledon in plants regulates hypocotyl elongation (Procko et al. 2014). Auxins play roles in vertical growth of plants, nutation processes and defoliation. These hormone groups are produced in the plant's terminal bud, and phototropism occurs with the help of these hormones. Cytokinins regulate root and body development by working with auxins in development of seeds, and lead to production of chloroplast, thus preventing yellowing of leaves and ageing. Growth is provided by stimulation of cell division, and growth of offshoots and leaves from buds is induced. Gibberellins induce rapid and abnormal elongation of the body and prevent defoliation. Abscisic acid blocks the pores of the plant in arid conditions and increases adaptation to dry environments by preventing loss of water from leaves. It is observed that young pine seedlings develop very weak in forests containing high numbers of plant species in the Ericaceae family showing allelopathic properties.

While the level of photosynthesis decreases in high temperatures for numerous tree species, resistance to high temperature is achieved by secretion of isoprene in birch leaves (Dong et al. 2016).

1.3.2 Volatile Compounds in Bloom and Fruit Ripening

Gibberellins stimulate bloom and early bloom and promote growth of fruits. Auxins are effective in fruit ripening. ABAs lead to abscission in flowers and fruits, which increases the ripening of the fruits.

Ethylene is one of the volatile compounds that play a role in plant development. Ethylene is secreted in the ripening of climatic fruits; the volatile ethylene regulates gene expression and softens the cell wall. While changes in volatile compounds increase the quality and attractiveness of the fruits, higher ethylene production leads to faster ripening. Chlorophyll loss in the fruit changes the colour, and secondary metabolites such as carotenoids and flavonoids regulated by ethylene and ABA are produced. From the development to the ripening of the fruit, glycose and fructose accumulate, citric and malic acids are created from organic acids, which in turn establish the taste of the fruit. Volatile compounds also determine the flavour of the fruit (Gómez et al. 2014).

About 900 different volatiles were found in fruits and vegetables. Ethylene also plays an important role in flavour creation in climacteric fruits. In a study on grape plants, ester, ketone and lactone compounds emerged after the veraison period, their levels increased up to the point of harvesting, and these compounds' scent activity values were higher than those of others (Chang et al. 2015).

Amount of ester increases starting at postpollination to the postharvest ripeness. Most aldehydes are produced in high amounts in early growth and harvest periods (Beaulieu and Grimm 2001). Aldehydes, hydrocarbons, alcohols, acids and ketones are volatile chemicals which play a part in the smell and flavour of numerous fruits. Volatile esters provide genuine flavour to the fruit.

1.3.3 Volatile Compounds in Promoting Pollinators

Fragrant secondary metabolites secreted in flowers are at least as attractive as colours for many insects, birds and mammals. The most important secondary metabolites responsible for scent in flowers are monoterpenes (such as linalool, limonene, citronellol and geraniol), sesquiterpenes (such as α-bisabolol, β-ionone and farnesol), aromatics (such as vanillin, eugenol and methyl eugenol), aliphatics (such as pentadecane and octanol), monoamines (such as methylamine, ethylamine and propylamine), diamines (such as putrescine and cadaverine) and indole alkaloids (such as indole and skatole). Pollinator insects are highly sensitive about different scent components. Using the scents that attract these insects as pheromones, for example, spraying the plants that are highly dependent on cross-pollination with these chemicals in time of flowering, may increase the amount of pollinator insects, and products can be obtained with higher efficiency and quality.

Plants attract pollinator insects and achieve reproduction with cross-pollination. More than 85% of flowery plants need insects for the pollination to happen. Studies focus on flower scents and plants that attract pollinators. Bees act as pollinators based on smell, rather than flower colour. They also communicate with other bees and find hives based on their sense of smell, instead of following flower colours.

Flowers on plants produce volatile compounds in varying and high amounts for pollination, and benzenoids are found in flower scents as VOCs.

Nectar, which contains volatile primary metabolites (sugars and amino acids) and secondary metabolites (alkaloids, phenolic amino acids that are not proteins with attractive and repulsive properties), is an important component.

1.4 Volatile Defence Against Herbivore Insects and Fungal Pathogens

Plants may be subjected to natural predators such as pests, nematodes, pathogens, fungi, bacteria and viruses, which decrease their yield and quality. Plants have developed various defence mechanisms against these biotic stresses.

While there are still not enough studies about the effective mechanisms of volatile oils and their components which are effective against pests in general, there are some studies about defence mechanisms that prevent feeding or have insecticidal effects (Roy and Das 2015; Shahab-Ghayoor and Saeidi 2015).

Secondary compounds found in plants (volatile oils, alkaloids, glycosides, etc.) are used as raw material for drugs, in addition to being found effective against pests. In the studies conducted (Wang et al. 2015; Degenkolb and Vilcinskas 2016), it was reported that some plants counteract and even kill some pests (nematodes) with the substances they secrete. Repellents which have an important place in biological struggle may be classified in two groups as natural and synthetic. It is reported that some significant natural repellents are *Pyrethrum*, *Artemisia* and *Mentha* species and secondary compounds like volatile oils. Secondary compounds in plants are studied in two groups as repellents and deterrents.

More than 2000 plant species have natural insecticide properties. For example, the pyrethrin substance found in Pyrethrum (*Chrysanthemum cinerariaefolium*) flowers has been used as insecticide material since the nineteenth century. Europeans learnt in the beginning of 1800s that dried flower powders of the Pyrethrum plant used by African natives as an insecticide was significant for Persians, and it was exported from Europe to the United States in the beginning of the twentieth century. These flower powders have been used by Westerners as a source of pyrethrin to produce fly spray for 160 years.

Today, another natural insecticide which is a tropical and subtropical plant is neem tree (*Azadirachta indica*), whose plants are used as a source of azadirachtin. Fruits, seeds and leaves of this plant contain various compounds which kill insects, prevent fungi from growing and developing and restrict the infective abilities of plant viruses.

Again, the volatile oil rich in β-asarone which is derived prom sweet flag plant (*Acorus calamus*) leads to infertility in insects and shows insecticidal properties. Methyl jasmonate is secreted from the wounded tissues of plants attacked by insects and animals, and it leads healthy plants which receive the smell to take precautions against possible attacks by insects and animals. The hordenine alkaloid found in barley (*Hordeum vulgare*), carvacrol found in oregano (*Oreganum* sp.) volatile oil and 1,8-cineole found in eucalyptus (*Eucalyptus* sp.) volatile oil are allomones which have strong allelopathic effects on other plant species.

Combined secretion of terpenoids, phenylpropanoid compounds and acid derivatives with volatile oils shows indirect defensive effects, while proteinase inhibitor, polyphenol oxidase activation and peroxidase play a direct role in the defence mechanism (Huang et al. 2015).

Terpenes play a defensive role in many plants against herbivores, and they are dissuasive for many herbivore mammals and insects because of their toxic effects. This is why they have important roles in plant defence. For example, monoterpene esters known as pyrethroids which are found in leaves and flowers of *Chrysanthemum* species show dramatic insecticidal effects. As they are unstable in nature and they have a negligible toxicity effect on mammals, both natural and synthetic pyrethroids are popular compounds used in commercial insecticides.

Monoterpenes accumulate in resin canals in needles, branches and bodies of coniferous plants like pines and firs, and these compounds have toxic effects on numerous insects including Scolytidae which give great harm to coniferous plants worldwide.

Certain mono- and sesquiterpenes in maize, cotton and wild tobacco are produced and secreted after the insects feed. These chemicals not only repel herbivorous insects that leave their eggs on the plants, but they also attract predatory and parasitic insects that feed on those herbivorous insects. Therefore, they minimise the potential harm, and their functions are not limited, in that they provide help for plants from other organisms.

While essential oil attracts insects and facilitates pollination, it also disperses sun beams by evaporating in the air to protect the plant from heat. In some cases, essential oils act as protective agents against harmful rodents, insects and bacteria which bring disease. Essential oils are chemically heterocyclic, hydroaromatic and fragrant combinations. Just as in fatty combinations, essential oils are composed of carbohydrates, thio acids, phenols, aldehydes, ketones, alcohols, carbonic acids, esters and especially terpenes (mono-, sesqui-, dipolyterpenes), besides causing their scent as oxygen combinations.

As essential oils are effective against pests in an inhalative way because they are composed of volatile compounds, studies focus on fumigant effect mechanisms and mostly storehouse pests (Altındal and Altındal 2011; Hamza et al. 2016; Ja-Eun et al. 2016). These implementations are achieved by fumigation of essential oils in an enclosed environment for a certain time.

Phytoncides are of the volatile essential oil branch (though some are not volatile) and they have a role to kill microbes. They carry antibiotic properties and eliminate many harmful and vector microbes and viruses in the air. They are found in almost all plants in different chemical structures. Particularly some vegetables like onions and cabbages carry significant amounts of phytoncides. Phytoncides are also found in plants such as garlic, horseradish (*Armoracia rusticana* L.), lemon and milfoil.

In contact effect implementations, essential oils are sprayed on pests, and they show their effects by the absorption of the substance by cuticula. Repellent effect occurs by the insect feeling the volatile oil in the air and fleeing the place. This effect may continue until the complete evacuation of the oil from the place based on the properties of volatile oil components.

Feeding inhibitors are compounds that temporarily or permanently prevent feeding based on their potential when they are consumed by insects. Essential oils affect the insect's peripheral nervous system by their dissuasive effect and prevent the continuation of feeding. Toxic effects disrupt physiological and biochemical activities after consumption by the insect.

Insect repellent effects of volatile oils are well known. There are found frequently in outward facing secretion hairs on plants, and they act as "warning signs" for toxicity as they repel the insect before it even feeds on the plant.

In many studies on phenolic substances (Büchel et al. 2015; Randriamanana et al. 2015), it was reported that phenols form a defensive mechanism in plants' protection against pathogens and herbivores.

Specific plant hormones such as ethylene are synthesised against most pathogens and pests. As a result of infection with the pathogen, signals which induce defence mechanisms are transmitted, and local and systemic antimicrobial defence is achieved. Ethylene is a signal compound which plays an important role in inducing defence reaction in plants by acting as a signal.

Considering the insect killing effects of plant volatile oils and compounds, their insecticidal effects were studied on *P. americana*, and the highest toxic effect was observed with the carvacrol compound (Ramírez-López et al. 2016). In another study, it was observed that allyl isothiocyanate and garlic volatile oil have high insecticide effect on *Blattella germanica* (L.) (Dictyoptera: Blattellidae) adults, and it may be used as a fumigant against this pest (Tunaz et al. 2009).

A significant portion of herbivore insects prefer plants with high nitrogen content. It is known that herbivore insects, as in Orthoptera, Coleoptera and Lepidoptera orders, usually require equal amounts of proteins and carbohydrates, while phloem and grain insects require higher amounts of carbohydrates. Based on their content of cellulose, plants develop mechanical resistance against digestive enzymes of insects, or the phenolic compounds in the food directly bind with digestive enzymes and affect the insect's digestion activities negatively.

Plants release VOCs against herbivores and attract the predators of herbivores. For instance, maize plant attacked by caterpillar releases terpenoid to repel herbivores and attract bees, which are enemies of herbivores (Fiers et al. 2013).

Defence signals are transmitted as soon as the herbivore insect starts feeding on the plant. Caterpillars affect the leaf tissue damage, quality and amount and remove plant tissue in about the size of leaves. Lima beans (*Phaseolus lunatus*) also release similar compounds against caterpillar damages.

There are an insufficient number of studies conducted on defence systems in different plant species. The studies conducted so far have mostly been on ants and *Spodoptera* larvae that feed on plants like *Arabidopsis*, maize, rice, tomato and tobacco. It was reported that jasmonic acid is produced in high quantities in the tobacco plant as a defence signal against caterpillars (Paré and Tumlinson 1999). It was indicated that *Fusarium* infection transmitted by the soil on leaves and/or roots in maize may induce secretion of volatiles (Piesik et al. 2011). Antifungal activity was found in volatile oils in birch leaves, and it was found to be highly effective against plant pathogens (Yaşar 2005). Conducted studies revealed that substances (β-caryophyllene, linalool) that attract EPNs (entomopathogenic nematodes) are secreted by infected or damaged plant roots (Laznik and Trdan 2016).

There is a very limited amount of information about the roles of plant volatile compounds against pathogens. VOCs make plants resistant to pathogens. Besides their signalling function, volatile compounds may also inhibit the development of pathogens and contribute in resistance (Quintana-Rodriguez et al. 2015). In terms of defence reactions, it was reported that ethylene production is in coordination with plant resistance and it provides warning about disease development (Martínez-Hidalgo et al. 2015).

References

Altındal D, Altındal N (2011) The effects of sage volatile oil (*Salvia officinalis*) and Turkish oregano volatile oil (*Origanum onites*) on stored cowpea (*Vigna sinensis* L.) seed. J Med Plants Res 5(20):5017–5020

Altındal N, Altındal D (2013) Allelopathic effect of sage and Turkish oregano volatile oils on in vitro in sainfoin (*Onobrychis viciifolia*). J. Essent Oil Bearing Plants 16(3):328–333. doi:10.1080/0972060X.2013.764180

Beaulieu JC, Grimm CC (2001) Identification of volatile compounds in cantaloupe at various developmental stages using solid phase microextraction. J Agric Food Chem 49:1345–1352

Büchel K, Fenning T, Gershenzon J, Hilker M, Meiners T (2015) Elm defence against herbivores and pathogens: morphological, chemical and molecular regulation aspects. Phytochem Rev. http://link.springer.com/article/10.1007/s11101-015-9442-0#/page-1. Accessed 21 May 2016

Byers KJRP, Bradshaw HD Jr, Riffell JA et al (2014) Three floral volatiles contribute to differential pollinator attraction in monkeyflowers (*Mimulus*). J Exp Biol 217:614–623. doi:10.1242/jeb.092213

Chang E-H, Jeong S-M, Hur Y-Y, Koh S-W, Choi I-M et al (2015) Changes of volatile compounds in *Vitis labrusca* 'Doonuri' grapes during stages of fruit development and in wine. Hortic Environ Biotechnol 56(2):137–144. doi:10.1007/s13580-015-0068-3

Degenkolb T, Vilcinskas A (2016) Metabolites from nematophagous fungi and nematicidal natural products from fungi as an alternative for biological control. Part I: metabolites from nematophagous ascomycetes. Appl Microbiol Biotechnol 100:3799–3812. doi:10.1007/s00253-015-7233-6

Dong F, Fu X, Watanabe N, Su X, Yang Z et al (2016) Recent advances in the emission and functions of plant vegetative volatiles. Molecules 21:124. doi:10.3390/molecules21020124

Farre-Armengol G, Filella I, Llusia J, Penuelas J et al (2015) Pollination mode determines floral scent. Biochem Syst Ecol 61:44–53

Fiers M, Lognay G, Fauconnier M-L, Jijakli MH et al (2013) Volatile compound-mediated interactions between barley and pathogenic fungi in the soil. PLoS One 8(6):1–18. doi:10.1371/journal.pone.0066805

Gómez MD, Vera-Sirera F, Pérez-Amador MA, et al (2014) Molecular programme of senescence in dry and fleshy fruits. J Exp Bot. http://jxb.oxfordjournals.org/. Accessed Apr 19 2016

Hamza AF, El-Orabi MN, Gharieb OH, El-Saeady A-HA, Hussein A-RE et al (2016) Response of *Sitophilus granarius* L. to fumigant toxicity of some plant volatile oils. J Radiat Res Appl Sci 9:8–14

Huang X-Z, Chen J-Y, Xiao H-J, Xiao Y-T, Wu J, Wu J-X, Zhou J-J, Zhang Y-J, Guo Y-Y et al (2015) Dynamic transcriptome analysis and volatile profiling of *Gossypium hirsutum* in response to the cotton bollworm *Helicoverpa armigera*. Sci Rep 5:11867. doi:10.1038/srep11867

Ja-Eun S, Jeong-Moon K, Na-Hyun L, Ji-Yeon Y, Hoi-Seon L et al (2016) Acaricidal and insecticidal activities of essential oils against a stored-food mite and stored-grain insects. J Food Protect 1:4–178. doi: http://dx.doi.org/10.4315/0362-028X.JFP-15-109

Kant MR, Bleeker PM, Van Wisk M, Schuurink RC, Haring MA, et al (2009) Plant volatiles in defence. In: Van Loon LC (ed) Advances in botanical research. Plant innate immunity: Elsevier Science, pp 614–651

Kim J-Y, Lee J-S, Kwon T-R, Lee S-I, Kima J-A, Leea G-M, Parka S-C, Jeonga M-J et al (2015) Sound waves delay tomato fruit ripening by negatively regulating ethylene biosynthesis and signaling genes. Postharvest Biol Technol 110:43–50

Laznik Ž, Trdan S (2016) The role of volatile substances emitted by cultivated plant's roots in indirect defense against soil herbivores. Insecticides Resist Intech, p 393–419

Li W, Yamaguchi S, Khan MA, An P, Liu X, Tran L-SP et al (2016) Roles of gibberellins and abscisic acid in regulating germination of *Suaeda salsa* dimorphic seeds under salt stress. Front Plant Sci 6. doi:10.3389/fpls.2015.01235

Louis J, Basu S, Varsani S, Castano-Duque L, Jiang V, Williams WP, Felton GW, Luthe DS et al (2015) Ethylene contributes to maize insect resistance1-mediated maize defense against the phloem sap-sucking corn leaf aphid. Plant Physiol 169:313–324

Martínez-Hidalgo P, García JM, Pozo MJ et al (2015) Induced systemic resistance against *Botrytis cinerea* by *Micromonospora* strains isolated from root nodules. Front Microbiol 6:1–11

Ogawa M, Hanada A, Yamauchi Y, Kuwahara A, Kamiya Y, Yamaguchi S et al (2003) Gibberellin biosynthesis and response during Arabidopsis seed germination. Plant Cell 15:1591–1604

Öz H. Doğal sinek kovucular (2013) http://www.Labmedya.com. Accessed 29 May 2016

Paré PW, Tumlinson JH (1999) Plant volatiles as a defense against insect herbivores. Plant Physiol 121:325–331

Peter KV (2012) Handbook of herbs and spices, 2nd ed, vol 2. Woodheat Publishing Limited, p 583

Piesik D, Lemnczykb G, Skoczekc A, Lamparskia R, Bocianowski J, Kotwica K, Delaney KJ et al (2011) Fusarium infection in maize: volatile induction of infected and neighboring uninfected plants has the potential to attract a pest cereal leaf beetle, Oulema melanopus. J Plant Physiol 168:1534–1542. doi:10.1016/j.jplph.2011.01.032

Procko C, Crenshaw CM, Ljung K, Noel JP, Chory J et al (2014) Cotyledon-generated auxin is required for shade-induced hypocotyl growth in brassica rapa. Plant Physiol 165:1285–1301

Quintana-Rodriguez E, Morales-Vargas AT, Molina-Torres J, Adame-Alvarez RM, Acosta-Gallegos JA, Heil M et al (2015) Plant volatiles cause direct, induced and associational resistance in common bean to the fungal pathogen *Colletotrichum lindemuthianum*. J Ecol 103:250–260. doi:10.1111/1365-2745.12340

Ramírez-López CB, García-Sánchez E, Martínez-Muñoz RE, Del Río RE, Martínez-Pacheco MM et al (2016) Chemical composition of the essential oil from *Ageratina jocotepecana* and its repellent effect on Drywood termite *Incisitermes marginipennis*. Bol Latinoam Caribb Plantas Medicinales Aromáticas 15(1):53–60

Randriamanana TR, Lavola A, Julkunen-Tiitto R et al (2015) Interactive effects of supplemental UV-B and temperature in European aspen seedlings: implications for growth, leaf traits, phenolic defense and associated organisms. Plant Physiol Biochem 93:84–93

Reisenman CE, Riffell JA (2015) The neural bases of host plant selection in a neuroecology frame work. Front Physiol 6:229. doi:10.3389/fphys.2015.00229

Roy A, Das S (2015) Molecular mechanism underlying the Entomotoxic effect of *Colocasia esculenta* Tuber agglutinin against *Dysdercus cingulatus*. Insects 6:827–846. doi:10.3390/insects6040827

Shahab-Ghayoor H, Saeidi K (2015) Antifeedant activities of essential oils of *Satureja hortensis* and *Fumaria parviflora* against Indian meal moth *Plodia interpunctella* Hübner (Lepidoptera: Pyralidae). Entomol Ornithol Herpetol 4:154. doi:10.4172/2161-0983.1000154

Shulaev V, Silverman P, Raskin I et al (1997) Airborne signalling by methyl salicylate in plant pathogen resistance. Nature 385:718–721. doi:10.1038/385718a0

Tunaz H, Er MK, Işıkber AA et al (2009) Fumigant toxicity of plant essential oils and selected Monoterpenoid components against the adult German cockroach, *Blattella germanica* (L.) (Dictyoptera: Blattellidae). Turk J Agric For 33:211–217. doi:10.3906/tar-0805-22

Wang Q, Eneji AE, Kong X, Wang K, Dong H et al (2015) Salt stress effects on secondary metabolites of cotton in relation to gene expression responsible for aphid development. PLoS ONE 10(6):1–14. doi:10.1371/journal.pone.0129541

Yaşar S (2005) Determination of fixed and essential oil contents and soil characteristic of some perennial medical plants that grow naturally in the campus of Çukurova University. Department of biology institute of natural and applied sciences university of Çukurova, MSc Thesis, pp 1–51

Zadeh SY, Ramin AA, Baninasab B et al (2015) Effect of gibberellic acid, stratification and salinity on seed germination of *Echinacea purpurea* cv. Magnus Herba Pol 61(3):13–22. doi:10.1515/hepo-2015-0019

Zhang D-W, Deng X-G, Fu F-Q, Lin H-H et al (2015) Induction of plant virus defense response by brassinosteroids and brassinosteroid signaling in Arabidopsis thaliana. Planta 241:875–885. doi:10.1007/s00425-014-2218-8

Characterization of Bacterial Volatiles and Their Impact on Plant Health Under Abiotic Stress

2

Anukool Vaishnav, Ajit Varma, Narendra Tuteja, and Devendra Kumar Choudhary

Abstract

Bacterial released volatile compounds (VOCs) in air enable bacteria to interact with their surrounding environment. Soil bacterial volatiles are known to contribute to plant interactions, and several studies also identified their influence on plant stress tolerance. Plant growth-promoting rhizobacterial (PGPR)-mediated VOCs are reported to increase seedling emergence, plant weight, crop yield, and stress resistance. The present chapter describes the characterization of different bacterial VOCs and their roles in enhancement of plant abiotic stress tolerance, a new research area, with potential agriculture applications.

Keywords

Abiotic stress • Bacterium • Plant growth-ptomoting bacterium • Volatile organic compounds • C4-bacterial volatiles

2.1 Introduction

Plants live naturally with many microorganisms, and the nutrient-rich environment of the rhizosphere is especially conducive to interactions between microorganisms and plants. Plants release different products through the roots into the surrounding

A. Vaishnav
ICAR-National Bureau of Agriculturally Important Microorganisms (NBAIM), Mau, Uttar Pradesh, India

A. Varma • N. Tuteja • D.K. Choudhary (✉)
Amity Institute of Microbial Technology (AIMT), Block 'E-3', 4th Floor, Amity University Campus, Sector-125, Noida, Gautam Buddha Nagar 201303, Uttar Pradesh, India
e-mail: dkchoudhary1@amity.edu

D.K. Choudhary et al. (eds.), *Volatiles and Food Security*,
DOI 10.1007/978-981-10-5553-9_2

area that attract a tremendous diversity of microorganisms (Perry et al. 2007). Some of these microorganisms have no observable effects on plant; others enhance or inhibit plant growth. Plant growth-promoting rhizobacteria (PGPR) can stimulate plant growth or increase tolerance by producing nonvolatile substances, such as the hormones auxin and cytokinin, as well as 1-aminocyclopropane-1-carboxylate (ACC) deaminase, which reduces plant ethylene levels, and siderophores, which facilitate root uptake of metal nutrients. In addition, certain PGPR promote plant growth by emitting volatile organic compounds (VOCs) (Vaishnav et al. 2017). Volatile compounds have low molecular weight (<300 Da) and high vapor pressure (0.01 kPa at 20 °C) in nature that can readily evaporate and diffuse through heterogeneous mixtures of solids, liquids, and gasses (Audrain et al. 2015). The spectrum of bacterial VOCs is influenced by heterogeneity of soil, which depends on bacterial/plant-secreted metabolites. Some VOCs are specific for a phylogenetic group and used for taxonomic purposes (Kai et al. 2016). The volatile compounds are generally produced as metabolic end products of anaerobic fermentation processes and extracellular degradation of complex organic molecules. The widely differing species of bacteria were capable of emitting a variety of volatile compounds, comprising of fatty acid derivatives, terpenoids, and aromatic, nitrogenous, and sulfurous compounds. Interestingly, many of the substances found in the bacterial scent spectra have not been identified yet, and their biological roles are also unknown. By referring to a few known examples and by comparison with known functions of scents from other groups of organisms, it is assumed that the bacterial scents serve as signal compounds for interspecies and intraspecies communication or from cell to cell, for the disposal of excess carbon compounds, or as substances that stimulate or inhibit growth (Wenke et al. 2010). Many investigations concerning VOC pattern of soil microorganisms were performed under different treatment conditions. Bolm et al. performed a large screening of volatile-mediated effects on *Arabidopsis thaliana*. The VOC effect was found highly dependent on cultivation medium and the inoculum quantity. The production of beneficial VOC compound butanediol was found higher on the nutrient-rich media LB and MR-VP and less pronounced on MS and Angel. In another study, *P. simiae* strain AU changed VOC pattern in the presence of soybean seeds and sodium nitroprusside (SNP, a nitric oxide donor) treatment. Some compound expression was enhanced in the presence of soybean seedlings, and few compounds like 4-nitroguaiacol and quinolone are newly expressed in the presence of SNP. These compounds were showed significant enhancement of seed germination and higher fresh weight of soybean under 100 mm NaCl stress (Vaishnav et al. 2016). As a result of the increasing interests in VOCs in mediating plant-microorganism interactions, the present chapter focuses on the chemical nature of microbial VOCs, as well as the effects of microbial VOCs on tolerance level of plants under abiotic stresses.

2.2 Chemical Side of VOCs

Solid-phase microextraction (SPME) and dynamic headspace volatile analyses have revealed that PGPR strains have the requisite machinery to synthesize a wide range of volatiles, including short-chain aliphatic aldehydes, esters, alcohols, organic acids, ethers, ketones, sulfur compounds, and hydrocarbons.

2.3 Hydrocarbons

These products are derived from fatty acid biosynthetic pathways. Short-chain alkanes (decane to tetradecane) are found in bacteria, while longer hydrocarbons such as hexadecane are particularly abundant in cyanobacteria, which are also known for their ability to synthesize branched hydrocarbons (Audrain et al. 2015). In a volatiles profiling study, *Paenibacillus polymyxa* strain E681 was exclusively released long-chain C13 tridecane which augments ISR and protected *Arabidopsis* seedlings against the biotrophic pathogenic bacterium *Pseudomonas syringae* pv. maculicola ES4326 (Lee et al. 2012). The priming with pure tridecane compound at 10 mM increased 4.7-fold in transcription of pathogenesis-related gene (PR1) before pathogen challenge and 3.3-fold increase at 3 h after pathogen challenge. Some other hydrocarbons like isoprene, acetylene, cyclohexane, 1-undecane, and dodecane were also found in *P. polymyxa* E681-derived VOCs blend. An n-decanal volatile compound derived from different *Pseudomonas* sp. was found to completely inhibited *Sclerotinia sclerotiorum* suggesting its role in biological control (Fernando et al. 2005). Recently, Vaishnav et al. (2016) demonstrated the effect of pure tridecane compound on soybean seed germination under salt stress. Seed germination was found higher in the presence of tridecane as compared to control.

2.4 Ketones/Alcohols

PGPR strains *B. amyloliquefaciens* IN937 and *B. subtilis* GB03 were found to produce acetoin (3-hydroxy-2-butanone) compound that triggers growth promotion and resistance in *Arabidopsis* seedlings. The acetoin and its oxidized form 2, 3-butanediol are derived from anaerobic fermentation of pyruvate. In vitro production of 2, 3-butanediol is also favored in the presence of sucrose as a major nutrient in growth media (Ryu et al. 2004). 2R, 3R-butanediol derived from *P. chlororaphis* O6 was found as major determinant in inducing resistance to drought in *Arabidopsis* through an SA-dependent mechanism (Cho et al. 2008). In addition, short-chain-branched alcohols such as 3-methyl-1-butanol and 2-methyl-1-butanol were also found to accumulate in *B. amyloliquefaciens* IN937a producing VOCs. These compounds are formed by enzymatic conversion of branched chain amino acids, i.e., leucine and isoleucine via the Ehrlich pathway (Marilley and Casey 2004). Long-chain aliphatic alcohols (i.e., 1-octanol, 1-decanol, and 1-dodecanol) are commonly associated with Enterobacteriaceae; they are produced through β- or α-oxidation of fatty acid derivatives, and thus, their concentration are markedly increased from cultures supplemented with fatty acids (Hamilton-Kemp et al. 2005). 1-Hexanol produced by *Pseudomonas* sp. (Fernando et al. 2005; Vaishnav et al. 2016) and *Bacillus* sp. (Chaurasia et al.) has been reported as a potential plant growth promoter and antifungal compound. During a large screening of VOCs, 1-hexanol was found most release compounds from 42 different bacterial strains and appeared to promote plant growth (Blom et al. 2011).

2.5 Acids

Short-chain fatty acids like acetic acid are derived from VOCs of an antagonist bacterium *Burkholderia tropica* which is found to inhibit the growth of four phytopathogenic fungi, *Colletotrichum gloeosporioides*, *Fusarium culmorum*, *Fusarium oxysporum*, and *Sclerotium rolfsii* (Tenorio-Salgado et al. 2013). A salt-tolerant PGPR bacterium *P. simiae* AU produced different types of acids in VOC blend, i.e., stearic acid, phthalic acid, acetic acid, oxalic acid, myristic acid, hexadecanoic acid, etc. The exposure of *P. simiae*-mediated VOCs is found to induce tolerance in soybean plants against salt stress (Vaishnav et al. 2016). These acids are major by-products of anaerobic metabolism; indeed, they are formed during bacterial fermentation of carbohydrates.

2.6 Sulfur Compounds

Dimethyl disulfide (DMDS) and dimethyl trisulfide (DMTS) are produced by most of PGPR strains. Benzothiazole compound was derived from different *Pseudomonas* spp. and showed potential in the inhibition of sclerotial activity and enhanced salt tolerance in soybean plants (Fernando et al. 2005; Vaishnav et al. 2016).

2.7 Inorganic Compounds

Hydrogen cyanide is mostly produced by virulent bacterial strains (e.g., *Pseudomonas* or *Chromobacterium* species). It is a potent inhibitor of cytochrome *c* oxidase and of other metal-containing enzyme, hence responsible for plant-killing effects (Blom et al. 2011). Cyanide production by rhizosphere bacteria is considered as a plant growth-promoting trait which is used as biocontrol against phytopathogenic fungi (Voisard et al. 1989).

2.8 Test Systems

Two types of systems are mainly used for volatiles studies as follows: (a) a closed setup (Petri dish or box) and (b) a directed airflow in order to transport the volatiles to the plant.

2.9 Closed Systems

More than half of all volatile studies were operated by passive diffusion using partite Petri dishes. A plastic border separates the dish into two (I-shaped) or three (Y-shaped) compartments (Fig. 2.1a). Bacteria and plants/pathogens are inoculated in individual compartment which make them physically separate to each other. The exchange of VOCs is facilitated solely via headspace. To prevent the escape of VOCs, Petri dishes are sealed with parafilm. In order to extract the volatile compounds, experiment is performed in tripartite Petri dishes. In this, one partition contained bacterial culture, second inoculated with plant seeds/pathogen, and in third sterile-activated charcoal placed. Activated charcoal has a good adsorbent quality; therefore, it is mainly used to collect the VOCs in a closed setup. After incubation period, the activated charcoal is collected and washed with any organic solvent according to the aim of the study. In many studies, dichloromethane (DCM) was used to extract all trapped volatile compounds, which were posteriorly analyzed by gas chromatography-mass spectrometry (GC-MS). This way, Fernando et al. (2005) found that mycelial growth was completely inhibited in the presence of bacterial VOCs which streaked in a different compartment. Mycelial growth was unaffected by the presence of volatile-producing bacteria, when the third compartment of the

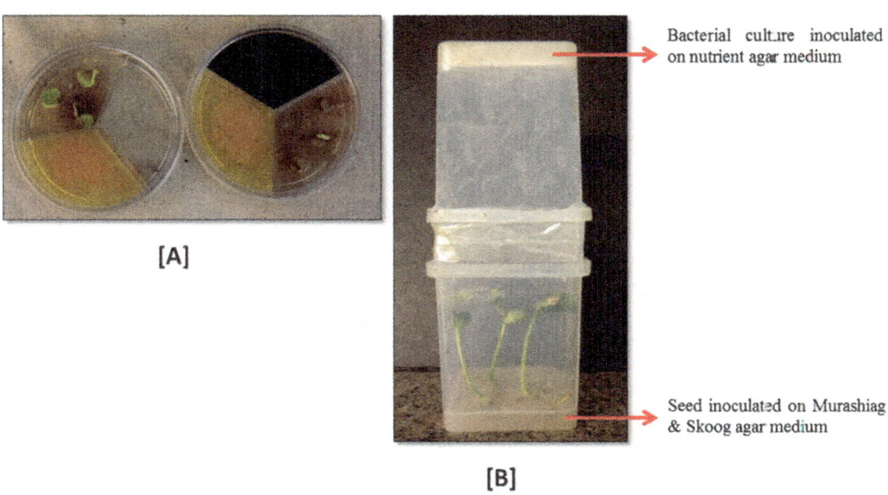

Fig. 2.1 Passive diffusion of VOCs in closed system (**a**) tripartite Petri plate with activated charcoal, bacterial culture and seed; (**b**) Magenta box

plates was amended with activated charcoal. Charcoal adsorbs volatiles as soon as they are produced, and hence no inhibitory effect was observed. Plant growth promotion was also observed with *partitie* Petri dish setup by Ryu et al. (2004), Blom et al. (2011), and Vaishnav et al. (2016). The limitation of this setup is narrow headspace; therefore, only young plant seedlings can be sampled. Most of the studies were performed with *A. thaliana* plants, because of its small size and its short life cycle. For other test plants and for attempts to use adult plants, bigger growth containers were designed (box systems) (Vaishnav et al. 2015) (Fig. 2.1b). In another approach, Park et al. (2015) inoculated *Pseudomonas fluorescens* SS101 in a plate which placed beneath the soil at the bottom of the pot. A filter is also placed above the plate to ensure that only VOCs could transfer between the plate and the soil. Tobacco seeds were sown onto the soil and after 4 weeks of sowing, the authors observed significant growth stimulation with an increase of fresh weight of tobacco.

2.10 Dynamic Air Stream Systems

In this system, a continue flow of purified air passed over bacterial culture plate and subsequently reached to targeted parts of plant (aerial/root) (Fig. 2.2). In this way, Kai and Piechulla (2009) performed experiment with *S. plymuthica*-derived VOCs by targeted aerial and roots of adult *A. thaliana* plants. For aerial part, they used a mini growth chamber in which VOCs enriched air directly reached from culture. For root, plants were kept in a perforated glass bowl and placed over a second glass bowl which attached with an air inlet. Air enriched with VOCs entered the lower compartment and reached to roots of the plants.

2.11 Characterization of Volatiles

The identification of volatiles is usually accomplished using gas chromatography coupled most often with mass spectrometry (GC-MS) in electron ionization mode (EI). The number of detectable volatiles in bacteria generally increases when

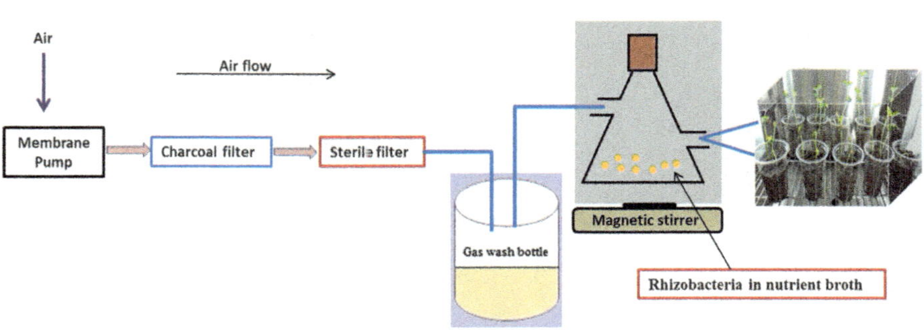

Fig. 2.2 Dynamic air stream system targeting the aerial plant parts

various techniques are applied. Hence, multiple approaches are used in combination with GC-MS., e.g., headspace airflow systems with GC-MS, trapping materials – static solid-phase microextraction (SPME) with GC-MS, and proton transfer reaction/mass spectrometry (PTR-MS). Headspace volatiles can be collected through air flow onto an absorbent filter and released by rinsing the filter with organic solvent. On the other hand, VOCs can be collected by SPME and directly released into a heated GC injector. SPME can extract volatiles from bacterial cultures in a relatively short amount of time and has been successfully used to collect them in several systems (Farag et al. 2013). In SPME, fiber plays role in absorbing of compounds based on their polarity and size. In the case of rhizobacterial volatiles, divinylbenzene/carboxen/PDMS fibers are mainly used. These fibers adsorb polar low molecular weight VOCs, which are the predominant VOCs released from rhizobacteria. Proton transfer reaction mass spectrometry (PTR-MS) technique allows online VOC measurements. This technique is also combined with GC or time of flight (TOF). After MS analysis, compounds are identified by comparison of their mass spectral fragmentation patterns with mass spectra libraries such as WILEY and NIST. Several times the identification cannot be possible on the basis of mass spectral libraries alone. In such condition, retention indices, derivatization, and comparison with reference compound are better alternate to elucidate the compound structure.

2.12 The Effects of Bacterial Volatile Emissions on Plant Abiotic Stress Tolerance

During the last few years, an increasing number of PGPR VOCs studies have demonstrated an effect on induced systemic tolerance (IST) against abiotic stresses. Under high salt concentration, excessive sodium (Na^+) creates both ionic and osmotic stresses in plants, leading to suppression of plant growth and reduction in crop yields. The entry of Na^+ ion is controlled by several transporters in which high-affinity K^+ transporter (HKT1) plays a major role. A soil bacterium *Bacillus subtilis* GB03-mediated VOCs has been reported to regulate HKT1 expression and conferred salt tolerance in *Arabidopsis thaliana*. The expression of HKT1 was tissue specific, GB03 concurrently down- and upregulates HKT1 expression in roots and shoots respectively, resulting lower Na+ accumulation in the plant as compared with controls (Zhang et al. 2008). Plants adjust their endogenous metabolism to cope with osmotic and ionic stress. Bacterial VOC-mediated salt tolerance was reported in soybean plants; *P. simiae* strain AU released such VOCs that not only decreased root Na^+ levels but also increased the accumulation of proline, which protect cells from osmotic stress. In addition, soybean plants showed higher level of vegetative storage protein (VSP), gamma-glutamyl hydrolase (GGH) and RuBisCo large-chain proteins that are known to help sustain plant growth under stress conditions (Vaishnav et al. 2015). *P. simiae* was found to emit 4-nitroguaiacol and quinolone compounds in the presence of nitric oxide donor sodium nitroprusside that were found to promote soybean seed germination under 100 mM NaCl. VOCs exposure induced antioxidative enzymes and nitrate reductase gene expression in soybean plants that relieve the negative effects of salt stress (Vaishnav et al. 2016). *Paraburkholderia phytofirmans* PsJN emitted 2-undecanone, 7-hexanol, 3-methylbutanol, and dimethyl disulfide compounds that directly effect on bacterial colonization, increasing plant growth rate and tolerance to salinity (Ledger et al. 2016).

Plants have their own mechanism to protect against osmotic stress. During drought condition, plant accumulates osmolytes and increase antioxidant activity to nullify the effect of osmotic stress outside the cell and reactive oxygen species inside the cell, respectively. *Bacillus thuringiensis* AZP2 and *Paenibacillus polymyxa* B primed wheat seeds showed enhanced tolerance against drought stress. Primed seeds were exhibited higher survivorship, dry mass, water use efficiency, and antioxidant enzyme activity. Three volatile compounds benzaldehyde, b-pinene, and geranyl acetone were found to emit from wheat seedlings and effective to mitigate early phases of stress development (Timmusk et al. 2014). In addition to adaptive responses, an increase in the plant hormones abscisic acid (ABA) and salicylic acid (SA) causes stomatal closure to minimize water loss through transpiration. In an experiment, root colonization of *Arabidopsis* plants with *P. chlororaphis* O6 induced tolerance to drought that was correlated with reduced water loss with stomatal closure. Drought tolerance was found to mediate by 2R, 3R-butanediol volatile compound produced by *P. chlororaphis* O6. In the lack of 2R, 3R-butanediol production, no induction of drought tolerance was found. Further study demonstrated free SA, NO, and hydrogen peroxide in *P. chlororaphis* O6-colonized drought-stressed plants which suggested a primary role of these signaling molecules in induced drought tolerance (Cho et al. 2008, 2013). In a VOC assay setup, *Bacillus subtilis* GB03 exposure increased an enzyme expression of choline synthesis phosphoethanolamine *N*-methyltransferase (PEAMT) resulting enhanced pool of choline and glycine-betaine in *Arabidopsis* plant under osmotic stress. Certain bacterial VOCs are involved in biofilm formation, which contain exopolysaccharides as major constituents, and these polysaccharides maintain soil moisture content and increase drought tolerance in plants (Naseem and Bano 2014).

Iron is a limiting nutrient for plants due to its minimal solubility in soils. Iron solubility decreased in the presence of high salt concentration. As a result, iron deficiency can occur in the plants and affect their metabolism. In a study, *Bacillus subtilis* GB03-mediated VOCs activate the plant's iron acquisition machinery to increase assimilation of metal ions in *Arabidopsis*. GB03 VOCs upregulates gene expression of Fe-deficiency-induced transcription factor 1 (FIT1), which is necessary for two iron transporter induction FRO2 and IRT1. On the other hand, VOCs also enhanced the acidification media by enhancing root proton release, which increased iron mobility in plants (Zhang et al. 2009). Higher salt concentration also reduces sulfur (S) availability in the soil. Sulfur is an essential element in cysteine and methionine amino acids, and its deficiency represses the photosynthesis activity and productivity. Plant uptakes sulfur through soil or airborne compounds, including volatile compounds emitted by soil bacteria like dimethyl disulfide (DMDS). Emission of DMDS was found from a PGPR *Bacillus* sp. B55, which enhanced the

S content in *Nicotiana attenuata* plants grown in S-deficient soils or impaired S uptake/assimilation/metabolism (Meldasu et al. 2013).

2.13 Conclusion and Future Prospective

This chapter reports the potential role of bacterial VOCs as airborne signals in plant interaction with beneficial effects. In many reports, bacterial VOCs enable plants to adapt to various environmental stresses and ultimately enhance plant growth. Despite that, many aspects of VOC interaction with plants are still poorly characterized. The combined analysis of metabolic and gene expression profiles will likely be an increasingly powerful approach to identifying the mechanism of plant perception for bacterial VOCs. Moreover, systematic use of radiolabeled VOC compounds will reveal how plants assimilate VOC components as metabolites. The question of how bacteria respond to diverse environment to produce VOCs could be addressed by screening of compounds based on different parameters, as well as by monitoring transcriptomic, proteomic, and metabolomic changes in response to different environments.

Acknowledgments Authors are very thankful to DBT and SERB for financial assistance.

References

Audrain B, Farag MA, Ryu C-M, Ghigo J-M (2015) Role of bacterial volatile compounds in bacterial biology. FEMS Microbiol Rev 39:222–233

Blom D, Fabbri C, Connor EC, Schiestl FP, Klauser DR, Boller T et al (2011) Production of plant growth modulating volatiles is widespread among rhizosphere bacteria and strongly depends on culture conditions. Environ Microbiol 13:3047–3058

Cho SM, Kang BR, Han SH, Anderson AJ, Park JY, Lee YH et al (2008) 2R,3R-butanediol, a bacterial volatile produced by *Pseudomonas chlororaphis* O6, is involved in induction of systemic tolerance to drought in *Arabidopsis thaliana*. Mol Plant-Microbe Interact 21:1067–1075

Cho SM, Kim YH, Anderson AJ, Kim YC (2013) Nitric oxide and hydrogen peroxide production are involved in systemic drought tolerance induced by 2R, 3R-butanediol in *Arabidopsis thaliana*. Plant Pathol J 29:427–434

Farag MA, Zhang H, Ryu CM (2013) Dynamic chemical communication between plants and bacteria through airborne signals: induced resistance by bacterial volatiles. J Chem Ecol 39:1007–1018

Fernando WGD, Ramarathnam R, Krishnamoorthy AS, Savchuk SC (2005) Identification and use of potential bacterial organic antifungal volatiles in biocontrol. Soil Biol Biochem 37:955–964

Groenhagen U, Baumgartner R, Bailly et al (2013) Production of bioactive volatiles by different *Burkholderia ambifaria* strains. J Chem Ecol 39:892–906

Hamilton-Kemp T, Newman M, Collins R et al (2005) Production of the long-chain alcohols octanol, decanol, and dodecanol by *Escherichia coli*. Curr Microbiol 51:82–86

Kai M, Piechulla B (2009) Plant growth promotion due to rhizobacterial volatiles – an effect of CO_2? FEBS Lett 583:3473–3477

Kai M, Effmert U, Piechulla (2016) Bacterial plant interactions: approaches to unravel the biological function of bacterial volatiles in the rhizosphere. Front Microbiol 7:108

Ledger T, Rojas S, Timmermann T, Pinedo I, Poupin MJ, Garrido T, Richter P, Tamayo J, Donoso R (2016) Volatile-mediated effects predominate in *Paraburkholderia phytofirmans* growth promotion and salt stress tolerance of *Arabidopsis thaliana*. Front Microbiol 7:18388

Lee B, Farag MA, Park HB, Kloepper JW et al (2012) Induced resistance by a long-chain bacterial volatile: elicitation of plant systemic defense by a C13 volatile produced by *Paenibacillus polymyxa*. PLoS One 7:1–11

Marilley L, Casey MG (2004) Flavours of cheese products: metabolic pathways, analytical tools and identification of producing strains. Int J Food Microbiol 90:139–159

Meldau DG, Meldau S, Hoang LH, Underberg S, Wunsche H, Baldwin IT (2013) Dimethyl disulfide produced by the naturally associated bacterium *Bacillus* sp B55 promotes *Nicotiana attenuata* growth by enhancing sulphur nutrition. Plant Cell 25:2731–2747

Naseem H, Bano A (2014) Role of plant growth-promoting rhizobacteria and their exopolysaccharide in drought tolerance of maize. J Plant Interact 9:689–701

Park YS, Dutta S, Ann M, Raaijmakers JM et al (2015) Promotion of plant growth by *Pseudomonas fluorescens* strain SS101 via novel volatile organic compounds. Biochem Biophys Res Commun 461:361–365

Perry LG, Alford ER, Horiuchi J, Paschke MW, Vivanco JM (2007) Chemical signals in the rhizosphere: root-root and root-microbe communication. In: Pinton R, Varanini Z, Nannipi P (eds) The rhizosphere. Biochemistry and organic substances at the soil-plant interface, 2nd edn. CRC press, Taylor and Francis Group, Boca Raton, pp 297–330

Ryu CM, Farag MA, Hu C-H, Reddy MS et al (2004) Bacterial volatiles induce systemic resistance in Arabidopsis. Plant Physiol 134:1017–1026

Tenorio-Salgado S, Tinoco R, Vazquez-Duhalt R, Caballero-Mellado J et al (2013) Identification of volatile compounds produced by the bacterium *Burkholderia tropica* that inhibit the growth of fungal pathogens. Bioengineering 4:236–243

Timmusk S, Abd El-Daim IA, Copolovici L, Tanilas T, Ka¨nnaste A et al (2014) Drought-tolerance of wheat improved by rhizosphere bacteria from harsh environments: enhanced biomass production and reduced emissions of stress volatiles. PLoS One 9(5):e96086

Vaishnav A, Kumari S, Jain S, Varma A, Choudhary DK (2015) Putative bacterial volatile-mediated growth in soybean (*Glycine max* L. Merrill) and expression of induced proteins under salt stress. J Appl Microbiol 119:539–551

Vaishnav A, Kumari S, Jain S, Varma A, Tuteja N, Choudhary DK (2016) PGPR mediated expression of salt tolerance gene in soybean through volatiles under sodium nitroprusside. J Basic Microbiol 56:1–15

Vaishnav A, Varma A, Tuteja N, Choudhary DK (2017) PGPR-mediated amelioration of crops under salt stress. In: Choudhary DK, Varma A, Tuteja N (eds) Plant-microbe interaction: an approach to sustainable agriculture. Springer, Singapore. ISBN:978-981-10-2854-0

Voisard C, Keel C, Haas D, Dèfago G (1989) Cyanide production by *Pseudomonas fluorescens* helps suppress black root rot of tobacco under gnotobiotic conditions. EMBO J 8(2):351–358

Wenke K, Kai M, Piechulla B (2010) Belowground volatiles facilitate interactions between plant roots and soil organisms. Planta 231:499–506

Zhang H, Kim MS, Sun Y, Dowd SE et al (2008) Soil bacteria confer plant salt tolerance by tissue-specific regulation of the sodium transporter HKT1. Mol Plant-Microbe Interact 21:737–744

Zhang H, Sun Y, Xie X, Kim MS, Dowd SE, Paré PW (2009) A soil bacteria regulates plant acquisition of iron via deficiency-inducible mechanisms. Plant J 58:568–577

Soil Reclamation Through Phytoextraction and Phytovolatilization

3

S.S. Arya, S. Devi, R. Angrish, I. Singal, and Kanta Rani

Abstract

Environmental pollution becomes most severe due to anthropologic actions including domestic waste generation and excessive utilization of fertilizers and pesticides to get better yield. Although the phenomenon of hyperaccumulation of metal ions in shoots of certain plants is known since long, the contemporary environmental concerns have prompted broad-based studies on hyperaccumulator plants that can phytoremediate contaminated soils. Phytoremediation is considered as an eco-friendly technology which is deployed to alleviate pollutants from environment components. The present chapter discusses phytoextraction and phytovolatilization mechanisms that are involved in the decontamination of the soil.

Keywords

Phytoremediation • Salinity • Hyperaccumulator • Pollutants

3.1 Introduction

Phytoremediation (phyto, plant, and *remedium*, restoring balance) is visualized as benign technology that depends upon the remarkable ability of some plants to remove or neutralize various chemicals (organics and metal ions) from the soil,

S.S. Arya (✉)
Department of Botany, Maharshi Dayanand University, Rohtak, Haryana, India
e-mail: aryasunder.hau@gmail.com

S. Devi • R. Angrish • I. Singal
Department of Botany and Plant Physiology, CCS HAU, Hisar, Haryana, India

K. Rani
Department of Botany, GGDSD College, Palwal, Haryana, India

D.K. Choudhary et al. (eds.), *Volatiles and Food Security*,
DOI 10.1007/978-981-10-5553-9_3

water, and air (Sarma 2011). It is very eco-friendly, cost-effective, aesthetically pleasing, and noninvasive to redress the alleviation of environmental hazards (Elizabeth 2005). Phytoremediation is brought about by plants having the ability to extract and accumulate the toxic metals or ions in the aboveground shoots (phytoaccumulation), or removing or decomposing various organic chemicals from the soil (phytodegradation), or acting as "filter" to remove toxic matter from an aqueous environment (rhizofilteration).

To complete the increasing demand of world population, there is excessive utilization of fertilizers and pesticides to get better yield. So, today environmental pollution becomes most severe also due to anthropologic actions including domestic waste generation (Kabata-Pendias and Pendias 1989). When excessive heavy metals/ions are present in the environment, a large amount absorbed by plant roots translocated to upright direction leading to reduced growth and metabolic disorder (Bingham et al. 1986). Buildup of soil salinity is one of the world's oldest and most serious agricultural problems in arid and semiarid regions (Tanji 1990). About 7.0 million hectares of agricultural land is infested with salinity worldwide, and these domains are expanding further. The chlorides and sulfates of sodium, calcium, and magnesium are the dominating soluble salts in them (Dahiya and Laura 1988). The existing technologies on farm salinity management that work well include surface and subsurface drainage. These are basically civil engineering technologies and are costly to install, are difficult to maintain, and have the problem of saline effluent management. Apart from that, under Indian conditions with fragmented landholdings, a wide application of such technology seems utopian. Phytoextraction is the natural ability of certain plants to accumulate unusually high amount of metal ions particularly in their leaves (Elizabeth 2005; Angrish and Devi 2014).

Phytoextraction and phytovolatilization occur simultaneously. Phytovolatilization is a diffusion process in which volatile organic compounds (VOCs) are absorbed by the plants and are released into the atmosphere. In recent years considerable research efforts have been made in the use of plants to remove inorganic or organic contaminants from the soil by the technique of phytoremediation (Devi et al. 2016). To improve the previous phytoremediation processes that are based on biological and engineering strategies, there is a need to know the physiological and molecular mechanism of different plants. This chapter represents the summarized work of the eminent scientists on phytoextraction and phytovolatilization processes that can be used in higher education teaching.

3.2 Origin

The US Environmental Protection Agency proposed the term phytoremediation in 1991 which was firstly reported by Raskin et al. But in open technical literature, the term phytoremediation was firstly used by Cunningham and Berti (1993). A German botanist works on the leaves of different plant species that are grown naturally on the soil which contained extraordinary high levels of zinc (Baumann 1885). It has been observed that 1% and 1.7% zinc accumulate in dry leaves of violet (*Viola calaminaria*) and the mustard (*Thlaspi calaminare*) species, respectively, whereas the plants growing in controlled condition accumulated zinc from 0.001% to 0.02% in their dry leaves. Half a century after, a word "alkali disease" was noted in animals in South Dakota. The cause of this disease was traced to the accumulation of selenium up to 0.6% in dry shoot/leaf mass (Byers 1935, 1936) of *Astragalus*. Shortly thereafter, two Italian botanists (Minguzzi and Vergnano 1948) reported 1% nickel in leaves of *Alyssum bertolonii* growing on nickel-enriched serpentine soils near Florence, Italy. The quest for using this unique hyperaccumulation ability of some plants was thus initiated.

3.2.1 Overview of Phytoremediation

Phytoremediation is also low cost over conventional methods for hazardous waste management (McCutchen and Schnoor 2003). Phytoremediation is a nondestructive cleanup method in which plants can be used as a tool to decontaminate the soil and water (Fig. 3.1).

There are five kinds of phytoremediation sub-techniques which exist (Salt et al. 1998):

- Phytoextraction: wherein plants accumulate pollutants (metals) in order to decontaminate soils
- Phytodegradation: wherein plants degrade organic pollutants directly via their own metabolic activities
- Phytostabilization: wherein plants stabilize pollutant in soil
- Phytovolatilization: deployment of plants to remove pollutants from air
- Rhizofilteration: deployment of plant roots or whole plant for filtration

3.3 Phytoextraction

Plants have been used to remove contaminants from soil, water, and air into harvestable plant biomass. Excessive amount of the contaminants are absorbed by the plants from the soil that are called as hyperaccumulators. The hyperaccumulator plants of Brassicaceae family (Kumar et al. 1995) have been deployed for phytoextraction. McCutchen and Schnoor (2003) observed that phytoextraction is becoming a more widely used remediation technology where field-level results have been shown (Brown et al. 1995). It includes the extent of contamination, metal bioavailability, and the plants' ability to intercept, absorb, and accumulate metals from soil which is becoming a challenge for researchers and managers of phytoextraction enterprises (Thangavel and Subbhuraam 2004). Salt et al. (1998) highlighted the remarkable ability of certain plants to hyperaccumulate metal ions (Cd, Ar, Pb, Ni, Co, etc.) in their harvestable parts (Salt et al. 1995; Cunningham and Ow 1996; Suresh and Ravisanker 2004). Hyperaccumulation of metal ions in plants takes place against concentration gradient at the expenditure of metabolic energy which

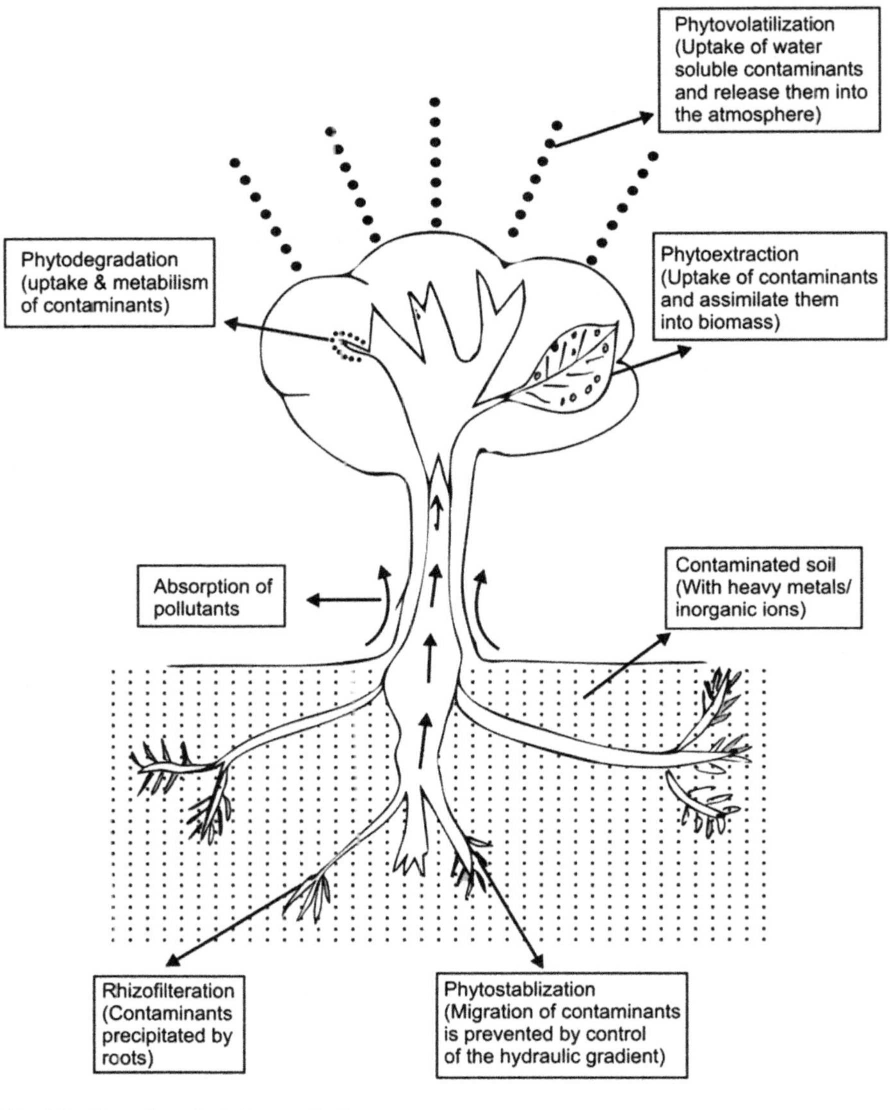

Fig. 3.1 Overview of phytoremediation

is obviously derived from the sun. Ion hyperaccumulation in harvestable parts is therefore acknowledgeably a "green" technology that is environmentally benign.

In contrast to heavy metal ions, phytoremediation of the component ions of salinity, i.e., Na^+, Ca^{2+}, Mg^{2+}, Cl^-, and SO_4^{2-}, has not received desired attention. There are scanty reports in literature (Yeo 1974; Williams 1960; Sairam and Tyagi 2004; Devi et al. 2016) where attempts to alleviate salinity using salt hyperaccumulating plants have been made. It is a matter of common knowledge that halophytes, which constitute the bulk of native flora of the saline soils, not only survive but also thrive on the saline milieu. This requires repeated cropping until the contaminated soil has reached acceptable levels for the farmers to cultivate their regular crops.

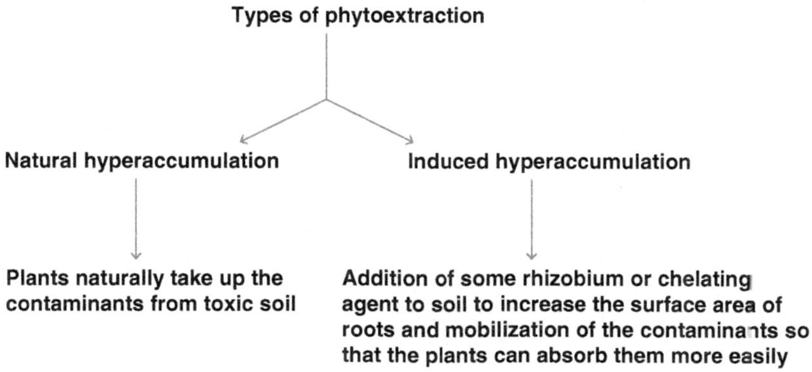

3.4 Mechanism of Hyperaccumulator Plants

Based on the availability of contaminants present in soil ecosystem, the hyperaccumulator plants can absorb. This capacity for accumulation is the result of the plants to that environment (Fig. 3.2). Sen et al. (1982) observed that despite aridity, the habitats of halophytes in arid regions are mostly wet. Dry and saline habitats do not have any vegetation. Halophytic roots are able to absorb water only from somewhat dilute soil solutions. As water is transpired, accumulation of salt takes place in shoot, particularly leaves. Under physiological drought conditions, the leaves of saline plants play an important role and develop a combination of xeromorphic and halophytic characteristics, viz., hair cover, salt excretory glands, and salt storage glands. Two of these features, i.e., salt glands and succulence vis-à-vis ion storage in cells, are important.

3.4.1 Salt Excretion

Salt excretion takes place through certain specialized glandular cells. Salt excretory glands have been reported in some of the non-succulent halophytes of the Indian arid zone, viz., *Aeluropus lagopoides*, *Sporobolus helvolus*, *Chloris virgata*, *Cressa cretica*, *Tamarix dioica*, *T. ericoides, and T. troupii.* As early as 1935, Frey-Wissling noted that important structures in the salt economy of some halophytes are salt glands. This fascinating trait has evolved convergently in many different families of angiosperms such as Plumbaginaceae, Tamaricaceae, Primulaceae, etc. including bladder trichomes of some Chenopodiaceae, e.g., *Atriplex* species. Excess salt may well be secreted by salt glands in some halophytes, e.g., *Spartina townsendii* (Skelding and Wintebotham 1939) and *Limonium* (Ziegler and Luttge 1967). Salt crystals secreted by glands are likely to fall again on the soil below due to gravity,

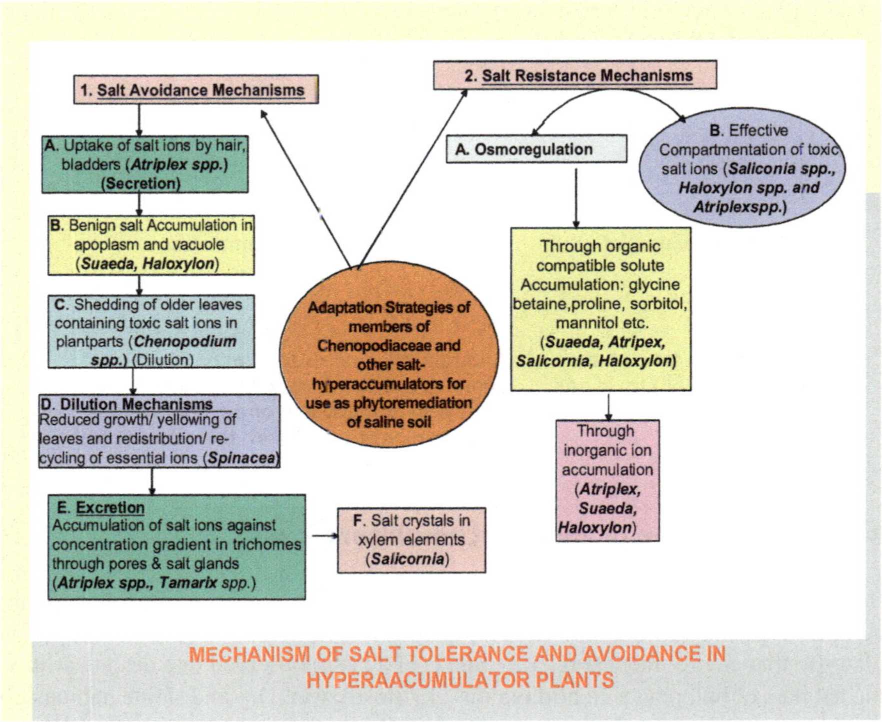

Fig. 3.2 A flowchart of some important salt tolerance strategies adopted by hyperaccumulator plants

dewdrops, or rainfall. The use of salt-excreter plants in saline soil remediation must, therefore, be critically assessed taking in account all these factors.

3.4.2 Succulence Mechanism

Ion accumulation in succulent halophytes, like *Haloxylon recurvum*, *H. salicornicum*, *Portulaca oleracea*, *Salsola baryosma*, *Sesuvium sesuvioides*, *Suaeda fruticosa*, *Trianthema triquetra*, *Zygophyllum simplex*, *Suaeda fruticosa*, *Salsola baryosma*, *Trianthema triquetra*, etc. where these are sequestered in high concentration in vacuolar sap is therefore an important mechanism of interest for salt hyperaccumulation from phytoremediation point of view (Sen et al. 1982).

3.5 Function

Because of several drawbacks, the older and traditional methods are not suitable for practical applications, and hence, the deployment of phytoremediation strategies to make soil heavy metal contamination-free is necessary (Lasat 2002). Potential for phytoremediation depends upon the interactions among soils, heavy metals, bacteria, and plants. Potential for phytoremediation depends upon the interactions among soils, heavy metals, bacteria, plants and their interactions are affected by a variety of factors, such as characteristics, activity of plants and rhizobacteria, climatic conditions, soil properties, fixation, mineralization, synthesis, and release of organic and inorganic compounds, root system etc.

1. *Role of mycorrhizae in phytoremediation*

Remediation of heavy metal contamination in soils is difficult as these cannot be destroyed biologically but are only transformed from one more toxic to less toxic form (Garbisu and Alkorta 2001). Phytoextraction is the use of plants to extract, sequester, and/or detoxify pollutants through physical, chemical, and biological processes (Wenzel et al. 1999). The process of metal uptake and accumulation in plants after mycorrhizal application increases the surface of the root (Fig. 3.3). Contaminants present in the soil are sorbed at root surface and enter into the root cells by crossing cellular membrane. Some of the contaminants absorbed are stored in the vacuole, and the rest of them enters into the root vascular tissue (xylem). Finally contaminants are translocated from root to shoot portion of the plant and dumped at a point sink for their incineration (Huang et al. 2005).

2. *Role of plant growth-promoting rhizobacteria (PGPR)*

Inoculation of PGPR in spermosphere and seed microbiolization with hyperaccumulator plants (Glick et al. 1999; Glick 2003) help in mitigating toxic effects of heavy metals on the plants (Belimov et al. 2004) and the release of chelating agents, acidification and phosphate solubilization (Abou-Shanab et al. 2003a). The use of PGPR with PGP ability in combination with plants is expected to provide high efficiency for phytoremediation (Whiting et al. 2001; Abou-Shanab et al. 2003a).

For example:

1. Size of Indian mustard increased by 50–100% upon inoculation with *K. ascorbata* SUD165/26 in Ni-contaminated soil (Burd et al. 1998).
2. In the presence of PGPR, toxicity level of nickel was significantly reduced in canola or tomato seeds (Burd et al. 1998).
3. PGPR enhanced accumulation of Se and Hg in wetland plant tissues (de Souza et al. 1999b).

3. *Plant-bacteria interactions*

During the process of symbiosis (plant and bacteria), the adaptation capabilities of both partners should be more in helping in alleviation of high level of heavy metals. Herein, bacteria help in augmentation of the contaminated soil and ameliorate functionality with improved soil and plan health (Elsgaard et al. 2001; Filip 2002).

Fig. 3.3 Increased uptake of contaminants on application of mycorrhiza

3.6 Phytovolatilization

Phytoextraction and phytovolatilization occur simultaneously. Phytovolatilization is one of the important processes of uptake and transpiration of water-soluble contaminants by the plants. Contaminants which are present in plants in soluble form undergo several processes and finally volatilize into the atmosphere along the stream of transpiration. Phytovolatilization has been widely used to remove mercury by converting its more toxic mercuric ion into less toxic elemental mercury. In the presence of VOCs, plants may help in alleviation and transportation of different types of organic compounds and thereby affect the fate of contaminants (Fig. 3.4) (Limmer and Burken 2016).

Direct Phytovolatilization In this process plant-mediated uptake and translocation of contaminants to the shoot portion to diffuse across hydrophobic barriers such as cutin in the epidermis or suberin in woody dermal tissues.

Indirect Phytovolatilization By deploying ample amount of soil plants, take vast quantities of water whereby activities of plant roots may increase the flux of volatile contaminants (Jasechko et al. 2013) from the subsurface through the following ways:

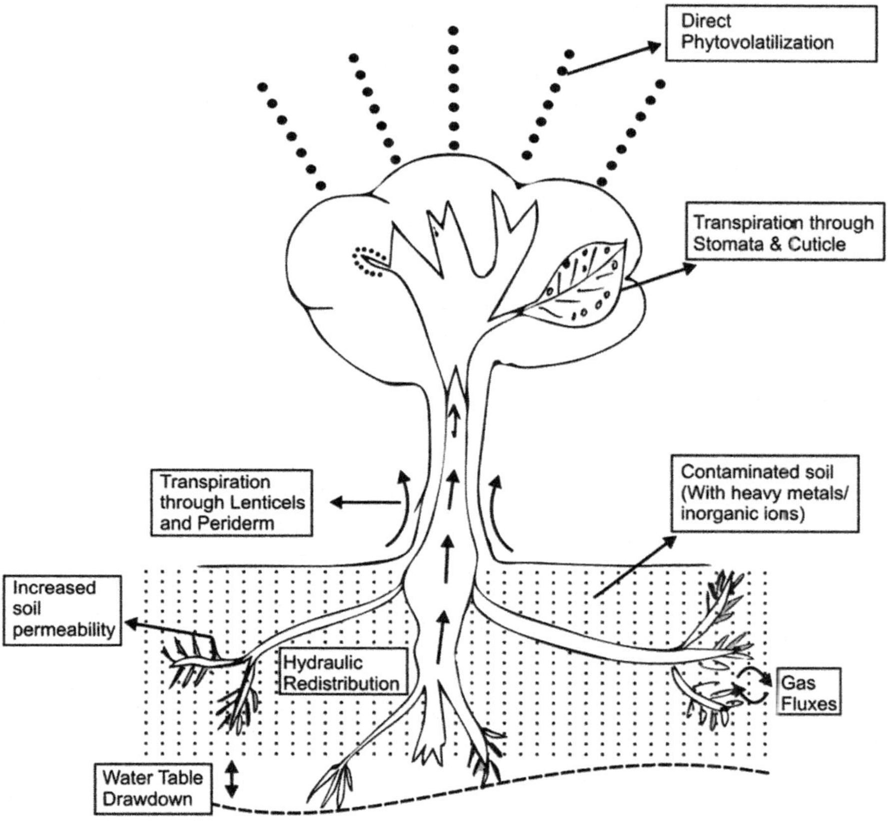

Fig. 3.4 Phytovolatilization processes: direct and indirect

- Lowering the water table.
- Water table fluctuations cause gas fluxes.
- Increased soil permeability.
- Hydraulic redistribution.
- Interception of rainfall that would otherwise infiltrate to dilute and advert VOCs away from the surface.

According to Negri et al. (2003), plant roots redistribute water throughout the subsurface by employing two types of hydraulic lift (Figs. 3.5 and 3.6) (Neumann and Cardon 2012). It is a pace of transportation stream wherein organic contaminants cross the cellular membrane of root passively and whereby volatilize (Dettenmaier et al. 2009).

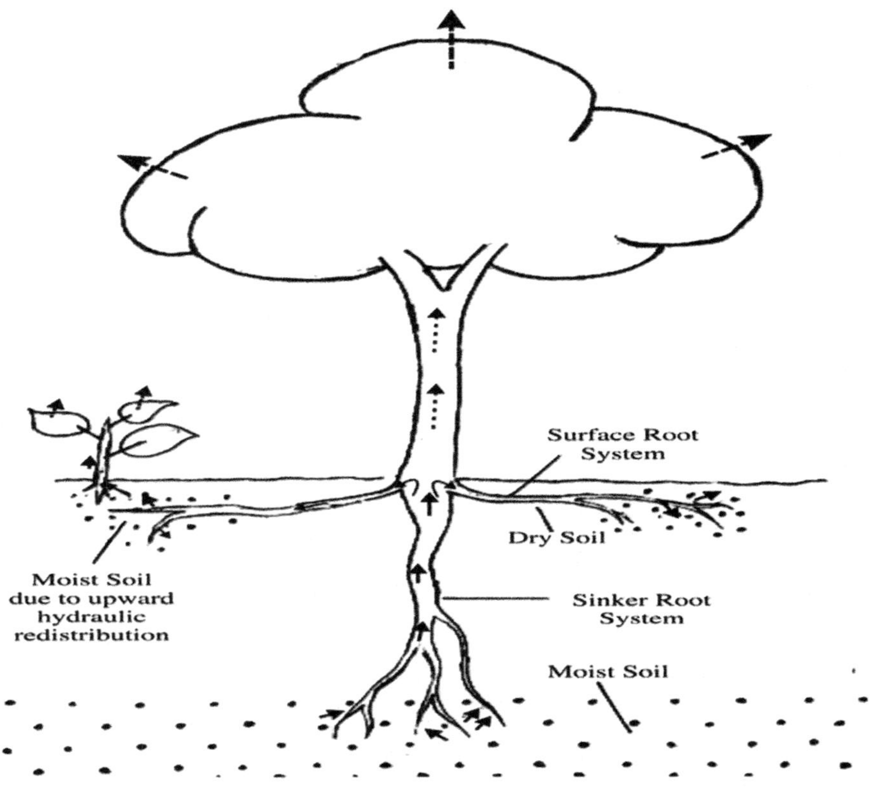

Fig. 3.5 Upward hydraulic redistribution in trees

3.7 Functions of Plant Volatiles

There are a number of functions implied through plant VOCs wherein several of them are:

1. *Plant Reproduction*: Floral scent is a signal for pollinators that can be used to pollinate wherein a diverse blend of plant VOCs attract pollinators and to ensure reproduction (Knudsen and Tollsten 1993). There is a large diversity of volatiles that may contain from 1 to 100 volatiles wherein amount varies from the low picogram range to more than 30 μg/h (Knudsen and Gershenzon 2006).
2. *Plant Defense*: To succumb the adverse effect, a blend of VOCs are produced upon attack of biotic stress (Vancanneyt et al. 2001; Dicke and van Loon 2000).

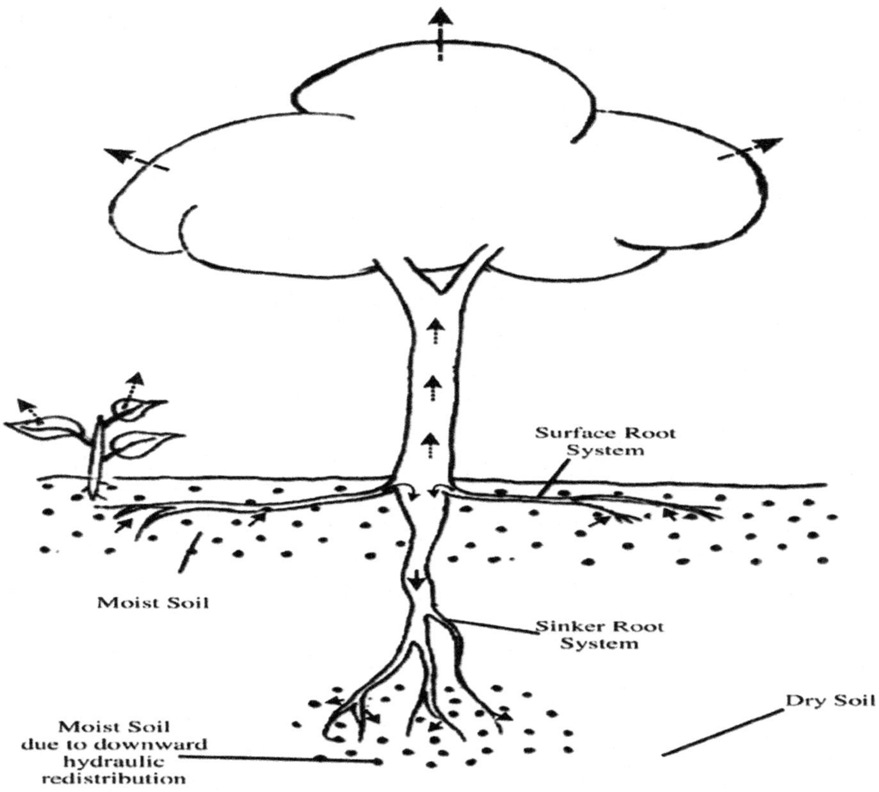

Fig. 3.6 Downward hydraulic redistribution in trees

The produced VOCs help in sustainability of plants in direct/indirect way by deploying lipoxygenase (LOX) pathway, the shikimic acid pathway, and products of the terpenoid pathway (Kessler and Baldwin 2001; Pichersky and Gershenzon 2002; Horiuchi et al. 2003; Gols et al. 2003; Heil 2004).

3. *Plant-Herbivore-Carnivore Interactions*: This type of tritrophic interaction induced by VOCs includes interactions between lima bean plants (*Phaseolus lunatus*), herbivorous spider mites (*Tetranychus urticae*), and carnivorous mites (*Phytoseiulus persimilis*) (Takabayashi and Dicke 1996). Merely lying of egg on plants produces VOCs which attract egg parasitoids (Hilker and Meiners 2002). Similarly, herbivore- and wound-induced VOCs attract predators/parasitoids in plant-caterpillar-parasitoid (Dicke and van Loon 2000) and plant-caterpillar-predatory bug interactions (Kessler and Baldwin 2001).
4. *Plant-Plant Interactions*: Herbivore-infested plants produce VOCs that also mediate plant-plant interactions and may induce the expression of defense genes

(Dicke et al. 1990; Arimura et al. 2002, 2004b). Release of herbivore-induced volatiles occurs both locally from damaged tissues and systemically from undamaged tissues and displays distinct temporal patterns (Arimura et al. 2004b). *Nicotiana tabacum*, for example, releases several herbivore-induced volatiles exclusively at night. These nocturnally emitted compounds repel female moths (*Heliothis virescens*), which search for oviposition sites during the night (De Moraes et al. 2001).

5. *Role of Plant Volatiles in Belowground Defense*: The emission of VOCs is not limited to aerial parts of the plants; rather it involves rhizosphere VOCs that help in plant defense against root-feeding enemies.

   For example:

   1. A bacterial (*Pseudomonas syringae* strain DC 3000) or fungal (*Alternaria brassicicola*) pathogen or rootfeeding insect (*Diuraphis noxia*) triggers the rapid emission of 1,8-cineole upon infection with *Arabidopsis* roots (Hammer et al. 2003; Chen et al. 2004; Ro et al. 2006) which enhances plant defense.
   2. Upon attack by weevil larvae *Otiorhynchus sulcatus*, emission of VOCs by roots of *Thuja occidentalis* was shown to attract the entomopathogenic nematode *Heterohabditis megidis* (Boff et al. 2001). Similarly, root-feeding larvae (*Delia radicum*) emit VOCs by turnip roots that attract the parasitoid *Trybliographa rapae* (Neveu et al. 2002).
   3. By deploying GC-MS, the emitted VOC was identified as the sesquiterpene (*E*)-*β*-caryophyllene that produced belowground upon root-insect-induced plant signal that strongly attracts an entomopathogenic nematode *Heterorhabditis megidis* (Rasmann et al. 2005).

6. *Abiotic Stresses*: Under abiotic stress it has been reported that plant VOCs maintained photosynthetic rate at elevated temperatures (Copolovici et al. 2005; Penuelas et al. 2005). Besides, fumigation with exogenous isoprene of fosmidomycin-fed leaves of red oak (*Quercus rubra*) and kudzu (*Pueraria lobata* [Willd.] Ohwi.) increased the ability of photosynthetic apparatus to recover from a brief high-temperature exposure (Sharkey et al. 2001). In addition, isoprenoids served as antioxidants to protect plants against a range of stresses including ozone-induced oxidative stress (Loreto et al. 2001, 2004) and singlet oxygen accumulation (Affek and Yakir 2002).

Some relevant reports regarding phytoextraction and phytovolatilization are tabulated (Table 3.1.).

Table 3.1. Some case studies involving different plants/halophytes in organic/inorganic ion remediation through phytoextraction and phytovolatilization

Sr. No	Hyperaccumulator plant(s) used and nature of studies	Quantification of remediation	Reference
1.	In a closed system with hybrid poplars after 7 days of exposure	Fraction of trichloroethylene (TCE) to the TCE taken up by the plant was 70–90%	Gordon et al. (1998)
2.	Phytovolatilization by alfalfa (*Medicago sativa*) treated with TCE and 1,1,1-trichloroethane (TCA) in soil mesocosms	Similar and notable fractions of TCE and TCA were volatilized	Narayanan et al. (1995)
3.	Poplar cuttings dosed with C-MTBE for 7 days	Volatilized 54% of the total MTBE	Rubin and Ramaswami (2001)
4.	In closed system with poplar and willow cuttings dosed with C-PCE or C-TCE	75% of the contaminant taken up by the plant was phytovolatilized	Jasechko et al. (2013)
5.	Hydroponic hybrid poplars dosed with C-MTBE for 10 days	Directly phytovolatilize 17% of the total (96%) MTBE	Jasechko et al. (2013)
6.	*Salicornia europaea* collected from the vicinity of Maharloo salt lake near Shiraz, Fars Province, Iran	High (30%) Na^+ accumulation corresponding to 31,500 μg g^{-1} dry weight. Amounts of Ni, Cr, Cd, Pb, and Hg were also found to be high in this plant	
7.	*Apocynum lancifolium, Chenopodium album*; Girlan, Khirezm region, northwest Uzbekistan	*Chenopodium album* produced 3.25 t ha^{-1} $year^{-1}$ dry biomass removing 569.6 kg ha^{-1} salt ions from 0.3 m of the soil profile amounting to 1.47% of the soil salts	Hamidov et al. (2007)
8.	*Suaeda fruticosa, Suaeda nudiflora, Salsola baryosma, Haloxylon recurvum*, and *Atriplex lentiformis* grown in salinity microplots at CCS Haryana Agricultural University, Hisar	These plants were found to be best salt hyperaccumulators and also had high biomass. These plants had the potential of desalinization of saline soils from 16 dSm^{-1} to 2 dSm^{-1} in 4.9–6.1 years	Datta and Angrish (2006) and Devi et al. (2008, 2016)
9.	Wild-growing *Tecticornia indica* and *Suaeda nudiflora*; Soliman sublake, northeast Tunisia	Both perennial plants exhibited high productivities and Na+accumulation, i.e., *T. indica* 7.4 t dry weight ha^{-1} and Na^+ 0.7 t ha^{-1} and *S. nudiflora* 0.75 t ha^{-1} and 0.22 t ha^{-1}, respectively. Soil salinity was lowered and microbial biomass was more in the remediated soil	Ouni et al. (2013)

3.8 Application

Phytoremediation processes may be applied near the industries or where the effluent has been reached. Worldwide phytoremediation projects have been carried out to mitigate the organic and inorganic contaminants that are released from different sources and excessive utilization of fertilizers and pesticides for agricultural purposes. Members of Chenopodiaceae and alpine pennycress, hemp, pigweed, etc. have proven to be successful for hyperaccumulating contaminants at toxic waste sites, i.e., abandoned metal mine workings and ongoing coal-mine discharges. This technology has become increasingly popular and employed at sites with soils contaminated with lead, uranium, and arsenic. Phytoremediation is a natural process which depends upon the rooting system and plants' ability to accumulate maximum contaminants in their aboveground biomass. These hyperaccumulator plants are also exposed to the herbivore animals, so it enters into the food web. Phytoremediation processes also have some advantages and disadvantages as follows:

Advantages

- Phytoremediation is less costly and easy to install both in situ and ex situ.
- Physiology of plants can be easily studied.
- It is very eco-friendly, aesthetically pleasing, and publicly acceptable.
- The process of phytomining increases the possibility of the reuse of valuable metals.
- It is more economically viable using the same tools and supplies as agriculture.
- It is less disruptive to the environment.
- It reduces the risk of spreading the contaminants by avoiding excavation.

Disadvantages

- Phytoremediation is limited up to the spreading of the roots.
- It requires a long-term commitment.
- It also requires the screening of hyperaccumulator plants.
- Toxicity of the contaminants leads to the death of the plant.
- There is recycling of the contaminants by entering into the food chain or released into the environment during autumn season.
- Environmental damage may be increased due to greater solubility or leaching of the contaminants.

3.9 Conclusion and Future Projections

On the basis of the literature reports, it is inferred conclusively beyond doubt that hyperaccumulator plants were able to phytoremediate the contaminated soils very efficiently and effectively (Fig. 3.7). These could provide proficient, sustainable, and low-cost plant-based technology for greening of contaminated wastelands and amelioration of physical and chemical nature of top layer of soil especially in arid

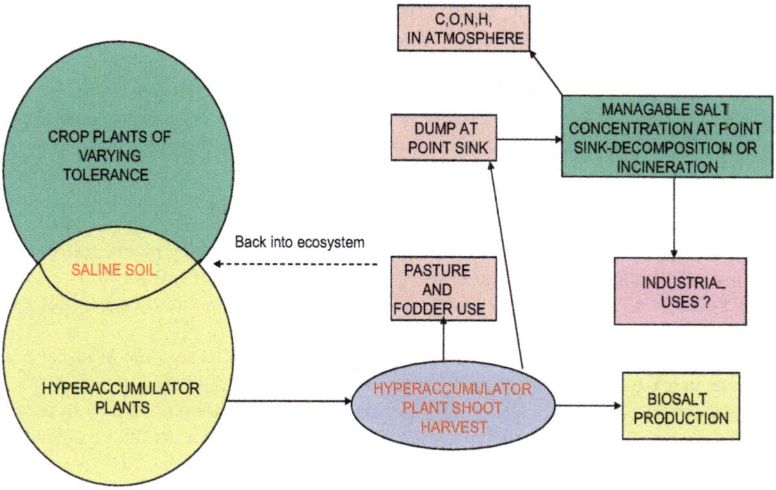

Fig. 3.7 Complementary use of hyperaccumulator plants and crop systems

and semiarid tracts of India. These hyperaccumulator plants also provide fodder, substituted vegetables, grain, fire (fuel) wood, and oil and hence are economically viable plants as well for livestock and rural people (Abbad et al. 2004). Another feasibility in the near future is the production of bio-salt or vegetable salt (CSMRI) from these hyperaccumulator plants.

In fact these plants use sun's energy to remove contaminants from soil. So transpiration-/translocation-mediated and active uptake and sequestration of contaminants are the core of hyperaccumulation technology. This phytoremediation technology involves the repeated cropping (harvestings) of these hyperaccumulator shoots until the soil contaminants have reached acceptable levels for the farmers to cultivate their regular crops. Further, these hyperaccumulator plants should always be harvested and dumped at a point sink or incinerated for further industrial uses as well.

References

Abbad A, El Hadrami A, El Hadrami I, Benchaabane A (2004) *Atriplex halimus* (Chenopodiaceae): a halophytic species for restoration and rehabilitation of saline degraded lands. Pak J Biol Sci 7:1085–1093

Abou-Shanab RA, Angle JS, Delorme TA, Chaney RL, van Berkum P, Moawad H, Ghanem K, Ghozlan HA (2003) Rhizobacterial effects on nickel extraction from soil and uptake by *Alyssum murale*. New Phytol 158:219–224

Affek HP, Yakir D (2002) Protection by isoprene against singlet oxygen in leaves. Plant Physiol 129:269–277

Angrish R, Devi S (2014) Potential of salt hyperaccumulation plants in salinity phytoremediation. Advances Plant Physiol 15:307–323

Arimura G, Ozawa R, Nishioka T, Boland W, Koch T, Kuhnemann F, Takabayashi J (2002) Herbivore-induced volatiles induce the emission of ethylene in neighboring lima bean plants. Plant J 29:87–98

Arimura G, Ozawa R, Kugimiya S, Takabayashi J, Bohlmann J (2004) Herbivore-induced defense response in a model legume: two-spotted spider mites, *Tetranychus urticae*, induce emission of (E)-β-ocimene and transcript accumulation of (E)-β-ocimene synthase in *Lotus japonicus*. Plant Physiol 135:1976–1983

Baumann A (1885) Das verhalten von zinksalzen gegen pffanzen and in boden

Belimov AA, Kunakova AM, Safroncva VI, Stepanok VV, Yudkin LY, Alekseev YV, Ozhemyakov AP (2004) Employment of rhizobacteria for the inoculation of barley plants cultivated in soil contaminated with lead and cadmium. Microbiology (Moscow) 73(1):99–106

Bingham FT, Pereyea FJ, Jarrell WM (1986) Metal toxicity to agricultural crops. Met Ions Biol Syst 20:119–156

Boff MIC, Zoon FC, Smits PH (2001) Orientation of *Heterorhabditis megidis* to insect hosts and plant roots in a Y-tube sand olfactometer. Entomol Exp Appl 98:329–337

Brown SL, Chany RL, Angle RS, Baker AJM (1995) Zinc and cadmium uptake by hyperaccumulator Thlaspi caerulescens and metal tolerant Silene vulgaris grown on sludge amended soils. Environ Sci Technol 25:1581–1585

Burd GI, Dixon DG, Glick BR (1998) A plant growth-promoting bacterium that decreases nickel toxicity in seedlings. Appl Environ Microbiol 64:3663–3668

Byers JN (1935) Selenium occurrence in certain soils in the United States, with a discussion of related topics. US Dep Agric Technol Bull 482:1–47

Byers HG (1936) Selenium occurrence in certain soils in the United States, with discussion of related topics. Second report US Dep. Agric Technol Bull 530:1–78

Chen F, Ro DK, Petri J, Gershenzon J, Bohlmann J, Pichersky E, Tholl D (2004) Characterization of root-specific Arabidopsis terpene synthase responsible for the formation of the volatile monoterpene 1,8-cineole. Plant Physiol 135:1956–1966

Copolovici LO, Filella I, Llusia J, Niinemets U, Penuelas J (2005) The capacity for thermal protection of photosynthetic electron transport varies for different monoterpenes in *Quercus ilex*. Plant Physiol 139:485–496

Cunningham SD, Berti WR (1993) Remediation of contaminated soils with green plants: an overview. Cell Dev Biol 29(4):207–212

Cunningham SD, Ow DW (1996) Promises and prospects of phytoremediation. Plant Physiol 110:715–719

Dahiya IS, Laura RD (1988) The recent advances in the characterization, reclamation and management of salt affected soils in India – A review. Intern J Trop Agric 6:157–117

Datta KS, Angrish R (2006) Selection, characterization and quantification of plant species for phytoremediation of saline soils. Final Progress Report, Ministry of Environment and Forests, Government of India, New Delhi, pp 1–166

De Moraes CM, Mescheer MC, Tumlinson JH (2001) Caterpillar induced nocturnal plant volatiles repel nonspecific females. Nature 410:577–580

de Souza MP, Huang CP, Chee N, Terry N (1999) Rhizosphere bacteria enhance the accumulation of selenium and mercury in wetland plants. Planta 209:259–263

Dettenmaier EM, Doucette WJ, Bugbee B (2009) Chemical hydrophobicity and uptake by plant roots. Environ Sci Technol 43(2):324–329

Devi S, Rani C, Datta KS, Bishnoi SK, Mahala SC, Angrish R (2008) Phytoremediation of soil salinity using salt hyperaccumulator plants. Indian J Plant Physiol 4:347–356

Devi S, Nandwal AS, Angrish R, Arya SS, Kumar N, Sharma SK (2016) Phytoremediation potential of some halophytic species for soil salinity. Int J Phytoremediation 18:693–696

Dicke M, van Loon JJA (2000) Multitrophic effects of herbivore-induced plant volatiles in an evolutionary context. Entomol Exp Appl 97:237–249

Dicke M, Abelis MW, Takabayashi J, Bruin J, Posthumus MA (1990) Plant strategies of manipulating predator-prey interactions through allelochemicals: prospects for application in pest control. J Chem Ecol 16:3091–3117

Elizabeth PS (2005) Phytoremediation. Annu Rev Plant Biol 56:15–39

Elsgaard L, Petersen SO, Debosz K (2001) Effects and risk assessment of linear alkylbenzene sulfonates in agricultural soil. 1. Short-term effects on soil microbiology. Environ Toxicol Chem 20(8):1656–1663

Filip Z (2002) International approach to assessing soil quality by ecologically-related biological parameters. Agric Ecosyst Environ 88(2):689–712

Garbisu C, Alkorta I (2001) Phytoextraction: a cost-effective plant-based technology for the removal of metals from the environment. Bioresour Technol 77(3):229–236

Glick BR (2003) Phytoremediation: synergistic use of plants and bacteria to clean up the environment. Biotechnol Adv 21(5):383–393

Glick BR, Patten CL, Holguin G, Penrose DM (1999) Biochemical and genetic mechanisms used by plant growth-promoting bacteria. Imperial College Press, London

Gols R, Roosjen M, Dijkman H, Dicke M (2003) Induction of direct and indirect plant responses by jasmonic acid, low spider mite densities, or a combination of jasmonic acid treatment and spider mite infestation. J Chem Ecol 29:2651–2666

Gordon M, Choe N, Duffy J, Ekuan G, Heilman P, Muiznieks I, Ruszaj M, Shurtleff BB, Strand S, Wilmoth J, Newman LA (1998) Phytoremediation of trichloroethylene with hybrid poplars. Environ Health Perspect 106:1001–1004

Hamidov A, Khaydarova V, Khamidov M, Neves A, Beltrao J (2007) Remediation of saline soils using Apocynum lancifolium and Chenopodium album. In: Proceedings of the 3rd IASME/WSEAS international conference on energy, environment, Agios Nikolas, Greece, July 24–26, pp 157–164

Hammer KA, Carson CF, Riley TV (2003) Antifungal activity of the components of *Melaleuca alternifolia* (tea tree) oil. J Appl Microbiol 95:853–860

Heil M (2004) Direct defense or ecological costs: responses of herbivorous beetles to volatiles released by wild Lima bean (*Phaseolus lunatus*). J Chem Ecol 30:1289–1295

Hilker M, Meiners T (2002) Induction of plant responses towards oviposition and feeding of herbivorous arthropods: a comparison. Entomol Exp Appl 104:181–192

Horiuchi JI, Arimura GI, Ozawa R, Shimoda T, Dicke M, Takabayashi J, Nishioka T (2003) Lima bean leaves exposed to herbivore-induced conspecific plant volatiles attract herbivores in addition to carnivores. Appl Entomol Zool 38:365–368

Huang Y, Tao S, Chen YJ (2005) The role of arbuscular mycorrhiza on change of heavy metal speciation in rhizosphere of maize in wastewater irrigated agriculture soil. J Environ Sci 17(2):276–280

Jasechko S, Sharp ZD, Gibson JJ, Birks SJ, Yi Y, Fawcett PJ (2013) Terrestrial water fluxes dominated by transpiration. Nature 496:347–350

Kabata-Pendias A, Pendias H (1989) Trace elements in the soil and plants. CRC Press, Boca Raton

Kessler A, Baldwin IT (2001) Defensive function of herbivore-induced plant volatile emissions in nature. Science 291:2142–2143

Knudsen JT, Gershenzon J (2006) The chemistry diversity of floral scent. In: Dudareva N, Pichersky E (eds) Biology of floral scent. CRC Press, Boca Raton, pp 27–52

Knudsen JT, Tollsten L (1993) Trends in floral scent chemistry in pollination syndromes: floral scent composition in moth-pollinated taxa. Bot J Linn Soc 113:263–284

Kumar PBA, Dushenkov V, Motto H, Raskin I (1995) Phytoextraction: the use of plants to remove heavy metals from soils. Environ Sci Technol 29(5):1232–1238

Lasat HA (2002) Phytoextraction of toxic metals: a review of biological mechanisms. J Environ Qual 31(1):109–120

Limmer M, Burken J (2016) Phytovolatilization of organic contaminants. Environ Sci Technol 50:6632–6643

Loreto F, Mannozzi M, Maris C, Nascetti P, Ferranti F, Pasqualini S (2001) Ozone quenching properties of isoprene and its antioxidant role in leaves. Plant Physiol 126:993–1000

Loreto F, Pinelli P, Manes F, Kollist H (2004) Impact of ozone on monoterpene emissions and evidence for an isoprene-like antioxidant action of monoterpenes emitted by *Quercus ilex* leaves. Tree Physiol 24:361–367

McCutchen SC, Schnoor JL (2003) Phytoremediation: transformation and control of contaminants. Wiley, Hoboken, pp 233–262

Minguzzi C, Vergnano O (1948) Il Contenuto di nichel nelle ceneri di Alyssum bertolonii Desu. Atti della Societa Tosoma di Scienze. Nat Mem Ser A 55:49–77

Narayanan M, Davis LC, Erickson LE (1995) Fate of volatile chlorinated organic compounds in a laboratory chamber with alfalfa plants. Environ Sci Technol 29:2437–2444

Negri MC, Gatliff EG, Quinn JJ, Hinchman RR (2003) Root development and rooting at depths. In Phytoremediation

Neumann RB, Cardon ZG (2012) The magnitude of hydraulic redistribution by plant roots: a review and synthesis of empirical and modeling studies. New Phytol 194:337–352

Neveu N, Grandgirard J, Nenon JP, Cortesero AM (2002) Systemic release of herbivore-induced plant volatiles by turnips infested by concealed root-feeding larvae *Delia radicum* L. J Chem Ecol 28:1717–1732

Ouni Y, Lakhdar A, Rabi M, Aoui AS, Maria AR, Chedly A (2013) Effects of the halophytes *Tecticornia indica* and *Suaeda fruticosa* on soil enzyme activities in a Mediterranean sabkha. Int J Phytoremediation 15:188–197

Penuelas J, Llusia J, Asensio D, Munne-Bosch S (2005) Linking isoprene with plant thermotolerance, antioxidants and monoterpene emissions. Plant Cell Environ 28:278–286

Pichersky E, Gershenzon J (2002) The formation and function of plant volatiles: perfumes for pollinator attraction and defense. Curr Opin Plant Biol 5:237–243

Rasmann S, Kollner TG, Degenhardt J, Hiltpold I, Toepfer S, Kuhlmann U, Gershenzon J, Turlings TCJ (2005) Recruitment of entomopathogenic nematodes by insect-damaged maize. Nature 434:732–737

Ro DK, Ehlting J, Keeling CI, Lin R, Mattheus N, Bohlmann J (2006) Microarray expression profiling and functional characterization of *AtTPS* genes: duplicated *Arabidopsis thaliana* sesquiterpene synthase genes *At4g13280* and *At4g13300* encode root-specific and wound-inducible $(Z)-\gamma-$ bisabolene synthases. Arch Biochem Biophys 448:104–116

Rubin E, Ramaswami A (2001) The potential for phytoremediation of MTBE. Water Res 35:1348–1353

Sairam RK, Tyagi A (2004) Physiology and molecular biology of salinity stress tolerance in plants. Curr Sci 86:407–421

Salt DE, Blaylock M, Kumar NPBA, Dushenkov V, Ensley BD, Chet I, Raskin I (1995) Phytoremediation: a novel strategy for the removal of toxic metals from the environment using plants. Biol Technol 13(5):468–474

Salt DE, Smith RD, Raskin I (1998) Phytoremediation. Ann Rev Plant Physiol Mol Biol 49:643–668

Sarma H (2011) Metal hyperaccumulation in plants: a review focusing on phytoremediation technology. J Environ Sci Technol 4:118–138

Sen DN, Rajpurohit KS, Wissing FW (1982) Survey and adaptive biology of halophytes in Western Rajasthan, India. Department of Botany and Geography, University of Jodhpur, vol 2. Dr. W. Junk Publishers, The Hague

Sharkey TD, Chen XY, Yeh S (2001) Isoprene increases thermotolerance of fosmidomycin-fed leaves. Plant Physiol 125:2001–2006

Skelding AD, Wintebotham J (1939) The structure and development of the hydathodes of *Spartina townsendii* groves. New Phytol 38:69–79

Suresh B, Ravisanker GA (2004) Phytoremediation – A novel and promising approach for environmental clean up. Critic Rev Biotech 24:97–124

Takabayashi J, Dicke M (1996) Plant-carnivore mutualism through herbivore-induced carnivore attractants. Trends Plant Sci 1:109–113

Thangavel P, Subbhuraam CV (2004) Phytoextraction: role of hyperaccumulators in metal contaminated soils. Proc Indian Natl Sci Acad 70:109–130

Tanji KK (1990) Agricultural Salinity Assessment and Management. Irrigation and Drainage Division, American Society of Civil Engineers, New York

Vancanneyt G, Sanz C, Farmaki T, Paneque M, Ortego F, Castanera P, Sanchez-Serrano JJ (2001) Hydroperoxide lyase depletion in transgenic potato plants leads to an increase in aphid performance. Proc Natl Acad Sci U S A 98:8139–8144

Wenzel WW, Lombi E, Adriano DC (1999) Biochemical processes in the rhizosphere: role in phytoremediation of metal-polluted soils. In: Prasad MNV, Hagemeyer J (eds) Heavy metal stress in plants: from molecules to ecosystems. Springer, Berlin, pp 273–303

Whiting SN, de Souza MP, Terry N (2001) Rhizosphere bacteria mobilize Zn for hyperaccumulation by Thlaspi caerulescens. Environ Sci Technol 35(15):3144–3150

Williams MC (1960) Effect of sodium and potassium salts on growth and oxalate content of halogeton. Plant Physiol 35:500–505

Yeo AR (1974) Salt tolerance in the halophyte Suaeda maritime L. Dum. D. Phil. Thesis Univ. Sussex, England, p 183

Ziegler H, Luttge U (1967) Die Salzdrusen von *Limonium vulgare*. II. Mitteilung, Die Lokisierung des chloride. Planta 74:1–17

Microbial Volatiles in Defense

4

Monika, Khan Mohd. Sarim, S.S. Arya, S. Devi, Vikender Kaur, and A. Singla

Abstract

Microbes are versatile, dynamic, and the most adaptive entities occurring in nature, and these properties enable them to strive in almost any conceivable environment. These are loaded with the variety of compounds having potential to cope up with such harsh conditions. Among the diversity of compounds, the important one is volatile compounds; these are light molecular weight, low vapor pressure compounds that easily disperse in environment, plant, and microbes and trigger metabolic and physiological responses that confer microbial defense and induce systemic resistance in plants. Basic chromatography and mass spectrometry techniques enable us to understand the chemical structure and function of these fascinating molecules. In addition to this, modern OMICS methods giving opportunity to deep insight of microbial diversity and strengthen the concept of volatile compounds function by providing real-time pictures of their expression and signaling. Incorporation of computational tools with molecular biology techniques incredibly creates a reservoir of knowledge-based database of volatile compounds' structure, function, diversity, signaling, and even prediction through statistical tools. Hence, we are closer to decipher the significance of microbial

Monika • K.M. Sarim
Department of Microbiology, CCS Haryana Agriculture University, Hisar, Haryana, India

S.S. Arya (✉)
Department of Botany, Maharshi Dayanand University, Rohtak, Haryana, India
e-mail: aryasunder.hau@gmail.com

S. Devi
Department of Botany and Plant Physiology, CCS HAU, Hisar, Haryana, India

V. Kaur
Germplasm Evaluation Division, ICAR-NBPGR, Pusa Campus, New Delhi, India

A. Singla
Sam Higginbottom Institute of Agriculture, Technology & Sciences, Allahabad, India

D.K. Choudhary et al. (eds.), *Volatiles and Food Security*,
DOI 10.1007/978-981-10-5553-9_4

volatile compounds in plants and microbes defense as well, and basic understanding, computational approaches, and intellectual input can definitely provide some fruitful findings.

Keywords
Volatile compounds • Microbes • Microbial volatile organic compounds (MVOCs) • Volatile defense

4.1 Volatile Compounds: Delighting Compounds from Microbes

Who can't love the smell coming from soil during rainy season? Yes it is "geosmin" a volatile compound frequently produced by an actinomycetes called *Streptomyces* (Gerber and Lechevalier 1965). Volatile organic compounds (VOCs) are carbon-structured solids or liquids that quickly turn to gas phase through vaporization at 0.01 kPa pressure and 20 °C (Pagans et al. 2006). Several of these diffusible metabolites act as antibiotics and antibiotic-like substances and also act as potential signal molecules that play important role in interactions between plant-bacteria or fungi-bacteria (Frey-Klett et al. 2011; Haq et al. 2014). Microbial volatiles are small molecules of low molecular weight, high vapor pressure, and low water solubility produced as intermediate compounds during microbial metabolism. MVC can be fetched from microbial culture during production of industrially important substances such as coolants, petroleum, from cleaners, and diverse pharmaceutical manufacturing. MVC are made up of fine architecture of amines, aldehydes, ketones, aromatic hydrocarbon, terpenes, sulfur, and, most importantly, alcohol. In a different aspect, these groups contribute to the distinct smell of variety of volatile compounds. Concentration of any volatile compound also poses great impact on the production and efficacy of these compounds. In an offbeat research, Rice and Koziel (2015) demonstrated the interplay of chemical concentration in product development of a legislative drug "marijuana" using multidimensional GC-MS and simultaneous sensory analysis. Result suggested that aroma profile from marijuana changed during fresh preparation to packaging and storage. This study will be helpful for dogs and drug officers to cope up with drug instances and monitoring.

4.2 Diversity of MOVC in Ecosystem

Microbial volatile organic compounds are extremely diverse in structure and functions like microbes themselves. Structural variation in species-/strain-specific MOVC has great importance in volatile compound studies. Bacteria produce a diverse array of organic and inorganic volatile compounds that facilitate acclimatization, pathogenesis, and signal transduction. Organic volatile compounds include:

1. *Hydrocarbons*: produced during fatty acid biosynthesis and classified into long-chain and short-chain hydrocarbons. The former are usually found in bacteria and cyanobacteria, whereas the latter generally occur in fungi (Tellez et al. 2001).
2. *Ketones and alcohols*: produced during fatty acid decarboxylation. Long chain alcohols (volatile compounds) frequently present in gram negative bacteria such as enterobacteriaceae family members (Hamilton-Kemp et al. 2005), and *Firmicutes* consist short-chain alcohol compounds (Faragh et al. 2013).
3. *Terpenes*: derivatives of dimethylallyl pyrophosphate and isopentenyl pyrophosphate produced through mevalonate pathway.
4. *Sulfur-containing compounds*: well known for distinct aroma in produce during catabolic degradation of methionine. Indoleacetic acid and other products such as trimethylamine (TMA) and 2-amino-acetophenone (2-AA) are the nitrogen-containing volatile compounds. HCN, H_2S, NO, etc., are the inorganic volatile compounds that also play important key role in bacterial and plant signaling and defense.

Many research findings also support the role of volatile structure on the efficiency of compounds action. Lee et al. (2012) described the impact of long-chain molecule structure on the potential of a volatile compound against phytopathogens. A bacterial VOC-C-13, produced by *P. polymyxa* that promote the plant growth via ethylene-mediated signaling and induced systemic resistance in *Arabidopsis* against *Pseudomonas syringae*. The study also suggested that long-chain compound are strain specific and, hence, can be used as marker for microbial systematic.

4.3 MOVC: A Potential Compound for Defense

Microbes and their volatile compounds perform several functions in microbes and plant biology. Many MOVC have been found to induce resistance and compete with pathogenic microorganisms and provide defense to plant and themselves too. Figure 4.1 depicts the volatile compounds production from rhizospheric microbes and their effect on plant's growth and defense.

Microbial volatile compounds protect plants from pathogen directly by inhibition of pathogenic organism or indirectly by eliciting plant defense response against pathogen. But microbial volatile compounds act in dose-dependent manner. For example, *Bacillus subtilis* GB03 single colony does not inhibit spore germination and growth of pathogenic fungi *Botrytis cinerea* and *Fusarium oxysporum.* When the number of colonies of *B. subtilis* gets increased, it inhibits the growth of pathogenic fungi. There is linear relationship between concentration of volatile compounds and growth inhibition of pathogens. Volatiles from *B. subtilis* of 100 mM concentration inhibit spore germination and growth inhibition in *Botrytis cinerea* and *Fusarium oxysporum* (Sharifi and Ryu 2016; Sharifi et al. 2013). Indirectly microbial compounds play role in plant defense by eliciting plant defense through

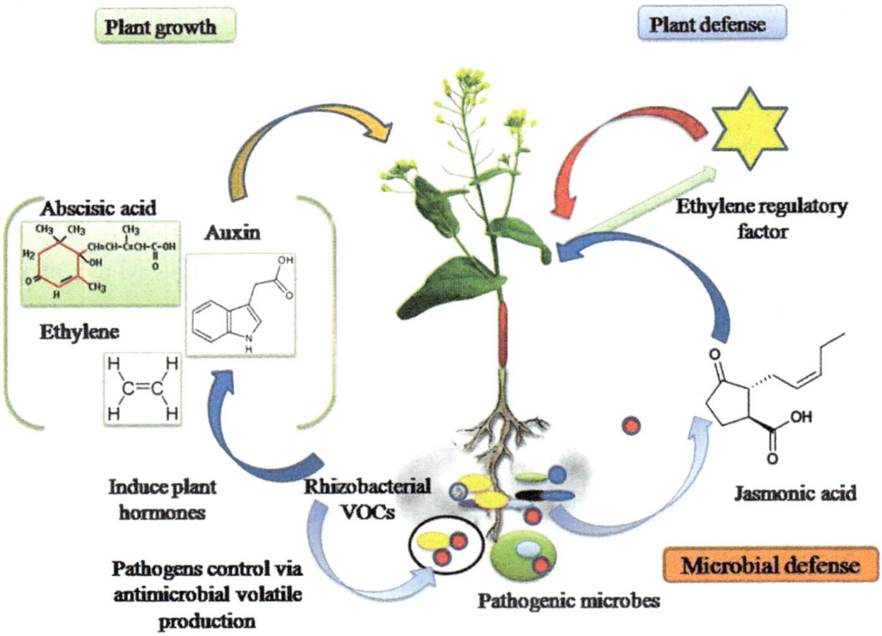

Fig. 4.1 Effects of rhizobacterial volatile compounds on plant growth and defense

induced systematic resistance. For example, long-chain volatile compounds like tridecane and hexadecane elicit induced systemic resistance against *Pseudomonas syringae* in *Arabidopsis*. The same dosages of different volatile compounds exhibit different responses in different pathogens. For example, a low dosage of butanediol inhibited the growth of *Microdochium nivale*, while similar dosage of acetoin induced resistance against *P. syringae* pv. (tomato in *Arabidopsis*). Microbial volatile compounds like methyl jasmonate, 3-pentanol, 2-hydroxy-3-pentanone, and acetoin do not inhibit the spore germination in growth inhibition of *B. cinerea* at 100 mM concentration, but this concentration is usually able to protect plants from disease by eliciting plant defense response against pathogen (Sharifi et al. 2013). In some reports it is evident that bacteria were able to produce 30–200 ng of volatiles depending on soil. The 2-200 ng of these compounds was efficiently activate systemic resistance against *Eriwinia carotovaora* (Ryu et al. 2004).

4.3.1 Volatile Compounds from Bacteria

Diverse array of bacteria have been reported as agent for pathogen control. Rhizobacteria have the ability to produce volatile compounds effective against various plant pathogens that help to strengthen the microbial as well as plant defense. *Sclerotinia sclerotiorum* a threatened soilborne fungus pathogen with broad host susceptibility is the causal agent of white mold in variety of economically important crops. Boland and Hall (1994) mentioned that around 408 species of 278 genera from 75 families are affected with this pathogen that makes it the most serious phytopathogen in irrigated crops. Antifungal effects of volatile compounds (alcohols, aldehydes, ketones, and sulfides) against sclerotia, ascospore germination, and mycelial growth of *Sclerotinia sclerotiorum* were mentioned by Fernandoa et al. (2004). Giorgio et al. (2015) evaluated biocide effects of volatile compound secreted by rhizobacteria against a well-known plant pathogen *Sclerotinia sclerotiorum.* Six rhizobacterial isolates were evaluated for antifungal compound, and VOCs were found effective against different pathogenic fungi and most effectively inhibited mycelial growth of *Sclerotinia sclerotiorum.* In another study bacterial volatile compounds (methanethiol, dimethyl disulfide (DMDS), and 1-undecene) produced by *Pseudomonas tolaasii* were identified by chromatography, and their bioactivity was evaluated against *Pleurotusostreatus and P. eryngii and Agaricus bisporus.* Earlier findings reveal that VOC of *P. talaasii* as virulent factor to control mycelial growth of harmful pathogen and confers plant protection from disease occurrence (Giorgio et al. 2015).

Plant growth-promoting rhizobacteria (PGPR) contribute to plant biomass increase, crop yield, and disease resistance through reduction of ethylene levels and by iron chelator and siderophore production. Besides this, many rhizobacteria produce volatile compounds that are reported as important compound in plant growth (Ryu et al. 2003). Chromatography techniques revealed the chemicals that trigger plant growth promotion are 2,3-butanediol and acetoin (Faraga et al. 2006). Later this group provided evidence that PGPR also play a pivotal role in plant metabolism and through volatile compounds and demonstrated their effects on the sulfur metabolism and defense from herbivores in *Arabidopsis*. *Bacillus amyloliquefaciens* (GB03) produce VOC-enhanced higher sulfur uptake, assimilation, and accumulation in *Arabidopsis*. It induced elevated production of glucosinolates that exhibited protection against herbivore *Spodoptera exigua*, a beet armyworm, BAW (Aziz et al. 2016). Another important aspect of microbial volatiles is the mechanism of action that modulates different physiological response. The characteristic feature of volatile compounds is their hydrophobicity that facilitates partition followed by increase permeability of lipid membrane. In a study, Trombetta et al. (2005) also mentioned the impact of monoterpene volatile compounds on the altered permeability of cell membrane of *Staphylococcus aureus* and *Escherichia coli* (Table 4.1).

4.3.2 Volatile Compound from Fungi

Many genera of fungi show a remarkable ability to produce and excrete volatile compounds. Fungi commonly produce volatiles composed of eight-carbon alcohols, short-chain alcohols, aldehyde, and ketones. Till now, 250 different volatile compounds have been detected and characterized from fungi. Most common VOC producing fungal genera are *Penicillium*, *Aspergillus*, *Mucor*, *Alternaria*, and *Ulocladium*. These volatile compounds have unique properties of plant defense, growth promotion, and antibacterial activity. Very little is known about the precursor and biosynthetic origin of most VOC's compounds. Bjurman et al. (1998) reported that compounds like 3-methyl-1-butanol, 2-methyl-1-propanol, and

Table 4.1 List of microbial volatile compounds with target pathogenic microbes

Bacteria	Compound	Pathogen	Effect	References
Pseudomonas P2 strain	Dimethyl disulfide, dimethyl trisulfide	*Rhizoctonia solani*	Inhibit mycelial growth	Elkahoui et al. (2014)
Streptomyces alboflavus	2-methylisoborneol	*Fusarium moniliforme, Aspergillus flavus, A. ochraceus, A. niger, and Penicillium citrinum*	Inhibit sporulation and mycelial growth	Wang et al. (2013)
B. subtilis B2g, B. cepacia 1S18, P. fluorescens L13-6-12		*Panagrellus redivivus, Bursaphelenchus xylophilus*	Inhibit growth	Vespermann et al. (2007) and Tarkka and Piechulla (2007)
Streptomyces griseus	Carvacrol, dimethyl sulfoxide (DMSO), cyclohexanol, naphthalene	*Penicillium chrysogenum and Botrytis cinerea*	Inhibited spore germination and mycelium growth	Danaei et al. (2014)
Bacillus subtilis		*A. alternata, C. oxysporum, F. oxysporum, P. lilacinus, P. variotii, and P. afertile*	Deformation of mycelial, hyphal, and conidial structures	Chaurasiaa et al. (2005)
Saccharomyces cerevisiae	Six VOC including ethanol and ester compounds	*Sclerotinia sclerotiorum*	Inhibited mycelial growth	Fialho et al. (2011)
Pseudomonas fluorescens and *Serratia plymuthica*	Volatile compound	*Agrobacterium tumefaciens* and *A. vitis*	Bacteriostatic effects	Dandurishvili et al. (2011)
Lysobacter enzymogenes ISE13	2,4-Di-tertbutylphenol	*Colletotrichum acutatum, Phytophthora capsici*	Inhibited spore germination and mycelial growth	Sang and Kim (2011)
Pseudomonas chlororaphis, P. corrugate	Cyclohexanol, 2-ethyl-1- hexanol	*Sclerotinia sclerotiorum*	Inhibition of mycelial growth	Fernando et al. (2005)
Bacterial and fungal volatile-mediated-induced systemic resistance in plants				
Mushroom	1-octen-3-olallo-ocimene and a C-6 aldehyde	*Botrytis cinerea*	*Arabidopsis thaliana*	Kishimoto et al. (2007)

(continued)

Table 4.1 (continued)

Bacteria	Compound	Pathogen	Effect	References
Ampelomyces sp. and *Cladosporium* sp.	m-cresol and methyl benzoate	*Pseudomonas syringae* pv. tomato DC3000 (Pst)	Induced systematic resistance (ISR) in *A. thaliana*	Naznin et al. (2014a, b)
Pseudomonas chlororaphis O6	2R,3R-butanediol		Tolerance to drought in *Arabidopsis thaliana*	Cho et al. (2008)
Bacillus amyloliquefaciens strain IN937a	3-pentanol			
	2,3-butanediol acetoin		Induced systemic resistance (ISR)	Zhou et al. (2010)

styrene are derived from metabolism of fatty acids. Lactones, isoprene nitrogen, and sulfur containing VOCs arise from amino acids. Monoterpenes (farnesol), isoprene, and sesquiterpenes arise from metabolism of acetyl-CoA. Several methyl ketones (acetone), alkanes, esters, and alkenes (1-undecene) also arise from fatty acids (Schmidt et al. 2015). Fungi are also the major contributor of plant defense by producing volatile compounds. A Japanese group isolated a plant growth-promoting fungus, *Talaromyces* sp., from agriculture field and evaluated for its ability to produce volatile compounds. They reported plant growth promotion and induced systematic resistance against *Colletotrichum higginsianum* due to β-caryophyllene produced by this fungus. VCs from *Cladosporium* sp. exhibited induced systematic resistance against plant pathogen *Pseudomonas syringae* (Naznin et al. 2014a, b). Eight-carbon compounds such as 1-octen-3-ol, 3-octanol, and 3-octanone (mushroom alcohol) are among the most common fungal VOCs. 1-octen-3-ol mixed in *Arabidopsis thaliana* inhibits the growth of plant fungus *Botrytis cinerea* (Kishimoto et al. 2007). Vinale et al. (2008) demonstrated that addition of 6-pentyl-2Hpyran-2-one (6PP), one of the earliest known volatile compounds from fungus to plants, reduced disease symptoms. Similarly, *Muscoder* produced different volatile compounds azulene, naphthalene, α-phellandrene, β-caryophyllene, tetrahydrofuran, and 2-pentylfuran growing on different host plant like *Ananas ananassoides*, *Myristica fragrans*, and *Ginkgo biloba*. These volatile compounds emitted by fungus used to control soilborne plant pathogen and reduce the need of methyl bromide fumigation (Fig. 4.1) (Yuan et al. 2012). Volatile compound produced by the culture of *Muscodor crispans* inhibited the growth of serious bacterial pathogen *Xanthomonas axonopodis* pv. *citri* and *Mycosphaerella fijiensis* causal organism of the black sigatoka disease of banana (Mitchell et al. 2010). Microbial volatile organic compounds, i.e., phenyl ethanol, ethyl acetate, and methylbutanol, excreted by the yeast *Saccharomyces cerevisiae* inhibit the growth of plant pathogenic fungus *Guignardia citricarpa* (causal agent of disease citrus black spot) (Fialho et al. 2010). Volatile compounds released by *Phoma* sp. inhibited a number of plant

Table 4.2 Fungal volatiles in plant defense

Species	Molecule	Target organism	Effect	References
Trichoderma virens	δ-Cadinene β-Caryophyllene	Fungi and plants	Growth enhancers and induction of defense response in plant (*Arabidopsis thaliana*) against fungus (*Botrytis cinerea*)	Angel Contreras-Cornejo et al. (2014)
Fusarium oxysporum MSA35	β-Caryophyllene	Plants	Compounds act as plant growth promoter and increased shoot length, root length, and fresh weight of lettuce seedling	Minerdi et al. (2011)
Trametes gibbosa	Octenol	Fungi	Acts as attractant for fungus to control beetles	Thakeow et al. (2008)
Trametes versicolor	β-Gualene γ-Patchoulene δ-Cadinene	Fungi	Acts as attractant for fungus to control beetles	Drilling and Dettner (2009)
Muscodor albus	Sesquiterpenoids, derivative of alcohols and naphthalene	Fungi	For control of postharvest plant diseases	Mercier and Jimenej (2004)
Trichoderma asperellum IsmT5	6-pentyl-α-pyrone (6PP)	Fungi and bacteria	Reduced spore germination	

pathogen like *Cercospora*, *Verticillium*, and *Sclerotinia* (Strobel et al. 2011) (Table 4.2).

4.4 Industrial Production of MVOCs

4.4.1 Response Surface Methodology (RSM) for Optimization of Culture Conditions

Response surface methodology is a statistical tool to optimize, build model, and evaluate effect of factors for most suitable culture conditions for the production of volatile or any other bioactive compounds (Usha et al. 2011). It can also be used in the product development and formulation. It works by minimizing the errors and determine effects of different parameters to show simultaneous, systematic, and efficient variation of all the parameters. RSM involves design, e.g., central composite design (CCD), or any other to fit a second-order polynomial by a least squares technique and use an equation to determine the effect of all variables (Hegde et al. 2013). RSM can also be used to design fermentation setup and media optimization for industrial production of microbial volatiles. In a relevant study RSM was used to optimize the HS-SPME-GC-MS for determination of volatile compounds in caprine coalho cheese and found three factors, viz., equilibration time, extraction time, and extraction temperature as 20 min, 40 min, and 45 °C temperature, respectively (Bezerra et al. 2015). Finally, application of RSM to get the real inference about the plant-microbe interactions and plant growth promotion is implemented. However, very little efforts have been made in this context. Three-dimensional response surface plots provide the real pictures that how a microorganism or combination of microorganisms is affecting the plant growth. There are several statistical packages such as SAS 9.1, Minitab, Design expert, R, Matlab, etc., to perform these simulations. Microbes can be also tested for their optimal performance under combined abiotic stresses.

4.4.2 Bioprocess Engineering for Efficient Recovery of Volatile Compounds from Fermentation Broth

Bioprospecting is increasing our knowledge about the microbe's ability for the production of diverse volatile compounds including microbial fuel, insecticides, antibacterial, etc. Many bacteria have the ability to produce such compounds in industrial reactor also, but the troublesome part is the extraction and recovery of compound. Schoen et al. (2016) developed an analytical system for the rapid quantification and total volatile organic compound extraction from reactor using platinum catalyst with a sensitive CO_2 detector. This platinum catalyst quantitatively oxidizes volatile compound to CO_2; precise oxidation of compound and respiratory CO_2 determination make this system accurate for industrial production of volatile compound. Besides this, other versatile approaches have been practiced such as online monitoring of microbial volatile compound using proton transfer reaction-mass spectrometry (PTR-MS). In four, *Escherichia coli*, *Shigella flexneri*, *Salmonella enterica*, and *Candida tropicalis*, tested strains, some marker ions were found in the headspace of microbial culture. This study indicated temporal emission of VOCs and established an effective way for online monitoring of VOCs during growth (Bunge et al. 2008).

4.5 Synthetic Biology for Redesigning MVOCs

Synthetic biology is a promising approach frequently being used to engineered microbial cell for the production of valuable compound. Synthetic biology aims to design and engineer biology-based parts, novel devices, and systems as well as redesign existing, natural biological systems. The basic concept behind this is the understanding of orchestration of cell, genes, and protein and designing the novel system that produces desired substrate in a required amount and of good quality. Efforts have been made to enhance production of different volatile compounds used for biofuel production that gives an indication that the approach can be implemented to engineer the microbial system for higher production of desired volatile compounds. A single bacterium or fungus some time might not be able to synthesize the entire component needed for the purpose. Hence, synthetic consortium development is a novel approach that targets inherent metabolic traits to produce high-quality and valuable amounts of desired volatile compounds.

4.6 OMICS Approaches Exploring New Concepts of Microbial Volatile-Mediated Plant Defense

Modern biology offers a wide platform to study biological mechanism and physiological responses with deeper understanding. Metagenomics gives liberty to identify complex microbial community without growing them on culture medium. Likewise, proteomics, transcriptomics, and metabolomics are the rapid methods to access the immense valuable information hidden in the microbial world. Species identification using 16S rRNA and ITS (internal transcribed sequences) analysis is commonly used technique for identification of bacteria (Handelsman 2004). Next-generation sequencing (NGS) of hypervariable regions of ITS or 16SrRNA such as V4–V5 facilitates the real analysis of whole microbial composition of any given sample and enables us to identify various potential strains that may possibly be not studied by any other traditional method (Fig. 4.2). Metagenomics deals with the

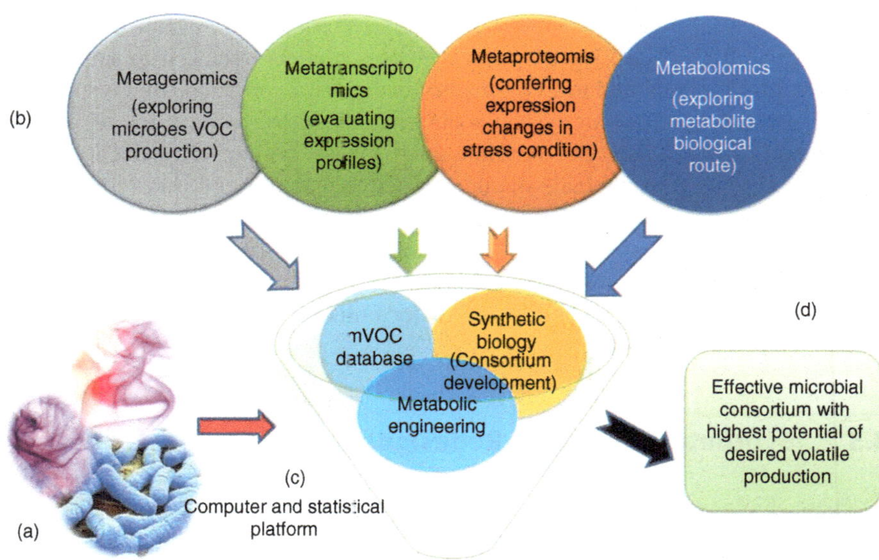

Fig. 4.2 Modern techniques deciphering potential of microbial volatiles: (**a**) evaporation of microbial volatile compounds, (**b**) OMICS technology unrevealing potential volatile producing microbes, and studying changes in gene expression related with respect to particular volatile, (**c**) advance web-based and computational tools offering search, prediction of right microbes for right volatile production, (**d**) outcome of modern technology implementation for highest exploration of microbial volatile compounds

nucleic acid of whole microbial community, purified and sequenced through Illumina platform. Operational taxonomic units (OTUs) generated systematically analyzed through statistical tools for the richness or evenness of a species. As individual strain has unique OTUs, therefore, its richness level in community can be easily determined. Hence, next-generation or Illumina sequencing provides more clear picture of microbial communities. Many web resources such as "GhostKOALA" have prepared for the rapid analysis of whole genome of plant and microbial genome which also strengthens the understanding of microbial molecular architecture of different metabolic pathways available in eukaryotic microbiome. In comparison with classical methods which were completely based on culturable methods for microbial identification using 16S rRNA gene sequencing and phylogenetic tree preparation, NGS is a little more expensive but affordable and also provides more precise information about the whole community. NGS data can be easily simulated through various available computational tools revealing the function by automatically annotating nucleotide sequences similar to those genes that have already been annotated and available in different databases.

4.6.1 Decoding Chemical Dialogues Between Microbes and Pathogens

Microbial interactions are the complex network of chemical signals that facilitate the biological information from one organism to another. The response of pathogen against antimicrobial compound produced by either fungi or bacteria needs to be studied; it will give a clear picture of kinds of responses, chemical signals, involved in any microbial interaction and chance to produce effective antimicrobial compounds than ever before. Raza et al. (2016) studied the effect of volatile organic compounds produced by *Bacillus amyloliquefaciens* SQR-9 against *Ralstonia solanacearum*. Proteomic analysis showed the expression pattern (down regulated 32 and upregulated 22) of different protein under infection process. Downregulated genes were related to virulence, carbohydrate and protein metabolism, antioxidant activity, protein folding, and translation.

4.7 Computational Approaches for Highest Exploration of MVOC in Human Welfare

4.7.1 MVOCs

A number of databases for various biological processes from pathogenesis (PHI-base) to bioremediation (EAWAG, UMBBD) are available. A database solely devoted to microbial volatile compounds has been created (Lemfack et al. 2014). One can search in database using PubChem ID, molecular weight, and name or even through the compound's property. Result shows compound's name, PubChem ID, synonyms, and overall structural information of the respective microbial volatile compounds. The web resource to access the MVOCs database information is http://bioinformatics.charite.de/mvoc/.

4.7.2 In Silico Approaches for Understanding and Modeling Desired Compounds

Rapid widespread of pathogenic variants weakened the resistant mechanism in any organism. To effectively cope up with this, there is an urgent need to develop computational tools to study molecular architecture and interaction modeling of molecules to predict the effective interaction of anti-infective compounds to pathogenic moieties. In silico approaches have been frequently used for analyzing and predicting microbial interaction. A novel approach to predict the missing link between microbial volatile pathways exploring valuable insight into volatile metabolomics and offer a platform for metabolite engineering to design desired compound in required quantity by chemoinformatics and bioinformatics tools. A new way, Reverse Pathway Engineering (RPE), generally uses small molecules that sense VOCs and suggests the reaction or enzymes that link to any known metabolic precursor, so the path can synthesize and aimed molecule can be synthesized. Liu et al. (2014) demonstrated this system biology mechanism nicely for tracing back the missing link in flavor volatile production in lactic acid bacteria. Other advance tools for exploring metabolic routes are ReBiT (Retro-Biosynthesis Tool Martin et al. 2009; Route Designer Law et al. 2009), etc.

4.7.3 Molecular Docking

Molecular docking is used to predict the structure and complex interactions that occur between two molecules, which ultimately affect their efficacy in an environment. It enables us to identify the target proteins which are being affected by a metabolite of microbial origin. The efficacy of a compound depends on the compatibility of a molecule with the other molecule or ligand communicates it to a biological system. Combined aspect of molecular physiology, bioinformatics, biophysics, and proteomics can decipher the real consequences behind the effects of microbial volatiles (Ferreira et al. 2015). In case of microbial volatile, the results obtained from chromatography techniques and further molecular docking of compound to a ligand can definitely provide good outcome. Molecular architecture of protein can be elucidated through docking on automated servers such as SWISS-MODEL. For chemical structure and molecular formula, Chemspider can be used; other docking tools are autodock, rosetta DOCK, ICM Pro, etc. Dharni et al. (2014) demonstrated the antifungal potential of novel isolated compound 2-4-di-tert-butylphenol against agriculturally important phytopathogenic fungus *Fusarium oxysporum*. The use of molecular docking depicted the binding of β-tubulin of *F. oxysporum* to the 2-4-di-tert-butylphenol and confirmed that novel compound is the potent inhibitor of compounds β-tubulin of *F. oxysporum*.

4.8 Conclusion

Microbial volatiles from primary and secondary metabolism are very diverse and have their different functions. Keeping in view the complexity of volatile compounds, a fresh term has recently proposed as "volatilome." VOCs from primary metabolism are the intermediate, whereas secondary metabolism produced VOC for specific purpose such as disease control, attractant, etc. Microbial volatile compounds (MOVCs) provide sustainable solution to overcome plant disease and boost up plant defense mechanism to directly or indirectly enable them to response, adopt, and survive in different conditions. Many reports incorporated in this chapter showed the successful rate of microbial volatile compounds in plant defense. But many of current insights are carried out under laboratory conditions.

More studies should be carried out to assess the cost-effectivity, ecofriendliness and sustainability of use of naturally produced microbial volatile compounds to better understand the role of the MOVCs at field level. In the coming future, we can expect more expanding knowledge about biodiversity of MOVCs, exploring underlying molecular mechanism of action, high-throughput plant responses to MOVCs, and production of a cost-effective delivery of these compounds to crop fields.

References

Angel Contreras-Cornejo H, Ias-Rodriguez L, Faro-Cuevas R, Lopez Bucio J (2014) Trichoderma spp. improve growth of Arabidopsis seedlings under salt stress through enhanced root development, osmolite production, and Na+ elimination through root exudates. Mol Plant Microbe Interact 27:503–514. doi:10.1094/MPMI-09-130265-R

Aziz M, Nadipalli RK, Xie X, Sun Y, Surowiec K, Zhang J-L, Paré PW (2016) Augmenting sulfur metabolism and herbivore defense in Arabidopsis by bacterial volatile signaling. Front Plant Sci 7:458

Bezerra TKA, Araújo ARR, Arcanjo NMO, da Silva FLH, Queiroga RCRE, Madruga MS (2015) Optimization of the HSSPME-GC/MS technique for the analysis of volatile compounds in caprine Coalho cheese using response surface methodology. Food Sci Technol Campinas 36(1):103–110

Bjurman J, Nordstrand E, Kristensson J (1998) Growth-phase-relatedproduction of potential volatile-organic tracer compounds by moulds on wood. Indoor Air 7:2–7

Boland G, Hall R (1994) Index of plant hosts to *Sclerotinia sclerotiorum*. Can J Microbiol 16:93–108

Bunge M, Araghipour N, Mikoviny T, Dunkl J, Schnitzhofer R, Hansel A, Schinner F, Wisthaler A, Margesin R, Mark TD (2008) On-line monitoring of microbial volatile metabolites by proton transfer reaction-mass spectrometry. Appl Environ Microbiol 74(7):2179–2186

Chaurasiaa B, Pandeya A, Palnib LMS, Trivedia P, Kumara B, Colvin N (2005) Diffusible and volatile compounds produced by an antagonistic Bacillus subtilis strain cause structural deformations in pathogenic fungi in vitro. Microbiol Res 160:75–81

Cho SM, Kang BR, Han SH, Anderson AJ, Park JY, Lee YH, Cho BH, Yang KY, Ryu CM, Kim YC (2008) 2R,3R-butanediol, a bacterial volatile produced by Pseudomonas chlororaphis O6, is involved in induction of systemic tolerance to drought in Arabidopsis thaliana. Am Phytopathol Soc eXtra 21(8):1067–1075

Danaei M, Baghizadeh A, Pourseyedi S, Amini J, Yaghoobi MM (2014) Biological control of plant fungal diseases using volatile substances of *Streptomyces griseus*. Eur J Exp Biol 4(1):334–339

Dandurishvili N, Toklikishvili N, Ovadis M, Eliashvili P, Giorgobiani N, Keshelava R et al (2011) Broad-range antagonistic rhizobacteria *Pseudomonas fluorescens* and Serratia plymuthica suppress Agrobacterium crown gall tumours on tomato plants. J Appl Microbiol 110:341–352

Dharni S, Sanchita MA, Samad A, Srivastava SK, Sharma A, Patra DD (2014) Purification, characterization, and in vitro activity of 2,4-di-*tert*-butylphenol from *Pseudomonas monteilii* PsF84: conformational and molecular docking studies. J Agric Food Chem., 2014 62(26):6138–6146

Drilling K, Dettner K (2009) Electrophysiological responses of four fungivorous coleoptera to volatiles of Trametes versicolor: implications for host selection. Chemoecology 19:109–115. doi:10.1007/s00049-009-0015-9

Elkahoui S, Djébali N, Karkouch I, Ibrahim AH, Kalai L, Bachkovel S, Tabbene O, Limam F (2014) Mass spectrometry identification of antifungal lipopetides from Bacillus sp. BCLRB2 against Rhizoctoniasolani and Sclerotiniasclerotiorum. Prikl Biokhim Mikrobiol 50:184–188

Faraga MA, Ryu CM, Sumner LW, Pare PW (2006) GC–MS SPME profiling of rhizobacterial volatiles reveals prospective inducers of growth promotion and induced systemic resistance in plants. Phytochemistry 67. (2006:2262–2268

Farag MA, Ryu C-M, Sumner LW, andParé, P. W. (2013) GC–MS SPME profiling of rhizobacterial volatiles reveals prospective inducers of growth promotion and induced systemic resistance in plants. Phytochemistry 67:2262–2268. doi:10.1016/j.phytochem

Fernandoa WGD, Ramarathnama R, Krishnamoorthy AS, Savchuk SC (2004) Identification and use of potential bacterial organic antifungal volatiles in biocontrol. Soil Biol Biochem 37:955–964

Fernando WGD, Ramarathnam R, Savchuk SC (2005) Identification and use of potential bacterial organic antifungal volatiles in biocontrol. Soil Biol Biochem 37:955–964

Ferreira GL, dos Santos DN, Oliva G, Andricopulo AD (2015) Molecular docking and structure-based drug design strategies. Mclecules 20:13384–13421

Fialho MB, Toffano L, Pedroso MP, Augusto F, Pascholati SF (2010) Volatile organic compounds produced by Saccharomyces cerevisiae inhibit the in vitro development of Guignardia citricarpa, the causal agent of citrus black spot. World J Microbiol Biotechnol 26:925–932. doi:10.1007/s11274-0090255-4

Fialho MB, de Moraes MHD, Tremocoldi AR, Pascholati SF (2011) Potential of antimicrobial volatile organic compounds to control *Sclerotinia sclerotiorum* in bean seeds. Pesq Agrop Bras Bras 46(2):137–142

Frey-Klett P, Burlinson P, Deveau A, Barret M, Tarkka M, Sarniguet A (2011) Bacterial-fungal interactions: hyphens between agricultural, clinical, environmental, and food microbiologists. Microbiol Mol Biol Rev 75:583–609

Gerber NN, Lechevalier HA (1965) Geosmin, an earthy-smelling substance isolated from actinomycetes. Appl Microbiol 13(6):935–938

Giorgio A, De Stradis A, Lo Cantore P, Iacobellis NS (2015) Biocide effects of volatile organic compounds produced by potential biocontrol rhizobacteria on Sclerotinia sclerotiorum. Front Microbiol 6:1056

Hamilton-Kemp T, Newman M, Collins R et al (2005) Production of the long-chain alcohols octanol, decanol, and dodecanol by Escherichia coli Curr. Microbiol 51:826

Handelsman J (2004) Metagenomics: application of genomics to uncultured microorganisms. Microbiol Mol Biol Rev 68(4):669–685

Haq IU, Zhang M, Yang P, Van Elsas JD (2014) The interactions of bacteria with fungi in soil: emerging concepts. Adv Appl Microbiol 89:185–215

Hegde S, Bhadri G, Narsapur K, Koppal S, Oswal P et al (2013) Statistical optimization of medium components by response surface methodology for enhanced production of bacterial cellulose by *Gluconacetobacter persimmonis*. J Bioprocess Biotech 4:142

Kishimoto K, Matsui K, Ozawa R, Takabayashi J (2007) Volatile 1-octen-3-ol induces a defensive response in *Arabidopsis thaliana*. J Gen Plant Pathol 73:35–37. doi:10.1007/s10327-006-0314-8

Law J, Zsoldos Z, Simon A, Reid D, Liu Y et al (2009) Route designer: a retrosynthetic analysis tool utilizing automated retrosynthetic rule generation. J Chem Inf Model 49:593–602

Lee B, Farag MA, Park HB, Kloepper JW, Lee SH et al (2012) Induced resistance by a long-chain bacterial volatile: elicitation of plant systemic defense by a C13 volatile produced by Paenibacillus polymyxa. PLoS One 7(11):e48744

Lemfack MC, Nickel J, Dunkel M, Preissner R, Piechulla B (2014) mVOC: a database of microbial volatiles. Nucleic Acids Res 42:D744–D748

Liu M, Bienfait B, Sacher O, Gasteiger J, Siezen RJ et al (2014) Combining chemoinformatics with bioinformatics: in silico prediction of bacterial flavor forming pathways by a chemical systems biology approach "Reverse Pathway Engineering". PLoS One 9(1):e84769

Martin CH, Nielsen DR, Solomon KV, Prather KL (2009) Synthetic metabolism: engineering biology at the protein and pathway scales. Chem Biol 16:277–286

Mercier J, Jimenez JI (2004) Control of fungal decay of apples and peaches by the biofumigant fungus Muscodoralbus. Postharv Biol Technol 31:1–8. doi:10.1016/j.postharvbio.2003.08.004

Minerdi D, Bossi S, Maffei ME, Gullino ML, Garibaldi A (2011) Fusariumoxysporum and its bacterial consortium promote lettuce growth andexpansin A5 gene expression through microbial volatile organic compound(MVOC) emission. FEMS Microbiol Ecol 76:342–351. doi:10.1111/j.1574-6941.2011.01051.x

Mitchell AM, Strobel GA, Moore E, Sears J (2010) Volatiles antimicrobials from Muscodor crispans, a novel endophytic fungus. Microbiology 156:270–277

Naznin HA, Kiyohara D, Kimura M, Miyazawa M, Shimizu M, Hyakumachi M (2014a) Systemic resistance induced by volatile organic compound semitted by plant growth-promoting fungi in Arabidopsis thaliana. PLoS One 9:e86882. doi:10.1371/journal.pone.0086882

Naznin HA, Kiyohara D, Kimura M, Miyazawa M, Shimizu M et al (2014b) Systemic resistance induced by volatile organic compounds emitted by plant growth-promoting fungi in Arabidopsis thaliana. PLoS One 9(1):e86882

Pagans E, Font X, Sanchez A (2006) Emission of volatile organic compounds from composting of different solid wastes: abatement by biofiltration. J Hazard Mater 131:179e186

Raza W, Ling N, Yang L, Huang Q, Shen Q (2016) Response of tomato wilt pathogen *Ralstonia solanacearum* to the volatile organic compounds produced by a biocontrol strain *Bacillus amyloliquefaciens* SQR-9

Rice S, Koziel JA (2015) Characterizing the smell of marijuana by odor impact of volatile compounds: an application of simultaneous chemical and sensory analysis. PLoS One 10(12):e0144160

Ryu CM, Farag MA, Hu CH, Reddy MS, Wei HX, Paré PW et al (2003) Bacterial volatiles promote growth in Arabidopsis. Proc Natl Acad Sci U S A 100:4927–4932

Ryu C-M, Farag MA, Hu CH, Reddy MS, Kloepper JW, Pare PW (2004) Bacterial volatiles induce systemic resistance in Arabidopsis. Plant Physiol 134:1017–1026. PMID:14976231; http://dx.doi.org/10.1104/pp.103.026583

Sang MK, Kim KD (2012) The volatile-producing Flavobacteriumjohnsoniae strain GSE09 shows biocontrol activity against Phytophthoracapsici in pepper. J. Appl. Microbiol. 113:383–398. doi:10.1111/j.1365-2672.2012.05330.x

Schmidt R, Cordovez V, de Boer W, Raaijmakers J, Garbeva P (2015) Volatile affairs in microbial interactions. ISME J. doi:10.1038/ismej.2015.42

Schoen HR, Peyton BM, Knighton WB (2016) Rapid total volatile organic carbon quantification from microbial fermentation using a platinum catalyst and proton transfer reaction-mass spectrometry. AMB Express 6:90

Sharifi R, Ryu C-M (2016) Are bacterial volatile compounds poisonous odors to a fungal pathogen Botrytis cinerea, alarm signals to Arabidopsis seedlings for eliciting induced resistance, or both? Front Microbiol 7:196. PMID:26941721; http://dx.doi.org/10.3389/fmicb.2016.00196

Sharifi R, Ahmadzade M, Behboudi K, Ryu C-M (2013) Role of *Bacillus subtilis* volatiles in induction of systemic resistance in Arabidopsis. Iran J Plant Prot Sci 44:91–101

Strobel, G., Singh, S. K., Riyaz-Ul-Hassan, S., Mitchell, A. M., Geary, B., and Sears, J. (2011). An endophytic/pathogenic phomasp from creosote bush producingbiologically active volatile compounds having fuel potential. FEMS Microbiol Lett 320, 87–94. doi: 10.1111/j.1574-6968.2011.02297.x

Tarkka MT, Piechulla B (2007) Aromatic weapons: truffles attack plants by the production of volatiles. New Phytol 175:383–386

Tellez MR, Schrader KK, Kobaisy M (2001) Volatile components of the cyanobacterium Oscillatoria perornata (Skuja). J Agric Food Chem 49:5989–5992

Thakeow P, Angeli S, Weissbecker B, Schuetz S (2008) Antennal and behavioral responses of *Cis boleti* to fungal odor of Trametes gibbosa. Chem Senses 33:379–387. doi:10.1093/chemse/bjn005

Trombetta D, Castelli F, Sarpietro MG, Venuti V, Cristani M, Daniele C et al (2005) Mechanisms of antibacterial action of three monoterpenes. Antimicrob Agents Chemother 49:2474–2478

Usha Rani M, Rastogi NK, Appaiah KA (2011) Statistical optimization of medium composition for bacterial cellulose production by Gluconacetobacter Hanenii UAC09 using coffee cherry husk extract-an agro industry waste. J Microb Biot 21:739–745

Vespermann A, Kai M, Piechulla B (2007) Rhizobacterial volatiles affect the growth of fungi and Arabidopsis thaliana. Appl Environ Microbiol 73:5639–5641

Vinale F, Sivasithamparam K, Ghisalberti EL, Marra R, Woo SL, Lorito M (2008) Trichoderma-plant pathogens interactions. Soil Biol Biochem 40:1–10

Wang C, Wang Z, Qiao X, Li Z, Li F, Chen M et al (2013) Antifungal activity of volatile organic compounds from Streptomyces alboflavus TD-1. FEMS Microbiol Lett 341:45–51. doi:10.1111/1574-6968.12088

Yuan Z, Chen Y, Xu B, Zhang C (2012) Current perspectives on the volatile-producing fungal endophytes. Crit Rev Biotechnol 32:363–373. doi:10.3109/07388551.2011.651429

Zhou C, Li Z, Yu D (2010) Bacillus megaterium strain XTBG34 promotesplant growth by producing 2-pentylfuran. J Microbiol 48:460–466. doi:10.1007/s12275-010-0068-z

Volatiles in the Rhizosphere: Bioprospecting for Sustainable Agriculture and Food Security

5

Shyamalina Haldar and Sanghamitra Sengupta

Abstract

Volatile organic compounds are low molecular weight lipophilic molecules with low boiling points and an appreciable vapor pressure under ambient conditions and constitute a small proportion of the total number of metabolites produced by living organisms. Volatiles are important aromatic compounds found in foods that evoke gustatory response in humans and animals. In addition they have important role in mediating communication between living organisms. Due to their roles as signaling molecules, within and between organisms, studies with regard to structural and functional diversities of these compounds are essential for an improved understanding of cellular and organismal communications in living systems. Since plant–microbe interactions are one of the most fascinating ecological phenomena that help to sustain the food cycle, ecological balance, and environmental stability, this chapter highlights the diversity of volatiles present in the plant rhizosphere. The rhizovolatiles discussed here include those produced by plants as well as by microorganisms inhabiting the rhizosphere. This chapter focuses on the role of these volatiles in the establishment of successful association between plants and other organisms and their beneficial effects on plant growth and development. This will value-add to our present understanding of the chemical cues defining the complexity and dynamism of rhizosphere functioning. At the end, this synthesis emphasizes on the potentiality of these volatiles for sustainable agriculture and food production to ensure food security.

S. Haldar
Department of Microbiology, Goa University, Taleigao Plateau, Goa 403206, India
e-mail: shyamalina@yahoo.com

S. Sengupta (✉)
Department of Biochemistry, University of Calcutta,
35, Ballygunge Circular Road, Kolkata 700019, West Bengal, India
e-mail: sanghamitrasg@yahoo.com

D.K. Choudhary et al. (eds.), *Volatiles and Food Security*,
DOI 10.1007/978-981-10-5553-9_5

Keywords
Ecology • Foods • Plant–microbe interactions • Rhizosphere • Signaling • Volatiles

5.1 Introduction

Volatiles are defined as the group of chemical elements or compounds with low boiling point. These compounds could be naturally occurring or produced by activities of microorganisms, plants, animals, and humans (Abdullah et al. 2015a). The definition of volatile organic compounds (VOCs) depends on the frame in which this term is used (http://www.eurofins.com/voc.aspx). According to the World Health Organization (WHO), VOC is any organic compound whose boiling point ranges from 50 to 260 °C with saturation vapor pressure greater than 100 kPa at 25 °C (ISO 16000-6). Compared to the anthropogenic sources which emit an estimated 142 Tg of carbon per year, biological sources emit in much higher amount (1,150 Tg of carbon per year) (Goldstein and Galbally 2007). The majority of naturally occurring VOC is produced by plants (1 pg of carbon in the form of VOC annually), and the rest is produced by animals, microbes, and fungi, such as molds (Pichersky and Gershenzon 2002). VOCs include a wide range of ubiquitous aromatic compounds such as hydrocarbons, alcohols, aldehydes, acids, esters, amines, thiols, xylene, and monoterpenes, for example, α-pinene and trace amounts of insect pheromones. These compounds chiefly correspond to compounds responsible for flavors detectable by humans at below part per trillion (ppt) levels and are endowed with diverse physical properties (Table 5.1; Rowan 2011). Few mammalian pheromones, androstenone (5α-androst-16-en-3-one) and fecal indolic compounds, and skatole (3-methylindole) which have higher molecular weights and sufficient vapor pressure that can be clearly perceived by humans are also considered as VOC (Liberles 2014; Moore et al. 1987). In particular, volatile compounds with low molecular weight (500 Daltons, Da) and high vapor pressure can diffuse in gaseous phase and hence serve as important signaling molecules (semiochemicals) within and between organisms. Although VOCs constitute only a small fraction of the total metabolites produced by an organism (1 % of the total metabolites), they take part in a wide range of biological processes (Dudareva et al. 2004; Rowan 2011). Thus measurement of the amount of the VOC produced and determination of their functional role are essential to understand the functioning of living systems.

The advent of technologies has led to the identification of volatile metabolites from different living organisms. Few databases have so far been developed that describe the volatiles and their biological functions (Abdullah et al. 2015a). The information on VOCs and their interactions with the organisms and contribution in the benefits of mankind pertaining to health, food, and industry are not well organized (Abdullah et al. 2015a; Dudareva et al. 2004, 2006). Although an extensive volume of literature exists on VOCs and their biosynthesis, our knowledge about their biological functioning and mechanism of action is still limiting (Holopainen

Table 5.1 Properties and source of different types of volatile organic compounds (VOCs)

Volatile	Source	Chemical class/ biosynthesis	Boiling point (°C) under 780 mm Hg
Androstenone	Mammalian pheromone	Steroid	372
α-Farnesene ($C_{15}H_{24}$)	Flower volatile and flavor precursor	Sesquiterpene	280
δ-Octalactone ($C_8H_{14}O_2$)	Microbial, dairy foods	Fatty acid oxidation	238
Dimethyl disulfide (CH_3SSCH_3)	Onion, garlic, coffee	Disulfide	109
Ethanol (C_2H_5OH)	Anaerobic respiration	Hydrocarbon	−102
Ethylene (C_2H_4)	Plant hormone	Alcohol	78
Hexanal ($C_6H_{12}O$)	Green leaf volatile	Aldehyde (lipoxygenase)	131
Isoprene (C_5H_8)	Plants (oxidative stress)	Terpenoid	265
Methanethiole (CH_3SH)	Biogenic	Sulfide	6
Skatole ($C_9H_9N_2$)	Feces	Aromatic heterocycle	78

and Blande 2012). This chapter highlights the biological activities of VOC associated with plant's rhizosphere which is an extremely ecologically important niche housing many of the complex interactions that occur between plant roots and microorganisms. A major part of this chapter provides a comprehensive description of the types of VOC produced by different plants and how these compounds mediate interactions between plants and microorganisms and macroorganisms residing in the vicinity of the roots. The second part of this chapter focuses on the functional relevance of the isolated rhizo-VOC from different plants with respect to agriculture and food production. Finally, we close this chapter by describing the limitations of the production of these volatiles and possible direction to solve the problems. In conclusion, we also highlight the plausible applications of these molecules in future economic benefits.

5.2 Volatiles Observed in the Rhizosphere

5.2.1 Rhizovolatiles of Plant Origin

An organism's interaction with its environment is fundamental to the survival of that organism and the performance of the ecosystem as a whole (Elton 1927). The association among the species includes four main types of two-way interactions: mutualism, commensalism, competition, and predation (herbivory and parasitism). Following the colonization in the terrestrial ecosystem, the "plants" as sessile organisms develop such interactions with innumerable micro- and macroorganisms for their survival and systemic functioning. The communication between plants and organisms is mediated through the exchange of chemicals which are synthesized and exuded by both plants and the partner-organisms. These chemicals lead to the development of a complex signaling network that culminates into the establishment of a strong beneficial association between them. Rhizosphere, the narrow zone surrounding the plant roots, represents one of the most diverse habitats on the planet and houses many such interactions (Trabelsi and Mhamdi 2013). The dynamicity of the rhizosphere ecology is defined by the physicochemical and biological events defined by the host plant itself through uptake of water and minerals and subsequent release of nutrients and carbon dioxide, exudation, and secretion of an array of chemical compounds altogether defined as "rhizosphere processes" (Philippot et al. 2013). An amount of 5–20 % of all photosynthetically fixed carbon is released by the plants in the rhizosphere to create a carbon-rich environment for numerous beneficial and pathogenic micro- and macroorganisms (Steeghs et al. 2004).

5.3 Structural Diversity

VOCs form an important component of the host of compounds released by the plant through their roots in the rhizosphere. The important characteristics of plant VOC include their structural diversity, species-specific property, and tissue specificity (Wei et al. 2016). Some of these secondary metabolites are restricted to certain plant taxa at high concentration and have no apparent roles in plants' primary metabolism (Schoonhoven et al. 2005). They are secreted only at certain stages of development or under specific circumstances to provide signals for the physiological state of plant cells. It has been shown that a given tissue has similar VOC fingerprints at a particular stage of developmental (Wei et al. 2016). VOC secreted from plants, in general, have molecular weight less than 300 Da and boiling points lower than 260 °C (Dong et al. 2016). Till date, more than 1700 volatiles have been recognized from the aerial and non-aerial parts of plants. These constitute more than 90 families, comprising approximately 1 % of all currently known plant specialized metabolites (Pichersky and Gershenzon 2002; Dudareva et al. 2004). Ethylene, the potent activator of plant defense responses, was the first gaseous hormone discovered in nature (Bleecker and Kende 2000).

Plant VOC has been classified on the bases of their chemical structures, physiological functions, and biosynthetic pathways. They can act as endogenous hormones (e.g., ethylene, methyl jasmonate (MeJA), and methyl salicylate (MeSA)) or can mediate the interactions between cospecific and heterospecific plant species and with organisms of higher trophic levels (herbivores; pollinators; enemies of herbivores). The following are few major biosynthetic pathways that produce plant VOC. These include (1) *mevalonate (MVA)/2-C-methyl-D-erythritol 4-phosphate pathway (MEP)* producing terpenoids and carotenoid derivatives, (2) *shikimate and phenylpropanoids pathway* producing benzenoids and phenylpropanoids from aromatic amino acids (such as phenylalanine), and (3) *lipoxygenase (LOX) pathway* producing green leaf volatiles (GLV), methyl jasmonate, and fatty acid derivatives like alcohols and aldehydes (Figs. 5.1 and 5.2; Wei et al. 2016). VOCs are produced by the removal of hydrophilic moieties and oxidation, hydroxylation, reduction, methylation, and acylation reactions (Pichersky et al. 2006; Tzin and Galili 2010; Becker et al. 2015).

Methanol (cell wall expansion & maintenance) → Formaldehyde (oxidative demethylation; dissociation of 5,10- methylene-THF) → Formic acid (photorespiration & oxidation of glyoxalic acid) → CO_2

Ethanol (stress) → Acetaldehyde (pyruvic acid decarboxylation; wound responce) → Acetic acid (carbohydrate & fat decomposition) → Fatty acids, steroids, terpenoids, carotenoids.

Isoprene (oxidative stress) → Methyl venyl ketone(oxidative stress – drought, radiation)

Stored monoterpenes (ionization of GDP) → Ethanol (anoxic condition – root flooding)

Fig. 5.1 Production and metabolism of plant VOC. *GDP* geranyl diphosphate, *THF* tetrahydrofolate, *VOC* volatile organic compound

Herbivore-induced plant volatiles (HIPV) and oviposition-induced plant volatiles (OIPV) are by far the most studied plant VOC, specifically produced in response to arthropod herbivore feeding, damage, and oviposition (Becker et al. 2015). In intact plants, VOC with glycosidic linkage and the enzyme beta-glycosidase which degrades the glycosidic bonds are located in vacuoles and cell walls, respectively. This enzyme–substrate compartmentalization inhibits the hydrolysis. However, the exogenous beta-glucosidase from the herbivores during piercing–sucking–chewing of plant parts can hydrolyze the glycosidically bound VOC and release the VOC (Dong et al. 2015). Herbivore attack generates superoxide by the action of nicotinamide adenine dinucleotide phosphate complex. This superoxide is converted to H_2O_2 which induces activation signaling by VOC like jasmonic acid, salicylic acid, and ethylene that stimulates production of phytohormone (Dong et al. 2015).

HIPV can be broadly classified into two categories. Some are emitted constitutively, and the remaining are induced only upon herbivore attack. Hemiterpenes like isoprene, linalool, and β-myrcene belong to monoterpenes, (E)- β-caryophyllene from sesquiterpenes, and eugenol from phenylpropanoids and (E)-2-hexenol; a type of GLV is constitutively expressed. However, the levels of these VOCs are increased upon attack by the herbivores (Becker et al. 2015). On the other hand, (E)- β-ocimene, (E)- β-farnesene, methyl salicylate, methyl jasmonate, and (Z)-3-hexenyl acetate belonging to monoterpenes, sesquiterpenes, benzenoids, fatty acid derivatives, and GLV, respectively, are induced only due to herbivore attack (Becker et al. 2015). HIPV attract parasitoids and predators which feed upon herbivore arthropods as a defense response in the neighboring plants in addition to exerting toxic effects on the herbivores directly (Ode 2013; Gols 2014; Becker et al. 2015). This indirect defense mechanism through HIPV is evident in as many as 49 plant species belonging to 25 different families and insects from five different orders till date (Mumm and Dicke 2010). Through protection from the insects, HIPV also increases plant strength and fitness (Becker et al. 2015).

Secretion of sulfur-containing VOC by plant following attacks by herbivores constitutes a multi-trophic interaction between plants and parasitoids. Roots emit

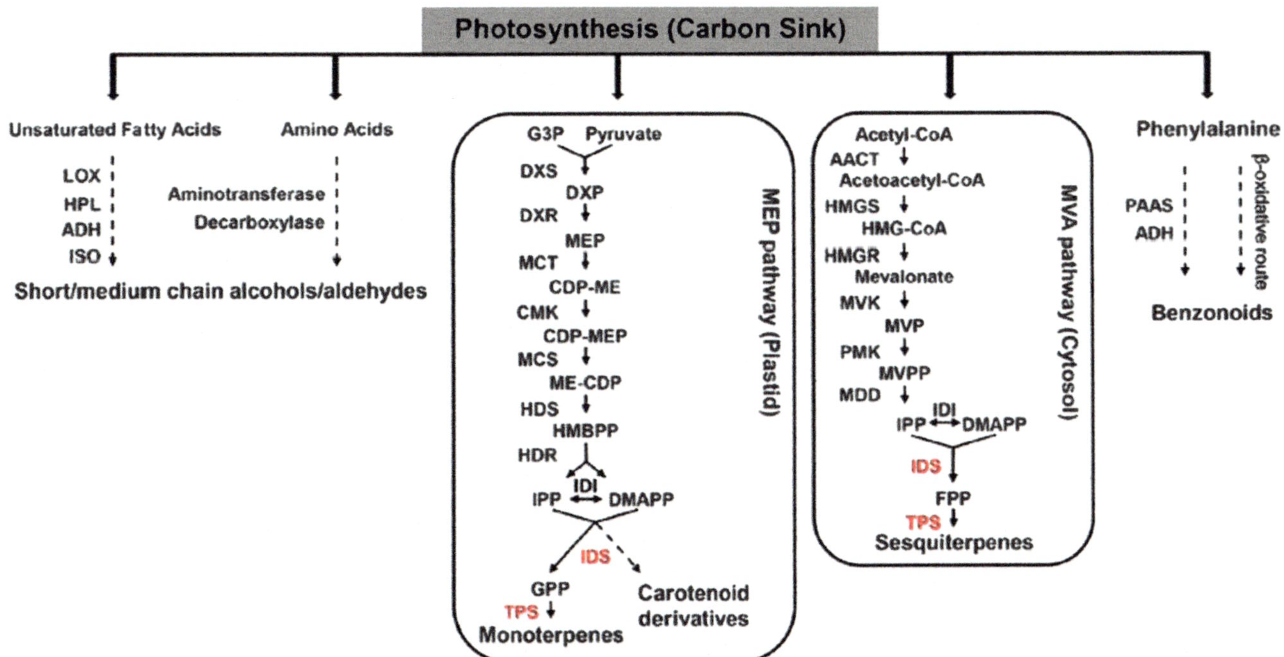

Fig. 5.2 Different pathways for synthesis of plant volatile compounds

higher amounts of sulfur-containing VOCs per gram dry mass than that of shoots (Kergunteuil et al. 2015). In this regard, the local emission of sulfur-containing VOC is more pronounced and temporarily dynamic in roots (Kergunteuil et al. 2015). Van Dam et al. analyzed the emission of sulfur-containing VOCs from herbivore-damaged roots of six *Brassica* species and correlated the findings with the composition of root glucosinolate which is a typical secondary metabolite involved in plant defense (van Dam et al. 2012). Upon root damage, glucosinolate is hydrolyzed to other VOCs such as thiocyanates, isothiocyanates (ITC), nitriles, and epithionitriles which act as signaling cues by the predators and parasitoids to trace the herbivorous hosts. This study showed that the emission of sulfide VOC is species and time dependent but not influenced by the type of damage or glucosinolate composition in the roots.

5.4 Functional Significance

VOCs are not directly involved in mediating growth, development, and reproduction in plant but participate in important ecological functions including inter- and/or intraspecies cross-talks, antimicrobial activities, and defense strategies against pests and pathogens (Abdullah et al. 2015a). They represent an important indicator for biological processes such as cell to cell communication, regulation of physiological processes, and modulation of interactions among a wide variety of organisms to maintain ecological equilibrium and environmental integrity (Patti et al. 2012; Fall et al. 1999). These semiochemicals function as hormones in the identification of mates, cospecifics, competitors, and predators (Abdullah et al. 2015b; Pichersky et al. 2006; Pichersky and Gershenzon 2002).

VOCs along with the nonvolatile compounds basically assist plants to adapt to the environment (Wei et al. 2016). They are emitted to act as attractants/repellants to/against the neighboring plants and other micro- and macroorganisms including herbivores, pathogens, pests, pollinators, and parasitoids, to assuage the heat and oxidative stresses, and to regulate the levels of systemic acquired resistance (SAR) to various diseases (Pichersky and Gershenzon 2002; Dicke et al. 2009; Frost et al. 2008; Schwarz et al. 2009). VOCs thus represent a key component in a plant's chemotype and have imperative roles in plant's ecosystem (Wei et al. 2016).

Application of high-throughput techniques such as proteomics and metabolomics has allowed to gather enormous information about the biochemical and molecular aspects of VOC, the mechanism of synthesis, and emission and explication of the probable functions of VOC. One of the frequently addressed questions is that why do plants emit volatiles. Investigations have proved that several factors such as endogenous circadian clock; various biotic stressors, e.g., pathogens and herbivore feeding; and abiotic stresses such as drought, heat stress, and ozone actually induce the release of VOC from plants (Dong et al. 2016; Holopainen and Blande 2012). The herbivorous-invertebrate feeding is considered to be one of the major triggers for VOC emission by plants (Holopainen and Blande 2012). The damaged roots transmit signals to distant parts and neighboring plants via VOC

(Holopainen and Blande 2012). The amount and proportion of VOC emitted allow the plant to send complex signals which then carry out assorted ecological functions including plant-plant communication and communication with second (herbivores and pollinators) and third (enemies of herbivores) trophic level organisms (Holopainen and Blande 2012).

In general, the communication in the root systems is based on nonvolatile hydrophilic exudates as water films shield the root surfaces and the soil particles, thereby allowing the free flow of air within the soil pores. Recent evidences imply that VOC are involved in root–root and root–microbial/pest interaction belowground as well (Schenkel et al. 2015). The alkaloids and phenolic compounds are the major signaling molecules found in the roots of the terrestrial plants grown in wet soil. These compounds mediate self-inhibition, avoidance of underground obstacles, and allelopathy. Studies with different varieties of barley cultivars demonstrated that VOC from one variety significantly allocated more biomass to the roots of the other varieties. This indicates that the phenomenon of coadaptation among different cultivars may be implicated for harnessing adaptation under various environmental conditions (Ninkovic 2003).

5.5 Mediators of Plant–Plant Communication

Peñuelas et al. have proposed plausible ecological roles of root-secreted VOC (Peñuelas et al. 2014). Volatiles emitted by underground roots regulate plant to plant communication mostly in the rhizosphere. This communication is important to distinguish between self-plants from the nonself ones. However the exact mechanism that differentiates different types of signals is not totally understood (Holopainen and Blande 2012). Experiments have shown that there is an increase in number and size of root branches when a number of nonself-plants grow together relative to that when the plants of same species are grown in the same place. This is possibly attributed to the increased availability of different kinds of nutrients for the plants growing in competition. Allogenetic recognition and physiological coordination among the roots also result in allocation of greater amount of resources for growth of the roots. This is believed to be mediated by VOC secreted by the plant roots. VOC, thus, functions in recognition of kins among the plants grown together (Holopainen and Blande 2012).

Volatiles such as sesquiterpene (E)-b-caryophyllene are induced in the roots to deal with heat stress and damage caused by insect feeding. (E)-b-caryophyllene forms 70 % of the total sesquiterpene secreted from the roots of *Copaifera officinalis* indicating the specificity of this compound in the legume's roots (Chen et al. 2009). In in vitro experiments in sand-filled olfactometer, (E)-b-caryophyllene was also produced from the damaged maize roots due to the rootworm (*Diabrotica virgifera*) feeding. It was also demonstrated that (E)b-caryophyllene evaporated and moved in humid sandy soil more swiftly. This latter observation indicates the rapid movement and transmission of signals by VOC among the neighboring roots.

5.6 Role in Plant Defense

In systemic acquired resistance (SAR) for various plants, viz., *Arabidopsis thaliana* and *Nicotiana tabacum* (tobacco), salicylic acid (SA)-binding proteins possessing esterase activity release the active defense volatile phytohormone SA through hydrolysis of methyl salicylate (MeSA) (Slaymaker et al. 2002; Vlot et al. 2008). The released SA is perceived by the receptors from the distant parts of the same plant or on the conspecific or heterospecific neighboring plants. In addition to this, neighboring plants might as well adsorb VOC and store them in the epidermis layers and release them to the atmosphere as and when needed.

In a study, Wei et al. threw light on diverse composition of VOC and chief VOC synthesizing enzymes from different tissues including roots in cucumbers. An abundance of aromatic benzenoids particularly oxidized monoterpenes was detected in the roots of the cucumbers. TPS11 and TPS14 were the two *terpene synthase* genes identified to be involved in the synthesis of terpenoids in the roots of cucumber. This study provides a solid foundation for future investigations of both the physiological functions of VOC with respect to improvement of flavor in cucumber plants (Wei et al. 2016). Proton-transfer-reaction mass spectrometry (PTR-MS) analyses revealed that ethanol, acetaldehyde, acetic acid, ethyl acetate, 2-butanone, 2,3,-butanedione, acetone and the monoterpene, and 1,8-cineole are the major VOCs released from *Arabidopsis* ecotype Columbia (Col-0) during interactions with *Pseudomonas syringae* DC3000 and the aphid *Diuraphis noxia* (Steeghs et al. 2004). The interactions between the pathogenic fungus *Alternaria brassicicola* with cultured *Arabidopsis* roots showed a distinct pattern of VOC emission with a release of high abundance of ethanol in addition to acetic acid, acetone, ethyl acetate, and 1,8-cineole indicating a rapid switching to alcoholic fermentation mode of metabolism induced by pathogen. This suggests that the plant roots are proficient enough to adapt to the changing scenario in the rhizosphere. Fingerprinting of VOC with functional genomics thus establishes a link between gene regulation with metabolic pathways and synthesis of bioactive volatile synthesis in response to pathogenic invasion in particular plant species (Steeghs et al. 2004). A few studies have been reported that VOCs are secreted from plant roots in the presence of insects (Neveu et al. 2002). Methyl jasmonate (MeJA) is an example of such insect-induced VOC that triggers the release of proteinase inhibitors and polyphenol oxidase in undamaged neighboring plants. In another study, in response to contact with parasites, accumulation of phytoalexin was observed in bean and barley plants (Farag et al. 2013).

5.7 Diversity of Rhizovolatiles of Bacterial Origin

Microorganisms colonizing in plant rhizosphere produce a large battery of VOC which manipulate the assembly and activity of rhizobacteria/fungi in the host and neighboring plants. Microbial activities include mineral sequestration, nutrient assimilation, metal sorption, substance solubilization, toxin removal, pathogen repulsion, and disease suppression (Bailly and Weisskopf 2012; Kai et al. 2016). Effects of microbial VOC on plants vary within a particular combination of plant and microbial species as well as other biotic and abiotic factors (Schmidt et al. 2015).

Plant-associated microbiome practically covers the entire plant surface. They are especially profuse in the nutrient-rich rhizosphere where the competition between the different organisms is high (De Vrieze et al. 2015). Approximately 400 of the 10,000 described microbial species have been shown to produce VOC (Ortíz-Castro et al. 2008). Till date, 1,088 VOCs identified from 517 microorganisms have been clustered in KNApSAcK Metabolite Ecology Database (Abdullah et al. 2015a). A total of 77 species of these are "pathogenic bacteria" that can be grouped into six clusters, of which again three clusters contain 100 % pathogenic species including *Pseudomonas aeruginosa*, *Klebsiella pneumoniae*, and *Escherichia coli*, while the remaining three clusters contain both pathogenic and nonpathogenic bacteria. Among 517 microorganisms, 92 species emit only one type of VOC, while 14 have been reported to emit 50 different types of VOC (Abdullah et al. 2015a). This statistical data confirms that a few microbial VOCs can act as its odor fingerprint. A total of 341 bacterial VOC from 1,044 species were reported to be associated with 11 types of biological activities. Of these, 57.3 % of the biological activities are related to chemical ecology which includes plant growth enhancement, antimicrobial, antifungal, attractant, defense, root growth inhibition, and repellent activities (Figs. 5.3 and 5.4; Table 5.2; Abdullah et al. 2015a). It is noteworthy to mention that the chemical structure of VOC determines their biological activities. Isoprenes, terpenes, alkanes, alkenes, alcohols, esters, carbonyls, and acids are volatiles that help in communication between microbes in a rhizo-consortium and the plants (Kai et al. 2007; Vespermann et al. 2007).

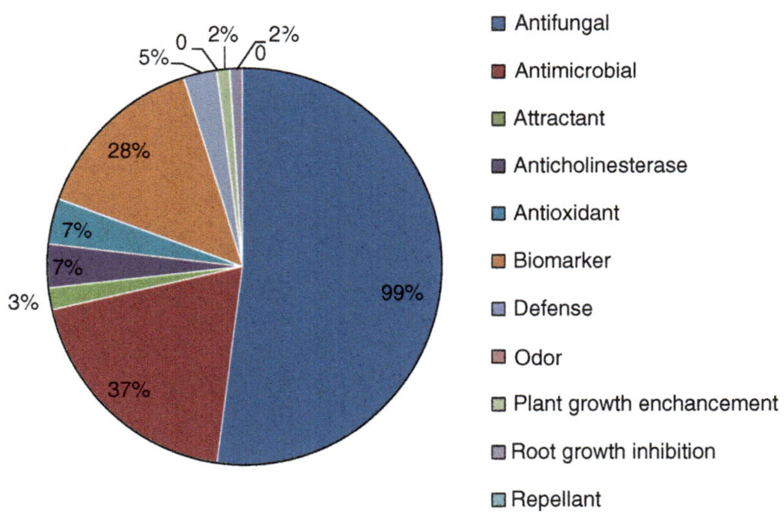

Fig. 5.3 Proportion of bacterial species producing different volatiles with different functions

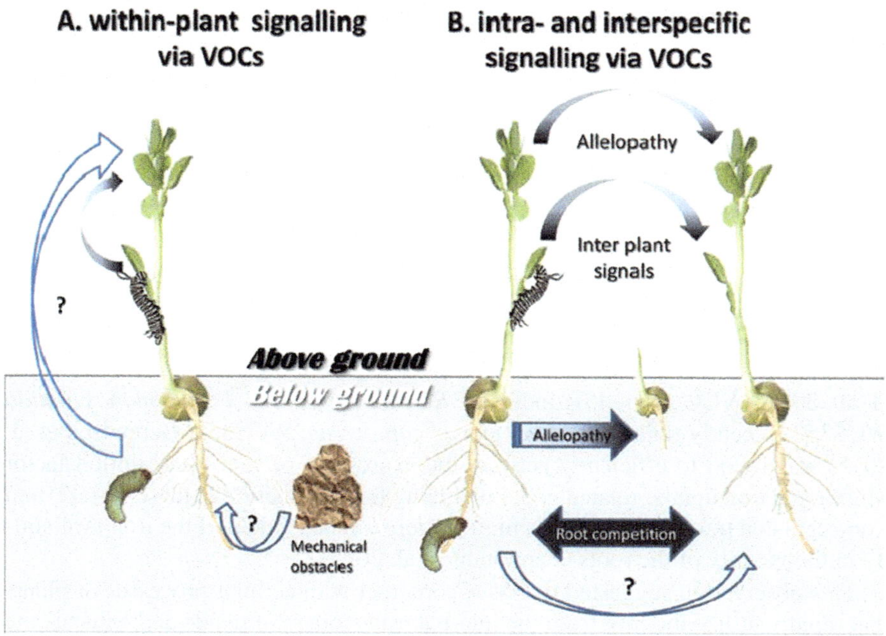

Fig. 5.4 Role of plant volatiles in aboveground and belowground signaling

Table 5.2 Bacterial volatiles and their functions

Cluster ID (count of VOCs)	Chemical structure	Related biological activities
Cluster 1 (55)	Terpenoids	Antimicrobial; anticholinesterase; defense; antioxidant
Cluster 2 (33)	Hydrocarbons	Antimicrobial; anticholinesterase; antioxidant
Cluster 3 (41)	Alkanes	Biomarker
Cluster 4 (18)	Alkenes	Antifungal
Cluster 5 (21)	Aldehydes, esters, carboxylic acid, ketone	Antimicrobial; antioxidant; biomarker; anticholinesterase; repellant
Cluster 6 (25)	Alcohol and alkanes	Plant growth enhancement; root elongation inhibition; odor
Cluster 7 (47)	Ester, carboxylic acid, ketone	Attractant; biomarker
Cluster 8 (15)	Epoxide, ether, ester, alcohol	
Cluster 9 (42)	Aromatic alcohol, carboxylic acid, ester, ketone, ether	Attractant
Cluster 10 (14)	Aromatic compound	Biomarker
Cluster 11 (30)	Diverse (C0-C6 molecules)	Biomarker

5.8 Beneficial Roles of Bacterial VOC

We are beginning to understand the effects of bacterial VOC on the growth and development of plants. The effects of bacterial VOC on plant growth (promotion and/or inhibition) have been nicely reviewed by Bailly and Weisskopf (2012). In general, bacterial VOCs mediate multiple functions of intra- and interspecies communication such as bacterial quorum sensing, growth and differentiation, inhibition of phytopathogenic fungal growth and differentiation, plant health and growth promotion (direct/indirectly), and antibiotic and stress resistance (De Vrieze et al. 2015).

Bacterial volatiles contribute in mineral utilization of plants. One notable example is during iron homeostasis in plants under iron-limiting conditions (Zamioudis et al. 2015). VOC from ISR-inducing *Rhizobacteria* like *Pseudomonas putida* WCS358 (recently renamed *Pseudomonas capeferrum* WCS358; Berendsen et al. 2015) was found to efficiently activate the expression of root transcription factor MYB7 and iron uptake-related genes including ferric reduction oxidase (FRO2) and iron-regulated transporter1 (IRT1) in *Arabidopsis* independent of the iron availability in the vicinity of the roots (Zamioudis et al. 2015).

This observation suggested that VOCs interact with cellular processes of plants that rapidly utilize the iron from the internal iron storage of plants and enhance the need for iron acquisition. Besides WCS358, seven other strains from the genus *Pseudomonas* and the phylum *Actinobacteria* could also potently activate the promoter of MYB72 gene suggesting that the members of the natural root microbiota play an important role in upregulating the capacity for iron uptake in the plant roots. To investigate the mechanisms of action of VOC from *Bacillus subtilis* in promotion of plant growth, Meldau et al. showed that B55 strain contributed to sulfur nutrition in *Nicotiana attenuata* (coyote tobacco) roots by increasing the sulfur absorption ability of the plants (Meldau et al. 2013). Other VOCs, viz., 13-tetradecadien-1-ol, 2-butanone, and 2-methyl-n-1-tridecene identified from *Pseudomonas fluorescens* SS101, were found to increase the number of lateral roots (Park et al. 2015).

VOC function is one of the chief determinants of induced systemic resistance [ISR] and induced systemic tolerance [IST] to abiotic stresses, such as drought and heavy metals in variety of plants including *Arabidopsis thaliana* (Zamioudis et al. 2015). With the help of VOC, plant growth-promoting rhizobacteria (PGPR) elicit ISR and plant growth promotion even in the absence of physical contact with plants. Initial research was conducted on the emission of VOC from plant growth-promoting bacteria (PGPB), *Bacillus* sp., which evoked ISR and also enhanced plant growth and development by improving plant mineral nutrition (Ryu et al. 2003, 2004a, b). VOC from *Bacillus subtilis* strain GB03 was shown to stimulate iron uptake mechanisms in *Arabidopsis*, thereby improving the iron content of the plants (Zamioudis et al. 2015). Following this discovery, numerous researches were undertaken to identify VOC from different species belonging to *Bacillus* and non-*Bacillus* genera and to study their role in plant growth promotion and disease suppression

(Yi et al. 2016). A number of in vitro culture-plate assays have been performed with different plant species incubated with PGPR, and the VOCs secreted from the bacterial species were identified. Petri dish bipartite studies have shown that acetoin and 2,3- butanediol secreted from *Enterobacter cloacae* JM22, *Pseudomonas fluorescens* 89B61, and *Pseudomonas chlororaphis* O6 could promote *Arabidopsis* growth through ISR (Kai et al. 2016). 2,3-Butanediol is one of the most well-characterized compounds isolated from a host of Gram-positive and Gram-negative bacteria, namely, *Aerobacter* sp., *Bacillus* sp., *Enterobacter* sp., *Klebsiella* sp., and *Serratia* sp. (Yi et al. 2016). Yi et al. demonstrated that 2,3-butanediol secreted by *Bacillus subtilis* has protective role both for the plants and the microbes with the subsequent development of ISR via ethylene and auxin homeostasis (Yi et al. 2016; Kanchiswamy et al. 2015). In another study, exogenous applications of 2,3-butanediol and its precursor 3-hydroxy-2-butanone (syn. acetoin) isolated from *Bacillus subtilis* GB03, *Bacillus pumilus* SE34, and *Bacillus amyloliquefaciens* IN937a were found to promote the growth of *Arabidopsis* plants (Farag et al. 2013; Kai et al. 2016). 2,3-Butanediol suppressed the growth of soilborne pathogen *Rhizoctonia solanacearum* and enhanced the growth of biocontrol *Pseudomonas protegens* Pf-5. The composition of root exudates is modified by 2,3-butanediol which in turn influenced the selectivity of the bacterial species in the rhizosphere. Thus in short 2,3-butanediol plays a role in the recruitment of the bacterial species at the junction of plant roots. It also activated the synthesis and secretion of defense molecules such as salicylic acid and ethylene from the plant roots and thereby enhanced the plant immunity. A study had shown that 2,3-butanediol promoted the robustness of the bacteria by protecting them against low pH of the rhizosphere by creating an alkaline environment during multiplication of bacterial cells. These together explained the role of 2,3-butanediol in promoting bacterial fitness and stability in the rhizosphere. In addition to this, 2,3-butanediol ensures optimal functions of the cell wall-degrading enzymes such as cellulase, pectinase, and protease and acts as an important factor responsible for virulence of *Pectobacterium* sp. and *Dicheya* sp., the causative agents for soft-rot disease (Kwan et al. 2013). VOC from *Pseudomonas chlororaphis* promoted growth by eliciting systemic resistance against the bacterial pathogen *Erwinia carotovora* in tobacco (*Nicotiana tabacum*) plants via GacS kinase-dependent production of 2,3-butanediol (Han et al. 2006). *Pseudomonas chlororaphis* also increased drought tolerance by stimulating the synthesis of jasmonic and salicylic acids in plants (Kanchiswamy et al. 2015). Numerous other bacterial VOCs such as those secreted from *Burkholderia cepacia* and *Staphylococcus* sp. have also been detected in the plant rhizosphere. These compounds have shown to improve plant growth. Along with *Bacillus* sp., volatiles from *Stenotrophomonas* and *Serratia* inhibited the growth of mycelia of pathogenic fungus *Rhizoctonia solani* and promoted growth in *Arabidopsis thaliana* seedlings, while the volatiles from *Burkholderia cepacia* and *Pseudomonas* sp. retarded the growth of the fungus to an extent.

5.9 Regulation of Secretion of Bacterial Volatiles

Secretion of VOC by the bacteria in the rhizosphere is greatly influenced by exact nature of multi-trophic interaction taking place in the rhizosphere. Following is an example of how quorum sensing may be modulated by VOC. It has been noticed that 2,3-butanediol stimulates GacS kinases which regulate the synthesis acyl-homoserine lactones (AHLs), a key molecule utilized by the bacteria to communicate with neighboring organisms (Han et al. 2006). Emission of the volatile metabolite 2-aminoacetophenone by *Burkholderia ambifaria*, *Pseudomonas aeruginosa*, and *Streptomyces* sp., on the other hand, is found to be regulated by the quorum sensing signals (Kai et al. 2016). This actually culminates into a nonrandom loss in bacterial communities and hence the pattern of metabolite secretion (Hol et al. 2015; Schulz-Bohm et al. 2015). Fungal VOCs have been shown to alter thermostability of *Collimonas pratensis* and *S. plymuthica* PRI-2C (Schmidt et al. 2015). In a study, de Vriez et al. identified a series of compounds such as dimethyl trisulfide (DMTS), S-methyl methanethiosulfonate (MMTS), nitropentane, isovaleric acid, undecanal, phenylpropanedione, propiophenone, and the simple ketones 3-hexanone, 2-decanone, and 2-undecanone secreted from different strains of *Pseudomonas* to elicit anti-oomycete potential against *Phytophthora infestans* in potato plants (De Vrieze et al. 2015). These compounds inhibited the growth of mycelium in *Phytophthora infestans* in addition to harboring sporicidal activities (De Vrieze et al. 2015).

5.10 Bioprospecting of Volatiles

A large number of VOC obtained from microbes and plants are used by humans as food additives or flavors, drugs, and fragrances and in agriculture (Schenkel et al. 2015). The use of VOC as an alternative novel strategy in sustainable agriculture and crop production has only recently been explored (Kanchiswamy et al. 2015). A large body of literature has already documented in vitro effects of VOC in plant growth promotion. Studies are ongoing to examine the suitability of several HIPVs to enhance the efficacy of biological control agents (reviewed in Peñaflor and Bento 2013). The strategies using plant VOC may be classified under four major categories. These include (1) selection and growing of plant cultivars that can release HIPV to attract natural enemies, (2) application of exogenous elicitors in the plants for induction of defense response, (3) release of synthetic HIPV to recruit natural enemies for the crops directly, and (4) manipulation of target genes of HIPV to enhance the production and release of volatiles and also to hasten the induced defense responses.

Studies with various inbred lines of corn and cotton crops have revealed the blends of HIPV vary among plant cultivars affect natural enemy foraging (Peñaflor and Bento 2013). Keeping in line, (E)-β-caryophyllene secreted by European varieties helps the host to find entomopathogenic nematodes more efficiently than the North American varieties, making the former a better biological control of the nematode (Rasmann et al. 2005). Therefore, selection of cultivars is important for improving biological control.

To develop crop plants with better pest resistance, plant metabolism may be engineered to target the pathways involved in VOC synthesis and regulation. The strategies usually involve manipulation of expression of genes related to terpenoid synthesis (TPs) and have been employed in model plants like *Arabidopsis* and crop plants like cotton and rice (Peñaflor and Bento 2013). The transgenic lines (Bt cotton and Bt rice) with manipulated synthesis and secretion of terpenoids are capable of attracting natural enemies better, which eventually affect the tripartite interactions of the plant with the insects and nematodes. Jasmonic acid, salicylic acid, and their metabolites such as methyl jasmonate, cis-jasmonate, and methyl salicylate are widely used as exogenous elicitors. These molecules function as fake signals to trigger the emission of a blend of toxins and HIPV even from the undamaged plants. As a result of these signals, the undamaged crops become more resistant to herbivore damage and begin to recruit natural enemies to both herbivore-infected and noninfected plants. Methyl salicylate (MeSA) is one of the strongest candidates often used as a synthetic plant volatile. It attracts a wide range of natural enemies in the field and aids to biological control by suppressing herbivore populations. Dimethylhexadecylamine, secreted by *Arthrobacter agilis* strain UMCV2, was found to modulate the morphogenesis in *Medicago sativa* in vitro by increasing the plant biomass and the length of the stem and the roots (Aviles-Garcia et al. 2016). In petri plate assays, it enhanced the formation of lateral roots indicating the role of this compound in plant growth and development. The racemic mixture of (RR) and (SS) isomers of 2,3-butanediol isolated from *Bacillus subtilis* GB03 and *Bacillus amyloliquefaciens* IN937a was found to trigger ISR when applied exogenously in *Arabidopsis* seedlings and also reduced the severity of the pathogenesis caused by *Erwinia carotovora* subsp. carotovora. Acetoin and 2,3-butanediol produced by different strains of *Bacillus subtilis* GB03 have been shown to affect the protective strategies of *Arabidopsis* upon infection. Both the transgenic and mutant lines of *Arabidopsis* showed stimulated defense response against *Bacillus subtilis* through increased production of ethylene. These studies highlight the role of bacterial VOC as plant defense stimulators, an observation which may be applied in agricultural field studies in future. In fact, VOCs have been applied in open-field studies very recently for sustainable protection of crops (Song and Ryu 2013). Application of VOC in open field thus remains a promising area to explore.

5.11 Conclusion

Small VOC emitted from microorganisms and the plants form important signaling molecules and have dramatic effects on plant development and thereby might be of help in better crop production and sustainable agriculture. However, the lack of techniques to isolate and identify VOC has resulted in gaps in knowledge regarding the functioning of VOC in rhizosphere. Therefore, to study the potentiality of VOC in situ, modeling of native rhizosphere system and developing rhizosphere platforms by combining various sets of microorganisms and plant species, is very much essential. This artificial system will mimic the functions and interactions of VOC actually occurring in an environmental niche and will provide a better understanding of the functionality of VOC. The elucidation of probable functions of VOC will provide solutions to exploit VOC toward developing cost-effective and eco-friendly strategies in agricultural practices worldwide by replacing the necessity for genetic modification and use of harmful pesticides, fungicides, and bactericides. The knowledge on VOC composition from plants will help to engineer hypersensitive genetically modified crops with better VOC secreting capabilities. The use of these plants in agricultural fields will confer enhanced protection in the neighboring plants through long-distance SAR via the released VOC. VOC may also be used to direct the plants toward the applied biological control agents (BCA) in field studies. In addition, VOC might have a great application toward development of trap crops to attract pests. Thus, volatiles may provide solutions for the development of ecologically feasible and environmental friendly approach toward crop protection in the future, thereby minimizing pest attacks and subsequent agricultural and economic losses. However, the strategies need to be developed for the appropriate use of volatiles in accordance to the geographical location and climatic conditions of a region, also considering the flora and the fauna constituency of the country. This can empower the biological control of harmful crop pests worldwide consequently leading to successful biological control programs.

Highlights

- Volatile organic compounds (VOCs) are small lipophilic compounds with low boiling points and high vapor pressure and produced as secondary metabolites from both micro- and macroorganisms.
- VOCs are important factors for aromatization of foods as well as key signaling molecules for inter- and intraspecies communications, taking part in plant–microbe and microbe–microbe interactions in rhizosphere.
- Plant VOCs have been classified on the basis of their chemical structures, physiological functions, and biosynthetic pathways and are mostly characterized by their structural diversity, species-specific property, and tissue specificity.
- Endogenous circadian clock, biotic stressors like pathogens and herbivore feeding, and abiotic stresses like drought, heat stress, and ozone actually induce the release of VOC from plants, the most important being the herbivore-induced plant volatiles (HIPV).
- VOCs from plants are the chief molecules for underground communication between neighboring plants and the other micro- and macroorganisms, helping in kin recognition, self-avoidance, systemic acquired resistance (SAR), long-distance signaling, defense response, and allelopathy.

- Bacterial VOCs (BVOCs) mediate multiple functions of intra- and interspecies communication such as bacterial quorum sensing, growth and differentiation, inhibition of phytopathogenic fungal growth and differentiation, direct or indirect plant health and growth promotion, induced systemic resistance (ISR), development of antibiotic, and stress responses.
- 2,3-Butanediol is one of the most well-characterized BVOCs isolated from a host of Gram-positive and Gram-negative bacterial groups including *Aerobacter* sp., *Bacillus* sp., *Enterobacter* sp., *Klebsiella* sp., *Pseudomonas* sp., and *Serratia* sp.
- The methods practiced for sustainable agriculture can be grouped under four major functional categories, namely, selection of plant cultivars with high defense response, application of exogenous elicitors and synthetic HIPV, and genetical manipulation of target genes of HIPV to enhance the production and release of volatiles and also to hasten the induced defense responses.
- In vitro and a few field-based studies on the use of BVOC in crop plants have elucidated upon the application of VOC as an alternative novel strategy for sustainable agriculture and crop production by replacing the chemical fertilizers and pesticides.

References

Abdullah AA, Altaf-Ul-Amin M, Ono N et al (2015a) Development and mining of a volatile organic compound database. Biomed Res Int 2015:139254. doi:10.1155/2015/139254

Abdullah AA, Altaf-Ul-Amin M, Nishioka T et al (2015b) Analysis of volatile metabolites emitted by various species to reveal their roles in chemical ecology and healthcare. Paper presented at 2nd international conference of biomedical engineering (ICoBE) at Penang, 30th–31st March, 2015. pp 1–6. doi:10.1109/ICoBE.2015.7235881

Aviles-Garcia ME, Flores-Cortez I, Hernández-Soberano C et al (2016) The plant growth-promoting rhizobacterium *Arthrobacter agilis* UMCV2 endophytically colonizes *Medicago truncatula*. Rev Argent Microbiol 48(4):342–346. doi:10.1016/j.ram.2016.07.004

Bailly A, Weisskopf L (2012) The modulating effect of bacterial volatiles on plant growth: current knowledge and future challenges. Plant Signal Behav 7(1):79–85. doi:10.4161/psb.7.1.18418

Becker C, Desneux N, Monticelli L et al (2015) Effects of abiotic factors on HIPV-mediated interactions between plants and parasitoids. Biomed Res Int 2015:342982. doi:10.1155/2015/342982

Berendsen RL, van Verk MC, Stringlis IA et al (2015) Unearthing the genomes of plant-beneficial Pseudomonas model strains WCS358, WCS374 and WCS417. BMC Genomics 16:539. doi:10.1186/s12864-015-1632-z

Bleecker AB, Kende H (2000) Ethylene: a gaseous signal molecule in plants. Annu Rev Cell Dev Biol 16:1–18. doi:10.1146/annurev.cellbio.16.1.1

Chen F, Al-Ahmad H, Joyce B (2009) Within-plant distribution and emission of sesquiterpenes from *Copaifera officinalis*. Plant Physiol Biochem 47(11–12):1017–1023. doi:10.1016/j.plaphy.2009.07.005

De Vrieze M, Pandey P, Bucheli TD et al (2015) Volatile organic compounds from native potato-associated *Pseudomonas* as potential anti-oomycete agents. Front Microbiol 6:1295. doi:10.3389/fmicb.2015.01295

Dicke M, van Loon JJ, Soler R (2009) Chemical complexity of volatiles from plants induced by multiple attack. Nat Chem Biol 5(5):317–324. doi:10.1038/nchembio.169

Dong L, Hou Y, Li F et al (2015) Characterization of volatile aroma compounds in different brewing barley cultivars. J Sci Food Agric 95(5):915–921. doi:10.1002/jsfa.6759

Dong F, Fu X, Watanabe N et al (2016) Recent advances in the emission and functions of plant vegetative volatiles. Molecules 21(2):124. doi:10.3390/molecules21020124

Dudareva N, Pichersky E, Gershenzon J (2004) Biochemistry of plant volatiles. Plant Physiol 135:1893–1902

Dudareva N, Negre F, Nagegowda DA et al (2006) Plant volatiles: recent advances and future perspectives. Crit Rev Plant Sci 25:417440

Elton CS (1927) Animal ecology. William Clowes and Sons Ltd, Great Britain

Fall R, Karl T, Hansel A et al (1999) Volatile organic compounds emitted after leaf wounding: online analysis by proton transfer-reaction mass spectrometry. J Geophys Res 104:15963–15974. doi:10.1029/1999JD900144

Farag MA, Zhang H, Ryu CM (2013) Dynamic chemical communication between plants and bacteria through airborne signals: induced resistance by bacterial volatiles. J Chem Ecol 39(7):1007–1018. doi:10.1007/s10886-013-0317-9

Frost CJ, Mescher MC, Carlson JE et al (2008) Why do distance limitations exist on plant-plant signaling via airborne volatiles? Plant Signal Behav 3(7):466–468

Goldstein AH, Galbally IE (2007) Known and unknown organic constituents in the Earth's atmosphere. Environ Sci Technol 41(5):1514–1521

Gols R (2014) Direct and indirect chemical defences against insects in a multitrophic framework: plant chemical defences against insect. Plant Cell Environ 37(8):1741–1752. doi:10.1111/pce.12318

Han SH, Lee SJ, Moon JH et al (2006) GacS-dependent production of 2R, 3R-butanediol by *Pseudomonas chlororaphis* O6 is a major determinant for eliciting systemic resistance against *Erwinia carotovora* but not against *Pseudomonas syringae* pv. tabaci in tobacco. Mol Plant-Microbe Interact 19(8):924–930. doi:10.1094/MPMI-19-0924

Hol WHG, Garbeva P, Hordijk C et al (2015) Non-random species loss in bacterial communities reduces antifungal volatile production. Ecology 96:2042–2048. doi:10.1890/14-2359.1

Holopainen JK, Blande JD (2012) Molecular plant volatile communication. Adv Exp Med Biol 739:17–31. doi:10.1007/978-1-4614-1704-0_2

Kai M, Effmert U, Berg G et al (2007) Volatiles of bacterial antagonists inhibit mycelial growth of the plant pathogen *Rhizoctonia solani*. Arch Microbiol 187(5):351–360. doi:10.1007/s00203-006-0199-0

Kai M, Effmert U, Piechulla B (2016) Bacterial-plant-interactions: approaches to unravel the biological function of bacterial volatiles in the rhizosphere. Front Microbiol 7:108. doi:10.3389/fmicb.2016.00108

Kanchiswamy CN, Malnoy M, Maffei ME (2015) Chemical diversity of microbial volatiles and their potential for plant growth and productivity. Front Plant Sci 6:151. doi:10.3389/fpls.2015.00151

Kergunteuil A, Dugravot S, Danner HJ et al (2015) Characterizing volatiles and attractiveness of five brassicaceous plants with potential for a 'push-pull' strategy toward the cabbage root fly, *Delia radicum*. Chem Ecol 41(4):330–339. doi:10.1007/s10886-015-0575-9

Kwan G, Charkowski AO, Barak JD (2013) *Salmonella enterica* suppresses *Pectobacterium carotovorum* subsp. carotovorum population and soft rot progression by acidifying the microaerophilic environment. MBio 4(1):e00557–e00512. doi:10.1128/mBio.00557-12

Liberles SD (2014) Mammalian pheromones. Annu Rev Physiol 76:151–175. doi:10.1146/annurev-physiol-021113-170334

Meldau DG, Meldau S, Hoang LH et al (2013) Dimethyl disulfide produced by the naturally associated bacterium *Bacillus sp.* B55 promotes *Nicotiana attenuata* growth by enhancing sulfur nutrition. Plant Cell 25:2731–2747

Moore JG, Jessop LD, Osborne DN (1987) Gas-chromatographic and mass-spectrometric analysis of the odor of human feces. Gastroenterology 93(6):1321–1329

Mumm R, Dicke M (2010) Variation in natural plant products and the attraction of bodyguards involved in indirect plant defense. Can J Zool 88(7):628–667

Neveu N, Grandgirard J, Nenon JP et al (2002) Systemic release of herbivore-induced plant volatiles by turnips infested by concealed root-feeding larvae *Delia radicum* L. J Chem Ecol 28(9):1717–1732

Ninkovic V (2003) Volatile communication between barley plants affects biomass allocation. J Exp Bot 54(389):1931–1939

Ode PJ (2013) Plant defenses and parasitoid chemical ecology. In: Wajnberg E, Colazza S (eds) Chemical ecology of insect parasitoids. Wiley-Blackwell, London

Ortíz-Castro R, Martínez-Trujillo M, López-Bucio J (2008) N-acyl-L-homoserine lactones: a class of bacterial quorum-sensing signals alter post-embryonic root development in *Arabidopsis thaliana*. Plant Cell Environ 31(10):1497–1509. doi:10.1111/j.1365-3040.2008.01863.x

Park YS, Dutta S, Ann M et al (2015) Promotion of plant growth by *Pseudomonas fluorescens* strain SS101 via novel volatile organic compounds. Biochem Biophys Res Commun 461(2):361–365. doi:10.1016/j.bbrc.2015.04.039

Patti GJ, Yanes O, Siuzdak G (2012) Innovation: metabolomics: the apogee of the omics trilogy. Nat Rev Mol Cell Biol 13(4):263–269. doi:10.1038/nrm3314

Peñaflor MF, Bento JM (2013) Herbivore-induced plant volatiles to enhance biological control in agriculture. Neotrop Entomol 42(4):331–343. doi:10.1007/s13744-013-0147-z

Peñuelas J, Asensio D, Tholl D et al (2014) Biogenic volatile emissions from the soil. Plant Cell Environ 37(8):1866–1891. doi:10.1111/pce.12340

Philippot L, Raaijmakers JM, Lemanceau P et al (2013) Going back to the roots: the microbial ecology of the rhizosphere. Nat Rev Microbiol 11(11):789–799. doi:10.1038/nrmicro3109 nrmicro3109

Pichersky E, Gershenzon J (2002) The formation and function of plant volatiles: perfumes for pollinator attraction and defense. Curr Opin Plant Biol 5(3):237–243

Pichersky E, Noel JP, Dudareva N (2006) Biosynthesis of plant volatiles: nature's diversity and ingenuity. Science 311(5762):808–811. doi:10.1126/science.1118510

Rasmann S, Kollner TG, Degenhardt J (2005) Recruitment of entomopathogenic nematodes by insect- damaged maize roots. Nature 434:732–737

Rowan DD (2011) Volatile metabolites. Metabolites 1(1):41–63. doi:10.3390/metabo1010041

Ryu CM, Farag MA, Hu CH et al (2003) Bacterial volatiles promote growth in Arabidopsis. Proc Natl Acad Sci U S A 100(8):4927–4932. doi:10.1073/pnas.0730845100

Ryu CM, Farag MA, Hu CH et al (2004a) Bacterial volatiles induce systemic resistance in Arabidopsis. Plant Physiol 134(3):1017–1026. doi:10.1104/pp.103.026583

Ryu CM, Murphy JF, Mysore KS et al (2004b) Plant growth-promoting rhizobacteria systemically protect *Arabidopsis thaliana* against cucumber mosaic virus by a salicylic acid and NPR1-independent and jasmonic acid-dependent signaling pathway. Plant J 39(3):381–392

Schenkel D, Lemfack MC, Piechulla B et al (2015) A meta-analysis approach for assessing the diversity and specificity of belowground root and microbial volatiles. Front Plant Sci 6:707. doi:10.3389/fpls.2015.00707

Schmidt R, Etalo DW, deJager V et al (2015) Microbial small talk: volatiles in fungal-bacterial interactions. Front Microbiol 6:1495. doi:10.3389/fmicb.2015.01495

Schoonhoven LM, van Loon JJA, Dicke M (2005) Insect-plant biology (2nd). Oxford University Press, Oxford. 9780198525950

Schulz-Bohm K, Zweers H, de Boer W (2015) A fragrant neighborhood: volatile mediated bacterial interactions in soil. Front Microbiol 6:1212. doi:10.3389/fmicb.2015.01212

Schwarz J, Gries R, Hillier K et al (2009) Phenology of semiochemical-mediated host foraging by the western boxelder bug, *Boisea rubrolineata*, an aposematic seed predator. J Chem Ecol 35(1):58–70. doi:10.1007/s10886-008-9575-3

Slaymaker DH, Navarre DA, Clark D et al (2002) The tobacco salicylic acid-binding protein 3 (SABP3) is the chloroplast carbonic anhydrase, which exhibits antioxidant activity and plays a role in the hypersensitive defense response. Proc Natl Acad Sci U S A 99:11640–11645. doi:10.1073/pnas.182427699

Song GC, Ryu CM (2013) Two volatile organic compounds trigger plant self-defense against a bacterial pathogen and a sucking insect in cucumber under open field conditions. Int J Mol Sci 14(5):9803–9819. doi:10.3390/ijms14059803

Steeghs M, Bais HP, de Gouw J et al (2004) Proton-transfer-reaction mass spectrometry as a new tool for real time analysis of root-secreted volatile organic compounds in Arabidopsis. Plant Physiol 135(1):47–58

Trabelsi D, Mhamdi R (2013) Microbial inoculants and their impact on soil microbial communities: a review. Biomed Res Int 2013:863240. doi:10.1155/2013/863240

Tzin V, Galili G (2010) New insights into the shikimate and aromatic amino acids biosynthesis pathways in plants. Mol Plant 3(6):956–972. doi:10.1093/mp/ssq048

van Dam NM, Samudrala D, Harren FJ et al (2012) Real-time analysis of sulfur-containing volatiles in Brassica plants infested with root-feeding Delia radicum larvae using proton-transfer reaction mass spectrometry. AoB Plants:pls021. doi:10.1093/aobpla/pls021

Vespermann A, Kai M, Piechulla B (2007) Rhizobacterial volatiles affect the growth of fungi and *Arabidopsis thaliana*. Appl Environ Microbiol 73(17):5639–5641. doi:10.1128/AEM.01078-07

Vlot AC, Liu PP, Cameron RK et al (2008) Identification of likely orthologs of tobacco salicylic acid-binding protein 2 and their role in systemic acquired resistance in *Arabidopsis thaliana*. Plant J 56:445–456. doi:10.1111/j.1365-313X.2008.03618.x

Wei G, Tian P, Zhang F et al (2016) Integrative analyses of nontargeted volatile profiling and transcriptome data provide molecular insight into VOC diversity in cucumber plants (*Cucumis sativus*). Plant Physiol 172(1):603–618. doi:10.1104/pp.16.01051

Yi HS, Ahn YR, Song GC et al (2016) Impact of a bacterial volatile 2,3-Butanediol on Bacillus subtilis rhizosphere robustness. Front Microbiol 7:993. doi:10.3389/fmicb.2016.00993

Zamioudis C, Korteland J, Van Pelt JA et al (2015) Rhizobacterial volatiles and photosynthesis-related signals coordinate MYB72 expression in Arabidopsis roots during onset of induced systemic resistance and iron-deficiency responses. Plant J 84(2):309–322. doi:10.1111/tpj.12995

Volatile Organic Compounds in Food Security: The Role of Neglected and Underutilized Legumes

6

J. Grebosz, S.U. Dandare, A.D. Ibrahim, and A.A. Aliero

Abstract

The significance of neglected and underutilized species in ameliorating micronutrient deficiency, food security, income generation and enhanced livelihood cannot be doubted. This chapter reviewed current knowledge on the expression patterns and functions of some leguminous plants proteome in response to volatile organic compounds (VOCs) released during biotic and abiotic stresses. The biogenesis of VOCs and their functional role in plant-plant signalling, environmental and biological stress responses were highlighted. Experimental evidences revealed that plant symbiont produce VOCs that induce resistance to phytopathological species and plant growth-promoting rhizobacteria (PGPR). The response to biotic and abiotic flux is achieved through protein production which confer resistance and survival ability to the plant through induced resistance patterns either by induced systemic resistance (ISR) or systemic acquired resistance (SAR). VOCs induced protein production under biotic stress related to bacterial and fungal infection. Bioprospecting of VOCs in legumes is advocated for improved food security as well as domestication and exploitation of their enor-

J. Grebosz
School of Biological Sciences, Queen's University Belfast, Belfast, UK

S.U. Dandare
School of Biological Sciences, Queen's University Belfast, Belfast, UK

Department of Biochemistry, Usmanu Danfodiyo University, Sokoto, Nigeria

A.D. Ibrahim (✉)
School of Biological Sciences, Queen's University Belfast, Belfast, UK

Department of Microbiology, Usmanu Danfodiyo University, Sokoto, Nigeria
e-mail: ibrahim.aliyu@udusok.edu.ng; aid4life@yahoo.com

A.A. Aliero
Department of Biological Sciences, Usmanu Danfodiyo University, Sokoto, Nigeria

D.K. Choudhary et al. (eds.), *Volatiles and Food Security*,
DOI 10.1007/978-981-10-5553-9_6

mous potential for sustainable agriculture. The practical application of VOCs as source of allelochemical in natural and agroecosystem towards improvement of crop productivity is emphasized.

Keywords
Legumes • Volatile organic compounds • Proteomics • Food security

6.1 Introduction

Numerous plant species that are cultivated for food across the globe are neglected and underutilized even though they play a significant role in human nutrition, food security and generation of income of the rural poor (Magbagbeola et al. 2010). These crops are continually maintained by preferences of culture and traditional practices, even though they are poorly characterized and neglected by research, conservationist (Dansi et al. 2012), plant breeders and policymakers and as such described as 'neglected and underutilized species (NUS)' (Padulosi et al. 2013). Despite huge potentials of NUS in reducing hunger, malnutrition and poverty, there is little attention paid to allow full harnessing of the possible values, thereby placing them under continual genetic erosion and disappearance danger which would further cause restriction of development options for the poor. Several neglected and underutilized crop species (NUCS) are very rich nutritionally (Padulosi et al. 1999; Johns and Eyzaguirre 2006; Ghane et al. 2010), making their erosion an immediate concern for global food security and nutritional status of the poor, and their improved use can result in better nutrition and reduce hidden hunger and poverty (Dansi et al. 2012). Many NUCS have been shown to adapt to difficult environments that are unfriendly for other conventional crops making capable of providing sustainable productions (Mal 2007) and meeting global food security needs.

Legumes (family: Fabaceae) are known as the second most valuable plant source for human and animal nutrition (Morris 2003). Legumes are said to be the third largest family amongst angiosperms and consist of approximately 650 genera and 20,000 species (Doyle 1994). Legumes are a staple food for subsistence, particularly in combination with cereals by a significant part of the human population. Although several other common proteinaceous edible legumes such as cowpea, soybean and other neglected and underutilized legumes that are occasionally used as food are available on the market, the consumption rate is not commensurate with production rate, and the demand is on the rise (Ali and Kumar 2000). Legumes are rich in nutrient content including starch, protein dietary fibre, oligosaccharides, phytochemicals and minerals, and they contribute to many health benefits to the humans due to their nutritional contents (Hangen and Bennink 2002; Vucenik and Shamsuddin 2006).

In the context of sustaining the global food security needs of humans, the exploitation of wild or neglected and underutilized legumes is certainly of great importance especially for meeting the food security and nutritional and agricultural development needs of developing nations (Awoyinka et al. 2016). Several of the wild or neglected and underutilized legumes are known to possess the required nutritional requirements to fight hidden hunger and malnutrition in developing nations. They are also known to cope with the increasing fluctuations in the climate around the world, making them a genetic reservoir and a sustainable food source for human consumption. Wild or neglected and underutilized legumes have recently been shown to possess adequate amounts of nutrients such as essential amino acids, protein, polyunsaturated fatty acids (PUFAs), beneficial bioactive compounds, fibre, vitamins and essential minerals which are comparable to other common consumed legumes (Awoyinka et al. 2016).

Some of the few wild or neglected and underutilized legumes include African yam bean or yam-pea (*Sphenostylis stenocarpa*), adzuki bean (*Vigna angularis*), Bambara bean or ground-bean (*Vigna subterranea*), mung bean (*Vigna radiata*), pigeon pea (*Cajanus cajan*), velvet bean (*Mucuna* spp.) and lima bean (*Phaseolus lunatus*). These neglected and underutilized legumes are shown to be genetically diverse in nature. Bambara groundnut (*Vigna subterranea* (L.) Verdc.) is an important neglected and underutilized legume crop which was believed to have originated from West Africa (Hepper 1963) and was originally cultivated in the warm tropics of Sub-Saharan Africa (Uguru and Ezeh 1997). Somta et al. (2011) showed high similarity in the genetic distance between West African and Cameroon/Nigerian species of Bambara groundnut. Consistency was found tight grouping in the majority of West African accessions with Cameroon/Nigerian accessions, away from Central African, East African and Southeast Asian accessions. It was shown that two subpopulations of Bambara groundnut exist with one subpopulation having most of the East African accessions and some Cameroon/Nigerian accessions, whereas the other subpopulation had most of the West African accessions including most of the Cameroon/Nigerian accessions (Somta et al. 2011). The diversity of *Cajanus cajan* has also been reported in Nigeria where about 12 *Mucuna* accessions exist (Ezeagu et al. 2003).

This chapter reviewed current knowledge on the expression patterns and functions of some leguminous plant's proteome in response to VOCs released during biotic and abiotic stresses. Mainly, it will focus on how bacterial and fungal VOCs elicit protein production and how this affects the survivability of the plant.

6.2 Volatile Organic Compounds: Elicitors of Plant's Defence Response

By virtue of their needs for normal growth, development and reproduction, plants utilize several primary metabolites and concomitantly accumulate a broad spectrum of natural products previously thought to be waste products for these organisms that lack defined excretory systems. This vast array of chemicals or by-products of primary metabolism is termed secondary metabolites. Though not directly involved in the immediate survivability of plants, secondary metabolites are said to play a crucial role in plant's interactions with their environment (Spinelli et al. 2011).

Essentially, they function in intra- (between plants) and inter-organism (between plants with microbes and animals) communications (Farag et al. 2013), as well as protection of the plant against biotic and abiotic stresses (Liu and Zhang 2015).

In general, greater than 100,000 natural products are known to be synthesized by plants (Dicke and Loreto 2010). Moreover, scientists have tried to classify them into groups. However, till date, there is no fixed, common agreement upon system for their classification. The most accepted classification is based on their peculiar structural characteristic which arises due to their biosynthetic nature and origin (Crozier et al. 2007). In this, the majority of secondary metabolites belong to one of these families, viz.:

- Alkaloids
- Terpenoid and steroids
- Polypropanoids
- Polyketides and fatty acids
- Others (specialized amino acids and carbohydrates)

At least, 1700 chemicals from these groups are known to be volatiles, the majority of which are terpenes (Niinemets et al. 2004; Dicke and Loreto 2010). Volatile organic compounds (VOCs) are carbon-based chemicals that have a high vapour pressure which causes their molecules to evaporate easily under normal room conditions. Their high volatility is due to the very low boiling point they possess (Dicke and Loreto 2010). The plethora of these chemically divergent compounds released annually into the atmosphere underscores their importance. For instance, biological sources are said to emit about 1150 Tg of VOCs excluding methane per annum (Goldstein and Galbally 2007), the majority (about 500 Tg) of which are produced by terrestrial plants as isoprenoids and probably a similar amount as monoterpenes (Dicke and Loreto 2010). These emissions occur almost exclusively from the stomata of the leaves. However, certain VOCs are also emitted by the other aboveground tissues (flowers and stems) and also below-ground from the roots. The rates of emission of all plant VOCs are dependent on environmental temperature and sunlight (Kesselmeier and Staudt 1999; Guenther et al. 2000). Supposedly, these excess emissions of carbon compounds will substantially have a direct effect on the amount of carbon fixed by plants and thus affect their physiology and productivity. Therefore, the question is why do plants release these relevant amounts of crucial carbon precursors?

Plant VOCs are involved in a wide range of functions which include plant-plant signalling, environmental and biological stress response, indirect plant defence by repulsion of polyphagous herbivores and pathogens, the attraction of herbivore's enemies and pollinators scavenge and removal of reactive oxygen species (ROS) and thermo-tolerance. Even though there is ample knowledge on the processes and metabolic pathways involved in the biosynthesis of several VOCs (Dudareva et al. 2004, 2006), very little is known about how plants can detect these signals and thus the mechanisms with which they exact their specific functions (Holopainen and Blande 2012). VOCs are not an exclusive preserve of plants; VOC emission is indeed recognized as a common property of a wide range of soil microbes since it was reported in 2003 by Ryu et al. Noteworthy are the beneficial plant symbionts which actively colonize the host's roots, exerting different effects on plant development (Pérez-Montaño et al. 2014; Liu and Zhang 2015). Collectively, they are called plant growth-promoting rhizobacteria (PGPR).

6.3 Rhizobacterial VOCs and Their Influence on Neglected and Underutilized Legumes

Due to the remarkable and central role, plants play a role in fixing the solar energy that powers all living processes, numerous microbiota surrounding the roots and rhizosphere soil to take advantage of the nutritious environment. Plants, especially legume roots, emit significant amounts of carbon- and nitrogen-containing compounds into the ground which microorganisms utilize for growth and multiplication (Van Loon and Glick 2004). Besides the symbiotic/mutualistic microbes that provide essential services to the plant, the soil environment also nurtures microbial pathogens and insect herbivores. This incredibly diverse microflora in the small zone around the roots (acclaimed to be one of the most energy-rich habitats on the Earth) interacts within the confines of the available nutrients, consequently leading to a vibrant interplay between the members of the microbial community (Choudhary et al. 2007; Pieterse et al. 2014). These interactions can be synergistic and antagonistic, for example, PGPR can suppress diseases through antagonism between bacteria and soil-borne pathogens.

There is scanty information on the volatile organic compound profile of neglected and underutilized legumes; information available in the literature have been shown that legumes formed close association with bacteria in their root zone (nodule), and the relationship formed is a nitrogen-fixing symbiotic relation with their leguminous plant hosts, while the bacteria are called root nodule bacteria, collectively termed rhizobia, and require high iron (Guerinot 1991). For example, *Bradyrhizobium* strain MAO 113 (isolated from *V. subterranea*) has been shown to infect *Vigna subterranea* (Gueye et al. 1998). *Bradyrhizobium* strain MAO 113 was originally isolated from the root nodules and rhizosphere soil of *Cajanus cajan* (Jadhav and Desai 1992). Clearly nitrogen-fixing symbiotic relation is evident in neglected and underutilized legumes like other common leguminous plants. There is also documented evidence of the role of volatile organic compounds (VOCs) in improving the effectiveness of nitrogen-fixing symbiotic relation in legumes. For example, the model legume plant species, *Medicago truncatula* roots, was shown to form a specific root nodule symbiotic relationship with the nitrogen-fixing bacterium *Sinorhizobium meliloti*, and the *Sinorhizobium meliloti* VOCs increased all plant growth parameters by the *M. truncatula* perceiving its symbiont through VOC emissions (Orozco-Mosqueda et al. 2013).

To date, eight bacterial volatiles possess a positive effect on plant growth, and these include 2,3-butanediol, 3-hydroxy-2-butanone, 2-pentylfuran, N,N-dimethylhexadecanamine, CO_2, 13-tetradecadien-1-ol, 2-butanone and 2-methyl-n-1-tridecene

(Ryu et al. 2003; Farag et al. 2006; Zou et al. 2010; Velázquez-Becerra et al. 2011; Bailly and Weisskopf 2012; Weise et al. 2013; Park et al. 2015). For example, the inoculation of the legume plant *Medicago sativa* with *Arthrobacter agilis* UMCV2 enhanced the growth and development of *M. sativa* seedlings (Velázquez-Becerra et al. 2011). The effect of *Arthrobacter agilis* UMCV2 inoculation observed in the roots system was linked to the bacterium VOCs' role in root development. The VOC was identified as N,N-dimethyl-hexadecanamine (dimethylhexadecylamine), an amino lipid that has a similar structure to the quorum-sensing signals of bacteria, which was found to modulate the bacterial growth and development of a plant in a dose-dependent way. It was concluded that legume plants can perceive bacterial VOCs to modulate growth and morphogenetic processes and identification of a novel signalling molecule potentially involved in the interactions between plant and rhizobacteria (Velázquez-Becerra et al. 2011).

6.4 Mechanisms of Plant Defence Responses

PGPR and its likes are agents that stimulate plant growth and manage soil and also plant health. The question is how do these agents exert this biological control and keep plants resistant to certain diseases?

Plants respond to various biotic stresses using their array of intrinsic defence apparatuses to counteract the deleterious effects of pathogens and parasites. When defence machineries are activated by a stimulus preceding infection by any biotic stress, the disease can be reduced, or even the plant becomes resistant to the disease. This phenomenon is called **induced resistance** – the situation of self-improved protective capacity developed by a plant when aptly excited (Choudhary et al. 2007). This resistance can be developed in response to insect herbivory, as a result of pathogenic infection, upon roots colonization by beneficial microorganisms or following treatment with certain chemicals (Pieterse et al. 2014). Induced resistance boosts the protective level of a plant against a vast array of attackers (Walters et al. 2013). Basically, the two most clearly defined forms of induced resistance are **induced systemic resistance** (ISR) and **systemic acquired resistance** (SAR), wherein the plant defensive machinery are prepared against future attacks from pathogens by prior infections or treatment from the same or different groups of organisms (Van Loon 2000).

6.4.1 Induced Systemic Resistance (ISR)

Upon an attack by a pathogen, induced resistance is not only expressed at the site of induction but also systemically in plant parts that are not in the same zone with the inducer, hence the term ISR (Walters et al. 2013). Therefore, ISR is not a confined property at the zone of induction but rather a global property distributed throughout the plant. ISR is acquired upon local induction by beneficial microbes such as the rhizobacteria (PGPR), and it is one of the widely recognized mechanisms of biocontrol including competition for an ecological niche/substrate and production of inhibitory allelochemicals by the host plant (Haas et al. 2002).

As mentioned earlier, ISR is potentiated by plant growth-promoting rhizobacteria (PGPR), and several studies have revealed the molecular mechanism driving this phenomenon. Typically, rhizobacteria-mediated ISR relies on jasmonic acid (JA) and ethylene (ET) that are synthesized by the plant as signalling molecules. These JA- or ET-induced pathways are characterized by the synthesis of a cascade of oxidative enzymes (peroxidases, polyphenol oxidases and lipoxygenases) and accumulation of low molecular weight compounds with antimicrobial properties (phytoalexins) (Pieterse et al. 2001; Yan et al. 2002). Specific *Pseudomonas* strains have been reported to induce systemic resistance in several plants including *Arabidopsis* and bean. An *Arabidopsis*-based model system was used to study plant-microbe interaction in which a non-pathogenic rhizobacterial strain *Pseudomonas fluorescens* WCS 417r was used as the inducer. Colonization of the model plant roots by the ISR inducer was found to elicit plant protection against several pathogens, including the bacterial leaf pathogens *Xanthomonas campestris* pv. *armoraciae* and *P. syringae* pv. *tomato*, the fungal leaf pathogen *Alternaria brassicicola*, the oomycete leaf pathogen *Peronospora parasitica* and the fungal root pathogen *Fusarium oxysporum* (Ton et al. 2002). In bean, an elevated level of hexanal (volatile antifungal compound) with a concomitant expression of proteins involved in hexanal synthesis was evident when ISR was elicited by *P. putida* BTP1 strain (Ongena et al. 2004).

Several studies have shown PGPR stimulating an SA-dependent ISR similar to pathogen-induced SAR even though ISR by beneficial microbes is triggered through JA and ET pathways which are salicylic acid (SA)-independent mechanisms. A typical example is an SA-producing *Pseudomonas aeruginosa* 7NSK2 shown to enhance disease resistance in wild-type bean (Audenaert et al. 2002). Even though several rhizobacteria possess the ability to synthesize SA, it is commonly not the key inducer of the observed systemic resistance (Ran et al. 2005; Djavaheri et al. 2012).

6.4.2 Systemic Acquired Resistance (SAR)

SAR was coined by Ross (1961) in an attempt to denote the phenomenon in which uninfected systemic plant parts develop immunity to a distant infection in the plant. SAR is acquired upon local induction by pathogens; it could be triggered by plant exposure to virulent or avirulent microbes (Pieterse et al. 2014). Pathogen-induced SAR is characterized by elevated levels of the plant hormone salicylic acid (SA) (Vlot et al. 2009). Furthermore, induction of SAR is associated with the synchronized activation of pathogenesis-related (PR) genes, several of which are responsible for the expression of a cascade of PR proteins with antimicrobial activity. These proteins include antifungals such as thaumatins, chitinases and glucanases. PR-1 is the best-characterized PR gene, and it is often used as a biomarker for SAR (Van Loon et al. 2006).

From the above explanations, it can thus be concluded that, although ISR and SAR are somewhat synonymous in which they are effective against a wide range of attackers, for practical reasons they are referred to differently. ISR is referred to when plant resistance is induced by a beneficial microbe (rhizobacteria mediated) or established to proceed via an SA-independent pathway and SAR when a pathogen triggers the induced resistance or confirmed to be SA dependent. It has also been reported that an amalgamation of ISR and SAR can intensify plant protection against attackers that are resisted through both routes, as well as broaden protection to a wider range of pathogens than ISR or SAR alone (Van Loon and Glick 2004; Haas and Defago 2005).

6.4.3 Hypersensitive Response (HR)

This is a form of resistance in plants characterized by deliberate cell death at the point of attempted pathogen entrance/infection. It occurs when the plant specifically recognizes the attacker during an incompatible interaction. The defence response proceeds with the accumulation of PR proteins, deposition of callose and lignin in the cell wall near the infection site and rapid death of plant cells at the location of the infection. These events essentially help prevent the spread of infection to other parts of the plant by limiting nutrient availability and thereby restricting growth and spread of microbial pathogens (Lamb and Dixon 1997). This line of defence is often used by certain plant species when pathogens suppress their basal resistance. HR is often accompanied by a broad spectrum of defence mechanisms in nearby cells; the onset of SAR is often a result of this response (Delaney 1997).

HR mechanism of elicitation involves the expression of a multitude of defence genes and the synthesis of antimicrobial secondary metabolites such as phytoalexins (Truman et al. 2006). It also triggers the synthesis and build-up of signal molecules such as SA, JA and protein kinases, activation of transcription factors and degradation of proteins by the polyubiquitin system (Pandey et al. 2005). HR is well characterized by the induction of the molecular biomarker *Athsr* in *Arabidopsis thaliana* following its leaves infiltration with the bacterial pathogen *Pseudomonas syringae* pv. *tomato* (Christopher-Kozjan and Heath 2003).

6.5 VOCs Links with the Proteome of Legumes

Proteomics is a study of structures and functions of the all proteins encoded by the organism's genome (called proteome) (Boersema et al. 2015; Marklund et al. 2015; Mozzi et al. 2016). The proteome is very labile and changes altogether with environmental conditions as well as with each organism's compartment. Moreover, proteome can be modified due to the post-translational modifications (PTMs), alternative splicing or protease activity (Mozzi et al. 2016). Nowadays, some advanced techniques have permitted analysis of the proteome in terms of activity, structure, degradation (degradomics), turnover, stoichiometry, localization and dynamics of the protein pathways (Boersema et al. 2015). The most popular techniques of proteomics are two-dimensional gel electrophoresis (2-DE) (including immobilized pH gradient – IPGs as the second dimension) followed by mass spectrometry (MS) (Görg et al. 2004). These methods allow quantitative and qualitative protein expression analysis (Görg et al. 2004; Boersema et al. 2015). Although proteomics is a highly used technique, there are just a few articles about the interaction between VOCs and the proteome of legumes. Thus, this chapter contains more general overview on the subject, also taking into account some studies beyond proteomics and legumes.

6.5.1 Biotic Stressors VOC-Connected Influence on the Proteome

Biotic stressors (insects and microorganisms) cause one of the following plant responses: systemic acquired resistance (SAR) or induced systemic resistance (ISR) – depending on the type of stressor and the regulatory pathway induced. SAR is regulated by salicylate (SA), while ISR is by jasmonate (JA) and ethylene. This results in the production of specific secondary metabolites such as pathogenesis-related proteins (PR proteins) and VOCs (Vallad and Goodman 2004). Both ways of response lead to the increased resistance to the biotic agents (Dicke et al. 2009).

6.5.2 Bacterial and Fungal VOCs' Influence on the Plant's Proteome

VOCs are produced by some microorganisms (bacteria and fungi) as well as by plants themselves. Researchers showed that there are some fungi (mycorrhizal fungi) and bacteria (plant growth-promoting rhizobacteria – PGPR) capable of producing VOCs somehow beneficial to plants. The PGPR bacteria living in the soil were proved to produce VOCs (i.e. 2,3-butanediol and acetoin by *Bacillus subtilis*, dimethyl disulphate by *Bacillus* sp., tridecane by *Paenibacillus polymyxa*) that protect plants from phytopathogenic (bacterial and fungal) infection. For instance, PGPR VOCs have been shown to protect *Arabidopsis* from infection by *Erwinia carotovora* spp. *carotovora* and *Pseudomonas syringae* pv. *tomato and* pv. *maculicola* as well as tobacco from *Botrytis cinerea* and *Cochliobolus heterostrophus* through induction of ISR (Ryu et al. 2004; Kai et al. 2008; Burketova et al. 2015). The very effective antifungal activity (reducing fungal growth) was also connected with the use of such bacteria as *Stenotrophomonas* spp., *Serratia* spp. and *B. subtilis*. The VOCs from the mentioned bacterial species limited the growth of some fungal strains, i.e. *P. carneus*, *M. bolleyi*, *P. betae*, *S. sclerotiorum and T. strictipile*. The bacterial strains were able to restrict the growth of *Sclerotinia*, *Verticillium* and *Rhizoctonia* that are well-known phytopathogens (Kai et al. 2008). The VOCs that limit the mycelia growth of *S. sclerotiorum* are benzothiazole, cyclohexanol, n-decanal, dimethyl trisulphide and 2-ethyl-1-hexanol, while the germination of this fungi is constrained by allyl alcohol (Huang et al. 1997; Fernando et al. 2005). Another PGPR bacterial, *Pseudomonas fluorescens* VOCs, have been observed to limit growth of such pathogenic fungi as *Rhizoctonia solani* and limit symptoms of *Botrytis cinerea* attack in *Medicago truncatula* (Kai et al. 2007; Hernandez-Leon et al. 2015). *P. fluorescens* is able to produce such VOCs as, i.e. 2,3-butanediol, butanal-3-methyl, isoprene, or acetone (Farag et al. 2006). However, the most important *P. fluorescens* VOCs regarding protection from phytopathogens are sulphur-rich compounds, i.e. dimethyl disulphide (DMDS), with proved antifungal activity through inducing ISR in plants (Huang et al. 2012; Hernandez-Leon et al. 2015). Moreover, the PGPR enhance plant's shoot and root growth as well as improve photosynthetic activity, nutrients uptake and biotic and abiotic stress tolerance (Kwon et al. 2010; Vaishnav et al. 2015).

PGPR bacteria living in the soil, namely, *Bacillus subtilis* GB03 and *Bacillus amyloliquefaciens* IN937a, were proved to induce photosynthesis, root and shoot growth and ISR as well (connected with the improved production of proteins involved in ethylene signalling pathway) (Ryu et al. 2004; Zhang et al. 2007; Kwon et al. 2010). The improved photosynthetic rates (reflected in the upregulation of plastid-specific proteins) are connected with the increased ion uptake (especially iron) caused by the *B. subtilis* VOC influence. Furthermore, VOCs of *Bacillus* sp. increase plant's growth rate probably due to the regulation of auxin homeostasis (Zhang et al. 2007). Also, they induce proteins involved in gluconeogenesis and TCA cycle – both pathways producing the energy needed for the growing at the fast pace plant (Kwon et al. 2010). In general, VOCs from *B. subtilis* upregulate 61 and downregulate 34 proteins in *Arabidopsis*. GB03 strain VOCs cause changes in expression of proteins responsible for the following: responsive to stimulus (32%), carbohydrate metabolism (18%), biosynthesis of plant hormones (17%), photosynthesis (10%), amino acid metabolism (10%), nitrogen metabolism (4%), oxidative stress (3%) and structure (2%) (Kwon et al. 2010). It was pointed out that the GB03 strain contributes to the improved resistance of a plant to the pathogenic bacteria, such as *Erwinia carotovora* spp. *carotovora* through turning on the ethylene signalling and ISR. The key role of ethylene was confirmed by the proteomic study of the influence of *B. subtilis* VOCs on *A. thaliana* where researchers observed upregulation of the enzymes participating in ethylene biosynthesis (aspartate aminotransferase, methionine adenosyltransferase 3, S-adenosylmethionine synthetase 2 – SAM-2) (Kwon et al. 2010). S-adenosylmethionine synthetase is a catalyser of the production of S-adenosylmethionine (AdoMet) which is connected with increased resistance to both biotic and abiotic stresses. Moreover, it was observed that bacterial VOCs cause the high expression of antioxidant and oxidative stress-related proteins, such as thioredoxin-dependent peroxidase 1, dehydroascorbate reductase, Fe superoxide dismutase 1 (SOD), glutathione S-transferase, structure of glutathione S-transferase, ascorbate peroxidase (APX) and glutathione reductase. These enzymes are capable of removing reactive oxygen species (ROS) which are produced during both biotic and abiotic stresses causing secondary oxidative stress. Thus, bacterial VOCs seem to have a potential role in improving the plants resistance not only to biotic but also abiotic stresses, such as drought or high salinity (discussed later on in this chapter). The former is also connected with the fact that VOCs might modify the negative influence of ROS on cells even during quite severe stress (Kwon et al. 2010). *Bacillus* sp. is already used in biofungicides (trade name – Kodiak for legumes) because of its antifungal activity (anti-*Fusarium*, anti-*Pythium*, anti-*Aspergillus*, anti-*Rhizoctonia*) (Perez-Garcia et al. 2011).

Research indicates that there are also some molecular mechanisms contributing to the enhanced tolerance of plants subjected to the PGPR VOC activity to abiotic stress. A study on PGPR bacteria *Pseudomonas simiae* strain AU and its VOCs influence on soybean (*Glycine max* L. Merrill; legume) sheds some light on the background of these mechanisms from proteomics point of view (Vaishnav et al. 2015). Despite some knowledge on the other PGPR from *Pseudomonas sp.*, *P. simiae* strain AU is very poorly investigated regarding VOCs identification or interaction with phytopathogens. However, it was shown that *P. simiae* VOCs have a positive influence on soybean's growth (even under abiotic, salt stress conditions) (Vaishnav et al. 2013). The bacterium contributes to the increased tolerance of soybean to salinity due to proline accumulation (playing a crucial role in osmotic adjustment and protection from proteins degradation) and reduction in Na^+ ions build-up. Moreover, it has been proved that the *P. simiae* strain AU VOCs induce soybean's ISR and upregulate proteins such as vegetative storage protein (VSP), gamma-glutamyl hydrolase (GGH) and RuBisCo large-chain protein. VSP influences Na^+ transporters and is responsible for the accumulation and maintaining the homeostasis of Na^+ ions. Due to the fact that high Na^+ ions uptake under salt stress conditions is a fundamental element in causing salt toxicity in the plant, it seems that VSP protein plays a major role in enhancing soybean's salt tolerance (which is supported by the other study on VSP2 gene that upregulated expression is known to increase salt tolerance). GGH is also crucial in raising the salt tolerance in developing plants (seedlings). Also, it participates in folate metabolism. Upregulation of RuBisCo (chloroplast-specific protein with a key role in photosynthesis) after *P. simiae* VOC exposure indicates that bacterial VOCs alleviate photosynthesis rate-limiting properties of the salt stress. Thus, VOCs' blend of *P. simiae* can improve the tolerance of soybean to osmotic/salt stress (Vaishnav et al. 2015). Other researchers confirmed that *P. simiae* can improve the tolerance of abiotic stress, such as drought also of the other legumes (mung bean – *Vigna radiata* L.) by inducing ISR (resulting in proline accumulation, reduction of stomata size, etc.) (Kumari et al. 2016). Other proteins with important functions in enhancing the resistance to salt stress are aquaporins which might be influenced by some mycorrhizal fungi and PGPR bacteria as well (Wani and Hossain 2016).

Furthermore, a few avirulent, plant-symbiont fungi species are also able to produce VOCs that induce plant resistance against some phytopathogens. These non-pathogenic fungi that colonize plants roots belong to the *Trichoderma* sp. just like PGPR, usually positively affects root growth, crop productivity, the uptake and use of nutrients as well as resistance to abiotic stresses (Harman et al. 2004; Stoppacher et al. 2010; Hung et al. 2013). Some of the positive effects of *Trichoderma* are connected with the production of VOCs, such as isobutyl alcohol, 3-methylbutanal and isopentyl alcohol (synthesized by *T. viride;* no 6PP identified in this species) which result in growth promotion of *Arabidopsis thaliana* (Hung et al. 2013). Moreover, *Trichoderma* can limit the growth of the other fungi by direct (mycoparasitism, antibiosis) or indirect action (induced plants resistance, SAR). Namely, *T. harzianum*, *T. asperellum* and *T. atroviride* were proved to induce plant resistance against pathogenic fungi, and due to this, the strains are used as biofungicides (against moulds) (Harman et al. 2004; Stoppacher et al. 2010; Hung et al. 2013). *Trichoderma asperellum* IsmT5 limits the severity of *Botrytis cinerea* and *Alternaria brassicicola* attack in *Arabidopsis thaliana* (although with the simultaneously reduced growth of *Arabidopsis*). Its antipathogenic effect is connected with the VOC called 6-pentyl-α-pyrone (6PP) which can act directly to lessen the germination of *Alternaria* spores and indirectly by inducing the *Arabidopsis* defence responses (ROS, anthocyanins accumulation, etc., similar to the response to oxidative stress) (Kottb et al. 2015).

As mentioned previously, *Trichoderma atroviride* was proved to act as an antagonist of some pathogenic organisms (i.e. *Botrytis cinerea*, *Rhizoctonia solani*). This species (strain P1) reduced the symptoms of *Xanthomonas campestris* pv. *phaseoli* infection in bean by 54% (Harman et al. 2004). Because *Trichoderma* sp. elicits SAR or ISR through various ways, such as by specific peptides, proteins and low molecular weight compounds production, still little is known about VOCs involved in this process (Harman et al. 2004). Interestingly, it has been proved that *Trichoderma* VOCs might have various effects on *Arabidopsis* growth depending on the age of *Trichoderma* culture and the plant itself. Despite the fact that *T. atroviride* 5-days and 14-day cultures were proved to produce 6PP (proved to induce plants growth and defence response previously), the former resulted in growth inhibition of young *Arabidopsis* seedlings (Lee et al. 2015). *T. harzianum* and *T. atroviride* VOCs, 6PP and harzianopyridone, positively influenced pea, tomato and canola growth and decreased the symptoms of *B. cinerea* and *L. maculans* on tomato and canola, respectively. 6PP activity is concentration dependent; it acts as a resistance stimulus in low concentrations and as a phytotoxin in high concentrations (Vinale et al. 2008). Furthermore, 6PP limits the growth of pathogenic fungi, such as *F. oxysporum* f. sp. *lycopersici*, and the production of deoxynivalenol – a mycotoxin synthesized by *Fusarium graminearum* (Scarseletti and Faull 1994; Cooney et al. 2001). 6PP, together with some VOCs from *T. atroviride*, has been proved to be limited by mycotoxin – fusaric acid (FA) produced by some *Fusarium* species (Stoppacher et al. 2010). Moreover, *T. atroviride* was shown to produce 1-octen-3-ol – a VOC which is a product of fatty acid metabolism, present in mushrooms and known to induce *Arabidopsis* defence response (Kishimoto et al. 2007; Stoppacher et al. 2010).

Unfortunately, to our knowledge, there is no evidence of proteomic study done on the influence of 6PP compound on legumes. However, a more general proteomic study about the three-way interactions (including the physical contact) between plant, *Trichoderma atroviride* and fungal pathogens (*B. cinerea* and *R. solani*) was carried out (Marra et al. 2006). The study indicates that *Trichoderma* influences fewer proteins in plants compared to the plant-pathogen or plant-pathogen-*Trichoderma* interaction proteome influence. *Trichoderma*, as well as *B. cinerea* interaction, resulted in the accumulation of homologues of PR proteins, such as

PR-4 and thaumatin-like protein, and proteins related to the activation of plant resistance mechanisms in beans. For example, nucleotide-binding site leucine-rich repeat (NBS-LRR)-type resistance protein was identified during interaction with *Trichoderma* as well as pathogens (in vast amounts in the presence of *R. solani*). This protein takes part in inducing apoptosis and hypersensitive response – HR through recognition of pathogenic metabolites. Noteworthy, PR proteins in beans are less upregulated in response to *Trichoderma* and pathogen than to pathogen alone. Interestingly, it was found that some defence-related proteins activated by pathogens were decreased in *Trichoderma*-bean interaction. Thus, *Trichoderma* presence can reduce protein-based resistance in beans. However, *Trichoderma* induces a specific defensive response in plants which was proved by identification of at least two PR proteins in bean leaves. *Trichoderma* itself caused 39 appeared ex novo, 87 absent, 27 upregulated and 38 downregulated proteins in the bean. *T. atroviride* alone decreased probable disease resistance of RPP8-like protein and SGT1-like protein (leading to the absence of the protein) while increased the production of thaumatin-like PR-5b protein (high upregulation) and pathogenesis-related protein 4B (slight increase) in bean leaves. Meanwhile, *B. cinerea* treatment influenced the mentioned proteins as well but in a different pattern. Namely, *B. cinerea* caused more significant reduction of RPP8-like protein but lower decrease of SGT1-like protein production compared to *Trichoderma*-treated plant proteome. Also, *B. cinerea* alone led to a decrease in production of pathogenesis-related protein 4B and the more limited upregulation of thaumatin-like PR-5b protein than *T. atroviride*. Interestingly, combined *T. atroviride* and *B. cinerea* treatment on the plants showed induced synthesis of all four mentioned plant defence proteins. Moreover, combined treatment showed higher protein induction compared to *T. atroviride* or *B. cinerea* treatment alone. *Trichoderma* and *R. solani* were also found to influence some defence-related proteins. *Trichoderma* treatment led to a higher accumulation of the NBS-LRR-type resistance protein, pathogenesis-related protein 10c, ribonuclease-like PR-10a protein as well as resistance RPP8-like protein and to a decrease in disease resistance protein production. On the other hand, *R. solani* stimulated all of the specified protein production, often in a more profound manner compared to *Trichoderma*-influenced plant. With the exception of disease resistance protein, all the aforementioned proteins were upregulated under simultaneous *T. atroviride* and *R. solani* treatment. However, the induction of the synthesis was lower compared to the upregulation occurring under *R. solani* treatment alone. Moreover, researchers found some homologues of known hydrophobins (avr proteins) in *T. atroviride* during the interaction with plants which might be involved in inducing a hypersensitive response (HR) and SAR. *T. atroviride* contributes to ROS production in the plant which is reflected in a stimulation of SOD production by *B. cinerea* when a three-way interaction with *Trichoderma* occurs (SOD removes ROS, produced by the plant as a defence response, and supports the pathogenic activity). In conclusion, *Trichoderma* influences different from pathogens groups of proteins in bean and modifies the synthesis pattern of proteins in a three-way interaction (plant-*Trichoderma*-pathogen) which shows that *Trichoderma* increase plants' resistance to pathogenic organisms at the molecular level (Marra et al. 2006).

Some mentioned proteins, called PR proteins, are produced by plants mostly under unfavourable conditions (with none or a minor production in the usual environmental state). These proteins are synthesized during both abiotic (cold, drought, heavy metals) and biotic (microbial infection, insect attack) stress. Some groups of PR proteins (PR-1, 2, 3, 4, 5, 6, 12, 13) were proved to limit fungi growth (Sudisha et al. 2011; Chassot et al. 2007; Sels et al. 2008). There are 17 PR protein classes known so far that possess various properties (i.e. chitinase, peroxidase, defensin, thionin, proteinase inhibitors). The size of these proteins varies between 5 and 75 kDa. PR proteins were proved to have antimicrobial and anti-herbivore properties. However, these abilities are limited to specific PR proteins against certain phytopathogens. Interestingly, not all PR proteins suppress phytopathogens' growth, although it is thought that this kind of proteins is still somehow involved in plant defence response. Different PR proteins (even belonging to the same family) are synthesized in various, specific plant organs (Sels et al. 2008). The PR-12 class includes plant defensins that have properties of alpha-amylase, ion channel blocking and antibacterial activity. The PR-13 class contains small (5 kDa), cysteine-rich proteins called thionins. Due to their ability to permeabilize cell membranes, thionins possess antifungal (i.e. *Fusarium oxysporum*) as well as antibacterial (i.e. *Pseudomonas syringae*) properties. Another class, PR-14, consists of small (7–9 kDa), cysteine-rich lipid transfer proteins. They owe their name to the ability to transfer phospholipids (phosphatidylinositol, phosphatidylcholine, galactolipids) non-specifically between membranes (Sels et al. 2008). They are usually produced in leaves and flowers and seldom in roots after some biotic or abiotic stress. It was proved that PR-14 proteins have antimicrobial activity against *Pseudomonas syringae* and *B. cinerea*, probably connected with the permeabilization of the cell membranes (Jung et al. 2005; Sels et al. 2008) (Table 6.1).

6.5.3 Herbivore-Legume Interaction and Its Influence on VOCs and Proteome

Another type of interaction that plants are subjected to is connected with the herbivores. There are different groups of herbivores able to affect plants in various manners, i.e. tissue-feeding caterpillars, cell-feeding thrips and phloem-feeding aphids (Choudhary et al. 2008). Aphids are phloem-feeding herbivores causing significant crop yield losses due to the phloem ingestion and virus transmission (Carillo et al. 2014; Babikova et al. 2014; Brechenmacher et al. 2015). Examples of aphids are green peach aphid, *Myzus persicae*, as well as pea aphid – *Acyrthosiphon pisum. Myzus persicae* is a pest that feeds on various plants species (c.a. 400 host plant species), vegetables, field crops and greenhouse crops (Feng and Isman 1995; Truong et al. 2015), whereas *A. pisum* can attack bean, soybean as well as pea and alfalfa crops (Carillo et al. 2014; Babikova et al. 2014; Brechenmacher et al. 2015). The interaction between plants and aphids has been investigated mostly by genetic and transcriptomic approaches with very limited number of proteomic studies (Ferry et al. 2011; Carillo et al. 2014).

Table 6.1 VOCs-induced proteins production under biotic stress related to bacterial and fungal infection

Plant of origin	Influencing organism	VOCs of the organism	Direction of proved enhancing tolerance activity	Proteins with increased expression	References
Mouse-ear cress (*Arabidopsis thaliana)*	*Bacillus subtilis* GB03	VOCs mixture, including well-known 2,3-butanediol and acetoin	*Erwinia carotovora* spp. *carotovora*	Thioredoxin-dependent peroxidase 1, dehydroascorbate reductase, Fe SOD, glutathione S-transferase, glutathione S-transferase, ascorbate peroxidase, glutathione reductase, aspartate aminotransferase, methionine adenosyltransferase 3, S-adenosylmethionine, SAM-2, photosystem II subunit P-1, RuBisCo activase, actin 8, chaperonin	Kwon et al. (2010)
Soybean (*Glycine max* L. Merrill)	*Pseudomonas simiae*	VOCs mixture	Salt and drought stress (abiotic stress)	VSP, gamma-glutamyl hydrolase, RuBisCo large chain	Vaishnav et al. (2015)
Bean (*Phaseolus vulgaris*)	*Trichoderma atroviride*	Probably VOCs mixture (including 6PP, harzianopyridone, 1-octen-3-ol) with a simultaneous physical contact	*R. solani, B. cinerea*	Thaumatin-like PR-5b protein, pathogenesis-related protein 4B, NBS-LRR-type resistance protein, pathogenesis-related protein 10c, ribonuclease-like PR-10a protein, RPP8-like protein	Marra et al. (2006), Vinale et al. (2008), Stoppacher et al. (2010)

Plants developed the ability to respond to herbivory stress through SAR or ISR. The various secondary metabolites are produced by plants depending on the plant and elicitor involved in the interaction as well as the species and developmental stage of the herbivore. The factors that can improve the resistance to biotic stressors are VOCs as well as PR proteins produced by plants during ISR (Choudhary et al. 2008; Truong et al. 2014). Plant VOCs are released as a response to herbivore wounding or herbivore-associated molecular patterns (HAMPS), i.e. as a reaction to fatty acid-amino acid conjugates (FACs) present in the herbivore saliva and contacting with the host plant. VOCs can act as a direct (discouraging to herbivorous behaviour) or an indirect defensive mechanism (by attracting predators of the insect herbivores) (Baldwin et al. 2002; Harmel et al. 2007; Huffaker 2015). Indeed, latest field experiments indicate that additional introduced portion of methyl salicylate – MeSa – limits the aphid infestation on plants and attracts natural enemies of the aphids as well (Cai et al. 2015). However, there is some evidence already of the ability of some aphids to avoid triggering VOCs emission in plants (Walling 2008; Schwartzberg et al. 2011). Interestingly, some mycorrhizal fungi, despite their positive role discussed earlier in this chapter, change plant VOCs' profile so that it appears more attractive to aphids. The attractiveness to aphids is strictly connected with low sesquiterpenes, especially (E)-β-farnesene (which is known aphids repellent and predators attractant) production (Babikova et al. 2014; Truong et al. 2014).

Proteomic analysis was carried out in order to examine the levels of protein synthesis in *Arabidopsis* in response to generalist sucking pest *Myzus persicae* Sulzer (Truong et al. 2015). *Myzus persicae* was proved to induce *Arabidopsis* VOC production. Interestingly, the production of two VOCs, 4-methylpentyl ITC and terpene, was positively correlated with the aphids' density on leaves (Truong et al. 2014). Proteomic analysis showed that 3 days of aphid infestation mostly increased the production of some photosynthetic proteins in *Arabidopsis*. However, the small chain of RuBisCo was downregulated in response to insect attack. Researchers explain this fact as a result of ROS activity that can damage the protein. The other hypothesis is based on the conviction of the need of carbon transfer in order to respond to the infection. Due to specific enzymes present in aphid saliva, aphids cause also changes in metabolism of sugars in plants. Moreover, aphid attack results in modification of amino acid metabolism. The former is reflected in the high production of alanine aminotransferase in aphid-infested plants. The sucking pests can also induce the production of ATP synthase (subunit beta) as well as elongation factor Tu. The translation-related factor Tu is connected with defence responses of plants. Interestingly, aphid attack causes downregulation of other proteins (receptor-like protein kinase, hydroxyproline-rich glycoprotein) involved in plants' response to stressors (Truong et al. 2015). The reason of the downregulation of these proteins is unknown, although it might be connected with energy conservation after aphid attack (Truong et al. 2015). In general, it was observed that more defence-related proteins are upregulated as a response to microorganisms than to insects (Stintzi et al.1993; Almagro et al. 2009; Liu et al. 2010).

As mentioned previously, the aphid infestation leads to the plant VOC production that are repellent to the insects. This finding has been supported by some papers focusing on various plants species (Harmel et al. 2007; James 2003; Zhu and Park 2005). However, the VOCs' profile seems not to change significantly due to the aphid infestation regarding some plants species, namely, for *Vicia faba* – bean. Researchers found that aphids *Acyrthosiphon pisum* L. don't influence significantly the VOCs production of bean (besides increasing the (E,E)-2–4-hexadienal, (Z)-3-hexenol and (E)-β-farnesene secretion) (Babikova et al. 2014; Schwartzberg et al. 2011). It has been proposed that this fact might be correlated with possibility of the adaptation of aphids to bypass triggering VOCs synthesis as well as to actively inhibit the plant VOCs production, thereby to suppress the plant's response (Walling 2008; Schwartzberg et al. 2011). Pea aphid is able to inhibit the production of (E)-β-ocimene and (E)-β-caryophyllene which are compounds that attract a parasitoid. It is noteworthy that other herbivore – beet armyworm caterpillars (chewing insect) – attack results in significant volatile (i.e. (E)-β-caryophyllene, (E,E)-4,8,12-trimethyl-1,3,7,11-tridecatetraene) release in bean (Schwartzberg et al. 2011).

Carillo et al. (2014) investigated a proteomic response of pea (*Pisum sativum*) to *A. pisum* attack. A resistant cultivar (P665) acts in a few pathways in order to overcome the negative effects of the aphid infestation. Namely, limiting the photosynthesis and biosynthesis of amino acids in the resistant line (with an increase of these in the susceptible Messire cultivar) is associated with its hostile properties due to the reduction of the nutrients inflow needed for aphids' feeding behaviour. Also, some wound signal molecules, such as lipoxygenases (LOXs) and leucine aminopeptidases (LAPs) as well as antioxidant ascorbate/glutathione cycle (ASC-GSH) proteins (monodehydroascorbate reductase, ascorbate peroxidase – APX; both upregulated in both cultivars), seem to be involved in the resistance to aphids. Interestingly, two heat shock proteins – Hsp70 proteins – were downregulated in resistant cultivar compared to non-changed accumulation in the susceptible line (Carillo et al. 2014).

Another legume, soybean (*Glycine max*; cultivar Syngenta NK S24-K4 RR), responds to aphid (*Aphis glycines*) attack by releasing VOCs as well. *A. glycine* leads to the increased production of such VOCs in field-grown soybean (reproductive stage) as cis-3-hexen-1-ol acetate (not induced in vegetative stage), methyl salicylate (MeSA) and farnesene (all of them are herbivore-induced plant volatiles – HIPVs). The mentioned volatiles act as an indirect defence response in soybean by attracting natural enemies of aphids. Interestingly, aphid attack resulted also in decrease of some of VOCs, like caryophyllene and humulene (known as HIPVs as well) (Cai et al. 2015). It has been shown that amongst soybean species, there are some cultivars less susceptible to the aphid invasion. The resistance of these cultivars was linked to the Rag1 and Rag2 genes which encode NBS-LRR proteins (conferring aphid resistance, e.g. in tomato). Despite the fact that a commercial aphid-resistant cultivar of soybean has been released, some aphids (biotype 3) able to live on aphid-resistant soybean (Rag1 and Rag2 plants) have appeared already (Brechenmacher et al. 2015). Brechenmacher et al. (2015) compared proteomes of two soybean cultivars differing in their resistance to aphids: resistant, Rag2, and susceptible, rag2, respectively. The resistant cultivar produces 132 proteins in modified manner compared to non-resistant line under non-stress conditions. When subjected to biotic stress (aphids), two soybean cultivars varied in the accumulation of 264 proteins. 86% of the stress-related proteins production was induced in the resistant soybean cultivar (Rag2) after aphid infestation. The most important stress-related proteins seemed to be as follows: MLO-like protein, thaumatin-like protein, two 70 kDa molecular chaperones and DREPP (developmentally regulated plasma membrane polypeptides) family that are accumulated significantly more intensive in Rag2 line (compared to rag2) when aphid attack occurs with no difference in synthesis between both cultivars under control conditions. The resistance to aphids might be also associated with water transport (three aquaporins production increase in Rag2), flux of energy and carbohydrate (raised accumulation of voltage-dependent anion-selective channel proteins, triose phosphate/phosphate translocator in Rag2), major and minor carbohydrate metabolism (high hexokinase, aldo-keto reductase, mitochondrial ADP/ATP carrier proteins synthesis in Rag2), amino acid metabolism (aspartate aminotransferase, threonine synthase, histidinol dehydrogenase, glutamate decarboxylase, isopropylmalate synthase accumulation) and secondary and hormone metabolism (high abundance of caffeic acid 3-O-methyltransferase, geranylgeranyl pyrophosphate synthetase, steroid 5-alpha reductase in Rag2). Also, some proteins connected with RNA processing are linked to the defence responses of Rag2. Rpb5 RNA polymerase, nucleolar protein NOP56, fibrillarin, spliceosomal proteins, RNA-binding protein, PR10 and HMG-box-related protein are proteins that accumulate in Rag2 line in response to aphids. Meanwhile, the other RNA-related proteins were less abundant in Rag2 than in rag2 which are PR10, two splicing factors, four RNA-binding proteins and bromodomain transcription factor. Resistant cultivars have also different signal transduction protein regulation than susceptible line. Namely, many proteins involved in signal transduction were downregulated in Rag2 – the only identified upregulated protein from this group was LRR serine/threonine protein kinase. Moreover, Rag2 cultivar produced decreased 2 xyloglucan fucosyltrasferases, putative LRR extensins and pectinesterase protein levels (cell wall metabolism proteins) compared to rag2 line. Interestingly, some proteins, like aminocyclopropanecarboxylate (ACC) oxidase, were more abundant in rag2 line under aphid attack (Brechenmacher et al. 2015).

Herbivore attack causes also the activation of some plants elicitor peptides. These peptides, called Peps, play an essential role in developing a defence response against herbivore as well as microorganisms by affecting other genes involved in defensive mechanisms (such as PR proteins, genes, etc.). Good example of such peptide is soybean's GmPep3 produced as a result of *S. exigua* attack and inducing VOC production. Peps genes orthologues are present in various plant species and can be used to induce defensive response in another closely related plants species, i.e. MtPep1 from *Medicago truncatula* (barrelclover) is capable of inducing VOC production in soybean (both legumes) (Huffaker 2015; Huffaker et al. 2013). Mentioned PR proteins (participating in defence response in plants) in plant-herbivore interaction are known to be activated due to the wounding caused by chewing insects. The PR-6 class consists of proteinase inhibitors that target such insect proteases, i.e. trypsin and chymotrypsin resulting in limited growth of the pests. However, PR-6 is also produced after microbial infection (fungal, *B. cinerea*, and bacterial – *Pseudomonas syringae*) (Jung et al. 2005; Sels et al. 2008) (Table 6.2).

Table 6.2 Changes in protein synthesis pattern during biotic herbivore-related stress

Plant of origin	Influencing organism	VOCs produced by plants	Upregulated proteins	Downregulated proteins	References
Mouse-ear cress (*Arabidopsis thaliana*)	*Myzus persicae*	4-methylpentyl ITC and terpene	Proteins involved in photosynthesis, alanine aminotransferase, ATP synthase subunit beta, elongation factor Tu, bifunctional enolase 2/transcriptional activator protein	RuBisCo small-chain protein, ribose-5-phosphate isomerase 3, ADP-ribosylation factor, receptor-like protein kinase, hydroxyproline-rich glycoprotein	Truong et al. (2014), Truong et al. (2015)
Pea (*Pisum sativum;* resistant cultivar P665)	Pea aphid (*Acyrthosiphon pisum*)	No information	LOXs, LAPs, monodehydroascorbate reductase, APX	Photosynthetic and amino acid metabolism proteins, HSP70	Carillo et al. (2014)
Soybean (*Glycine max*; Rag2-resistant line)	*Aphis glycines*	cis-3-hexen-1-ol acetate, MeSA and farnesene	MLO-like protein, thaumatin-like protein, two 70 kDa molecular chaperones, DREPP plasma membrane polypeptide, aquaporins, voltage-dependent anion-selective channel proteins, triose phosphate/phosphate translocator, hexokinase, aldo-keto reductase, mitochondrial ADP/ATP carrier proteins, aspartate aminotransferase, threonine synthase, histidinol dehydrogenase, glutamate decarboxylase, isopropylmalate synthase, caffeic acid 3-O-methyltransferase, geranylgeranyl pyrophosphate synthetase, steroid 5-alpha reductase, Rpb5 RNA polymerase, nucleolar protein NOP56, fibrillarin, spliceosomal proteins, RNA-binding protein, HMG-box-related protein, Tu elongation factor, ascorbate peroxidase	PR-10, two splicing factors, four RNA-binding proteins, bromodomain transcription factor, 2 xyloglucan fucosyltrasferases, putative LRR extensins, pectinesterase proteins	Brechenmacher et al. (2015)

6.5.4 Abiotic Stress Influence on the VOCs and Proteome

Plants are exposed to many different abiotic stressors, such as drought, salinity, heavy metals, etc. (very often coupled with biotic stresses as well). They result in photosynthetic activity decrease, damage and lowering the economic value of the crops and decreasing the yields. In general, abiotic stressors modify primary and secondary metabolism in plants, amongst which biogenic volatile organic compounds, BVOCs, production exists (Loreto and Schnitzler 2010). It has been proved that VOC (i.e. isoprenoids) production is triggered not only by biotic but also during some abiotic stresses and has some role in mitigating the mentioned. Amongst these stressors are drought, salinity, extreme temperatures, mechanical damage, pollutants (i.e. ozone and other oxidants) and high light intensity (Vickers et al. 2009; Loreto and Schnitzler 2010; Pinto et al. 2010; Winter et al. 2012; Griesser et al. 2015). Volatiles' role in reducing the negative influence of abiotic stress (protection from damage of photosynthetic apparatus) is explained by 'single biochemical mechanism for multiple physiological stressors' and might be connected either with their direct participation in ROS (created during contact with each abiotic stressor and causing oxidative damage) removal and affecting oxidative cellular state, indirect influence on ROS signalling or through membrane stabilization (membrane hydrophobic interactions stabilization) (Vickers et al. 2009).

Interestingly, PGPR bacteria and other soil microbiota (able to produce VOCs) influence plants in a way which leads to a higher tolerance not only to biotic (as discussed previously in this chapter) but also to abiotic stresses, as follows: water, temperature, salinity and heavy metal stresses. Microbes, i.e. *Azospirillum* sp. (i.e. *Azospirillum brasilense* reduces negative influence of salt stress in pea), *Pseudomonas syringae*, *P. fluorescens*, *P. simiae*, *Bacillus* sp., *Trichoderma harzianum* and *Trichoderma* sp., can alleviate salinity and drought stress, while *Bacillus polymyxa*, *Pseudomonas alcaligenes* and *Glomus* sp. enhance tolerance to nutrient deficiency in plants (Chakraborty and Chakraborty 2015; Vaishnav et al. 2015; Kumari et al. 2016). One of the significantly influencing crop abiotic stresses is salt stress. It is estimated that up to 20% of the agricultural (cultivated) lands might be affected by excessive salinity. Salt stress interferes ion homeostasis and leads to the secondary oxidative stress associated with the damage and growth inhibition of plants. Salt stress results in plant growth inhibition (or even death) and crop yield decrease (Chen et al. 2016). Studies have shown that salt stress ($CuCl_2$) induces the production of MeSa, monoterpenes (linalool), sesquiterpenes (alpha-curcumene, alpha-zingiberene) and (Z)-3-hexen-3-ol in (excised) rice (*Oryza sativa*) leaves (Obara et al. 2002). However, the study has been carried out on excised leaves – leaves with mechanical damage which may cause VOCs emission by itself.

As stated in this chapter, PGPR bacteria are able to alleviate salt stress. An example is *Bacillus amyloliquefaciens* SQR9 which works through inducing ISR, increasing total soluble sugars (TSS) with a protective function on cells, inducing antioxidative enzymes activity (scavenging ROS), maintaining the photosynthetic activity and maintaining Na^+ homeostasis. The maintenance of Na^+ balance under salt stress is probably associated with the enhanced ability of Na^+ sequestration to vacuoles, limiting Na^+ uptake by roots or expelling Na^+ excess. The reduction of Na^+ intake by maize roots in SQR9 presence can be connected with the production of exopolysaccharides (EPS) by the bacteria (Chen et al. 2016). Also, *Bacillus subtilis* GB03 VOCs alleviate salinity stress in *Arabidopsis* through modifying expression (downregulation) of HKT1 protein which is responsible for Na^+ uptake and transport (Chakraborty and Chakraborty 2015). Moreover, *P. simiae* VOCs also mitigate salt stress in soybean by maintaining water homeostasis and photosynthetic activity (Vaishnav et al. 2015). Plants are often subjected also to heavy metal stress (high concentrations of metals in soil) leading to the secondary oxidative stress, photosynthesis and growth decrease and damage. The sources of the heavy metal contamination are industrial wastes, pesticides and anthropogenic activities (Tuteja and Gill 2016).

Heavy metals, like Cd and Cu (introduced into the soil in, i.e. agricultural fertilizers), in excess amounts limit plant growth and photosynthesis. Researchers found that heavy metals in moderate concentrations (concentrations similar to the occurring in polluted soils) do not influence maize intact plant VOC production (Winter et al. 2012). Meanwhile, maize subjected to both abiotic and biotic herbivore stresses (combined stressors are very common in natural environment) induces VOC production in maize but with no significant induction of volatiles by metals. However, the high concentration of heavy metals (80 μM Cu for maize) primes the release of plants anti-herbivore VOCs (i.e. green leaf volatiles, (E)-hexenal; terpenes, (E)-β-farnesene). This kind of VOCs is produced in response to abiotic stress due to some overlapping of signalling pathways (triggered also by biotic stress), probably caused by ROS generated during both abiotic and biotic stresses (Vickers et al. 2009; Winter et al. 2012). It was proposed that Cu excess in environment leads to priming, the production of JA, and to signalling cascade which results in VOC production that might cause the attraction of parasitoids to plants. The produced VOCs negatively influence the growth of herbivores feeding on plants. Moreover, Cu treatment leads to faster plant response to herbivory (Winter et al. 2012).

Another abiotic stress – ozone stress – can affect plants resulting in VOC emissions as well. Ozone is becoming a serious abiotic stressor which causes damage to crops resulting in yield decrease. This air pollutant is high especially in urban areas and can reach the peak concentrations as high as 120–500 nL/L (Vuorinen et al. 2004). Emitted VOC blend as a response to ozone seems to be dependent on the cultivar and its tolerance to abiotic stress. Tobacco plants that are more tolerant (less visible damage) produce less amounts of VOCs as well as different bouquets of VOCs than more vulnerable cultivar. Namely, both cultivars produce methyl salicylate and sesquiterpenes, although susceptible cultivar produces additionally also C6 volatiles (Heiden et al. 1999).

High amounts of O_3 induce the production of homoterpene (E)-4,8-dimethyl-1,3,7-nonatriene (DMNT) in lima bean plants (Vuorinen et al. 2004; Pinto et al. 2010). The mentioned homoterpene is emitted also as a result of herbivore attack (Pinto et al. 2010). It was observed that lima bean plants subjected to O_3 stress produce (E)-β-ocimene, limonene, 1,8-cineole, linalool, 2-butanone, (E,E)-4,8,12-trimethyl-1,3,7,11-tridecatetraene (TMTT) and (Z)-3-hexenyl acetate and nonanal as well (Vuorinen et al. 2004; Souza et al. 2013). However, researchers are not unanimous in their opinions on VOCs emitted by lima bean under ozone stress, probably due to differences in the experimental approach that might be connected either with the analysis of excised leaves or intact plants of stem-excised plants. For instance, some researchers state the lima bean emits only DMNT, TMTT and (Z)-3-hexenyl acetate after ozone treatment, while the mentioned VOC synthesis is observed also during spider mite attack with the difference in (E)-β-ocimene emission (which occurs exclusively in case of the herbivore attack) (Vuorinen et al. 2004).

Anti-herbivore VOCs (synthesized also after abiotic stress occurrence) are expected to attract some herbivore predators. However, it has been showed that VOC bouquet emitted by lima bean after O_3 exposure doesn't influence tritrophic signalling and predatory mites are able to distinguish between spider mite-infested and non-infested plants even under O_3 co-treatment conditions. Moreover, natural enemy of spider mite is not appealed more to O_3-treated plants in any significant manner than to non-treated plants (Vuorinen et al. 2004). Furthermore, not only the particular volatiles' presence but also the share of individual compounds in VOC blend seems crucial for natural enemies of spider mite to arrive with feeding-focused behaviour to specific plants (Vuorinen et al. 2004).

As a consequence of the exposition to ozone, soybean accumulates ROS and peroxidate lipids (oxidative burst) before any visible negative features occur (Ahsan et al. 2010). In order to cope with the stress, soybean induces antioxidant enzymes, secondary metabolites, sucrose as well as kinases (involved in signal transduction resulting in modified genes expression) production. There are 20 and 32 proteins with modified expression when subjected to ozone treatment in leaves and chloroplasts, respectively (Ahsan et al. 2010). The most abundant group of identified proteins is connected with photosynthesis (almost half of them, 45%) which shows that photosynthesis is the most affected biochemical pathway by stress. Soybean's photosynthetic activity and carbon assimilation seem to drop as an effect of the ozone action. The discussed issue is reflected in a downregulation of the following proteins participating in photosynthesis, i.e. RuBisCo small and large subunits, RuBisCo activase and sedoheptulose-1,7-bisphosphatase, and downregulation of proteins of photosystem I/II, i.e. oxygen-evolving enhancer protein 1, cyclophilin, photosystem I reaction centre subunit VIII, chlorophyll a/b-binding protein, ferredoxin-NADP reductase, cytochrome c-553-like and cytochrome b6-f complex iron-sulphur subunit. However, researchers also found out that photosystem I subunit PsaD altogether with chloroplastic RuBisCo large subunit-binding protein was upregulated which might be associated with their ability to control the other stress-affected protein folding. In contrary, proteins involved in antioxidant defence (APX, catalase, superoxide dismutase both in leaves and chloroplast) and carbon metabolism (i.e. invertase, fructokinase, fructose-bisphosphate aldolase 1) are upregulated after ozone treatment. The latter is strictly connected with starch degradation (and sucrose synthesis) that is used in TCA cycle as a source of energy (compensating disruption of photosynthesis). Short acute ozone stress triggers some soybean metabolic signalling pathways which are reflected in accumulation of kinase family protein in leaf and putative receptor-like protein kinase 2 and serine/threonine-protein kinase SAPK10 in chloroplast. Kinases are proteins with a key role in perception of O_3 stress and signal transduction of it. Moreover, ROS and sugars released after O_3 exposure trigger some signal transduction pathways as well resulting in differentially expressed genes under stress conditions (e.g. sucrose activates some genes, also connected with secondary metabolism). Also, sucrose (highly accumulated in stressed soybean) was proved to be involved in such plant responses to stress as ROS production as well as ROS scavenging (Ahsan et al. 2010).

A proteomic response to ozone stress has been investigated also in the case of bean (*Phaseolus vulgaris* L. cv. IDIAP R-3). The study showed modifications in antioxidant enzymes activity: decrease in SOD and increase in APX accumulation. Furthermore, O_3 stress induced synthesis of small heat shock protein (HSP 33 kDa) as well as naringenin-7-O-methyltransferase (NOMT). Also, researchers proposed pathogenesis-related protein 2 as a potential marker of ozone stress (Torres et al. 2007) (Table 6.3).

Table 6.3 Proteins synthesis pattern in ozone-stressed legumes

Plant of origin	Influencing factor	VOCs produced by plant	Upregulated proteins	Downregulated proteins	References
Soybean (*Glycine max*)	Ozone	Not known	Photosystem I subunit PsaD, chloroplastic RuBisCo large subunit-binding protein, APX, catalase, superoxide dismutase, invertase, fructokinase, fructose-bisphosphate aldolase 1, kinase family protein, putative receptor-like protein kinase 2, serine/threonine-protein kinase SAPK10	RuBisCo small and large subunits, RuBisCo activase, sedoheptulose-1,7-bisphosphatase, proteins of photosystem I/II, i.e. oxygen-evolving enhancer protein 1, cyclophilin, photosystem I reaction centre subunit VIII, chlorophyll a/b-binding protein, ferredoxin-NADP reductase, cytochrome c-553-like, cytochrome b6-f complex iron-sulphur subunit, carbon metabolism proteins	Ahsan et al. (2010)
Bean (*Phaseolus vulgaris*)	Ozone	Not known	APX, HSP 33 kDa, NOMT, PR-2	SOD	Torres et al. (2007)

6.5.5 VOC Plant-Plant Interaction

As mentioned earlier in this chapter, plants release VOCs as a result of the pathogen attack. To our knowledge, no study has detected a VOC receptor in plants yet, but it seems that plants can interact with each other through VOCs. Some studies indicate that VOCs produced by the infected plant can be transferred by wind to the neighbouring plants resulting in the changes of some genes encoding specific defensive proteins. In other words, this phenomenon, known as 'talking trees', indicates that plants can develop a defensive mechanism to prevent an infection that can be spread from another plant (Baldwin et al. 2002).

It is thought that VOCs might act as biocontrol compounds in agriculture. Bean (*Phaseolus vulgaris*) shows higher resistance to *Colletotrichum lindemuthianum* (pathogenic fungi) when subjected to VOCs produced by resistant cultivar. A direct influence on the limiting germination of fungi conidia was connected with such VOCs as limonene, linalool, nonanal, methyl salicylate and methyl jasmonate (Quintana-Rodriguez et al. 2015). Also lima bean (*Phaseolus lunatus*) indicated increased resistance to *Pseudomonas syringae* (pathogenic bacterium) when treated with benzothiadiazole (BTH) (chemical compound inducing SAR but limiting growth of plants) or VOCs produced by resistant plants. Special role seems to be associated with nonanal which primes resistance to pathogenic bacteria (by triggering PR-2) (Yi et al. 2009). Moreover, jasmonic acid (JA) treatment (which leads to VOCs emission) might be considered as one of the ways of biocontrol due to the fact that lima bean plants exposed to JA treatment are less attractive to some herbivorous insects (Ballhorn et al. 2013).

6.6 Conclusion

Plants demonstrate various molecular responses to different environmental stimuli. PGPR bacterial VOCs enhance plant photosynthetic, nutrient uptake abilities and help in maintaining ion and water balance (even during abiotic stresses) as well as protect from biotic stressors, mycorrhizal fungi stimulate defence-related proteins (PR-4, PR-5, PR-10) and protect from phytopathogenic infection. Aphid infestation leads to increase of photosynthetic activity and amino acid metabolism in non-resistant plants (they modify plant metabolism due to the feeding on phloem). In contrary, resistant to aphids, lines of plants synthesize many antioxidant and defence proteins as well as some proteins involved in maintaining water homeostasis with simultaneous decrease in photosynthetic and amino acid metabolism proteins and release VOCs repellent to herbivores (or attracting herbivore enemies). Moreover, VOC called MeSa deserves special attention when considering VOCs with plants-protecting potential due to the fact that it has been found in plants subjected to both biotic and abiotic stresses and has been proved to possess some anti-herbivore activity. Based on the aforementioned, bioprospecting of VOCs in legumes is advocated for the improvement of food security as well as domestication and exploitation of their enormous potential for sustainable agriculture. The practical application of

VOCs as source of allelochemical in natural and agroecosystem towards improvement of crop productivity is emphasized.

References

Ahsan N, Nanjo Y, Sawada H, Kohno Y, Komatsu S (2010) Ozone stress-induced proteomic changes in leaf total soluble and chloroplast proteins of soybean reveal that carbon allocation is involved in adaptation in the early developmental stage. Proteomics 10:2605–2619

Ali M, Kumar S (2000) Problems and prospects of pulses research in India. Indian Farm 50:4–13

Almagro L, Gomez Ros LV, Belchi-Navarro S, Bru R, Ros Barcelo A, Pedreno MA (2009) Class III peroxidases in plant defence reactions. J Exp Bot 60:377–390

Audenaert K, Pattery T, Cornelis PH, Ofte M (2002) Induction of systemic resistance to *Botrytis cinerea* in tomato by *Pseudomonas aeruginosa* 7NSK2: the role of salicylic acid, pyochelin, and pyocyanin. Mol Plant-Microbe Interact 15:1147–1156

Awoyinka OA, Ileola AO, Imeoria CN, Tijani TD, Oladele FC, Asaolu MF (2016) Comparison of phytochemicals and anti-nutritional factors in some selected wild and edible bean in Nigeria. Food Nutr Sci 7:102–111

Babikova Z, Gilbert L, Bruce T, Dewhirst SY, Pickett JA, Johnson D (2014) Arbuscular mycorrhizal fungi and aphids interact by changing host plant quality and volatile emission. Funct Ecol 28:375–385

Bailly A, Weisskopf L (2012) The modulating effects of bacterial volatiles on plant growth. Current knowledge and future challenges. Plant Signal Behav 7:1–7

Baldwin IT, Kessler A, Halitschke R (2002) Volatile signalling in plant-plant-herbivore interactions: what is real? Curr Opin Plant Biol 5:1–3

Ballhorn DJ, Kautz S, Heil M (2013) Distance and sex determine host plant choice by herbivorous beetles. PLoS One 8(2):e55602. doi:10.1371/journal.pone.0055602

Boersema PJ, Kahraman A, Picotti P (2015) Proteomics beyond large-scale protein expression analysis. Curr Opin Biotechnol 34:162–170

Brechenmacher L, Nguyen THN, Zhang N, Jun TH, Xu D, Rouf Mian MA, Stacey G (2015) Identification of soybean proteins and genes differentially regulated in near isogenic lines differing in resistance to aphid infestation. J Proteome Res 14:4137–4146

Burketova L, Trda L, Ott PG, Valentova O (2015) Bio-based resistance for sustainable plant protection against pathogens. Biotechnol Adv 33:994–1004

Cai L, Koziel JA, O'Neal ME (2015) Studying plant–insect interactions with solid phase microextraction: screening for airborne volatile emissions response of soybeans to the soybean aphid, *Aphis glycines* Matsumura (Hemiptera: Aphididae). Chromatography 2:265–276

Carillo E, Rubiales D, Castillejo MA (2014) Proteomic analysis of pea (*Pisum sativum* L.) response during compatible and incompatible interactions with the pea aphid (*Acyrthosiphon pisum* H.) Plant Mol Biol Rep 32:697–718

Chakraborty U, Chakraborty B (eds) (2015) Abiotic stresses in crop plants, vol 232. CPI Group Ltd, Croydon, pp 241–242

Chassot C, Nawrath C, Metraux JP (2007) Cuticular defects lead to full immunity to a major plant pathogen. Plant J 49:972–980

Chen L, Liu Y, Wu G, Njeri KV, Shen Q, Zhang N, Zhang R (2016) Induced maize salt tolerance by rhizosphere inoculation of *Bacillus amyloliquefaciens* SQR9. Physiol Plant 157 10.1111/ppl.12441

Choudhary DK, Prakash A, Johri BN (2007) Induced systemic resistance (ISR) in plants: mechanism of action. Indian J Microbiol 47(4):289–297

Choudhary KD, Johri NB, Prakash A (2008) Volatiles as priming agents that initiate plant growth and defence responses. Curr Sci 94:598–604

Christopher-Kozjan R, Heath MC (2003) Cytological and pharmacological evidence that biotrophic fungi trigger different cell death execution processes in host and nonhost cells during the hypersensitive response. Physiol Mol Plant Pathol 62:265–275

Cooney JM, Lauren DR, Di Menna ME (2001) Impact of competitive fungi on trichothecene production by *Fusarium graminearum*. J Agric Food Chem 49:522–526

Crozier A, Clifford MN, Ashihara H (2007) Phenols, polyphenols and tannins: an overview. Plant Sec Metab Occur Struct Role Hum Diet. doi:10.1002/9780470988558

Dansi A, Vodouhe R, Azokpota P, Yedomonhan H, Assogba P, Adjatin A, Loko YL, Dossou-Aminon I, Akpagana K (2012) Diversity of the neglected and underutilized crop species of importance in Benin. Sci World J 2012:932947., 19 pages. doi:10.1100/2012/932947

Delaney TP (1997) Genetic dissection of acquired resistance to disease. Plant Physiol 113(1):5–12

Dicke M, Van Loon JJA, Soler R (2009) Chemical complexity of volatiles from plants induced by multiple attack. Nat Chem Biol 5(5):317–324

Dicke M, Loreto F (2010) Induced plant volatiles: from genes to climate change. Trends Plant Sci 15(3):115–117

Djavaheri M, Mercado-Blanco J, Versluis C, Meyer J-M, Van Loon LC, Bakker PAHM (2012) Iron-regulated metabolites produced by *Pseudomonas fluorescens* WCS374rare not required for eliciting induced systemic resistance (ISR) against *Pseudomonas syringae* pv. tomato in Arabidopsis. Microbiol Open 1:311–325

Doyle J (1994) Phylogeny of the legume family: an approach to understanding the origins of nodulation. Annu Rev Ecol Evol Syst 25:325–349

Dudareva N, Pichersky E, Gershenzon J (2004) Biochemistry of plant volatiles. Plant Physiol 135:1893–1902

Dudareva N, Negre F, Nagegowda DA, Orlova I (2006) Plant volatiles: recent advances and future perspectives. Crit Rev Plant Sci. 2006 25:417–440

Ezeagu IE, Maziya-Dixon B, Tarawali G (2003) Seed characteristics and nutrient and antinutrient composition of 12 *Mucuna* accessions from Nigeria. Trop Subtrop Agroecosyst 1:129–140

Farag MA, Ryu CM, Sumner LW, Pare PW (2006) GC-MS SPME profiling of rhizobacterial volatiles reveals prospective inducers of growth promotion and induced systemic resistance in plants. Phytochemistry 67:2262–2268

Farag MA, Zhang H, Ryu CM (2013) Dynamic chemical communication between plants and bacteria through airborne signals: induced resistance by bacterial volatiles. J Chem Ecol 39:1007–1018. doi:10.1007/s10886-0130317-9

Feng R, Isman MB (1995) Selection for resistance to azadirachtin in the green peach aphid, *Myzus persicae*. Experientia 51:831–833

Fernando WG, Ramarathnam R, Krishnamoorthy AS, Savchuk SC (2005) Identification and use of potential bacterial organic antifungal volatiles in biocontrol. Soil Biol Biochem 37:955–996

Ferry N, Stavroulakis S, Guan W, Davison GM, Bell HA, Weaver RJ, Gatehouse AM (2011) Molecular interactions between wheat and cereal aphid (*Sitobion avenae*): analysis of changes to the wheat proteome. Proteomics 11:1985–2002

Ghane SG, Lokhande VH, Ahire ML, Nikam TD (2010) *Indigofera glandulosa Wendl.* (Barbada) a potential source of nutritious food: underutilized and neglected legume in India. Genet Resour Crop Evol 57(1):147–153

Goldstein AH, Galbally IE (2007) Known and unexplored organic constituents in the earth's atmosphere. Environ Sci Technol 41(5):1514–1521

Görg A, Weiss W, Dunn MJ (2004) Current two-dimensional electrophoresis technology for proteomics. Proteomics 4(12):3665–3685

Griesser M, Weingart G, Schoedl-Hummel K, Neumann N, Becker M, Varmuza K, Liebner F, Schuhmacher R, Forneck A (2015) Severe drought stress is affecting selected primary metabolites, polyphenols, and volatile metabolites in grapevine leaves (*Vitis vinifera* cv. Pinot noir). Plant Physiol Chem 88:17–26

Guerinot ML (1991) Iron uptake and metabolism in the rhizobia/legume symbioses. Plant Soil 130(1-2):199–209

Gueye M, James MK, Kierans M, Sprent JI (1998) The development and structure of root nodules of bambara groundnut [Voandezeia (Vigna) subterranea]. World J Microbiol Biotechnol 14(2):177–184. doi:10.1023/A:1008869710607

Guenther A, Geron C, Pierce T, Lamb B, Harley P, Fall R (2000) Natural emissions of non-methane volatile organic compounds, carbon monoxide, and oxides of nitrogen from North America. Atmos Environ 34(12–14):2205–2230

Haas D, Defago G (2005) Biological control of soilborne pathogens by fluorescent pseudomonads. Nat Rev Microbiol 4:1–13

Haas D, Keel C, Reimann C (2002) Signal transduction in beneficial plant rhizobacteria with bio-control properties. Anton van Leeuw 81:385–395

Hangen L, Bennink MR (2002) Consumption of black beans and navy beans (*Phaseolus vulgaris*) reduced azoxymethane-induced colon cancer in rats. Nutr Cancer 44:60–65

Harman GE, Howell CR, Viterbo A, Chet I, Lorito M (2004) *Trichoderma* species – oppurtinistic avirulent plant symbionts. Nat Rev Microbiol 2:43–56

Harmel N, Almohamad R, Fauconnier M, Du Jardin P, Verheggen F, Marlier M, Haubruge E, Francis F (2007) Role of terpenes from aphid-infested potato on searching and oviposition behavior of *Episyrphus balteatus*. Insect Sci 14:57–63

Heiden AC, Hoffmann T, Kahl J, Kley D, Klockow D, Langebartels C, Mehlhorn H, Sandermann H, Schraudner M, Schuh G, Wildt J (1999) Emission of volatile organic compounds from ozone-exposed plants. Ecol Appl 9:1160–1167

Hepper FN (1963) Plants of the 1957–58 West Africa expedition II: the bambara groundnut (Voandzeia subterranea) and Kersting's groundnut (*Kerstingiella geocarpa*) wild in West Africa. Kew Bull 16(3):395–407

Hernández-León R, Rojas-Solís D, Contreras-Pérez M, Orozco-Mosqueda MC, Macías-Rodríguez LI, Reyes-de la Cruz H, Valencia-Cantero E, Santoyo G (2015) Characterization of the antifungal and plant growth-promoting effects of diffusible and volatile organic compounds produced by *Pseudomonas fluorescens* strains. Biol Control 81:83–92

Holopainen JK, Blande JD (2012) Molecular plant volatile communication. Adv Exp Med Biol 739:17–31

Huang H, Huang JW, Saidon G, Erickson R (1997) Effect of allyl alcohol and fermented agricultural wastes on carpogenic germination of sclerotia of *Sclerotinia sclerotiorum* and colonization of *Trichoderma* spp. Can J Plant Pathol 9:43–46

Huang C-J, Tsay J-F, Chang S-Y, Yang H-P, Wu W-S, Chen C-Y (2012) Dimethyl disulfide is an induced systemic resistance elicitor produced by Bacillus cereus C1L. Pest Manag Sci 68(9):1306–1310

Huffaker A (2015) Plant elicitor peptides in induced defense against insects. Curr Opin Insect Sci 9:44–50

Huffaker A, Pearce G, Veyrat N, Erb M, Turlings TCJ, Sartor R, Shen Z, Briggs SP, Vaughan MM, Alborn HT, Teal PEA, Schmelz EA (2013) Plant elicitor peptides are conserved signals regulating direct and indirect antiherbivore defense. PNAS 110:5707–5712

Hung R, Lee S, Bennett JW (2013) *Arabidopsis thaliana* as a model system for testing the effect of *Trichoderma* volatile organic compounds. Fungal Ecol 6:19–26

Jadhav RS, Desai AJ (1992) Isolation and characterization of siderophore from cowpea *Rhizobium* (peanut isolate). Curr Microbiol 24:137–141

James DG (2003) Field evaluation of herbivore-induced plant volatiles as attractants for beneficial insects: methyl salicylate and the green lacewing, *Chrysopa nigricornis*. J Chem Ecol 29:1601–1609

Johns T, Eyzaguirre PB (2006) Symposium on "wildgathered plants: basic nutrition, health and survival" linking biodiversity, diet and health in policy and practice. Proc Nutr Soc 65:182–189

Jung HW, Kim KD, Hwang BK (2005) Identification of pathogen-responsive regions in the promoter of a pepper lipid transfer protein gene (CALTPI) and the enhanced resistance of the CALTPI transgenic *Arabidopsis* against pathogen and environmental stresses. Planta 221:361–373

Kai M, Effmert U, Berg G, Piechulla B (2007) Volatiles of bacterial antagonists inhibit mycelial growth of the plant pathogen *Rhizoctonia solani*. Arch Microbiol 187:351–360

Kai M, Vespermann A, Piechulla B (2008) The growth of fungi and *Arabidopsis thaliana* is influenced by bacterial volatiles. Plant Singal Behav 3:482–484

Kesselmeier J, Staudt M (1999) Biogenic volatile organic compounds (VOC): an overview on emission, physiology and ecology. J Atmos Chem 33:23–88

Kishimoto K, Matsui K, Ozawa R, Takabayashi J (2007) Volatile 1-octen-3-ol induces a defensive response in *Arabidopsis thaliana*. J Gen Plant Pathol 73:35–37

Kottb M, Gigolashvili T, Groβkinsky DK, Piechulla B (2015) *Trichoderma* volatiles effecting *Arabidopsis*: from inhibition to protection against phytopathogenic fungi. Front Microbiol 6:1–14

Kumari S, Vaishnav A, Jain S, Varma A, Choudhary DK (2016) Induced drought tolerance through wild and mutant bacterial strain *Pseudomonas simiae* in mung bean (*Vigna radiate* L.) World J Microbiol Biotechnol 32:4

Kwon YS, Ryu CM, Lee S, Park HB, Han KS, Lee JH, Lee K, Chung WS, Jeong MJ, Kim HK, Bae DW (2010) Proteome analysis of Arabidopsis seedlings exposed to bacterial volatiles. Planta 232:1355–1370

Lamb C, Dixon RA (1997) The oxidative burst in plant disease resistance. Annu Rev Plant Physiol Plant Mol Biol 48(1):251–275

Lee S, Hung R, Yap M, Bennett JW (2015) Age matters: the effects of volatile organic compounds emitted by *Trichoderma atroviride* on plant growth. Arch Microbiol 197:723–727

Liu X-M, Zhang H (2015) The effects of bacterial volatile emissions on plant abiotic stress tolerance. Front Plant Sci 6:774. doi:10.3389/fpls.2015.00774

Liu JJ, Sturrock R, Ekramoddoullah AK (2010) The superfamily of thaumatin-like proteins: its origin, evolution and expression towards biological function. Plant Cell Rep 29:419–436

Loreto F, Schnitzler JP (2010) Abiotic stresses and induced BVOCs. Trends Plant Sci 15:154–166

Magbagbeola JAO, Adetoso JA, Owolabi OA (2010) Neglected and underutilized species (NUS): a panacea for community focused development to poverty alleviation/ poverty reduction in Nigeria. J Econ Int Finance 2(10):208–211

Mal B (2007) Neglected and underutilized crop genetic resources for sustainable agriculture. Indian J Plant Genet Res 22(1):1–16

Marklund EG, Degiacomi MT, Robinson CV, Baldwin AJ, Benesch JLP (2015) Collision cross sections for structural proteomics. Structure 23:791–799

Marra R, Ambrosino P, Carbone V, Vinale F, Woo SL, Ruocco M, Ciliento R, Lanzuise S, Ferraioli S, Soriente I, Gigante S, Turra D, Fogliano V, Scala F, Lorito M (2006) Study of the three-way interaction between *Trichoderma atroviride*, plant and fungal pathogens by using proteomic approach. Curr Genet 50:307–321

Morris B (2003) Legumes. In: Katz SH (ed) Encyclopedia of food and culture, vol 3. Charles Scribner & Sons, New York, pp 1–2

Mozzi F, Raya RR, Vignolo GM (2016) Biotechnology of lactic acid bacteria: novel applications. Wiley, Second edition, 159

Niinemets Ü, Loreto F, Reichstein M (2004) Physiological and physicochemical controls on foliar volatile organic compound emissions. Trends Plant Sci 9(4):180–186

Obara N, Hasegawa M, Kodama O (2002) Induced volatiles in elicit or treated and rice blast fungus inoculated rice leaves. Biosci Biotechnol Biochem 66:2549–2559

Ongena M, Duby F, Rossignol F, Fauconnier M, Dommes J, Thonart P (2004) Stimulation of the lipoxygenase pathway is associated with systemic resistance induced in bean by a nonpathogenic Pseudomonas strain. Mol Plant-Microbe Interact 17(9):1009–1018

Orozco-Mosqueda M, Macías-Rodríguez LI, Santoyo G, Farías-Rodríguez R, Valencia-Cantero E (2013) *Medicago truncatula* increases its iron-uptake mechanisms in response to volatile organic compounds produced by *Sinorhizobium meliloti*. Folia Microbiol 58:579–585

Padulosi S, Eyzaquirre P, Hodgkin T (1999) In: Janick J (ed) Challenges and strategies in promoting conservation and use of neglected and underutilized crop species. ASHS Press, Alexandria

Padulosi S, Thompson J, Rudebjer P (2013) Fighting poverty, hunger and malnutrition with neglected and underutilized species (NUS): needs, challenges and the way forward. Bioversity International, Rome

Pandey A, Ger M, Huang H, Yip M, Zeng J, Feng T (2005) Expression of the hypersensitive response-assisting protein in Arabidopsis results in hairpin-dependent hypersensitive cell death in response to Erwinia carotovora. Plant Mol Biol 59(5):771–780

Park Y-S, Dutta S, Ann M, Raaijmakers JM, Park K (2015) Promotion of plant growth by *Pseudomonas fluorescens* strain SS101 via novel volatile organic compounds. Biochem Biophys Res Commun 461(2):361–365

Perez-Garcia A, Romero D, de Vincente A (2011) Plant protection and growth stimulation by microorganisms: biotechnological applications of *Bacilli* in agriculture. Curr Opin Biotechnol 22:187–193

Pérez-Montaño F, Alías-Villegas C, Bellogín RA, Del Cerro P, Espuny MR, Jiménez-Guerrero I, López-Baena FJ, Ollero FJ, Cubo T (2014) Plant growth promotion in cereal and leguminous agricultural important plants: from microorganism capacities to crop production. Microbiol Res 169(5–6):325–336

Pieterse CMJ, Ton J, van Loon LC (2001) Cross-talk between plant defense signaling pathways: boost or burden? Agri-Biotech Net 3:1–18

Pieterse CMJ, Zamioudis C, Berendsen RL, Weller DM, van Wees SCM Bakker PAHM (2014) Induced systemic resistance by beneficial microbes

Pinto DM, Blande JD, Souza SR, Nerg AM, Holopainen JK (2010) Plant volatile organic compounds (VOCs) in ozone (O3) polluted atmospheres: the ecological effects. J Chem Ecol 36:22–34

Quintana-Rodriguez E, Morales-Vargas AT, Molina-Torres J, Adame-Alvarez RM, Acosta-Gallegos JA, Heil M (2015) Plant volatiles cause direct, induced and associational resistance in common bean to the fungal pathogen *Colletotrichum lindemuthianum*. J Ecol 103:250–260

Ran LX, Van Loon LC, Bakker PAHM (2005) No role for bacterially produced salicylic acid in rhizobacterial induction of systemic resistance in Arabidopsis. Phytopathology 95:1349–1355

Ross AF (1961) Systemic acquired resistance induced by localized virus infections in plants. Virology 14:340–358

Ryu C-M, Farag MA, Hu C-H, Reddy MS, Wei H-X, Paré PW, Kloepper JW (2003) Bacterial volatiles promote growth in *Arabidopsis*. Proc Natl Acad Sci U S A 100:4927–4932. doi:10.1073/pnas.0730845100

Ryu CM, Farag MA, Hu CH, Reddy MS, Kloepper JW, Pare PW (2004) Bacterial volatiles induce systemic resistance in Arabidopsis. Plant Physiol 134:1017–1026

Scarselletti R, Faull JL (1994) In vitro activity of 6-pentyl-α-pyrone, a metabolite of *Trichoderma harzianum*, in the inhibition of *Rhizoctonia solani* and *Fusarium oxysporum f. sp. lycopersici*. Mycol Res 98:1207–1209

Schwartzberg EG, Boroczky K, Trumlinson JH (2011) Pea aphids, *Acyrthosiphon pisum*, suppress induced plant volatiles in broad bean, *Vicia faba*. J Chem Ecol 37:1055–1062

Sels J, Mathys J, De Coninck BMA, Cammue BPA, De Bolle MFC (2008) Plant pathogenesis-related (PR) proteins: a focus on PR peptides. Plant Physiol Biochem 46:941–950

Somta P, Chankaew S, Rungnoi O, Srinives P (2011) Genetic diversity of the Bambara groundnut (*Vigna subterranea* (L.) Verdc.) as assessed by SSR markers. Genome 54(11):898–910

Souza SR, Blande JD, Holopainen JK (2013) Pre-exposure to nitric oxide modulates the effect of ozone on oxidative defences and volatile emissions in lima bean. Environ Pollut 179:111–119

Spinelli F, Cellini A, Marchetti L, Nagesh KM, Piovene C (2011). Emission and function of volatile organic compounds in response to abiotic stress, abiotic stress in plants mechanisms and adaptations, Prof. Arun Shanker (ed), ISBN: 978–953–307-394-1, InTech, Available from http://www.intechopen.com/books/abiotic-stress-in-plants-mechanisms-and-adaptations/emission-andfunction-of-volatile-organic-compounds-in-response-to-abiotic-stress

Stintzi A, Heitz T, Prasad V, Wiedemann-Merdinoglu S, Kauffmann S, Geoffroy P, Legrand M, Fritig B (1993) Plant 'pathogenesis-related' proteins and their role in defense against pathogens. Biochimie 75:687–706

Stoppacher N, Kluger B, Zeilinger S, Krska R, Schuhmacher R (2010) Identification and profiling of volatile metabolites of the biocontrol fungus *Trichoderma atroviride* by HS-SPME-GC-MS. J Microbiol Methods 81:187–193

Sudisha J, Sharathchandra RG, Amruthesh KN, Kumar A, Shetty HS (2011) Pathogenesis related proteins in plant defense response. Prog Biol Control 12:379–403

Ton J, van Pelt JA, van Loon LC, CMJ P (2002) Differential effectiveness of salicylate-dependent and jasmonate/ethylene-dependent induced resistance in Arabidopsis. Mol Plant-Microbe Interact 15:27–34

Torres NL, Cho K, Shibato J, Hirano M, Kubo A, Masuo Y, Iwahashi H, Jwa NS, Agrawal GK, Rakwal R (2007) Gel-based proteomics reveals potential novel protein markers of ozone stress in leaves of cultivated bean and maze species of Panama. Electrophoresis 28:4369–4381

Truman W, de Zabala MT, Grant M (2006) Type III effectors orchestrate a complex interplay between transcriptional networks to modify basal defense responses during pathogenesis and resistance. Plant J 46:14–33

Truong DH, Delory BM, Vanderplanck M, Brostaux B, Vanderyecken A, Heuskin S, Delaplace P, Francis F, Lognay G (2014) Temperature regimes and aphid density interactions differentially influence VOC emissions in *Arabidopsis*. Arthropod Plant Interact 8:317–327

Truong DH, Bauwens J, Delaplace P, Mazzucchelli G, Lognay G, Francis F (2015) Proteomic analysis of *Arabidopsis thaliana* (L.) Heynh responses to a generalist sucking pest (*Myzus persicae* Sulzer). Plant Biol 17:1210–1217

Tuteja N, Gill SS (2016) Abiotic stress response in plants. Wiley-VCH Verlag GmbH and Co. KGaA, Weinheim, pp 219–221

Uguru MI, Ezeh NE (1997) Growth, nodulation and yield of bambara groundnut (Vigna subterranea (L) Verdc) on selected Nigerian soils. J Sci Food Agric 73(3):377–382

Vaishnav A, Jain S, Kasotia A, Kumari S, Gaur RK, Choudhary DK (2013) Effect of nitric oxide signaling in bacterial-treated soybean plant under salt stress. Arch Microbiol 195(8):571–577

Vaishnav A, Kumari S, Jain S, Varma A, Choudhary DK (2015) Putative bacterial volatile-madiated growth in soybean (*Glycine max* L. Merrill) and expression of induced proteins under salt stress. J Appl Microbiol 119:539–551

Vallad GE, Goodman RM (2004) Systemic acquired resistance and induced systemic resistance in conventional agriculture. Crop Sci 44(6):1920–1934

Van Loon LC (2000) Systemic induced resistance. In: Slusarenko AJ, Fraser RSS, van Loon LC (eds) Mechanisms of resistance to plant diseases. Kluwer, Dordrecht, pp 521–574

Van Loon LC, Glick BR (2004) Increased plant fitness by rhizobacteria. In: Sandermann H (ed) Molecular ecotoxicology of plants. Springer, Berlin/Heidelberg, pp 177–205

Van Loon LC, Rep M, Pieterse CMJ (2006) The significance of inducible defense-related proteins in infected plants. Annu Rev Phytopathol 44:135–162

Velázquez-Becerra C, Macías-Rodríguez LI, López-Bucio J, Altamirano-Hernández J, Flores-Cortez I, Valencia-Cantero E (2011) A volatile organic compound analysis from Arthrobacter agilis identifies dimethylhexadecylamine, an amino-containing lipid modulating bacterial growth and *Medicago sativa* morphogenesis in vitro. Plant Soil 339:329–340

Vickers CE, Gershenzon J, Lerdau MT, Loreto F (2009) A unified mechanism of action for volatile isoprenoids in plant abiotic stress. Nat Chem Biol 5:283–291

Vinale F, Sivasithamparam K, Ghisalberti EL, Marra R, Barbetti MJ, Li H, Woo SL, Lorito M (2008) A novel role for Trichoderma secondary metabolites in the interactions with plants. Physiol Mol Plant Pathol 72:80–86

Vlot AC, Dempsey DA, Klessig DF (2009) Salicylic acid, a multifaceted hormone to combat the disease. Annu Rev Phytopathol 47:177–206

Vucenik I, Shamsuddin AM (2006) Protection against cancer by dietary IP6 and inositol. Nutr Cancer 55:109–125

Vuorinen T, Nerg AM, Holopainen JK (2004) Ozone exposure triggers the emission of herbivore-induced plant volatiles, but does not disturb tritrophic signalling. Environ Pollut 131:305–311

Walling LL (2008) Avoiding effective defences: strategies employed by phloem-feeding insects. Plant Physiol 146:859–866

Walters DR, Ratsep J, Havis ND (2013) Controlling crop diseases using induced resistance: challenges for the future. J Exp Bot 64:1263–1280

Wani SH, Hossain MA (eds) (2016) Managing salt tolerance in plants: molecular and genomic perspectives, vol 271–273. CRC Press, Boca Raton, pp 277–278

Weise T, Kai M, Piechulla B (2013) Bacterial ammonia causes significant plant growth inhibition. PLoS One 8:e63538. doi:10.1371/journal.pone.0063538

Winter TR, Borkowski L, Zeier J, Rostas M (2012) Heavy metal stress can prime for herbivore-induced plant volatile emission. Plant Cell Environ 35:1287–1298

Yan Z, Reddy MS, Ryu CM, Mc Inroy JA, Wilson M, Kloepper JW (2002) Induced systemic protection against tomato late blight elicited by PGPR. Phytopathology 92:1329–1333

Yi HS, Heil M, Adame-Alvarez RM, Ballhorn DJ, Ryu CM (2009) Airbourne induction and priming of plant defences against a bacterial pathogen. Plant Physiol 151:2152–2161

Zhang H, Kim MS, Krishnamachari V, Payton P, Sun Y, Grimson M, Farag MA, Ryu CM et al (2007) Rhizobacterial volatile emissions regulate auxin homeostasis and cell expansion in *Arabidopsis*. Planta 226:839–851

Zhu JW, Park KC (2005) Methyl salicylate, a soybean aphid induced plant volatile attractive to the predator *Coccinella septempunctata*. J Chem Ecol 31:1733–1746

Zou C, Li Z, Yu D (2010) Bacillus megaterium strain XTBG34 promotes plant growth by producing 2-pentylfuran. J Microbiol 48:460–466. doi:10.1007/s12275-010-0068-z

Plant Volatiles and Defense

7

Kanta Rani, S.S. Arya, S. Devi, and Vikender Kaur

Abstract

In nature, plants are adversely affected by environmental conditions like abiotic and biotic stresses. In addition to mechanical or physical defensive ways, plants have an important chemical defensive mechanism to cope up with these adverse conditions. Plant volatile organic compounds (VOCs) have various direct and indirect defense roles against various abiotic stresses (like temperature, water stress, ozone, salt stress, and heavy metals) and biotic stresses like herbivores and pathogen (above ground and below ground). In this chapter we review the defensive role of plant VOCs against abiotic and biotic stresses.

Keywords

Plant defense • Plant volatile compounds • Temperature • Ozone • Heavy metals • Water stress • Salt stress • HIPVs and pathogens

K. Rani (✉)
Department of Botany, GGDSD College, Palwal, Haryana, India
e-mail: drkantachhokar@gmail.com

S.S. Arya
Department of Botany, Maharshi Dayanand University, Rohtak, Haryana, India

S. Devi
Department of Botany and Plant Physiology, CCS HAU, Hisar, Haryana, India

V. Kaur
Germplasm Evaluation Division, ICAR-NBPGR, Pusa Campus, New Delhi, India

D.K. Choudhary et al. (eds.), *Volatiles and Food Security*,
DOI 10.1007/978-981-10-5553-9_7

7.1 Introduction

Being sessile organisms, plants have to face the various unfavorable environmental stress factors (abiotic and biotic stresses). Plant growth under these abiotic and biotic stresses resulted in decreased crop production throughout the world. There are various physical or mechanical and chemical defensive ways by which plants protect themselves from these stresses. Chemical defense in plants includes various volatile compounds that release from different parts like their leaves, bark, flowers, fruits, and even from roots. Till date about 1700 volatile organic compounds (VOCs) have been identified and isolated from different plant families. Plant volatiles can be classified into hydrocarbons, aldehydes, alcohols, ethers, ketones, and esters on the basis of their chemical structure. These plant volatiles are represented by phenylpropanoids/benzoids, terpenoids, fatty acid, and amino acid derivatives which constitute around 1% of plant secondary metabolites (Dudareva et al. 2006).

The plant emits constitutive volatile compounds under nonstressed conditions and induced volatile and constitutive volatile compounds under stressed conditions (Loreto and Schnitzler 2010; Niinemets 2010; Niinemets et al. 2010b). Induced volatile organic compound emissions under abiotic stress could enhance the defense response of plants during biotic stress (Cornath et al. 2006; Copolovici et al. 2014; Heil and Kost 2006; Niinemets 2010). Similarly biotic stress-affected plants protect themselves against abiotic stress either by increasing emission of these volatile compounds or by their direct effect on abiotic stress resistance (Copolovici and Niinemets 2016; Fujita et al. 2006; Owen and Penuelas 2005). This is known as priming effect of volatile compounds. Among the constitutive volatiles, isoprenoids, including isoprene and terpenes such as monoterpenes and sesquiterpenes, are important volatile compounds that are released from plants and have important role during respiration, photosynthesis, and fluidity of cell membrane and in biotic interactions (Fineschi et al. 2013; Harrison et al. 2013; Vranova et al. 2012).

Constitutive volatiles are stored in some specialized structures like resin ducts in case of coniferous plants, and in Labiatae these are stored in glandular trichomes (Hare 2011). These stored VOCs help to deter and reduce the attack of herbivory and also reduce the biological activity of pathogens due to their toxic nature (Popp et al. 1995; Ward et al. 1997; Litvak and Monson 1998; Baier et al. 2002). On the other hand, constitutively released non-stored volatiles can help directly in defense mechanism against abiotic stress by acting as antioxidants and make membrane more stable (Sharkey et al. 2008; Chen et al. 2009; Vicker et al. 2009; Possell and Loreto 2013). Loreto et al. (2006) observed the emission of GLV (green leaf volatile) example of induced volatile that is released within minutes after affected by stress (Hare 2011). Here in this chapter we discuss various roles of plant volatile compounds in plant defense under biotic and abiotic stresses.

7.2 Abiotic Stress and Plant Volatile Compounds

Major abiotic stresses like flood, drought, salinity, cold, high temperature, heavy metals, and ozone have negative effect on food crop agriculture production globally (Doupis et al. 2011). Primary and secondary metabolisms of plants are adversely affected by these abiotic stresses in different ways, thereby reducing the crop production (Loreto and Schnitzler 2010). Plants are sedentary organism, so to complete their life cycle they have to adjust with change in environment. Plants have many morphological adaptations that help to cope up with abiotic stress. For instant, plants growing under low temperature go through cold acclimation, causing change in composition of cell membrane and concentration of protein that help to reduce formation of intracellular ice crystal and dehydration under low temperature (Thomashow 1998). The susceptibility or tolerance to any type of stress is a very elaborate process because different stages of plant development are affected by the stress or several stresses (Chinnusamy et al. 2004).

However, the plants have internal chemical defense mechanism using volatile organic compounds to fight against these different abiotic stresses.

7.3 Temperature

High temperature causes the generation of hydrogen peroxide by enhanced photorespiration and reactive oxygen species (Vallelian-Bindschedler et al. 1998; Vacca et al. 2004) that have direct effect on structure and function of thylakoid and thylakoid leakiness and stimulate cyclic electron flow and photochemical reactions (Pastenes and Horton 1996; Bukhov et al. 1999; Schrader et al. 2004). High temperature often affects stomatal behavior either per se or because high temperature can lead to drought stress. Stomatal opening under mild heat stress helps to dissipate latent heat via transpiration. Whereas stomatal closure reduces water loss by improving water use efficiency, i.e. a ratio between net CO assimilation and transpiration (Loreto and Schnitzler 2010).

Plants have many mechanisms by which they are able to tolerate high temperature (Sharkey and Schrader 2006). One of the ways is heat shock proteins (Nover et al. 2001; Vierling 1991).

Temperatures directly affect the emission of volatile organic compounds. As with increasing temperature, more of VOCs enter into the gas phase and are released in the atmosphere. At high temperature the concentration of stored VOCs in specialized tissues like ducts and glands reaches a very high concentration; these stored VOCs are released in a very less amount unless their pools are not injured either by herbivory (Litvak and Monson 1998) or by strong wind, forest fire, etc., whereas non-stored volatiles are constitutively released from the leaves' mesophylls according to concentration gradients (Alessio et al. (2004).

There are many VOCs that are released in response to high temperature, act as the thermoprotectant, and help in stabilizing chloroplast membranes (Sharkey and Singsaas 1995). Release of VOC isoprene helps to protect plants under heat stress. Its emission may be light dependent or using carbon from the Calvin cycle during photosynthesis (Monson and Fall 1989; Loreto and Sharkey 1990; Affek and Yakir 2003; Schnitzler et al. 2004; Ferrieri et al. 2005). Experimental results of Sharkey et al. (1996), Singsaas et al. (1999), and Lehning et al. (2001) show that canopy position has a direct correlation with the isoprene emission (top leaves release about four times more isoprene than bottom leaves due to more exposure to heat stress) that is related with thermotolerance. Moreover ISP activity also shows a difference with respect to canopy position. Jardine et al. (2015) observed in a study that during morning, there is increase in photosynthesis and emission of terpenoids, and in the afternoon terpenoids are replaced by green leaf volatiles. This might be due to increase in temperature during afternoon as compared to morning, which leads to collapse of photosynthesis and isoprene antioxidant system; moreover high temperature is associated with desiccation which causes peroxidation of membrane and release of green leaf volatiles.

The experiment by Loreto et al. (2006) showed release of many volatile organic compounds under high temperature. Isoprene emission may be due to the reason that requirement of optical temperature for isoprene and isoprene synthase activity is around 42 °C that results into its rapid stimulation (Monson et al. 1992). The decrease rate of isoprene emission at high temperature was observed by Singsaas and Sharkey (2000); this decrease may be due to the role of isoprene in maintaining leaf at high temperature rather than to enzyme denaturation. On the contrary Loreto et al. (2006) also observed total inhibition of isoprene at high-temperature treatment might be due to insufficient availability of photosynthetic carbon for isoprene formation and that may be due to reduction of photosynthesis at high-temperature treatment.

Copolovici et al. (2005) blocked the monoterpene emission by using fosmidomycin, inhibitor of isoprenoid biosynthetic pathways in *Quercus ilex* (L.) that reduces the photosynthetic thermotolerance. But if fumigated externally with low concentration of monoterpene, it helps to restore heat stress resistant partly. Sharkey et al. (2001) observed similar findings by fumigating isoprene externally on leaves of red oak and kudzu that were fed by fosmidomycin. This suggests that monoterpene- and isoprene-non-emitting plants can also become heat tolerant by fumigation from nearby growing plants that emit these volatiles in warm windless days (Copolovici et al. 2005). Bukhov et al. (1999) and Pastens and Horton (1996) proposed the mechanism by which isoprene and monoterpenes help in thermotolerance that thylakoid membrane becomes leaky at high temperature and increases the hydrophobic interactions of thylakoid membrane, thereby stabilizing interactions between lipids and membrane protein under high-temperature stress condition (Sharkey and Yeh 2001). Thermotolerance of thylakoid membrane is due to lipophilic isoprene that is partitioned into membrane and checking water channels at high temperature (Singsaas et al. 1997; Sharkey et al. 2001).

Moreover, to study the function of isoprene as well as monoterpene within plant defense against high temperature, many transgenic plants have been genetically engineered that release isoprene or monoterpene or those in which their biosynthetic pathways have been inhibited. It was observed that transgenic plants that emit isoprene are comparatively more resistant against heat stress as compared to wild types (Loivamaki et al. 2007; Sasaki et al. 2007; Behnke et al. 2007). The emission of other VOCs such as methyl butenol, ethanol, and acetaldehyde from conifers under high temperature was observed by Schade and Goldstein (2001).

7.4 Ozone and Heavy Metal Stress Generate Reactive Oxygen Species (ROS)

Among the different pollutants like ozone, SO, HF, or NOx, ozone (O_3) is a major phytotoxic pollutant. Ozone is generated when anthropogenic hydrocarbons and oxides of nitrogen (NO, NO_2) and sulfur (SO_x) react with solar UV radiation. Ozone present in the stratosphere is beneficial as it shields the Earth from UV rays; however, tropospheric ozone is harmful to our plant life as it acts as highly reactive oxidant. The effects of ozone on plants are decreased rates of photosynthesis, leaf injury, reduced growth of roots and shoots, accelerated senescence, and reduced crop productivity. Ozone enters the plant through stomata and is converted to different forms of ROS. There ROS cause oxidation of proteins, RNA, and DNA, lipid peroxidation of cell membrane, and activation of programmed cell death (PCD).

There are many experimental evidences that show that plant volatile compounds play a significant role against ozone stress. Increase emission of ethylene has been observed under high levels of O_3 (Francesco et al. 2011; Mehlhorn and Wellburn 1987). Similarly in *Arabidopsis*, Vahala et al. (1998) observed that ACS_6 gene level of ACS (ACC synthase generate limiting enzyme in ethylene biosynthesis pathway) family elevated under ozone treatment. On the contrary, Calfapietra et al. (2007) observed that isoprene emission was reduced; this might be due to that just after entering into leaves, O_3 reacts with different compounds. One of the compounds is isoprene that is found in high concentration (Loreto et al. 1998). So it might react with O_3 and results into its reduced emission and produces new products like methyl vinyl ketone and methacrolein (Fares et al. 2008 in a study, by fumigation of O_3 on isoprene-emitting trees).

Within the plant and environment, volatile compound terpene plays the same dual role; in the environment in the absence of NO_X, it reacts with ozone, while in the presence of NO_X, it forms ozone by reacting with it (Fehsenfeld et al. 1992; Loreto and Schnitzler 2010). Similarly within plants, terpene reacts with ozone and ROS species, which protect plants from O_3 stress effect (Loreto et al. 2001; Hewitt et al. 1990). These results were experimentally proved by Vickers et al. (2009) by using non-emitting transgenic tobacco plant, which were severely affected by ozone treatment and isoprene-emitting plants that show resistance against ozone stress.

In *Quercus ilex* (L.) leaves, the inhibition of monoterpene emission by fosmidomycin makes it ozone susceptible and results into reduced photosynthesis and generation of ROS that causes membrane peroxidation (Loreto et al. 2004). Beauchamp et al. (2005) observe increased emissions of C6 compounds in ozone-stressed leaves, proportionate to ozone dose. Heiden et al. (1999) reported elevated levels of sesquiterpene production under ozone stress. The rate of sesquiterpene emission is directly proportional to plant resistance against O_3 and shows its direct role in O_3 stress defense.

Isoprene helps in ozone defense by acting as antioxidants in leaves. This is on the basis that it reacts with ozone (different concentrations with variable time exposure) and ROS like hydroxyl radicals as studied by Loreto et al. (2001) and Loreto and Velikova (2001). This reduced damage may be due to the action of isoprene in strengthening the membrane.

7.5 Heavy Metals

Heavy metals (HMs) like Fe, Mn, Zn, and Cu play an important role as many enzymes require the presence of heavy metals for their functioning (Terry 1980; Hansh and Mendel 2009). Sources of HMs may be natural or artificial; naturally these are existing in the Earth's crust and become the part of soil during the process of weathering. Wind dust coming from Sahara desert generally carries HMs in different proportions, e.g., less amount of Zn, Cr, Ni, and Pb and high amount of Fe (Ross 1994). Artificially HMs may be released by various agricultural activities (pesticides, fertilizers, fungicides, etc.), vehicle transportation, landfills, and coal burning that produce Cd, Hg, Ni, Fe, Al, Mn, and Ti. Fire also adds some volatile HMs (Hg and Se); few more HMs like Cr, Cu, Pb, and Zn are added to the environment by the corrosion of commercial waste products (Verkleji 1993).

The uptake of heavy metals can lead to inhibition of photosynthesis (by binding of HMs to different active sites of photosynthetic apparatus and inhibiting RuBisCO and PEP carboxylase enzyme activity; Mallick and Mohan 2003), disruption of membrane structure (causing fluidity of membrane lipid), ion homeostasis, and activation of PCD programmed cell death. One of the reasons that HMs (like Cd, As, and Al) are so toxic is that they can mimic other essential metals, e.g., Ca and Mg, and take their place in essential reactions, thereby disrupting these reactions. HMs can also lead to increase of reactive oxygen species in the chloroplast, mitochondria, and peroxisomes (Apel and Hirt 2004).

Heavy metals are generally categorized in two groups on the basis of mode of action in plant cell: (i) redox active HMs (e.g., Fe, Cu, Cr, Co) that generate ROS by direct redox reaction in cell and (ii) redox inactive HMs (e.g., Cd, Zn, Ni, Al, etc.) that also lead to ROS formation but by indirect methods (by interacting with antioxidant system, by disrupting electron transport chain, or by enhancing LOX (lipoxygenase) activity that leads to lipid peroxidation). Visual effect of HM stress is reduced plant growth (Sharma and Dubey, 2007), chlorosis of leaf, necrosis, senescence, etc.

7.6 Water and Salt Stress

Water and salt stress shows a direct effect on stomatal opening and closing and photosynthesis process (Fall et al. 1999), which reduce the emission of volatile compounds by varying the carbon supply in MEP pathways and by increasing resistance to emission of volatile compounds (Loreto and Schnitzler 2010). Sharkey and Loreto (1993) in kudzu leaves founded that isoprene emission is not reduced until photosynthesis is completely inhibited under drought stress.

Effect of drought stress includes closing of stomata, increase amount of osmaticum, and change in size and shape of leaf and characteristics of root system that help to improve conservation and increase water use efficiency (Chaves et al. 2002; Beis and Patakas 2010). Drought also causes the generation and accumulation of ROS and free radicals (Fig. 7.1) that adversely affect the membrane (Ionenko and Anisimov 2001; Munns 2002).

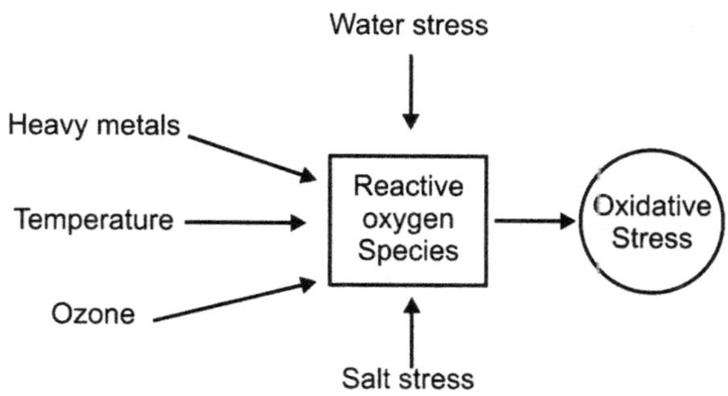

Fig. 7.1 Abiotic stress causes the increase concentration of reactive oxygen species

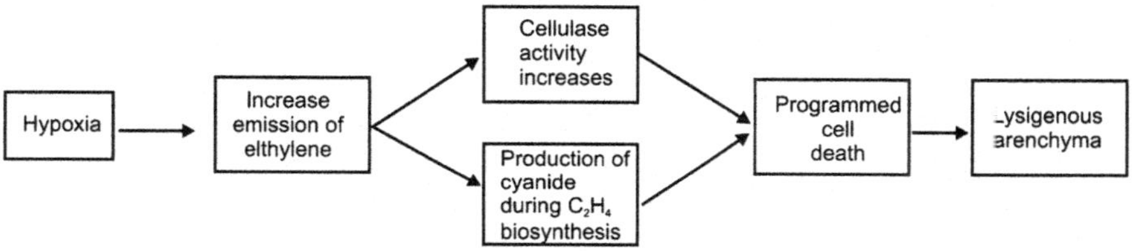

Fig. 7.2 Formation of lysigenous aerenchyma under hypoxia condition

Under drought stress, VOC isoprene plays an important role which was proved experimentally by Ryan et al. (2014) by using isoprene-emitting transgenic *Nicotiana tabacum* (where concentration of ROS and lipid peroxidation level shows no change with respect to stress) and non-isoprene-emitting plants (showed that concentration of ROS and lipid peroxidation level increases under drought stress). Thus, isoprene emission might help in maintaining membrane integrity under drought stress.

Similarly Timmusk et al. (2014) observed the emission of other VOCs (like monoterpenes, geranyl acetone, and benzaldehyde) in wheat plant (*Triticum aestivum*) under drought condition, and their emissions were smaller in those plants which were resistant against drought.

Water logged or flooding leads to low oxygen condition known as hypoxia. Under this condition, emission of ethylene increases that helps in avoiding stress by elongating submerged plant parts such as stem and leaf petioles and in developing new adventitious roots, leaf epinasty, aerenchyma tissues, etc. (Francesco et al. 2011). Peng et al. (2005) reported the upregulation of ACS family genes (e.g., ACS2, ACS6, and ACS9) in *Arabidopsis* under flooding. Aerenchyma provides a conduit for gas diffusion between roots and aerial organs; they can be formed by lysigeny, by schizogeny, or by combination of both, i.e., schizo-lysigeny. Lysigenous aerenchyma is formed due to the cyanide (causing cell death) that is formed during the biosynthetic pathway of ethylene (Fig. 7.2); moreover formation of aerenchyma by means of ethylene is coupled with increased activity of cellulose (He et al. 1994).

Flooding too increases the release of other volatile compounds like ethanol, acetaldehyde, methanol, GLV, C6 GLV, and C9 oxylipin nonanal-lipid membrane degradation indicator in Quercus robur and *Prunus serotina* (Bourtsoukidis et al. 2014; Copolovici and Niinemets 2010; Rottenberger et al. 2008). Under drought stress, wheat plants lead to emission of GLVs (like (Z)-3-hexenal, (E)-2-hexenal, and (Z)-3-hexen-1-yl acetate) and very small amounts of terpenoids as compared to control plants that produce smaller amounts of VOCs.

Salinity is also somehow related with water availability to plants due to osmotic stress. This stress is due to salt ion like chloride and sulfate that reduces the intake of water from soil to plant root. This will cause the physiological drought resulting into almost the same morphological, anatomical, and physiology changes as that of under drought condition. Due to their common effect, i.e., reduced availability or intake of water, plants use similar defense mechanism to cope up with these stresses (Tanou et al. 2009; Zhang et al. 2007; Zhao et al. 2007).

The resistant emission of terpene under salt as well as water stresses was reported by Loreto and Schnitzler (2010). This resistant emission of isoprene to salt and drought stress may be due to the increasing source of carbon probably from respiration (Teuber et al. 2008; Loreto et al. 2004) or starch breakdown (Gray et al. 2006). In some studies there is downregulation or repression of the isoprene emission that might lead to induction of large amount of other antioxidant molecules (Behnke 2009).

Osmotic stress is also related to emission of ethylene. Moreover increase activity of ACS in tomato cells has been reported under osmotic stress (Felix et al. 2000).

7.7 Biotic Stress and Plant Volatile

Plants being a sedentary organism cannot escape themselves from attackers in their life cycle. These attackers include the plant pathogens and herbivorous arthropods. These herbivorous species approximately account for half of the estimated six million insects (Schoonhoven et al. 2005). The pathogenic microbes cause the equal threat to plants and the diversity of which has not yet been quantified (Strange and Scott 2005). Plants have developed various defense approaches during the evolution, which include direct and indirect defenses. The plant exterior such as trichomes, hairs, thorns, spines, and thicker leaves is the major characteristic and influences the herbivore's biology which acts as direct defense mechanisms. The synthesis of toxic compounds like anthocyanins, terpenoids, alkaloids, phenolic compounds, and quinones either retards or kills the growth of the herbivores and acts as the first line of defense for many plant species against pathogens and herbivores (Hanley et al. 2007). Indirect defense mechanisms, on the other hand, comprise a second line of defense against pathogens and herbivory insect, conciliate by the synthesis of volatile compounds that attract the herbivores' enemies by supplying food derived from the extra floral nectar, and accommodate to boost the efficacy of natural enemies (Arimura et al. 2009). VOCs (volatile organic compounds) are released from plant parts, especially the aboveground (Zhang and Schlyter 2004; Fineschi et al. 2013) and into the surrounding environment and even from the roots into the water and soil air cavities (Hiltpod et al. 2011; Turling et al. 2012). So by using such abilities plants can be well defended and be represented in better way in the upcoming generations as compared to those which fails to resist against their killer.

7.8 Plant Defense Response to Insect Herbivores

The herbivores depend upon the green plants which are the main source of foods primarily. In nature, there are varieties of major threats to plants which are quite magnificent and vast. The herbivores such as mammals, reptiles, amphibians, birds, mollusks, arthropods, worms, bacteria, fungi, virus, and microbes are the significant threats to the plant (Karban and Baldwin 1997).

Based on their feeding behaviors, three major types of insect herbivores can be distinguished:

1. *Phloem Feeders* – whiteflies and aphids that cause little damage to cells of epidermis and mesophyll layer. Phloem feeders that insert stylet into the phloem tissue especially sieve tubes of leaves and stems. Although the amount of direct injury is low, when these insects serve as vectors for plant viruses, then they can cause great damage.
2. *Cell Content Feeders* – such as thrips and mites are piercing and sucking insect that cause an intermediate amount of physical damage to plant cells.
3. *Chewing Insects* – such as caterpillars, grasshoppers, and beetles that cause the most significant damage to plants.

VOCs synthesized in counter to these herbivore infestations are familiar as the herbivore-induced plant volatiles (HIPVs). These volatile compounds (HIPVs) emerge from the herbaceous damaged part (most probably the flowers and leaves) and furnish airborne signals in the form of chemical information which either attract the carnivorous arthropods (predators and parasitoids) and/or resist the ovipositing female from the host plants (Arimura et al. 2005; Turlings and Tons 2006; Zakir et al. 2009; Dicke and Baldwin 2010; Hare 2011). The carnivorous arthropods face many problems because of their victims that are under natural selection and seem to reduce the release of cues for carnivores. Although herbivores induced plant volatiles, ooze in relatively large quantities and which is herbivore specific and released from plants from which herbivores feed can resolve this problem. These HIPVs are detectable and authentic cues relatively that can be best utilized by carnivorous arthropods (Vet and Dicke 1992). The emissions of volatile compounds, present in glands on the leaf surface, are not induced constitutively. The emission is due to consequence of cell rupture and not under the direct controls of plants itself. Their toxicity helps in direct defense by deterring and reducing the food and feeding activities of herbivores and, thus, inhibiting biological activities of pathogens (Popp et al. 1995; Ward et al. 1997; Baier et al. 2002). The release of stored volatile compounds also serves as signals in host plant selection (Kelsey and Joseph 1997; Mita et al. 2002). Mostly, the plant species are induced volatile emitters, while few are constitutive emitters. After the onset of stress on plants, the green leaves release volatile compounds such as C_5 alcohols, aldehydes, and esters that are synthesized through lipoxygenase pathways (Loreto et al. 2006). Terpenoids come in light later on (Turling et al. 1998). The metabolic changes in plants in response to herbivory damage are expressed usually even in hour to days which in consequence release the HIPVs (Kunert et al. 2002), and the attraction was initiated within an hour in carnivorous arthropods (Scascighini et al. 2005). Most of the plant responses to insect herbivores involve both a wound response and recognition of certain compounds present in insect saliva which is known as elicitors (also called herbivore-associated molecular patterns – HAMPs). Many HAMPs of induced volatiles from plants have been identified (Dong et al. 2016; Zakir 2011) including volicitin that is separated from oral secretion of *Spodoptera exigua* larva (beet army worm). *Cotesia marginiventris*, female parasitic wasps, the natural enemies of *S. exigua*, were highly attractive to volatile mixture that is secreted from mechanically destroyed maize plants after volicitin application (Alborn et al. 1997).

The first claimed elicitors, β-galactosidase, are separated from regurgitate of the cabbage butterfly, *P. brassicae* caterpillar. The compound (gut regurgitate) of *P. brassicae* caterpillar attracts the parasitic wasps (*Cotesia glomerata*), which when applied to the vigorous leaves of cabbage plants release a specific admix of volatile which is alike to that of the herbivore-destroyed plants (Mattiacci et al. 2006); inceptin was from the oral secretion of *Spodoptera frugiperda*, the fall army worm after feeding them on cowpea leaves (Schmelz et al. 2006), caeliferins from regurgitant of grasshopper (*Schistocerca americana*), and an unidentified heat-labile constituent (Allmann et al. 2010). The elicitors are reported in chewing herbivores, while in piercing-sucking herbivore are not reported even now. Although it is not an easy task to calculate the elicitors quantity from small piercing -sucking herbivores, yet β-glucosidase (dominant elicitor) was identified from rice brown plant hopper that induce emission of volatiles in rice plants (Du et al. 2005).

Dong et al. (2016) described briefly the incident that elicits the response within the first second to minutes (Maffei et al. 2007), which is responsible for the recognition and activation of signal transduction pathways for the plant–insect interaction (Fig. 7.3). The initial occurrence that takes place within the seconds or minutes after attack measurable is plasma membrane potential, momentarily followed by change in cytosolic Ca^{2+} concentration and H_2O_2 formation, and is mediated by HAMPs. The activation of NADPH oxidase complex is induced by herbivore strike that produces the anion superoxide which instantly changed to H_2O_2 under the effect of superoxide dismutase. Hydrogen peroxide assembles in the cell which provokes diversity of late events detectable within minutes or hours afterward like kinase activation and plant hormones, i.e., jasmonic acid (JA) and salicylic acid (SA) signaling. These kinases then increase transcript levels of genes involved in JA, SA, JA-Ile and ethylene biosynthesis, which in turn enhance level of these phytohormones (Wu et al. 2007) that activate the genes involved in formation of plant volatiles, especially JA that lead to emission of Herbivore-induced volatiles within minutes and days.

7.9 Role of Herbivore-Induced Volatiles

7.9.1 Above Ground

Volatile compounds like terpenoids, phenylpropanoids, and fatty and amino acid derivatives are emitted in response to herbivore damage (Dudareva et al. 2006). Herbivore-induced plant volatiles (HIPVs) include a number of VOCs, but methyl salicylate (MeSA) is found frequently and has been reported in lima bean (Arimura et al. 2009) and *Arabidopsis* (Chen et al. 2003). MeSA produced sticky baits in many leaf and floral parts to attract the insect predators, e.g., big-eyed bugs, ladybird beetle, green lacewing, etc. Green leafy volatiles (GLVs) comprise of C_6-aldehydes (Arimura et al. 2009), and they help in plant defense by attracting their natural enemies (Maffei 2010). MeSA can be detected by attacking herbivore (Abdul et al. 2012) and is also used to inhibit *Mamestra brassicae* (cabbage moth) oviposition (Ulland et al. 2008). Volatile plant hormone, i.e., methyl jasmonate (MeJa), is produced by the damaged plants that help in the production of phenolics and glucosinolates (defensive compounds). MeJa can also be used as indirect defense compound of plants in organic agriculture, e.g., feeding damage can be completely avoided by the moth larvae of diamondback moth (*Plutella xylostella*), and experimental evidences showed that even if we spray low concentration of MeJa, i.e., 0.1% solution, it can still be helpful in the emission of homoterpenes and sesquiterpenes (Holopainen 2005).

Figure 7.3 represents the production of volatile compounds by the attacked plants that warned their neighbors also. Several physiological and molecular mechanisms have been discovered, but some of the studies also showed that plants growing under natural condition can also benefit from these signals (Heil and Bueno 2007; Kost and Heil 2006). Arimura et al. (2000) revealed the elevated expression of defensive genes on exposure of undamaged *Phaseolus vulgaris* to volatile compounds released from the neighbored plant. On application of a volatile compound, i.e., jasmonate, elevated level of monoterpene β-ocimene and mRNA was also noticed in *Vicia faba*. Experimental evidence of Sugimoto et al. (2014) also revealed how plants receive volatiles and start emission of defensive compounds, e.g., tomato plant [conversion of alcohol to (Z)-3-hexenylvicianoside after absorption of airborne (Z)-3- hexenol].

7.9.2 Below Ground

Beyond the aboveground parts, plant roots also produced HIPVs in counter to the overspread by underground insect pest that attracts the predator of infesting herbivore. A root-feeding pest *Diuraphis noxia* (Steegh et al. 2004) prompts the release of 1,8-cineol, a monoterpene volatile that is very toxic and acts as repellent for some insect pest (Tripathi et al. 2001). Roots of *Thuja occidentalis* (van Tol et al. 2001)

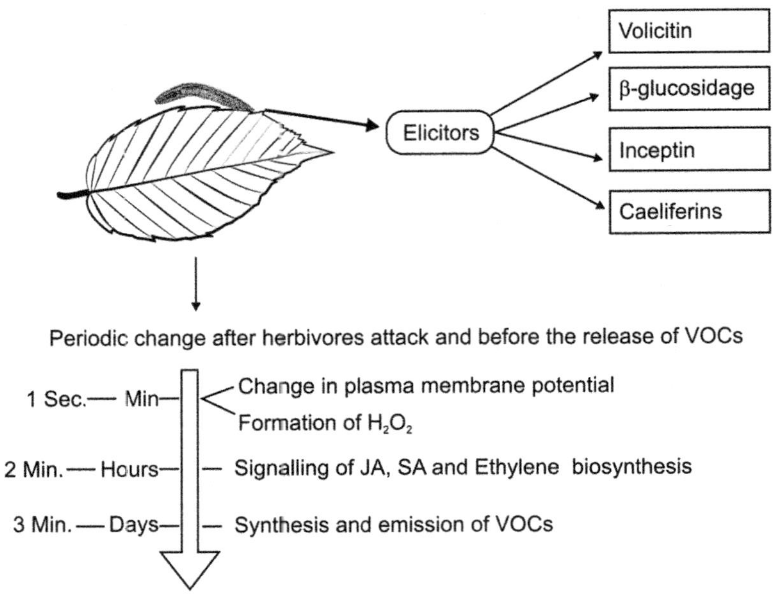

Fig. 7.3 Periodic changes during release of VOCs after herbivore attack (Modified from Dong et al. 2016)

release volatiles after infesting *Otiorhynchus sulcatus* which attract the pathogenic nematode (*H. megidis*). Moreover root exudates of broad bean plants infested by aphid have been found to make nonspecific nearby growing plants more attractive to foraging parasitoid (Guerrieri et al. 2002). Similarly, Rasmann et al. (2005) observed in maize plants that roots release a sesquiterpene (E)-β-caryophyllene when invaded by the larvae of *D. virgifera* that attracts *H. megidis*. The volatile cis-3-hexenyl acetate from leaves of both lima bean and hybrid poplar plant plays an important role in plant communication (Frost et al. 2008). This compound is also released by blueberry (Rodriguez-Saona et al. 2009) and sagebrush (Kessler et al. 2006) and is released within 5 min after the attack of herbivores (Arimura et al. 2009), and it acts as an important candidate of signals from damaged plant part to undamaged one. Emission of this compound both under mechanical damage and in response to herbivore attack shows that it acts as a general signal that has been detected in many plant species under multiple stimuli (Holopainen and James 2012). However few studies recommended that volatile compounds released by root in the soil minimize germination and growth of surrounding plants (Singh et al. 2001). Allelopathic effects (Fig. 7.4) of root-emitted volatile compounds, like monoterpenoid a-pinene, have been observed in many plant species, e.g., holm oak (Asensio et al. 2007), and purple sage (*Salvia leucophylla*, Nishida et al. 2005) emits a-pinene belowground (Wouter and Ronald 2009). Singh et al. (2006) studied five different plant species and observed inhibition of seed germination of three species after application of a-pinene. However, there was inhibition of early root growth in all

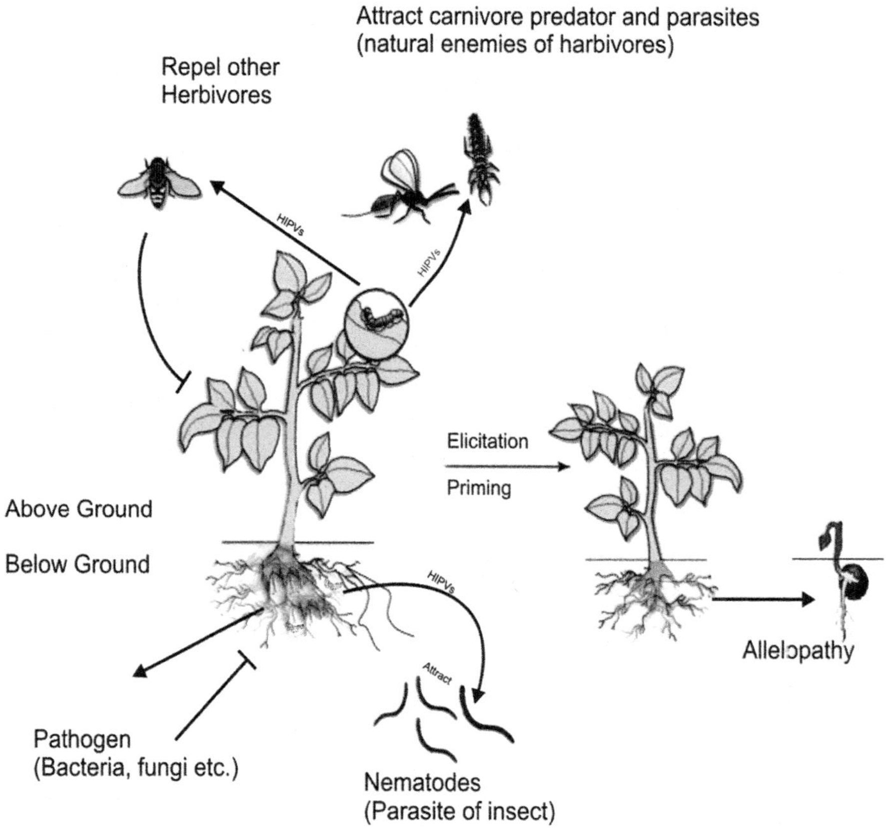

Fig. 7.4 An overview of role of volatile compounds in plants under biotic stress, where these volatiles can act as cues for natural enemies of the insect herbivore and repellent for other herbivores. Similarly belowground part also releases VOCs when attacked by herbivores that in turn attract insect-parasitizing nematodes. Moreover these released VOCs help in elicitation or priming of defense responses in healthy undamaged leaves of the same plant or in the neighboring plant and also have antimicrobial activities.

five plant species, and increased level of oxidative stress was observed in root. Additionally, a range of other *S. leucophylla* root-emitted volatile compounds are camphor, camphene, 1,8-cineole, and b-pinene that inhibited germination and growth of rapeseed (*Brassica campestris*), after its exogenous application, indicating the allelopathic effects of these volatile compounds. Nishida et al. (2005) suggested that this allelopathic effect of VOCs helps to prevent other species to enter in Salvia vegetation that helps to avoid interspecific competition. VOCs emitted from root as well as from shoot have inhibiting effects on nearby plant, e.g., snapdragon (*Antirrhinum majus*) flowers, emitted volatile compounds that inhibit the root growth in neighboring *Arabidopsis* seedlings under laboratory conditions (Horiuchi et al. 2007). Karban (2007) showed that upon clipping of leaves of sagebrush

(*Artemisia tridentata*), the germination of neighbouring seeds of different species is inhibited. When experimental condition was such that no air was in contact except its transmittance through soil, then there was no inhibitory effect on germination. von Dahl and Baldwin (2009) reported that when working with coyote tobacco (*Nicotiana attenuate*), ethylene emitted by plant exerts allelopathic effect on nearby grown plant roots by reducing its length, whereas transgenic plants which produce less amount of ethylene did not show any effect on root growth.

7.10 Plant Defense Response to Pathogen

Plants throughout their lives come in contact with a wide array of pathogens, continuously exposed to a diverse array of pathogens. Once inside the plant, pathogens generally employ one of three main attack strategies to use the host plant as a substrate for their own proliferation.

(a) Necrotrophic pathogens – attack their host by secreting cell wall-degrading enzymes and toxins, which eventually kill the affected plant cells. These dead tissue is then colonized by the pathogens and serves as food source.
(b) Biotrophic pathogens – attack the living tissue of plant and feed on substrates provided by their host.
(c) Hemibiotrophic pathogens – initially act as a biotroph and then become a necrotroph.

Biotic stress due to pathogen results into emission of various GLVs. For example, Lima bean leaves release E-2-hexenal and Z-3-hexenol (Croft et al. 1993) upon infection by pathogenic bacteria *Pseudomonas syringae* pv. *phaseolicola* being a hemibiotrophic pathogen (Rico and Preston 2008); E-2-hexenal is released when tobacco plants get infected by *Pseudomonas syringae* (Heiden et al. 2003). In both cases GLV emission starts within 18–20 h after infection; still the bacteria are in its exponential phase and that remains for 3 days (Whalen et al. 1991).

Many data have explained n-hexenal antimicrobial activity of GLVs against bacteria (Croft et al. 1993; Prost et al. 2005; Hamilton-Kemp 1992; Nakamura and Hatanaka 2002). Farmer and Davoine (2007) observed the maximum antimicrobial activity of E-2-hexenal of C6 aldehydes. Fungal attack also results into emission of GLVs like Z-3-hexenal, E-2-hexenal, Z-3 hexenol, E-2-hexenol, and Z-3 hexenyl acetate that are released from *Fusarium* species-infected maize plants (Piesik et al. 2011). Similarly *Botrytis cinerea* causes the emission of Z-3-hexenal, E-2-hexenal, and n-hexanal in *Arabidopsis* plants (Kishimoto et al. 2008). Early studies reported that C6 and C9 aldehydes emitted from crushed leaves of tomato have repressive effect on *Alternaria and Botrytis cinerea* spores (Hamilton-Kemp et al. 1992). Depending upon the different kinds of pathogen, their pathogenicity is affected by GLVs, e.g., in case of bacteria such as *Xanthomonas* and *Pseudomonas*, green leaf volatiles help in its pathogenicity by affecting phytohormone balance, whereas in the *Botrytis* fungus, it has direct toxicity by GLVs. Moreover, for proper defense response against *Botrytis cinerea*, JA is necessary as it has been reported that JA-insensitive mutant is comparatively more susceptible to the fungus (Thomma et al. 1998). Brown et al. (1995) reported that another GLV such as C6-aldehydes helps to defend the plant from plant pathogen by influencing their conidia germination (Shiojiri et al. 2006; Yi et al. 2009). Experimentally in green house study, it has been proved that silencing of HPL gene (hydroperoxide lyase gene responsible for synthesis of GLVs) enhances the susceptibility of *Arabidopsis thaliana* plants against *Botrytis cinerea*, a fungus pathogen, whereas overexpression of this gene leads to strengthening the protection against the fungus. Shiojiri et al. (2006) observed influence of fungal infection on synthetic (Z)-3-hexen-1-ol. Therefore, emission of some HIPVs might have a positive role against pathogens, such as *Botrytis cinerea*, that invade through wounds caused by herbivores (Fig. 7.4).

References

Abdul RW, Michael GP, Tariq A, Ahad AB, Barkat H, Savarimuthu I, Sharma HC (2012) Mechanisms of plant defense against insect herbivores. Plant Signal Behav 7(10):1306–1320

Affek HP, Yakir D (2003) Natural abundance carbon isotope composition of isoprene reflects incomplete coupling between isoprene synthesis and photosynthetic carbon flow. Pl Physiol 131:1727–1736

Alborn HT, Turlings TCJ, Jones TH, Stenhagen G, Loughrin JH, Tumlinson JH (1997) An elicitor of plant volatiles from beet armyworm oral secretion. Science 276:945–949

Alessio GA et al (2004) Direct and indirect impacts of fire on the isoprenoids emission from mediterranean vegetation. Funct Ecol 18:357–364

Apel K, Hirt H (2004) Reactive oxygen species: metabolism, oxidative stress, and signal transduction. Annu Rev Plant Biol 55:373–399

Argueso CT, Hansen M, Kieber JJ (2007) Regulation of ethylene biosynthesis. J Plant Growth Regul 26(2):92–105

Arimura G, Ozawa R, Shimoda T, Nishioka T, Boland W, Takabyashi J (2000) Herbivory-induced volatiles elicit defence genes in lima bean leaves. Nature 406:512–515

Arimura G, Kost C, Boland W (2005) Herbivore – induced, indirect plant defences. Biochim Et Biophys Acta Mol Cell Biol Lipids 1734(2):91–111

Arimura GI, Matsui K, Takabayashi J (2009) Chemical and molecular ecology of herbivore-induced plant volatiles: proximate factors and their ultimate functions. Plant Cell Physiol 50:911–923

Asensio D, Peñuelas J, Filella I, Llusià J (2007) On-line screening of soil VOCs exchange responses to moisture, temperature and root presence. Plant Soil 291:249–261

Baier P, Führer E, Kirsitis T, Rosner S (2002) Defense reactions of Norway spruce against bark beetles and the associated fungus *Ceratocystis polonica* in secondary pure and mixed species stands. For Ecol Manag 159:73–86

Beauchamp J, Wisthaler A, Hansel A, Kleist E, Miebach M, Niinemets Ü et al (2005) Ozone induced emissions of biogenic VOC from tobacco: relations between ozone uptake and emission of LOX products. Plant Cell Environ 28:1334–1343

Behnke K, Ehlting B, Teuber M, Bauerfeind M, Louis S, Hänsch R, Polle A, Bohlmann J, Schnitzler JP (2007) Transgenic, non-isoprene emitting poplars don't like it hot. Plant J 51:485–499

Behnke K, Kleist E, Uerlings R, Wildt J, Rennenberg H, Schnitzler JP (2009) RNAi-mediated suppression of isoprene biosynthesis in hybrid poplar impacts ozone tolerance. Tree Physiol 29:725–736

Beis A, Patakas A (2010) Differences in stomatal responses and root to shoot signalling between two grapevine varieties subjected to drought. Func Plant Biol 37:139–146

Bourtsoukidis E, Kawaletz H, Radacki D, Schuetz S, Hakola H, Hellen H, Noe S, Moelder I, Ammer C, Bonn B (2014) Impact of flooding and drought conditions on the emission of volatile organic compounds of Quercus robur and Prunus serotina. Trees Struct Funct 28:193–120

Brown GC et al (1995) Green leaf volatiles inhibit conidial germination of the entomopathogen *Pandora neoaphidis* (Entomophthorales: Entomophthoraceae). Environ Entomol 24:1637–1643. 66

Bukhov NG, Wiese C, Neimanis S, Heber U (1999) Heat sensitivity of chloroplasts and leaves: leakage of protons from thylakoids and reversible activation of cyclic electron transport. Photosyn Res 59:81–93

Calfapietra C, Wiberley AE, Falbel TG, Linskey AR, Mugnozza GS, Karnosky DF, Loreto F, Sharkey TD (2007) Isoprene synthase expression and protein levels are reduced under elevated O3 but not under elevated CO2 (FACE) in field-grown aspen trees. Plant Cell Environ 30:654–661

Cao Y, Wu Y, Zheng Z, Song F (2006) Overexpression of the rice EREBP-like gene OsBIERF3 enhances disease resistance and salt tolerance in transgenic tobacco. Physiol Mol Plant Pathol 67(3):202–211

Chaves MM, Pereira JS, Maroco J, Rodrigues ML, Ricardo CPP et al (2002) How plants cope with water stress in the field: photosynthesis and growth. Ann Bot 89:907–916

Chen F, Tholl D, D'Auria JC, Farooq A, Pichersky E, Gershenzon J (2003) Biosynthesis and emission of terpenoid volatiles from *Arabidopsis* flowers. Plant Cell 15:481–494

Chen JW, Bai KD, Cao KF (2009) Inhibition of monoterpene biosynthesis accelerates oxidative stress and leads to enhancement of antioxidant defenses in leaves of rubber tree (Hevea brasiliensis). Acta Physiol Plant 31:95–101. doi:10.1007/s11738-008-0205-z

Chinnusamy V, Schumaker K, Zhu J (2004) Molecular genetic perspectives on cross-talk and specificity in abiotic stress signalling in plants. J Exp Bot 55:225–236

Copolovici L, Niinemets Ü (2010) Flooding induced emissions of volatile signalling compounds in three tree species with differing waterlogging tolerance. Plant Cell Environ 33:1582–1594

Copolovici L, Niinemets Ulo. Springer International Publishing Switzerland (2016), J.D. Blande, R. Glinwood (eds.), Deciphering Chemical Language of Plant Communication, Signaling and Communication in Plants, Ch. Environmental Impacts on Plant Volatile Emission

Copolovici LO, Filella I, Llusià J, Niinemets Ü, Peñuelas J (2005) The capacity for thermal protection of photosynthetic electron transport varies for different monoterpenes in Quercus ilex. Plant Physiol 139:485–496

Copolovici L, Kannaste A, Remmel T, Niinemets U (2014) Volatile organic compound emissions from Alnus glutinosa under interacting drought and herbivory stresses. Environ Exp Bot 100:55–63

Copolovici L, Niinemets Ulo. Springer International Publishing Switzerland (2016), J.D. Blande, R. Glinwood (eds.), Deciphering Chemical Language of Plant Communication, Signaling and Communication in Plants, Ch.Environmental Impacts on Plant Volatile Emission

Croft K, Juttner F, Slusarenko A (1993) Volatile products of the lipoxygenase pathway evolved from *Phaseolus vulgaris* (L.) leaves Inoculated with *Pseudomonas syringae* pv *phaseolicola*. Plant Physiol 101:13–24

Dicke M, Baldwin IT (2010) The evolutionary context for herbivore induced plant volatiles: beyond the 'cry for help'. Trends in Plant Sci 15(3):167–175

Dong F, Xiumin F, Watanabe N, Xinguo S, Yang Z (2016) Recent advances in the emission and functions of plant vegetative volatiles. Molecules 21(2):124

Doupis G, Chartzoulakis K, Beis A, Patakas A (2011) Allometric and biochemical responses of grapevines subjected to drought and enhanced ultraviolet-B radiation. Aus J Grape Wine Res 17:36–42

Du MH, Yan XC, Lou YG, Cheng JA (2005) Studies on active chemicals in the saliva of the rice brown planthopper (Nilaparvata lugens) that elicit the production of rice volatiles (in Chinese). J Zhejiang Univ (Agric Life Sci) 31:237–244

Dudareva N, Negra F, Nagegowda D, Orlova I (2006) Plant volatiles : recent advances and future perspectives. Cri Rev Plant Sci 25(5):417–440

Fall R, Karl T, Hansel A, Jordan A, Lindinger W (1999) Volatile organic compounds emitted after leaf wounding: on-line analysis by proton-transfer-reaction mass spectrometry. J Geophys Res 104:15963–15974

Fares S, Loreto F, Kleist E, Wildt J (2008) Stomatal uptake and stomatal deposition of ozone in isoprene and monoterpene emitting plants. Plant Biol 10:44–54

Farmer EE, Davoine C (2007) Reactive electrophile species. Curr Opin Plant Biol 10:380–386

Fehsenfeld F, Calvert J, Fall R, Goldan P, Guenther A, Hewitt N, Lamb B, Liu S, Trainer M, Westberg H, Zimmerman P (1992) Emissions of volatile organic compounds from vegetation and the implications for atmospheric chemistry. Global Biogeochem 4:389–430

Felix G, Regenass M, Boller T (2000) Sensing of osmotic pressure changes in tomato cells. Plant Physiol 124:1169–1179

Ferrieri RA, Gray DW, Babst BA, Schueller MJ, Schlyer DJ, Thorpe MR et al (2005) Use of carbon-11 in Populus shows that exogenous jasmonic acid increases biosynthesis of isoprene from recently fixed carbon. Plant Cell Environ 28:591–602

Fineschi S, Loreto F, Staudt M, Peñuelas J (2013) Diversification of volatile isoprenoid emissions from trees: evolutionary and ecological perspectives. In: Niinemets Ü, Monson RK (eds) Biology, controls and models of tree volatile organic compound emissions. Springer Science+Business Media BV, Berlin, pp 1–20

Francesco Spinelli AC, Livia M, Karthik MN, Chiara P (2011) Emission and function of volatile organic compounds in response to abiotic stress, Abiotic stress in plants - mechanisms and adaptations

Frost CJ, Mescher MC, Dervinis C et al (2008) Priming defense genes and metabolites in hybrid poplar by the green volatile cis-3-hexenyl acetate. New Phytol 180:722–734

Fujita M, Fujita Y, Noutoshi Y, Takahashi F, Narusaka Y, Yamaguchi-Shinozaki K, Shinozaki K (2006) Crosstalk between abiotic and biotic stress responses: a current view from the points of convergence in the stress signaling networks. Curr Opin Plant Biol 9:436–442

Gray DW, Goldstein AH, Lerdau M (2006) Thermal history regulates methylbutenol basal emission rate in Pinus ponderosa. Plant Cell Environ 29:1298–1308

Guerrieri E, Poppy GM, Powell W, , Rao, R. Pennacchio, F., (2002) Plant-to-plant communication mediating in -flight orientation of *Aphidius ervi*. J Chem Ecol 28: 1703–1715

Hamilton-Kemp TR, McCracken CT, Loughrin JH, Andersen RA, Hildebrand DF (1992) Effects of some natural volatile compounds on the pathogenic fungi *Alternaria alternata* and *Botrytis cinerea*. J Chem Ecol 18:1083–1091

Hanley ME, Lamont BB, Fairbanks MM, Rafferty CM (2007) Plant structural traits and their role in antiherbivore defense. Perspect Plant Ecol Evol Syst 8:157–178

Hansh R, Mendel RR (2009) Physiological functions of mineral micronutrients (Cu, Zn, Mn, Fe, Ni, Mo, B, Cl). Curr Opin Biotech 12(3):59–266

Hare JD (2011) Ecological role of volatiles produced by plants in response to damage by herbivorous insects. Annu Rev Entomol 56:161–180

Harrison SP, Morfopoulos C, Dani KGS, Prentice IC, Arneth A, Atwell BJ, Barkley MP, Leishman MR, Loreto F, Medlyn BE, Niinemets U, Possell M, Penuelas J, Wright IJ (2013) Volatile isoprenoid emissions from plastid to planet. New Phytol 197:49–57

He CJ, Drew MG, Morgan PW (1994) Induction of enzymes associated with lysigenous aerenchyma formation in roots of Zea mays during hypoxia or nitrogen starvation. Plant Physiol 105:861–865

Heiden AC, Hoffmann T, Kahl J, Kley D, Klockow D, Langebartels C, Mehlhorn H, Sandermann H, Schraudner M, Schuh G, Wildt J (1999) Emission of volatile organic compounds from ozone-exposed plants. Ecol Appl 9:1160–1167

Heiden AC, Kobel K, Langebartels C (2003) Emissions of oxygenated volatile organic compounds from plants part I: emissions from lipoxygenase activity. J Atmos Chem 45:143–172

Heil M, Bueno JCS (2007) With plant signaling by volatiles leads to induction and priming of an indirect plant defense in nature. Proc. Natl. Acad. Sci. USA 104:5467–5472

Heil M, Kost C (2006) Priming of indirect defenses. Ecol Lett 9:813–817

Hewitt CN, Kok GL, Fall R (1990) Hydroperoxides in plants exposed to ozone mediate air pollution damage to alkene emitters. Nature 344:56–58

Hiltpold I, Erb M, Robert CAM, Turlings TCJ (2011) Systemic root signalling in a belowground, volatile mediated tritrophic interaction. Plant Cell Environ 34:1267–1275

Holopainen JK (2005) Improvement of biological control by volatile plant compounds. FORSKNINGSNYTT 1:1–2

Holopainen JK, James D Blande (2012). Molecular plant volatile communication. Sensing in Nature, edited by Carlos López-Larrea. ©2012 Landes Bioscience and Springer Science+Business Media pg no 17–31

Horiuchi JI, Badri DV, Kimball BA, Negre F, Dudareva N, Paschke MW, Vivanco JM (2007) The floral volatile, methyl benzcate, from snapdragon (Antirrhinum majus) triggers phytotoxic effects in Arabidopsis thaliana. Planta 226:1–10

Huang Z, Zhang Z, Zhang X, Zhang H, Huang D, Huang R (2004) Tomato TERF1 modulates ethylene response and enhances osmotic stress tolerance by activating expression of downstream genes. FEBS Lett 573(1–3):110–116

Ionenko IF, Anisimov AV (2001) Effect of water deficit and membrane destruction on water diffusion in the tissues of maize seedlings. Biol Plant 44:247–252

Jardine KJ, Chambers JQ, Holm J, Jardine AB, Fontes CG, Zorzanelli RF, Meyers KT, de Souza VF, Garcia S, Gimenez BO, Piva LR (2015) Green leaf volatile emissions during high temperature and drought stress in a central Amazon rainforest. Plants 4:678–690

Karban R (2007) Experimental clipping of sagebrush inhibits seed germination of neighbours. Ecol Lett 10:791–797

Karban R, Baldwin IT (1997) Induced responses to herbivory. University of Chicago Press, Chicago

Kelsey RG, Joseph G (1997) Ambrosia beetle host selection among logs of Douglas fir, western hemlock, and western red cedar with different ethanol and α-pinene concentrations. J Chem Ecol 23:1035–1051

Kessler A, Halitschke R, Diezel C, Baldwin IT (2006) Priming of plant defense responses in nature by airborne signaling between Artemisia tridentate and *Nicotiana attenuata*. Oecologia 148:280–292

Kishimoto K, Matsui K, Ozawa R, Takabayashi J (2008) Direct fungicidal activities of C6-aldehydes are important constituents for defense responses in Arabidopsis against *Botrytis cinerea*. Phytochemistry 69:2127–2132

Kost C, Heil M (2006) Herbivore-induced plant volatiles induce an indirect defence in neighbouring plants. J Ecol 94:619–628

Kunert M, Biedermann A, Koch T, Boland W (2002) Ultrafast sampling and analysis of plant volatiles by a hand-held miniaturised GC with pre-concentration unit: kinetic and quantitative aspects of plant volatile production. J Sep Sci 25:677–684

Lehning A, Zimmer W, Zimmer I, Schnitzler JP (2001) Modeling of annual variations of oak (Quercus robur L.) isoprene synthase activity to predict isoprene emission rates. J Geo Res 106D:3157–3166

Litvak ME, Monson RK (1998) Patterns of induced and constitutive monoterpene production in conifer needles in relation to insect herbivory. Oecologia 114:531–540

Loivamäki M, Gilmer F, Fischbach RJ, Sörgel C, Bachl A, Walter A, Schnitzler JP (2007) Arabidopsis, a model to study biological functions of isoprene emission? Plant Physiol 144:1066–1078

Loreto F, Schnitzler J (2010) Abiotic stresses and induced BVOCs. TrendsPlant Sci 15(3):154–166

Loreto F, Sharkey TD (1990) A gas-exchange study of photosynthesis and isoprene emission in Quercus rubra L. Planta 182:523–531

Loreto F, Velikova V (2001) Isoprene produced by leaves protects the photosynthetic apparatus against ozone damage, quenches ozone products, and reduces lipid peroxidation of cellular membranes. Plant Physiol 127:1781–1787

Loreto F, Förster A, Dürr M, Csiky O, Seufert G (1998) On the monoterpene emission under heat stress and on the increased thermotolerance of leaves of Quercus ilex L. fumigated with selected monoterpenes. Plant Cell Enviro 21:101–107

Loreto F, Mannozzi M, Maris C, Nascetti P, Ferranti F, Pasqualini S (2001) Ozone quenching properties of isoprene and its antioxidant role in leaves. Plant Physiol 126:993–1000

Loreto F, Pinelli P, Manes F, Kollist H (2004) Impact of ozone on monoterpene emissions and evidences for an isoprene-like antioxidant action of monoterpenes emitted by Quercus ilex (L.) leaves. Tree Physiol 24:361–367

Loreto F, Barta C, Brilli F, Nogues I (2006) On the induction of volatile organic compound emissions by plants as consequence of wounding or fluctuations of light and temperature. Plant Cell Environ 29:1820–1828

Maffei ME (2010) Site of synthesis, biochemistry and functional role of plant volatiles. S Afr J Bot 76:612–631

Maffei ME, Mithöfer A, Boland W (2007) Before gene expression: early events in plant–insect interaction. Trends Plant Sci 12(7):310–316

Mallick N, Mohn FH (2003) Use of chlorophyll fluorescence in metal-stress research: a case study with green microalga Scenedesmus. Ecotoxicol Environ 55:64–69

Mehlhorn H, Wellburn AR (1987) Stress ethylene formation determines plant sensitivity to ozone. Nature 327:417–418

Mita E, Tsitsimpikou C, Tsiveleka L, Petrakis PV, Ortiz A, Vagias C, Roussis V (2002) Seasonal variation of oleoresin terpenoids from *Pinus halepensis* and *Pinus pinea* and host selection of the scale insect *Marchalina hellenica* (Homoptera, Coccoidea, Margarodidae, Coelostonidiinae). Holzforschung 56:572–578

Monson RK, Fall R (1989) Isoprene emission from aspen leaves: the influence of environment and relation to photosynthesis and photorespiration. Plant Physiol 90:267–274

Monson RK, Jaeger CH, Adams WW III, Driggers EM, Silver GM, Fall R (1992) Relationships among isoprene emission rate, photosynthesis, and isoprene synthase activity as influenced by temperature. Plant Physiol 98:1175–1180

Munns R (2002) Comparative physiology of salt and water stress. Plant Cell Environ 25:239–250

Nakamura S, Hatanaka A (2002) Green-leaf-derived C6-aroma compounds with potent antibacterial action that act on both gram-negative and gram-positive bacteria. J Agric Food 50:7639–7644

Niinemets U (2010) Mild versus severe stress and BVOCs: thresholds, priming and consequences. Trends Plant Sci 15:145–153

Niinemets U, Monson RK, Arneth A, Ciccioli P, Kesselmeier J, Kuhn U, Noe SM, Penuelas J, Staudt M (2010) The leaf level emission factor of volatile isoprenoids: caveats, model algorithms, response shapes and scaling. Biogeo 7:1809–1832

Nishida N, Tamotsu S, Nagata N, Saito C, Sakai A (2005) Allelopathic effects of volatile monoterpenoids produced by Salvia leucophylla : inhibition of cell proliferation and DNA synthesis in the root apical meristem of Brassica campestris seedlings. J Chem Ecol 31:1187–1203

Nover L, Bharti K, Döring P, Mishra SK, Ganguli A, Scharf KD (2001) Arabidopsis and the heat stress transcription factor world: how many heat stress transcription factors do we need? Cell Stress Chaperones 6:177–189

Owen SM, Penuelas J (2005) Opportunistic emissions of volatile isoprenoids. Trends Plant Sci 10:420–426

Pastenes C, Horton P (1996) Effect of high temperature on photosynthesis in beans. 2. CO2 assimilation and metabolite contents. Plant Physiol 112:1253–1260

Peng HP, Lin TY, Wang NN, Shih MC (2005) Differential expression of genes encoding 1-aminoc yclopropane-1-carboxylate synthase in *Arabidopsis* during hypoxia. Plant Mol. Biol 58:15–25

Piesik D, Lemńczyk G, Skoczek A, Lamparski R, Bocianowski J, Kotwica K, Delaney KJ (2011) Fusarium infection in maize: volatile induction of infected and neighboring uninfected plants has the potential to attract a pest cereal leaf beetle, *Oulema melanopus*. J Plant Physiol 168:1534–1542

Popp MP, Johnson JD, Lesney MS (1995) Changes in ethylene production and monoterpene concentration in slash pine and loblolly pine following inoculation with bark beetle vectored fungi. Tree Physiol 15:807–812

Possell M, Loreto F (2013) The role of volatile organic compounds in plant resistance to abiotic stresses: responses and mechanisms. In: Niinemets Ü, Monson RK (eds) Biology, controls and models of tree volatile organic compound emissions. Springer Science+Business Media B.V, Berlin, pp 209–235

Prost I, Dhondt S, Rothe G, Vicente J, Rodriguez MJ, Kift N, Carbonne F, Griffiths G, Esquerré-Tugayé MT, Rosahl S, Castresana C (2005) Evaluation of the antimicrobial activities of plant oxylipins supports their involvement in defense against pathogens. Plant Physiol 139(4):1902–1913

Rasmann S, Köllner TG, Degenhardt J, Hiltpold I, Toepfer S, Kuhlmann U, Gershenzen J, Turling JCV (2005) Recruitment of entomopathogenic nematodes by insect-damaged maize roots. Nature 434:732–737

Rico A, Preston GM (2008) Pseudomonas syringae pv. tomato DC3000 uses constitutive and apoplast-induced nutrient assimilation pathways to catabolize nutrients that are abundant in the tomato apoplast. Mol. Plant Microbe Interact 21:269–282

Rodriguez-Saona CR, Rodriguez-Saona LE, Frost CJ (2009) Herbivore-induced volatiles in the perennial shrub, Vaccinium corymbosum, and their role in inter-branch signaling. J Chem Ecol. 35:163–175

Ross SM (1994) Toxic metals in soil – plant systems, vol 469. Wiley, Chichester

Rottenberger S, Kleiss B, Kuhn U, Wolf A, Piedade MTF, Junk W, Kesselmeier J (2008) The effect of flooding on the exchange of the volatile C2-compounds ethanol, acetaldehyde and acetic acid between leaves of Amazonian floodplain tree species and the atmosphere. Biogeosciences 5:1085–1100

Ryan AC, Hewitt CN, Possell M, Vickers CE, Purnell A, Mullineaux PM, Davies WJ, Dodd IC (2014) Isoprene emission protects photosynthesis but reduces plant productivity during drought in transgenic tobacco (*Nicotiana tabacum*) plants. New Phytol 201:205–216

Sasaki K, Saito T, Lämsä M, Oksman-Caldentey KM, Suzuki M, Ohyama K, Muranaka T, Ohara K, Yazaki K (2007) Plants utilize isoprene emission as a thermotolerance mechanism. Plant Cell Physiol 48:1254–1126

Scascighini N, Mattiacci L, D'Alessandro M, Hern A, Rott AS, Dorn S (2005) New insights in analysing parasitoid attracting synomones: early volatile emission and use of stir bar sorptive extraction. Chemoecology 15:97–104

Schade GW, Goldstein AH (2001) Fluxes of oxygenated volatile organic compounds from a ponderosa pine plantation. J Geophys Res 106:3111–3124

Schmelz EA, Carroll MJ, LeClere S, Phipps SM, Meredith J, Chourey, PS, Alborn HT, Teal PEA (2006) Fragments of ATP synthase mediate plant perception of insect attack. Proc Natl Acad Sci. USA 103: 8894–8899

Schnitzler JP, Graus M, Kreuzwieser J, Heizmann U, Rennenberg H, Wisthaler A et al (2004) Contribution of different carbon sources to isoprene biosynthesis in poplar leaves. Plant Physiol 135:152–160

Schoonhoven LM, van Loon JJA, Dicke M (2005) Insect plant biology, 2nd edn. Oxford University Press, Oxford

Schrader SM, Wise RR, Wacholtz WF, Ort DR, Sharkey TD (2004) Thylakoid membrane responses to moderately high leaf temperature in Pima cotton. Plant Cell Environ 27:725–735

Sharkey TD, Loreto F (1993) Water stress, temperature, and light effects on the capacity for isoprene emission and photosynthesis of kudzu leaves. Oecologia 95:328–333

Sharkey TD, Schrader SM (2006) High temperature stress. In: Physiology and molecular biology of stress tolerance in plants. Springer, Netherlands, pp 101–129

Sharkey TD, Singsaas EL (1995) Why plants emit isoprene. Nature 374:769

Sharkey TD, Yeh S (2001) Isoprene emission from plants. Annu Rev Plant Physiol Plant Mol Biol 52:407–436

Sharkey TD, Singsaas EL, Vanderveer PJ, Geron CD (1996) Field measurements of isoprene emission from trees in response to temperature and light. Tree Physiol 16:649–654

Sharkey TD, Chen X, Yeh S (2001) Isoprene increases thermotolerance of fosmidomycin-fed leaves. Plant Physiol 125:2001–2006

Sharkey TD, Wiberley AE, Donohue AR (2008) Isoprene emission from plants: why and how. Ann Bot 101:5–18. doi:10.1093/aob/mcm240

Shiojiri K, Kishimoto K, Ozawa R, Kugimiya S, Urashimo S, Arimura G, Horiuchi J, Nishioka T, Matsui K, Takabayashi J (2006) Changing green leaf volatile biosynthesis in plants: an approach for improving plant resistance against both herbivores and pathogens. Proc Natl Acad Sci U S A 103:16672–16676

Singh HP, Batish DR, Kaur S, Ramezani H, Kohli RK (2001) Comparative phytotoxicity of four monoterpenes against Cassia occidentalis. Ann Appl Biol 141:111–116

Singh HP, Batish DR, Kaur S, Arora K, Kohli RK (2006) Alpha-pinene inhibits growth and induces oxidative stress in roots. Ann Bot 98:1261–1269

Singsaas EL, Sharkey TD (2000) The effects of high temperature on isoprene synthesis in oak leaves. Plant Cell Environ 23:751–757

Singsaas EL, Lerdau M, Winter K, Sharkey TD (1997) Isoprene increases thermotolerance of isoprene-emitting species. Plant Physiol 115:1413–1420

Singsaas EL, Laporte MM, Shi JZ, Monson RK, Bowling DR, Johnson K, Sharkey TD (1999) Kinetics of leaf temperature fluctuation affects isoprene emission from red oak (Quercus rubra) leaves. Tree Physiol 19:917–924

Steeghs M, Bais HP, de Gouw J, Goldan P, Kuster W, Northway M, Fall R, Vivanco JM (2004) Proton-transfer-reaction mass spectrometry as a new tool for real time analysis of root- secreted volatile organic compounds in Arabidopsis. Plant Physiol 135:47–58

Strange RN, Scott PR (2005) Plant disease: a threat to global food security. Annu Rev Phytopathol 43:83–116

Sugimoto K, Matsui K, Iijima Y, Akakabe Y, Muramoto S, Ozawa R, Uefune M, Sasaki R, Alamgir KM, Akitake S, Nobuke T (2014) Intake and transformation to a glycoside of (Z)-3-hexenol from infested neighbors reveals a mode of plant odor reception and defense. Proc Natl Acad Sci U S A 111:7144–7149

Tanou G, Job C, Rajjou L, Arc E, Belghazi M, Diamantidis G, Molassiotis A, Job D (2009) Proteomics reveals the overlapping roles of hydrogen peroxide and nitric oxide in the acclimation of citrus plants to salinity. Plant J 60:795–804

Terry N (1980) Limiting factors in photosynthesis. 1. Use of iron stress to control photochemical capacity in vivo. Plant Physiol 65:114–120

Teuber M, Zimmer I, Kreuzwieser J, Ache P, Polle A, Rennenberg H, Schnitzler JP (2008) VOC emission of grey poplar leaves as affected by salt stress and different N sources. Plant Biol 10:86–96

Thomashow MF (1998) Role of cold responsive genes in plant freezing tolerance. Plant Physiol 118:1–8

Thomma BP, Eggermont K, Penninckx IA, MauchMani B, Vogelsang R, Cammue BP, Broekaert WF (1998) Separate jasmonate-dependent and salicylate-dependent defense-response pathways in Arabidopsis are essential for resistance to distinct microbial pathogens. Proc Natl Acad Sci U S A 95:15107–15111

Timmusk S, Abd El Daim IA, Copolovici L, Tanilas T, Kannaste A, Behers L, Nevo E, Seisenbaeva G, Stenstrom E, Niinemets U (2014) Drought tolerance of wheat improved by rhizosphere bacteria from harsh environments: enhanced biomass production and reduced emission of stress volatiles. PLoS One 9(5)

Tripathi AK, Prajapati V, Aggarwal KK, Kumar S (2001) Toxicity, feeding deterrence, and effect of activity of 1,8-cineole from Artemisia annua on progeny production of Tribolium castaneum (Coleoptera: Tenebrionidae). J Econ Entomol 94:979–983

Turlings TC, Ton J (2006) Exploiting scents of distress: the prospect of manipulating herbivore-induced plant odours to enhance the control of agricultural pests. Curr Opin Plant Biol 9:421–427

Turlings TCJ, Lengwiler UB, Bernasconi ML, Wechsler D (1998) Timing of induced volatile emissions in maize seedlings. Planta 207:146–152

Turlings TCJ, Hiltpold I, Rasmann S (2012) The importance of root-produced volatiles as foraging cues for entomopathogenic nematodes. Plant Soil 358:51–60

Ulland S, Ian E, Mozuraitis R, Borg-Karlson AK, Meadow R, Mustaparta H (2008) Methyl salicylate, identified as primary odorant of a specific receptor neuron type, inhibits oviposition by the moth Mamestra brassicae L. (Lepidoptera, noctuidae). Chem Senses 33:35–46

Vacca RA, de Pinto MC, Valenti D, Passarella S, Marra E, De Gara L (2004) Production of reactive oxygen species, alteration of cytosolic ascorbate peroxidase, and impairment of mitochondrial metabolism are early events in heat shock-induced programmed cell death in tobacco bright-yellow 2 cells. Plant Physiol 134(3):1100–1112

Vahala J, Schlagnhaufer CD, Pell EJ (1998) Induction of an ACC synthase cDNA by ozone in light-grown Arabidopsis thaliana leaves. Physiol Plant 103:45–50

Vallelian-Bindschedler L, Schweizer P, Mösinger E, Metraux JP (1998) Heat-induced resistance in barley to powdery mildew (Blumeria graminis f.sp. hordei) is associated with a burst of active oxygen species. Physiol Mol Plant Pathol 52:185–199

Van Tol RW, Van Der Sommen AT, Boff MI, Van Bezooijen J, Sabelis MW, Smits PH (2001) Plants protect their roots by alerting the enemies of grubs. Ecol Lett 4:292–294

Verkleji JAS (1993) The effects of heavy metals stress on higher plants and their use as biomonitors. In: Markert B (ed) Plant as bioindicators: indicators of heavy metals in the terrestrial environment. VCH, New York, pp 415–424

Vet LEM, Dicke M (1992) Ecology of infochemical use by natural enemies in a tritrophic context. Annu Rev Entomol 37:141–172

Vickers CE, Possell M, Cojocariu CI, Velikova VB, Laothawornkitkul J, Ryan A, Mullineaux PM, Nicholas Hewitt C (2009) Isoprene synthesis protects transgenic tobacco plants from oxidative stress. Plant Cell Environ 32:520–553

Vierling E (1991) The roles of heat shock proteins in plants. Ann Rev Plant Physiol Plant Mol Biol 42:579–620

von Dahl CC, Baldwin IT (2009) Use of silenced plants in allelopathy bioassays: a novel approach. Planta 229(3):569–575

Vranova E, Coman D, Gruissem W (2012) Structure and dynamics of the isoprenoid pathway network. Mol Plant 5:318–333

Ward BB, Courtney KJ, Langenheim JH (1997) Inhibition of *Nitrosomonas europaea* by monoterpenes from coastal redwood (*Sequoia sempervirens*) in whole cell studies. J Chem Ecol 23:2583–2598

Whalen MC, Innes RW, Bent AF, Staskawicz BJ (1991) Identification of Pseudomonas syringae pathogens of Arabidopsis and a bacterial locus determining a virulence on both Arabidopsis and soybean. Plant Cell 3:49–59

Wouter K, Ronald P (2009) Biogenic volatile organic compounds and plant competition. Trends Plant Sci 3(15):126–132

Wu JQ, Hettenhausen C, Melda S, Baldwin IT (2007) Herbivory rapidly activates MAPK signaling in attacked and unattacked leaf regions but not between leaves of Nicotiana attenuate. Plant Cell 19:1096–1122

Yi HS, Heil M, Adame-Alvarez RM, Ballhorn DJ, Ryu CM (2009) Airborne induction and priming of plant defenses against a bacterial pathogen. Plant Physiol 151:2152–2161

Zakir A (2011) Inducible defenses in herbivore plant interactions: functions mechanisms and manipulations. Introductory paper at the faculty of landscape planning. Hortic Agri Sci 2:1–25

Zakir A, Bengtsson M, Hansson BS, Witzgall P, Anderson P (2009) Herbivore induced volatiles reduce egglaying in Egyptian cotton leafworm. Poster presentation in the 25th annual meeting of International Society of Chemical ecology (ISCE), Neuchatél, Switzerand

Zhang QH, Schlyter F (2004) Olfactory recognition and behavioural avoidance of angiosperm nonhost volatiles by conifer-inhabiting bark beetles. Agri For Entomol 6(1):1–20

Zhang F, Wang YP, Yang YL, WuH WD, Liu JQ (2007) Involvement of hydrogen peroxide and nitric oxide in salt resistance in the calluses from Populus Euphratica. Plant Cell Environ 30:775–785

Zhao J, Ren W, Zhi D, Wang L, Xia G (2007) Arabidopsis DREB1A/CBF3 bestowed transgenic tall rescue increased tolerance to drought stress. Plant Cell Rep 26:1521–1528

Microbial Volatile in Defense

8

Khaled A. Selim, Dina E. El Ghwas, Rehab M. Selim, and Mohamed I. Abdelwahab Hassan

Abstract

Microbial volatile organic compounds (MVOCs) produced by microorganisms like bacteria and fungi are eco-friendly and considered a cost-effective sustainable strategy. There is potency for microbial VOCs in biotechnological applications as agriculture, industry, and medicine. Fungal and bacterial microbial volatile organic compounds (MVOCs) can be used instead of harmful pesticides, fungicides, and bactericides to protect plants from pathogens and increase crop productivity. MVOCs can be used to improve human health by

K.A. Selim (✉)
Interfaculty Institute of Microbiology and Infection Medicine, Eberhard Karls Universität Tübingen, Tübingen, Germany

Chemistry of Natural and Microbial Products Department, Pharmaceutical and Drug Industries Research Division, National Research Centre, Cairo, Egypt
e-mail: khaled.selim@uni-tuebingen.de

D.E. El Ghwas
Chemistry of Natural and Microbial Products Department, Pharmaceutical and Drug Industries Research Division, National Research Centre, Cairo, Egypt

R.M. Selim
Chemistry of Natural and Microbial Products Department, Pharmaceutical and Drug Industries Research Division, National Research Centre, Cairo, Egypt

Chemistry Department, University of Winnipeg-Mb, Winnipeg, Canada

M.I. Abdelwahab Hassan (✉)
Jena Microbial Resource Collection, Leibniz Institute for Natural Product Research and Infection Biology – Hans Knöll Institute, Jena, Germany

Faculty of Biology and Pharmacy, Institute of Microbiology, Department Microbiology and Molecular Biology, Friedrich, Schiller University Jena, Jena, Germany

National Research Centre (NRC), El Bohouth st, Dokki, Giza, 12622, Egypt
e-mail: mohamed.iasmael@gmail.com

D.K. Choudhary et al. (eds.), *Volatiles and Food Security*,
DOI 10.1007/978-981-10-5553-9_8

characterization of volatiles released by lower respiratory tract bacterial pathogens and their effective role as diagnostic markers in patient breathe testing. VOCs emitted by the human body, such as those which are released from the gut, have a great potential for the diagnostic and therapeutic inspections and help in the identification of liver enzyme activities and consequently investigating the influence of the metabolites on the liver function during disease development. This gives credit/point for the microbes as a promising tool for biological control that manages the disease and in the meantime saves the humanity and surrounding environment. In this chapter, we would like to draw the attention to the importance of microbial volatile organic compounds in the defense as general mechanism, and the chemistry of some MVOCs which has great role in the defense. Chat Conversation EndType a message.

Ke ywords
MVOCs • Fungi • Bacteria • Agriculture compound

8.1 Introduction

Microorganisms have an indispensable role in our life and are the most valuable creatures on our planet. They are of immense importance to the environment, human health, and our economy. Microbes affect every aspect of life on our planet; without microbes life cannot be maintained on earth. Majority of the microbes have profound beneficial effects such as fixing atmospheric gases and converting dead animal and plant materials into simpler substances that can be recycled again into the environment and reused in the beginning of the food chain. This example confirms that we cannot stand without the microorganisms in our life. These organisms have the same society and community like the human or may be more sophisticated. Moreover, we can find the treatment for the pathogenic species from other species or the same species but with different strain, so a study about the behavior of these intelligent organisms is interesting and comes back with a lot of profit. Microbes have an amazing variety of shapes and sizes. They have a wide diversity and live in every part of the biosphere. Volatile organic compounds (VOCs) have perverse characters on earth because of their physical and chemical properties which include low boiling point, high vapor pressure, and low molecular mass. Clearly volatiles are one of the constituents of the climate; however, they additionally reveal in permeable materials (rocks, plants, and soil pores) and are dissolved in liquid media. To clarify the complexity of VOCs, the expression "volatilome" has been proposed (Maffei et al. 2011). Volatile organic compounds (VOCs) can go a long way from the purpose of creation through the environment, permeable soils and fluid to give whole picture from chemical side for both short- and long-distance organismic interactions. Therefor, VOCs can be consider as a remarkable tool for investigation the mechanistic mode of action for the intracellular organisms from the biochemical view (Maffei et al. 2011). The main sources of volatile compounds are animal and

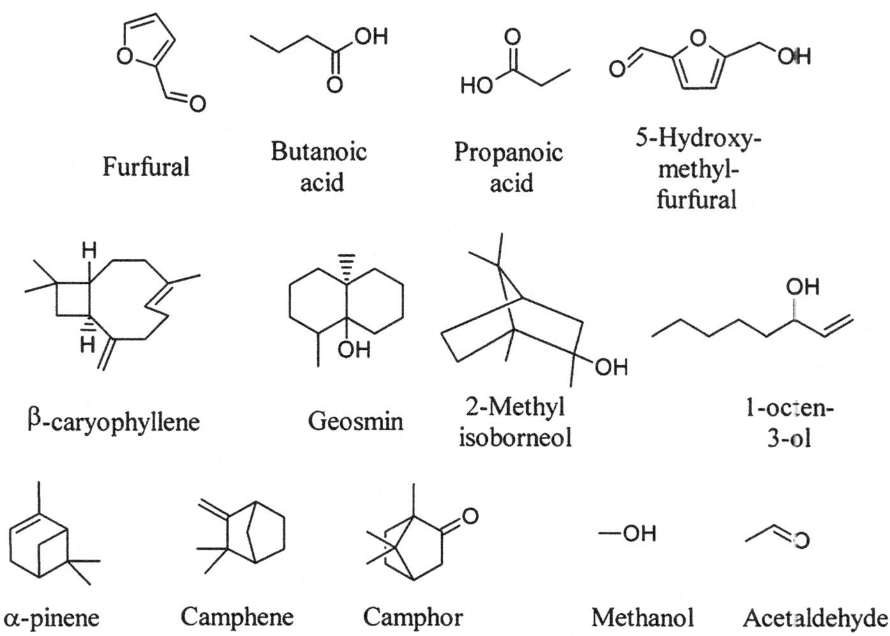

Fig. 8.1 Most frequently released MVOCs (Adapted from Kanchiswamy et al. 2015)

plant kingdoms, and they have been completely examined previously, whereas examination of volatile compounds of microorganisms (e.g., bacteria and fungi) was not properly investigateed for a while (Junker and Tholl 2013; Weisskopf 2013; Peñuelas et al. 2014). Microbial volatile organic compounds (MVOCs) are liquids and carbon-based solids that come into gas phase by vaporizing at 0.01 kPa at a temperature of roughly 20 °C (Pagans et al. 2006). Thus, MVOC production is influenced by the diversity of the microbial species, phases of the microbial growth, and other conditions such as pH, temperature, and nutrient compositions (Whillans and Lamont 1995). On the other hand, the compounds originating from microbial metabolism hardly exist because they have other environmental sources besides microbial metabolism (von Reuss et al. 2010). Among the most commonly released VOCs are furfural, acetaldehyde, butanoic acid, methanol, propanoic acid, geosmin, camphene, camphor, 2-methylisobor-neol, 5-hydroxy-methyl-furfural, α-pinene, β-caryophyllene, and 1-octen-3-ol as illustrated in Fig. 8.1 (Perl et al. 2011; Juenger et al. 2012; Sundberg et al. 2013; Kanchiswamy et al. 2015). On the other hand, they open the door for chemical research through which the fundamental and important details about the molecular basis of microbial exercises are rendered upon (Thorn and Greenman 2012). This can supply the real-world potency for better control and use of microorganisms (Liang et al. 2008). Additionally, MVOCs act as marker compounds to identify the fungal and bacterial species on our planet (Fiedler et al. 2001). Moreover, the significant number of the distinguished and identified MVOCs is produced by soil microorganisms (Insam and Seewald 2010).

8.2 Identification of Microbial Volatile Compounds

Volatile organic compounds (VOCs) are low-molecular-weight lipophilic compounds produced by different biosynthetic pathways. VOCs are responsible for the interactions among plants and have a role in antagonists and mutualistic symbionts above and below the soil surface (Garbeva et al. 2014b; Lemfack et al. 2014; Kanchiswamy et al. 2015). MVOCs provide basic information about microbial activities at molecular levels (Thorn and Greenman 2012). All living organisms can produce volatiles, but not all have a remarkable role in ecological interactions (Garbeva et al. 2004). The amount and composition of volatiles released by microorganisms depend on cultivation conditions and the physiological conditions of the microorganism such as moisture rates, oxygen presence, pH changes, and temperature (Romoli et al. 2014). Bacterial volatiles are typically dominated by alcohols, alkenes, ketones, benzenoids, terpenes, pyrazines, esters, and acids, whereas fungal volatiles are dominated by aldehydes, alcohols, ketones, alkenes, acids, benzenoids, and esters (Piechulla and Degenhardt 2014). More than 100 species of fungi and bacteria inhabiting the soil were described as MVOC producers (Effmert et al. 2012). Volatile metabolites are the main players in microbe interactions, whereas nonvolatile ones inhibit and sometimes stop the vital functions of bacteria (Tirranen and Gitelson 2006). MVOCs which are produced mainly by bacteria and fungi are synthesized and released during the primary metabolism (the synthesis and translation of DNA, amino acids, fatty acids) and the secondary metabolism (intermediates of the primary metabolism), essentially in the metabolic oxidation of glucose from different precursors like acetic acid, keto acids, fatty acids, and amino acids as illustrated in Fig. 8.2 (Korpi 2001).

However, the difference between primary and secondary metabolism is not clear, and the MVOCs are formed during both two metabolic pathways. Several studies manifested that MVOCs are defined as by-products from the primary metabolism and mycotoxins are the end products of the secondary metabolism. The primary metabolism is essential in all microorganisms, because it produces metabolites which have significant influence on metabolic processes like growth, development, and reproduction of microorganisms, while the secondary metabolism is still poorly understood and has a negligible advantage because it stops the production of these vital metabolites after the active growth (Bentley and Bennett 1988; Vining 1990). The function of secondary metabolism is mysterious, but it has different functions according to their diversity and differences in chemical structures. Most of these metabolites are produced either intracellularly or extracellularly; therefore, they could be used for detoxification of microorganisms or waste material. Among the well-famed secondary metabolites are antibiotics, toxins, and dyes (Bentley and Bennett 1988). Furthermore, nutritional disturbances and disorders such as deficiency in nitrogen and carbon sources causes the production of the secondary metabolites; also, modifications in the nutritional status can lead to enhancement of MVOC production (Korpi 2001). Moreover, the production of certain fungal MVOCs is associated with mycotoxin production. Also, secondary metabolites in some cases act as inhibitors of the primary metabolism (Sunesson 1995), and the

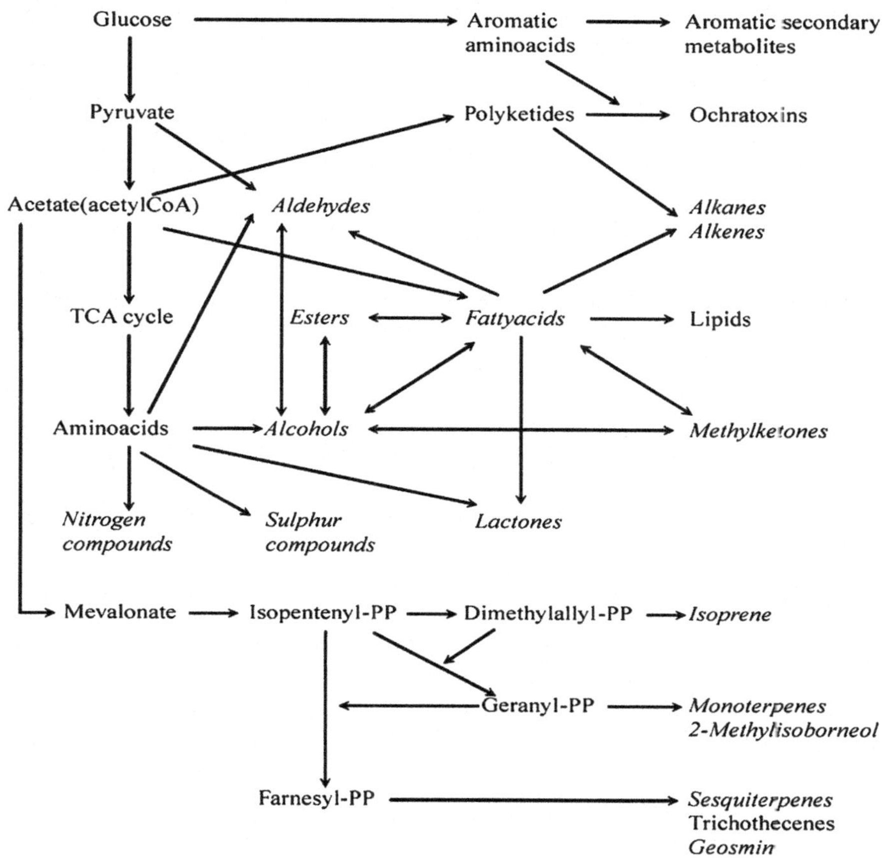

Fig. 8.2 Main metabolic pathways for the production of some MVOCs and mycotoxins. Volatile compounds are in italics. Abbreviations: *CoA* coenzyme A, *PP* pyrophosphate, *TCA* tricarboxylic acid

volatile compounds produced by some bacterial species can enhance the production of mycotoxin (Barr 1976). Such combinations have been described between the monoterpenes, sesquiterpenes, and trichothecenes, sesquiterpenes and aflatoxins, and ketones and ochratoxins (Demyttenaere et al. 2003, 2004; Wilkins et al. 2003).

8.3 Environmental Microbial Volatile Organic Compounds

The spontaneous environmental chemical reaction may alter the production of MVOCs into other compounds. For instance, alcohols are oxidized easily to carboxylic acids and aldehydes (Wilkins et al. 1997). Ketones in the air could interact with hydroxyl radicals to form aldehydes (Atkinson et al. 2000). On the other hand, the atmosphere chemical reactions may produce organic acids, ketones, and aldehydes, but the intermediate products formed during the environmental reactions have been suggested to be much more irritating than the corresponding original reactants and the end products (Wolkoff et al. 1999, 2000; Weschler 2000). In instances under extreme humidity conditions, the interaction between isoprene and ozone produces methacrolein, hydrogen peroxide, and methyl vinyl ketone (Sauer et al. 1999), all of which are defined as irritants. Also, human activities, building materials, foodstuffs, smoking, and traffic can be considered as sources of environmental MVOCs (Schleibinger et al. 2002). Till now, more than 250 compounds have been identified in laboratory studies (Wilkins and Larsen 1995b and Jelen and Wasowicz 1998); MVOCs produced by mixed cultures on the building materials have been examined in a few studies (Claeson et al. 2002a, b). Most of the laboratory research have been done with selected pure cultures of individual species, which are cultivated from cereals, agar, and other foodstuffs, bedding materials (like peat, straw, and shavings) or building materials (like wall paper, wood, and gypsum), house dust, and degradable household waste (Schleibinger et al. 2003; Meruva et al. 2004; Claeson and Sunesson 2005 and Claeson 2006).

Microorganisms, including algae, protozoa, fungi, bacteria, and viruses, are found in an all-around space and in both indoor and open-air environments. Fungi and heterotrophic bacteria use simple sugars and amino acids as organic sources for carbon and energy. Molds are able to grow rapidly because of the moisture conditions like suitable temperature, pH, and nutrient that are available in indoor environments. Also, some different species from bacteria and fungi are able to grow in any humid indoor conditions. The growth of molds on different types of materials is accompanied by the growth of bacteria in spite of researchers giving more attention to fungi in their studies more than bacteria. For example, primary colonizers, such as *Aspergillus*, *Eurotium*, and *Penicillium*, can be cultivated when the humidity percentage is more than 75%, secondary colonizers grow when humidity is at 80–90%, and tertiary colonizers like *Fusarium*, *Stachybotrys*, and actinomycetes are able to grow at a humidity above 90% (Pasanen et al. 1992; Batterman 1995). Furthermore, approximately 70% of the investigated moldy buildings which belong to the ascomycetes are *Streptomyces*. It was found in the indoor air of buildings; therefore, it has been considered as an indicator of destruction of water buildings (Hyvärinen et al. 2002; Suutari et al. 2002). On the other hand, several scientists demonstrated that actinomycetes are substantial for the degradation of multiple materials including plastics and rubber and materials that are hard to decay (Suutari et al. 2002). Bacteria and fungi have the ability to produce a wide diversity of compounds which have a role in biochemical processes. Majority of these compounds are volatile ones such as esters, alcohols, hydrocarbons, terpenes, sulfur-containing compounds, ketones, and carboxylic acids; some of the most common microbial volatiles are shown in Table 8.1 (Fischer et al 2000; Wilkins et al. 2000; Fiedler et al. 2001). The attention in using MVOCs as a signal of biocontamination was firstly boosted by the food-processing industry in the 1970s, when unpleasant odors of MVOCs were detected during the storage or processing of foodstuffs (Dainty et al. 1984, 1989; Börjesson et al. 1989, 1992; Wilkins and Scholl 1989). After that, MVOCs were investigated to recognize, identify, and isolate the microbes which are mainly fungi

Table 8.1 Examples of some MVOC compounds produced by some microorganisms

Compound group	Subgroups	Examples of common compounds
Hydrocarbons	Alkanes, alkenes, dienes, trienes	Octane, 1-octene
Terpenes	Hemiterpenes (C_5 hydrocarbons, alcohols, ketones)	Isoprene, limonene, geosmin
	Monoterpenes (C_{10} hydrocarbons, alcohols, ethers, ketones)	
	Sesquiterpenes (C_{15}, C_{11}, C_{12} hydrocarbons, alcohols, ketones)	
	Diterpenes (C_{20} hydrocarbons)	
Alcohols	Saturated, unsaturated, branched	1-Octen-3-ol, 2-methyl-2-propanol
Carboxylic acids and esters	Saturated, unsaturated, branched, diols, ketols	Acetic acid, ethyl acetate
Ketones	Methyl(2-) ketones (saturated, branched)	2-Butanone, 3-methyl-2-pentanone, 2-hexanone, 3-hexanone, cyclopentanone
	Ethyl(3-) ketones (saturated, unsaturated)	
Sulfur derivatives	Thiols, monosulfides, disulfide, trisulfides, S-methyl thioesters, thioethers	Dimethyl disulfide
Aromatic compounds	Hydrocarbons, alcohols, ethers, ketones, phenols	Styrene
Nitrogen-containing heterocyclics	Alkoxypyrazines, indoles, pyrroles, alkylfurans, γ- and δ-lactones	3-Methylfuran

(Wilkins et al. 2000; Karlshøj and Larsen 2005). Fungi and bacteria may affect the human health through the production of fungal mycotoxins, bacterial exo- and endotoxins, components of cell wall, and MVOCs. There are three ways in which bacteria and fungi could affect human health: allergic reactions, infections, and toxic responses (Stetzenbach 1998; Górny et al. 2002). Also, the exposure to MVOCs is one of the main reasons of a number of health problem symptoms like infection of the nose and eye, throat irritation, and tiredness (Samson 1985; Stetzenbach 1998).

On the other hand, individuals invest a greater part of their time inside, and as these buildings become older, the number of people experiencing negative health effects has been increasing since the 1970s (Redlich et al. 1997). However, MVOCs in indoor air environments were detected in the 1990s for the first time (Bayer and Crow 1994; Morey et al. 1997; Wilkins et al. 1997). These problems found in buildings are always named sick building syndrome (SBS). Many of MVOCs have been identified in indoor air, but indoor concentrations of many pollutants are always higher than those of outdoor (Jones 1999). Building products are always the reasons of the pollution of indoor air, but MVOCs resulted from diversity of other sources such as microorganisms, human activities, furniture, humidity, and solvents (Jones 1999). Nowadays there is a general agreement on the connection between humidity in buildings and health effects such as respiratory disease, asthma, wheezing, and coughing and general health problems such as headache and tiredness (Bornehag et al. 2001, 2004). These side effects have been characterized by the World Health Organization (WHO) which also include irritation (nose, eye, and throat), asthma, asthma-like symptoms (wheezing and chest tightness), neurotoxin effects (headaches, irritability, and fatigue), and skin symptoms (irritation and dryness) (WHO 1986).

There is no single or group of environmental factors that cause SBS, but there are many suggestions such as air contaminants, bacteria, fungi, dust, humidity, and poor ventilation which may be working individually or together (Redlich et al. 1997). The clear reasons for the previously mentioned symptoms have not yet been illustrated; however, organic chemical compounds and microbial agents have been submitted as one of causative agents. Humidity of building materials undergoes degradation processes resulting in the emission of chemical compounds, and the water content of the materials also supports the growth of microbes. The growth of fungi has been considered as the major reason that causes health problems in buildings; however, the literature observed that indoor air spores are poorly found (Bornehag et al. 2004 and Nevalainen and Seuri 2005).

The microbial development is always holed up behind ceilings or carpets which can be the reason behind the absence of connection between health symptoms and measured spore levels in indoor air. However, MVOCs are able to spread into the building structures, in this manner adding to the total combinations of VOCs to which those humans remaining inside are exposed. Thus, there is no relationship between MVOCs and a health problem which has been found that the interference of MVOCs in the SBS complex of problem is still flawed. On the other hand, due to many sources, the concentrations of indoor MVOCs vary, but both nonproblem and problem building have a total count of MVOC concentrations of less than 0.5 mg/m^3; however, the concentration of a single compound rarely exceeds 50 μg/m^3 (Brown 1999). However, a lot of scientists failed to find the relation between the total accounts of MVOC concentrations and health problems (Berglund and Johansson 1996; Mølhave et al. 1997; Ten Brinke et al. 1998). Furthermore, the composition of the substrate and microbial species is the most important factor to produce MVOCs (Sunesson et al. 1997). Also, there are other factors effecting MVOC production pattern such as temperature, growth phase, light, moisture, pH of the substrate, and availability of different levels of O_2 or CO_2 (Sunesson et al. 1996). The substrate composition has great effect on both quantitative and qualitative excretion of volatile compounds, for example, nutrient-rich media enhanced more quantities and different types of metabolites than do media with poor nutrient source (Wilkins et al. 2000). Moreover, in a damp moldy building, there are odors of fungal VOCs released from microscopic fungi that are easily identified, and the investigators study the negative effects of these odors on human health under the name of sick building syndrome (SBS) as mentioned before. On the other hand, about 250 of volatile organic compounds have been recognized from different fungal species, and it produced a mixture of simple phenols, heterocycles, thioalcohols, aldehydes, hydrocarbons, alcohols, ketones, thioesters, and their derivatives (Chiron

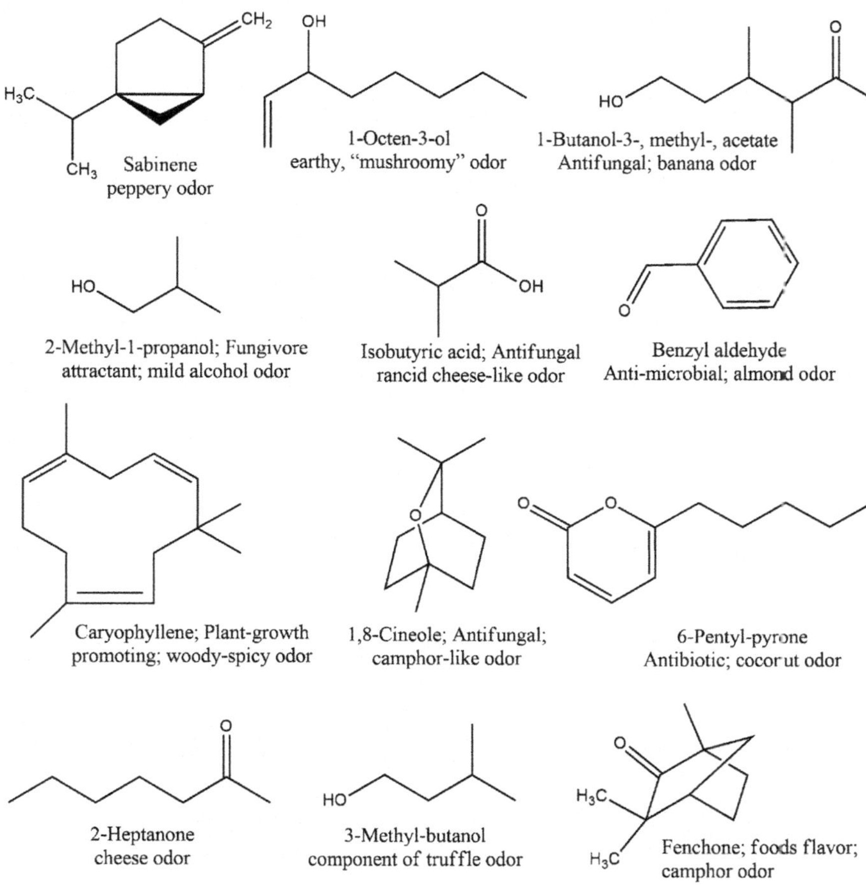

Fig. 8.3 Structure, functions, and odors of some VOCs produced by fungi

and Michelot 2005; Korpi et al. 2009; Ortiz-Castro et al. 2009). Fungal VOCs are released from both primary and secondary metabolism pathways (Korpi et al. 2009), and they are ideal (info-chemicals) because they can diffuse from the atmosphere and soil as illustrated in Fig. 8.3 (Morey et al. 1997). Also, during the past few years, unexpected high numbers of bacterial volatiles have been investigated (about 1000 of bacterial VOCs) including ketones, esters, alkenes, alcohols, alkanes, terpenoids, and sulfur compounds. A lot of investigations proved that bacteria emit their volatile compounds during their interactions with other microorganisms, in turn affecting communities and populations (Kai et al. 2009; Romoli et al. 2014) like interaction between bacteria and bacteria, bacteria and fungi, bacteria and plants, and bacteria and vertebrates. The appearances of these volatile compounds are due to a specific metabolism pathway in the bacteria (Kai et al. 2009). These volatile compounds are also ideal info-chemicals (like fungi) where it happened in the biosphere at different concentrations and over long distances (Wheatley 2002).

8.4 Fungal Volatile Organic Compounds

The actual roles of fungal VOCs in different aspects in daily activities as in the food industry are illustrated in Fig. 8.4. The VOCs are an alternative method for harmful chemical pesticides for the management plant disease during growth and a safety method to control the post-harvest disease by preventing the growth of any bacterial or fungal disease which is known as "mycofumigation" or organizing the phenomena of plant–microbe interaction. Also, VOCs are used in the classification of fungi because some species of fungi produce their own compounds. On the other hand, fungal VOCs have been studied as source of fuel or "mycodiesel." Finally, fungal VOCs are the key factor in food and flavor research, and volatile compounds from fungus have a role in human diseases which help the organism to invade the immune systems and cause severe diseases (Styger et al. 2011).

Lipid-soluble compounds are the most known of the microbial volatile organic compounds (MVOCs). Therefore, it has low water solubility also; numerous MVOCs have special odors, and this does not surprise the people who are working with fungal VOCs. For instance, the macro-fungi-like mushrooms which are highly important in the culinary arts have different combinations of VOCs like terpenes,

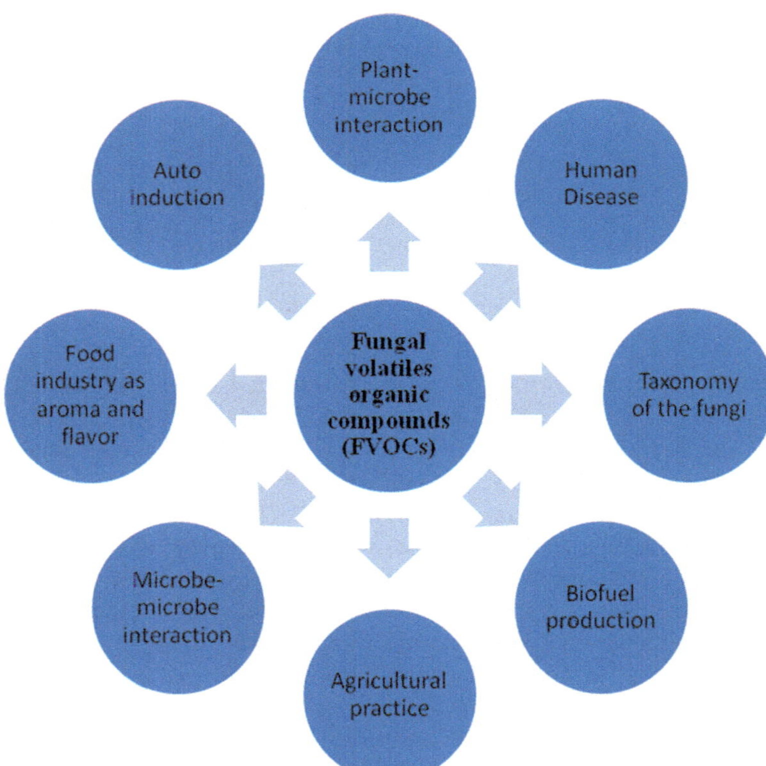

Fig. 8.4 The scheme representing the use of fungal VOCs in a different application

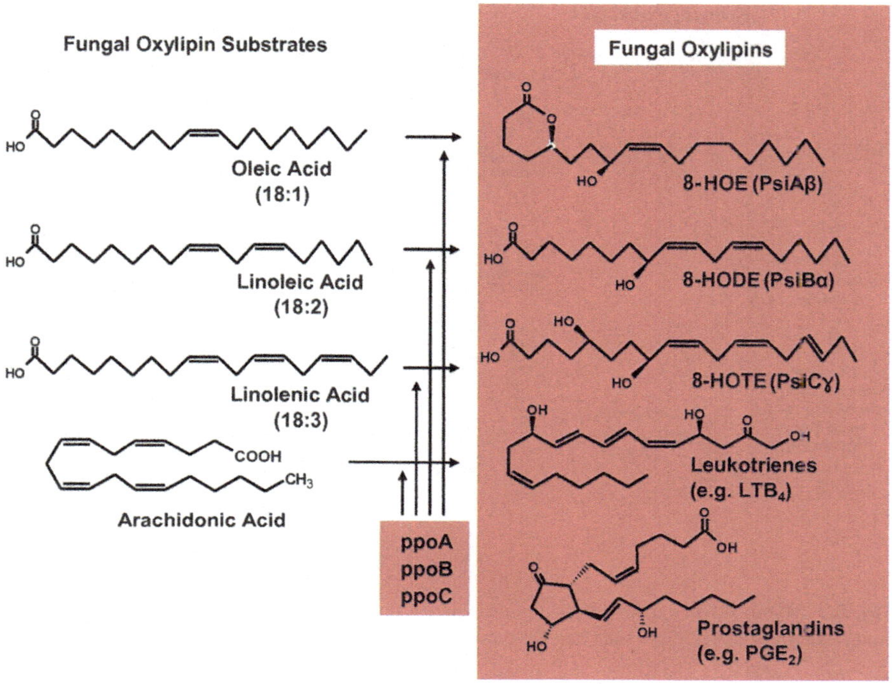

Fig. 8.5 Biosynthesis of fungal oxylipins in *Aspergillus nidulans* (Adapted from Christensen and Kolomiets 2011)

aldehydes, aromatics, and alcohols (Splivallo et al. 2007; Cho et al. 2008a, b; Fraatz and Zorn 2010). The most famous example is oxylipins which are a distinct group of lipids and defined as polyunsaturated fatty acids (PUFA) which gain oxygen either by enzymatic or nonenzymatic processes. Oxylipins have various characters in all living organisms such as mammals, microbes, and plants (Kachroo and Kachroo 2009; Shea and Del Poeta 2006 and Howe 2007). Fungal oxylipins participate obviously in various metabolic processes, for example, cell growth, apoptosis, sexual and asexual spore, differentiation, and pathogenesis (Noverr et al. 2003; La Camera et al. 2004). The discovery of the Psi factor (for precocious sexual inducer) opened the door to determine the role of fungal oxylipins in physiological processes (Champe et al. 1987). Phospholipids and acylglycerides are origins of fungal oxylipins that are hydrolyzed by lipases to produce oleic acid (18:1), linoleic acid (18:2), linolenic acid (18:3), and arachidonic acid (20:4) as illustrated in Fig. 8.5 (Sakuradani et al. 2009).

Several reports mentioned that several species of fungi can produce oxylipins such as *Aspergillus* spp., *Fusarium* spp., *Gaeumannomyces graminis*, *Laetisaria arvalis*, and *Cercospora zeae-maydis* (Shim and Dunkle 2002; Tsitsigiannis and Keller 2007). For example, dioxygenase (DOX) and lipoxygenase (LOX) from *Gaeumannomyces graminis* were the first enzymes characterized and participated in fungal oxylipin formation. The DOX and linoleate 8-DOX (lds) which are

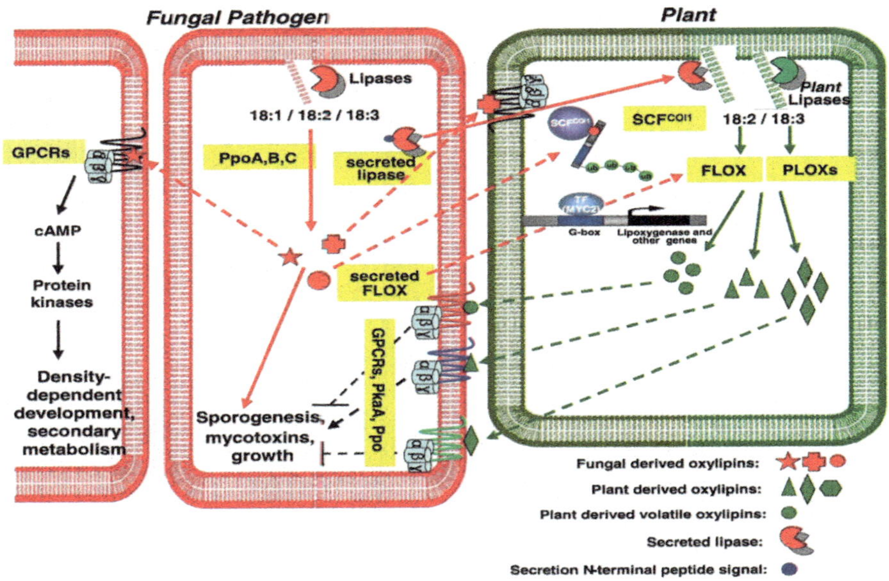

Fig. 8.6 Hypothetical model of oxylipin-mediated signal communication between plants and fungi (Adapted from Christensen and Kolomiets 2011)

homologous to the mammalian cyclooxygenase (COX) produce linoleic acid-derived 8-hydroxy linoleic acid (Hamberg et al. 1998; Su et al. 1998; Hörnsten et al. 1999). On the other hand, as shown in Fig. 8.6, when the fungus invades the plant, oxylipins from both sides interact with each other, and each side would like to dominate on the other side from oxylipin mechanism. One hypothesis postulated that oxylipins produced from plant can control the fungal oxylipin biosynthesis. Other hypotheses are controversial, and they could prove that oxylipins from fungus can alter the metabolism of lipids in the plant during plant–fungus interactions (Brodhagen et al. 2008). Invading fungi can produce lipase and LOX inside of the host cell to change metabolism of the lipid in the host by breaking down free fatty acids and oxidizing to produce oxylipins. The oxylipin-derived fungi used host-derived oxylipins to invade growth and produce the spores and mycotoxins (Voigt et al. 2005). All these studies support the ability of fungi to use the host oxylipins to increase fungal pathogenicity.

8.5 MVOCs Produced by Endophytic Fungi and Mycofumigation Process

The research on endophyte was concentrated on endophytic fungi, which appear to be ubiquitous among plants. All plant species inspected to the date have at least one species of endophyte (Petrini 1991; Arnold et al. 2001), and a lot of plant species may contain different endophytic species within apparent health tissues (Stone et al.

2000). An uncommon endophytic fungus was isolated and identified from *Cinnamomum zeylanicum* growing in a rainforest in Honduras (Strobel et al. 2001 and Worapong et al. 2001). This fungus lacked spores and spore-producing structures that have whitish hyphae (rope-like) which widely vary in diameter. Moreover, the result of 18S rDNA sequence data indicated that this organism was a *Xylariaceae* sp. This organism was able to produce a mixture of volatile organic compounds (VOCs), and according to the available chemical, molecular, and structural knowledge, this organism was given another genus name *Muscodor albus*. During the past 30–40 years, a lot of VOCs from fungi have been isolated and identified. However, no one has described the production of VOCs from the genus Muscodor albus which has been counted as effective tool against wide range of fungi, bacteria, and some nematodes and arthropod species as the genus *Muscodor albus* (Strobel et al. 2001; Mercier and Jimenez 2004; Mercier and Manker 2005; Strobel 2006; Riga et al. 2008). This inhibition effect was due to combinations of VOCs produced by *M. albus* in the culture medium (Strobel et al. 2001). The VOCs produced by *M. albus* consist of different organic acids, esters, short-chain alcohols, aromatic hydrocarbons, and ketones (Strobel et al. 2001), and this varies according to the type of nutrients, duration of incubation, substrate, temperature, and other environmental parameters (Nilsson et al. 2004 and Fiedler et al. 2005). The first entophytic fungus to be identified was *M. albus* which is found in the inner tissues of plants (Wilson 1995). Despite the fact that the collaborations between host plant and endophytes are regularly asymptomatic either for all or part of their life cycle, these can also be considered as communistic relationships. There are a lot of models which demonstrated that endophytic fungi improved the host plant growth, fitness, and response to stress (Waller et al. 2005; Rodriguez et al. 2009; Yuan et al. 2010). VOCs of *Muscodor albus* can be used as a defense barrier for the plant against potential pathogens or insects, besides helping the microorganisms to live (Strobel et al. 2001). On the other hand, the toxicity of VOCs released by *M. albus* was used to inhibit the growth of other microorganisms in agricultural seeds, plants, and soil. This process is called mycofumigation (Stinson et al. 2003; Mercier and Manker 2005; Strobel 2006). Recently, mycofumigation is used in transportation and storage of different fruits and vegetables due to the presence of *M. albus* which increases their shelf life till used (Mercier and Jimenez 2004). Therefore, *M. albus* is the promising strain for controlling this malady (Stinson et al. 2003). In vitro, the VOCs emitted from *M. albus* can inhibit the growth of *Monilinia fructicola*, *Penicillium expansum, and B. cinerea* which cause disease for peach. Moreover, the volatiles can reduce the incidence of post-harvest disease for peaches for more than 7 days of storage (Mercier and Jimenez 2004). Volatiles of *M. albus* also can prohibit the formation of molds (Mercier and Jimenez 2007). The volatiles of *Oxyporus latemarginatus* EF069 can be deemed as a promising tool for controlling the post-harvest diseases affecting negatively several crops and fruits as the VOCs secreted by *O. latemarginatus* EF069 diminish the rotten apples, which is caused by *B. cinerea* and *Rhizoctonia* moth orchid, after harvesting (Lee et al. 2009).

Also, the production of VOCs by endophytic fungi permits the host plant to fight the other pathogen which attacks them as well as harsh environments like salinity and drought (Verma et al. 2007). This motivated a lot of researchers analyzing the endophytic fungi to investigate the use of endophytic VOCs in medicine, agriculture and biotechnological industry (Banerjee et al. 2010; Tomsheck et al. 2010; Ahamed and Ahring 2011). Other examples for endophytic fungi which release volatile organic compounds were that isolated from *Odontoglossum* sp. (Orchidaceae) which was collected from a cloud forest in the Ecuadorian ecosystem. Orchidaceae is one of the largest plant family in the world, and it contains many of endophytes that release various of VOCs. An interesting peppery aroma was produced by the organism, so it was further selected for studying its biology, taxonomy, and VOCs (Singh et al. 2011).

Endophytic fungi of *Xylariaceae* family are a rich source of VOCs. 2-Phenylethanol is among VOCs produced by Aspergillus niger which has been isolated from the host plant *Rosa damascena* (Wani et al. 2010); 2-methyl butane, β-butyrolactone, and 2-butene dinitrile produced by *Botrytis* sp. BTF21 on the host plant *Musa* sp. (Ting et al. 2010); sabinene, 3-methylbutan-1-ol, 2-methylpropan-1-ol, and acetone produced by *Phomopsis* sp. on the host plant *Odontoglossum* sp. (Singh et al. 2011); and β-elemene, α-selinene, β- selinene, and 2,5-dihydrotoluene produced by *Nodulisporium* sp. on the host plant *Cinnamomum loureirii* (Park et al. 2010) as illustrated in Fig. 8.7.

8.6 MVOCs Produced by Bacteria

Bacteria release highly diverse inorganic and organic volatile compounds, which consist of different chemical classes, like nitrogen- and sulfur-containing compounds, acids, terpenes, and fatty acid derivatives (hydrocarbons, alcohols, and ketones), as shown in Fig. 8.8 (Ladygina et al. 2006). The short-chain alkanes are located in the microorganisms, while longer hydrocarbons like hexadecane are found in *Cyanobacteria*, which also have the ability to synthesize branched hydrocarbons (Tellez et al. 2001; Ladygina et al. 2006). Also, decarboxylation of fatty acids resulted in methyl ketones. Under anaerobic conditions, the fermentation of pyruvate resulted in acetoin (3-hydroxy-2-butanone) and its oxidized form 2,3-butanedione (Ryu et al. 2003), and this happens when acetolactate synthase enzyme catalyzes the condensation of two pyruvate molecules into acetolactate, then decarboxylates to form acetoin, and then oxidizes to form 2,3-butanedione. The processes which regulate the synthesis of acetoin and 2,3-butanedione in the environment are still unclear. *Enterobacteriaceae* produced long-chain aliphatic alcohols (e.g., 1-decanol, 1-dodecanol, and 1-octanol); they are produced through β- or α-oxidation of fatty acid derivatives when the medium was supplemented by it (Hamilton-Kemp et al. 2005). In contrast, *Proteobacteria* and *Firmicutes* produce short-chain alcohols (like 2,3-butanediol) during minimum oxygen conditions to supply a substitutional electron sink for the regeneration of NAD^+ when aerobic respiration is limited from pyruvate (Farag et al. 2013; Whiteson et al. 2014). On the other hand, by the Ehrlich pathway, enzymatic conversion of branched-chain amino

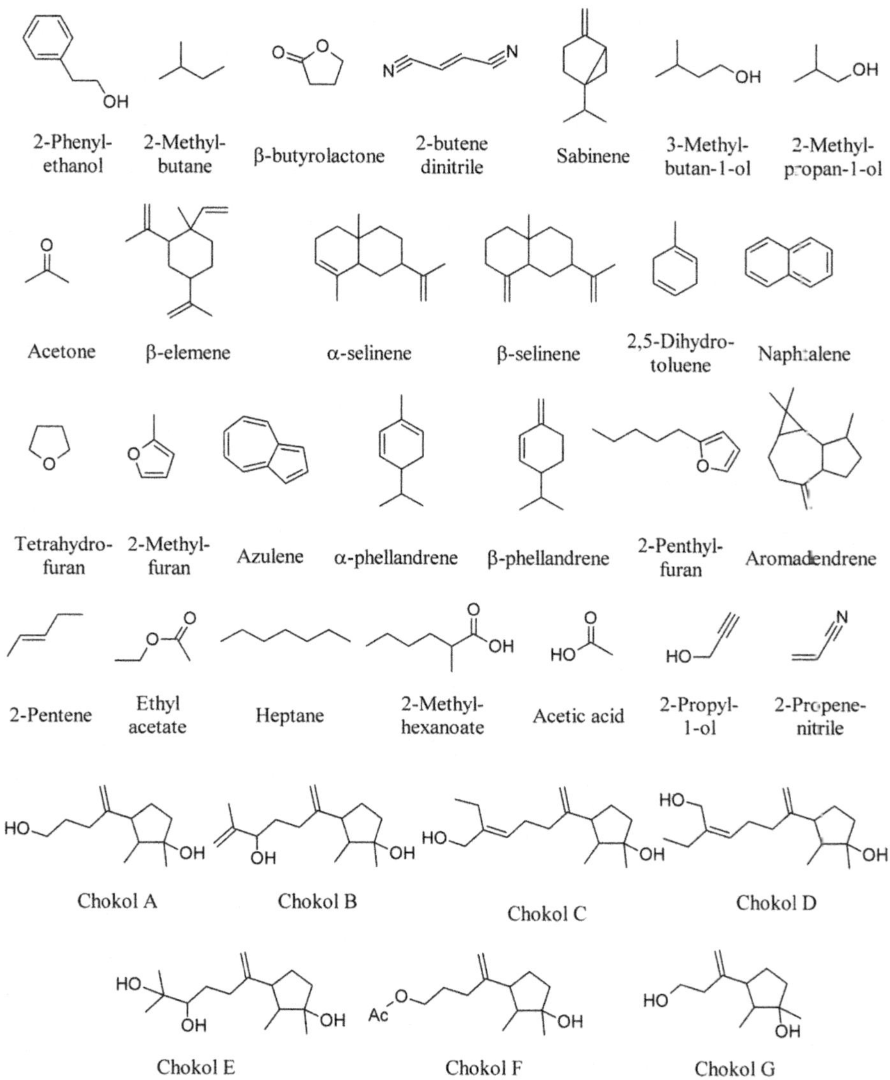

Fig. 8.7 Some volatile organic compounds of endophytic fungi (Adapted from Kanchiswamy et al. 2015)

acids such as isoleucine and leucine resulted in short-chain-branched alcohols *like* 3-methyl-1-butanol *and 2-methyl-1-butanol* (Marilley and Casey 2004).

Organic acids are produced less than alcohols and ketones, probably; many of the described short-chain fatty acids have been emitted by bacteria-like propionic acid, acetic acid, or butyric acid. These compounds exemplify the by-products of the anaerobic metabolism, and they are composed during fermentation processes of carbohydrates in bacteria. Also, anaplerotic pathway of the tricarboxylic acid cycle

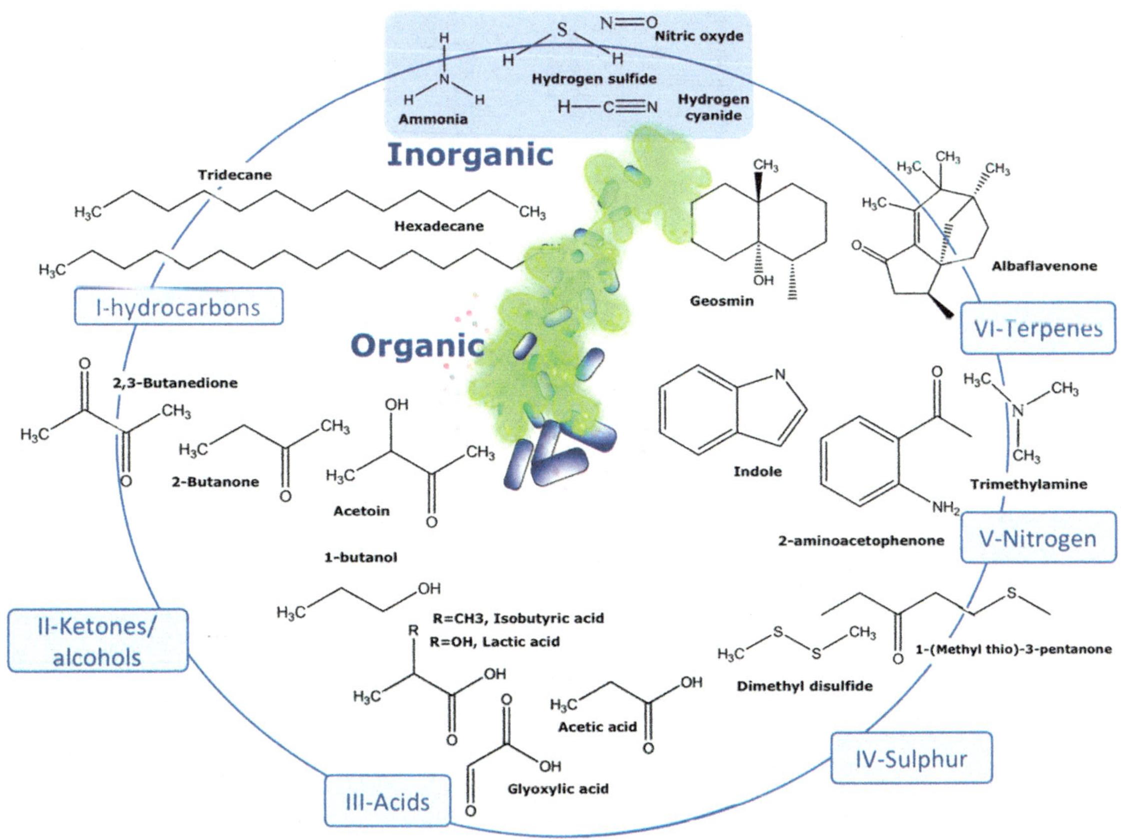

Fig. 8.8 Chemical classes of volatile compounds released by bacteria (Adapted from Audrain et al. 2015)

which allows cells to use the carbon compounds leads to the production of glyoxylic acid (Lorenz and Fink 2002; Schulz and Dickschat 2007). However, other metabolic pathways like amino acid metabolism or ethylene glycol can produce glyoxylic acid (Muckschel et al. 2012). On the other hand, sulfur compounds confer fermentation of food aroma like wine and cheese. The biosynthesis of methionine-derived volatiles, like 1-(methyl thio)-3-pentanone and dimethyl sulfide, is always intermediated by bacteria. Also, most volatile sulfur compound pathways from L-methionine include cleavage of 3-dimethylsulfoniopropionate which is created by marine algae and higher plants (Stefels 2000). Trimethylamine which is responsible for the odor of spoiling fish and also found in animal and human intestines is a tertiary volatile amine, and it is released upon the reduction of trimethylamine oxide (TMAO). This occurred under anaerobic conditions, while bacterial conversion occurred under both aerobic and anaerobic conditions. Another example is 2-amino-acetophenone (2-AA) which is an aromatic compound responsible for the grape-like odor of *Pseudomonas aeruginosa* cultures and could be used to detect *P. aeruginosa* infections (Que et al. 2013). Indole excretion has been reported in more than 85 species of bacteria (Lee and Lee 2010). Biosynthetic pathway of these compounds has been well studied in *Escherichia coli*, where it relies on tryptophanase encoded by *tnaA* that can reversibly modify tryptophan into indole, pyruvate, and ammonia. Regulation of indole biosynthesis is controlled by several environmental conditions such as cell population density, existence of extracellular tryptophan, catabolite repression, pH, and temperature (Lee and Lee 2010). Furthermore, terpenes are derived from isopentenyl pyrophosphate and dimethylallyl pyrophosphate building units, which can arise either from mevalonate pathway or deoxyxylulose phosphate pathway. Bacterial strains can produce monoterpenes (C_{10}), sesquiterpenes (C_{15}), and their derivatives (Schulz and Dickschat 2007). The earthy odorant geosmin and antibiotic albaflavenone are both terpenoid compounds formed by bacteria, while geosmin, a degraded sesquiterpene, is produced in several bacterial species, and albaflavenone is exclusively found in *Streptomyces* (Schulz and Dickschat 2007). Also, bacteria released inorganic volatile compound like hydrogen sulfide (H_2S), nitric oxide (NO), hydrogen cyanide (HCN), or ammonia. Hydrogen sulfide (H_2S)-producing bacteria produce this gas during cysteine degradation (Mattila and Thomas 2014). Nitric oxide (NO) is released from L-arginine by nitric oxide synthases (Mattila and Thomas 2014). Ammonia is released via L-aspartate catabolism from the metabolism of peptide and amino acid (Bernier et al. 2011). It is observed that, as a by-product in *E. coli*, K12 ammonia is released by an aspartate ammonia lyase when aspartate was converted to fumarate (Bernier et al. 2011). On the other hand, a few bacterial species such as some species of *Chromobacterium*, *Rhizobium*, and *Pseudomonas* have been detected to produce hydrogen cyanide (HCN). The HCN biosynthesis is catalyzed by HCN synthase and encoded by *hcnABC* genes, which forms HCN and CO_2 from glycine at the end of the exponential phase and under oxygen limiting conditions (Blom et al. 2011).

8.7 MVOCs in the Field for Plant Productivity and Crop Sustainability

The remarkable application of fertilizers and synthetic pesticides supplies immediate solutions for the crop yield problems and plant disease, respectively, but in the end, they drastically affect human health and our environment. Although biocontrol agents are produced by diverse microorganisms, they are becoming suitable substituents for the chemical fertilizers and pesticides, and their reduced efficiency and still highly expensive and incompatible field performance relegate them to niche product (Glare et al. 2012). Researchers who focused on MVOC–plant interactions have led to an increasingly conceptual understanding of the complicated and dynamic character of MVOCs, through confirming their indispensable role for enhancing crop productivity and protection. Plant treatment with MVOCs leads to remarkable modification of the metabolism, physiology, and transcriptional status of the plant, which strengthens the hypothesis that plants are able to respond to and perceive MVOCs (Cortes-Barco et al. 2010a, b; Song and Ryu 2013). Many studies clearly proved the necessity for application of MVOCs in open field and insist on their diverse roles for increasing resistance to manage pathogen, providing the protection from herbivores, so in general they act as biocontrol agents. Although, MVOCs play an important role in antibiosis and signaling also, they may act as regulators of growth and development of plant. But, on the molecular and ecological basis, the microbial volatile is not completely clear. Several mechanisms such as defense against other microorganisms, inter- and intraspecies interactions, and promotion of plant growth have been proposed. Therefore, to understand the role of VOCs in plant productivity, we must study the genome, metabolome, and proteome of plants when treated with MVOCs (Park et al. 2013; Song et al. 2013). Microbial volatiles affect the resistance to different pathogens and regulation of phytohormones and influence plant growth and development (Piechulla and Degenhardt 2014; Peñuelas et al. 2014). Plant growth is promoted by the production of GacS kinase-dependent production of 2,3-butanediol in presence of MVOCs provides the optimum condition for the growth of the plant (Han et al. 2006). 2,3-Butanediol produced by *Bacillus subtilis* helps the *Arabidopsis* plant to tolerate salt and enhance the induction systemic resistance (ISR) which regulate auxin and ethylene homeostasis, while *Pseudomonas chlororaphis* can produce the same compound which enables the plant to tolerate drought and increase resistance to *Erwinia carotovora* but not against *Pseudomonas syringae* pv. *tabaci* (Ryu et al. 2004; Cho et al. 2008a, b; Zhang et al. 2008). Furthermore, the volatiles of *Chromobacterium*, *Burkholderia*, *Serratia*, *Pseudomonas*, and *Stenotrophomona* showed adverse impacts on plant growth and development (Bailly and Weisskopf 2012). On the other hand, the analysis of proteomes and metabolomes of *Arabidopsis* plant exposed to *Bacillus subtilis* volatiles indicates the presence of different signaling pathways to increase plant growth such as cell wall modification, hormonal modulations, stress responses pathways, enzyme activity, antioxidant activity, and photosynthetic pathways (Kwon et al. 2010). Moreover, *Trichoderma viride* volatile compounds induce lateral root elongation and enhance flowering which indicates the importance of volatile compounds for the growth (Hung et al. 2013). Also, 1-octen-3-ol is produced by many fungi and contributes to enhanced plant resistance to the necrotrophic fungal pathogen *Botrytis cinerea* through the induction of defense signaling pathways (Contreras-Cornejo et al. 2014; Kishimoto et al. 2007). The volatiles from *Alternaria alternata* and *Penicillium aurantiogriseum* augment the growth and starch formation in various plant species (Ezquer et al. 2010). Interestingly, volatiles from non-pathogenic strains of *Fusarium oxysporum* MSA35 promote the growth of the lettuce (*Lactuca sativa*) (Minerdi et al. 2009; Minerdi et al. 2011). These examples conclude that MVOCs have potential effects on plant growth and increase plant productivity and welfare; MVOCs can be used instead of costly and unsustainable chemical pesticides and fertilizers that might harm human health. Spreading the concept of MVOCs as an alternative and promising way still remains a challenge and needs more effort.

8.8 VOCs as Biocontrol Agents

In disease control strategies, it is very important to select effective biocontrol agents. *Bacillus subtilis* has antifungal properties, so that a number of plant and animal diseases can be controlled (Whipps 2001). Volatile compounds are important in the functioning of both atmospheric ("aboveground") and soil ("belowground") ecosystems. Fungal VOCs have potential for biotechnological applications in agriculture, industry, and medicine. In agriculture, the potential of fungal VOCs as biological control (biocontrol) agents is of interest to control fungal pests and to employ a more environmentally sound pest management strategy by reducing fungicide use on crop plants.

8.8.1 Aboveground Actions (Aerial Part)

Fungi which associate to plants and cause no negative influence on their hosts called endophytic fungi (Bacon and White 2000). These fungi secrete a mixture of VOCs, and it is clear that from an ecological view, it is much more complicated than researchers have previously estimated. Furthermore, the volatile organic compounds of endophytes may provide beneficial influence on the host plant through forming supplementary lines of defense against diseases that may cause disease for the host (Macias-Rubalcava et al. 2010). For instance, VOCs emitted from *Muscodor albus* inhibit pathogenic fungi and bacteria attacking the plants as mentioned before (Strobel et al. 2001). Moreover, the VOCs released from *Muscodor fengyangensis* and *Muscodor yucatanensis* can inhibit the growth of pathogenic species of bacteria and fungi (Zhang et al. 2010). Also, volatile compounds released from *Muscodor crispans* can suppress multiple pathogen diseases such as *Mycosphaerella fijiensis* which infect bananas, as well as the *Xanthomonas axonopodis* pv. *Citri*, which constitute incurable problem for citrus crops which is considered as one of the main exports of different countries (Mitchell et al. 2010). Endophytic fungi can help the plant to survive under severe conditions. For example, VOCs released from the fungus *Phoma* sp. (from creosote bush) help the shrub to tolerate and survive in a desert (Strobel et al. 2011). It has the ability to produce unique VOCs which can damp or kill a wide variety of pathogens such as *Sclerotinia*, *Cercospora*, *Verticillium*, and *Ceratocystis*, while for *Aspergillus*, *Colletotrichum*, and *Trichoderma*, there is not any observable effect (Strobel et al. 2011).

8.8.2 Belowground Action (Soil Region)

Fungistasis is defined as the growth of the fungal hyphae that lose the ability to grow under optimum moisture conditions and temperature due to either competition for nutrients with other microorganisms or secreting diverse volatile compounds that effect negatively on other organisms in the soil (Watson and Ford 1972). Soils which are dominated by fungi are highly filled with 3-methyl-2-pentanone, trimethylamine, methyl pyrazine, dimethyl disulfide, 2,5-dimethyl-pyrazine, nonadecane, and N-dimethyloctylamine, but soils without fungistasis do not (Xu et al. 2004). Volatile fungistatic compounds can inhibit the growth of three fungal species, namely, *Paecilomyces lilacinus*, *Pochonia chlamydosporia*, and *Clonostachys rosea*; these volatile compounds do not need competition among soil microbes for emission, and sometimes they may have a distinguished mechanism to manage the fungistatic effect in the soil (Xu et al. 2004). The pores of the soils can be replenished by VOCs, and according to the characteristic properties of the surrounding area, VOCs can distribute (Aochi and Farmer 2005). Also, volatile compounds emitted by nonpathogenic *Fusarium oxysporum* strain-associated bacteria inhibited the growth of the pathogenic *F. oxysporum* which is disease to the plants (Minerdi et al. 2009). Furthermore, *Muscodor albus* can manage the rapid death of broccoli seedlings caused by *Rhizoctonia solani* and the root rot of bell pepper caused by *Phytophthora capsici* (Mercier and Manker 2005). On the other hand, the expression of gene responsible for biocontrol in *Pseudomonas fluorescens* is highly expressed during the presence of the volatiles emitted by *Trichoderma atroviride* (Lutz et al. 2004). Therefore, the treatment of the plants with volatiles of soilborne fungi can provide protection from pathogen attack, e.g., treatment of *Arabidopsis thaliana* with 1-octen-3-ol (mushroom alcohol) inhibits the growth of *Botrytis cinerea* through the regulated production of defense gene (Kishimoto et al. 2007). Moreover, allo-ocimene and an aldehyde which contain six carbon atoms have similar defense actions in *A. thaliana* (Kishimoto et al. 2006a, b). Also, other previous literature could reveal antagonist effect of fungal VOCs on plant growth, for example, fungal volatile compounds secreted by *Tuber* spp. prevent the growth of *A. thaliana* plant and mycorrhizal fungi that can form dead zones, consequently rubbing the competitors of their host (Splivallo et al. 2007a, 2011).

8.9 VOCs and Antibiotics

Antibiotic resistance, biofilm formation, and virulence could be influenced by bacterial volatile compounds since they play an important role in airborne interactions between bacteria. During food transformation and fermentation, bacteria release many different volatile compounds (Bos et al. 2013). Volatile bacterial secondary metabolites have influence on the growth of exposed plants, fungi, and animals (Ma et al. 2012). *E. coli* can produce more than 14 volatile compounds that have tangible role in mortality, resistance to different antibiotics, and adhesion processes. One of these promising compounds is TMA which can coordinate antibiotic tolerance in bacteria with different mechanisms (Létoffé et al. 2014). On the other hand, fungi produce antibiotics used in managing human disease; *Muscodor crispans* produce different types of VOCs such as propanoic acid, 2-methylbutyl ester, and 3-methylbutyl ester which help in managing several pathogens affecting humans such as *Staphylococcus aureus*, *Mycobacterium tuberculosis, and Yersinia pestis* which are drug resistant (Mitchell et al. 2010). These volatile compounds released have bactericidal criteria against *Escherichia coli* (Zhang et al. 2010).

8.10 VOCs Role in Bacterial–Fungal Interactions

Volatile compounds can serve as signaling molecules in the communication among different soil microbes although their physicochemical properties enable diffusion through soil particles. With respect to the functioning of soil microbial volatiles, most attention has been given to the suppressive effects of bacterial volatiles on soil eukaryotes that are harmful to agricultural crops, e.g., plant pathogenic fungi (Zou et al. 2007; Verginer et al. 2010; Garbeva et al. 2014a, b). *S. plymuthica*, *S. odorifera*, *Stenotrophomonas maltophilia*, *Stenotrophomonas rhizophila*, *P. fluorescens*, and *P. trivialis* which are rhizobacterial isolates produce complex MVOCs that inhibit the growth of many phytopathogenic and non-phytopathogenic fungi (Vespermann et al. 2007; Kai et al. 2010).

To our knowledge, there is very little research concerning with the investigation of the role of volatile compounds in fungal–bacterial interactions. The volatile compounds released by ascomycete fungi have indispensable role in interactions between fungi and bacteria. Volatiles influence phonotypical responses such as biofilm formation, growth of the microorganism, antimicrobial activity, and motility (Ruth Schmidt et al. 2016).

The role of MVOCs in the antagonistic interactions between mycorrhizal fungi and rhizobacteria and their ecological role have been described recently in a review by Effmert et al. 2012. For example, pyrrolnitrin (PRN) is a chlorinated phenylpyrrole antibiotic that is firstly identified from *Burkholderia pyrrocinia* and finally found in other genera, like *Serratia*, *Myxococcus*, *Enterobacter, and Pseudomonas* (Garbeva et al. 2004). It has a broad spectrum against a lot of fungi belonging to the *Ascomycota*, *Deuteromycota*, and *Basidiomycota* including a lot of phytopathogens with economical importance like *Botrytis cinerea*, *Sclerotinia sclerotiorum*,

Rhizoctonia solani, and *Verticillium dahlia*. Also, *Xanthomonas campestris* pv. *vesicatoria* emit more than 50 compounds; the majority of them are methyl ketones and ketones, while the dominant compounds are 10-methylundecan-2-one followed by 8-methylnonan-2-one, 9-methylundecan-2-one, geranylacetone, pentadecen-2-one, and 12-methyltridecan-2-one as illustrated in Fig. 8.9. When some of these compounds were tested on the fungus *R. solani*, contradictory results were found with either promoting or inhibiting effects (Weise et al. 2012).

8.11 MVOCs and Human Health

8.11.1 Bacterial Volatiles as Markers for Respiratory Diseases Diagnosis

Bacteria cause approximately a third to a half of adult community-acquired respiratory infections (Johansson et al. 2009). In respiratory tract infections, antibiotics are prescribed according to a patient history or self-report of sputum characteristics and length of symptoms (Tonkin-Crine et al. 2011). A broad spectrum and wide range of antibiotic activity are desirable, and this leads to antibiotic resistance (Costelloe et al. 2010). So the novel trend right now in the treatment is doing an analysis of a patient's breath which is known as "breath testing" which will provide us with the exact disease and can help us to avoid anticipating the inappropriate professional of antibiotics in the treatment of respiratory infections. Breath testing is defined as determining the contents of volatile metabolites in the breath to detect patients with *Pseudomonas aeruginosa* infection and tuberculosis. On the other hand, gas chromatography (GC) together with electron impact (EI) and mass spectrometry (MS) is the most established reliable and widely used approach to recognize and characterize gases and volatile compounds. Resolving power for different isomers in GC, the existence of large reference databases and efforts at standardization will increase the probability for similar results to be obtained from different laboratories (Zhang et al. 2011; Phillips et al. 2012). Although volatiles from different types of bacterial species have been characterized and well known, bacterial VOCs do not appear in the major databases (i.e., Wiley, NIST, and Massfinder) used for initial hypothetical identification of compounds compared to the mass spectra (Kai et al. 2009). Characterization of bacterial volatiles encompasses the growth in laboratory cultures where production of diverse volatiles is dependent on the type of media, amount of aeration, incubation time, and other environmental factors (Nawrath et al. 2012). A few types of bacterial species are of highly specific interest with regard to the common adult community-acquired lower respiratory infections. These include *Streptococcus pneumonia* and *Haemophilus influenzae*, both of which are always present in the upper respiratory tract of healthy individuals *Moraxella* (formerly *Branhamella*). *Moraxella catarrhalis* is only normally present in children, frequently causing ear infections, and can be recognized as a contributor in adult pneumonia (Ramirez and Anzueto 2011). *Mycobacterium tuberculosis* is estimated to currently reside in about a third of the world's population most often

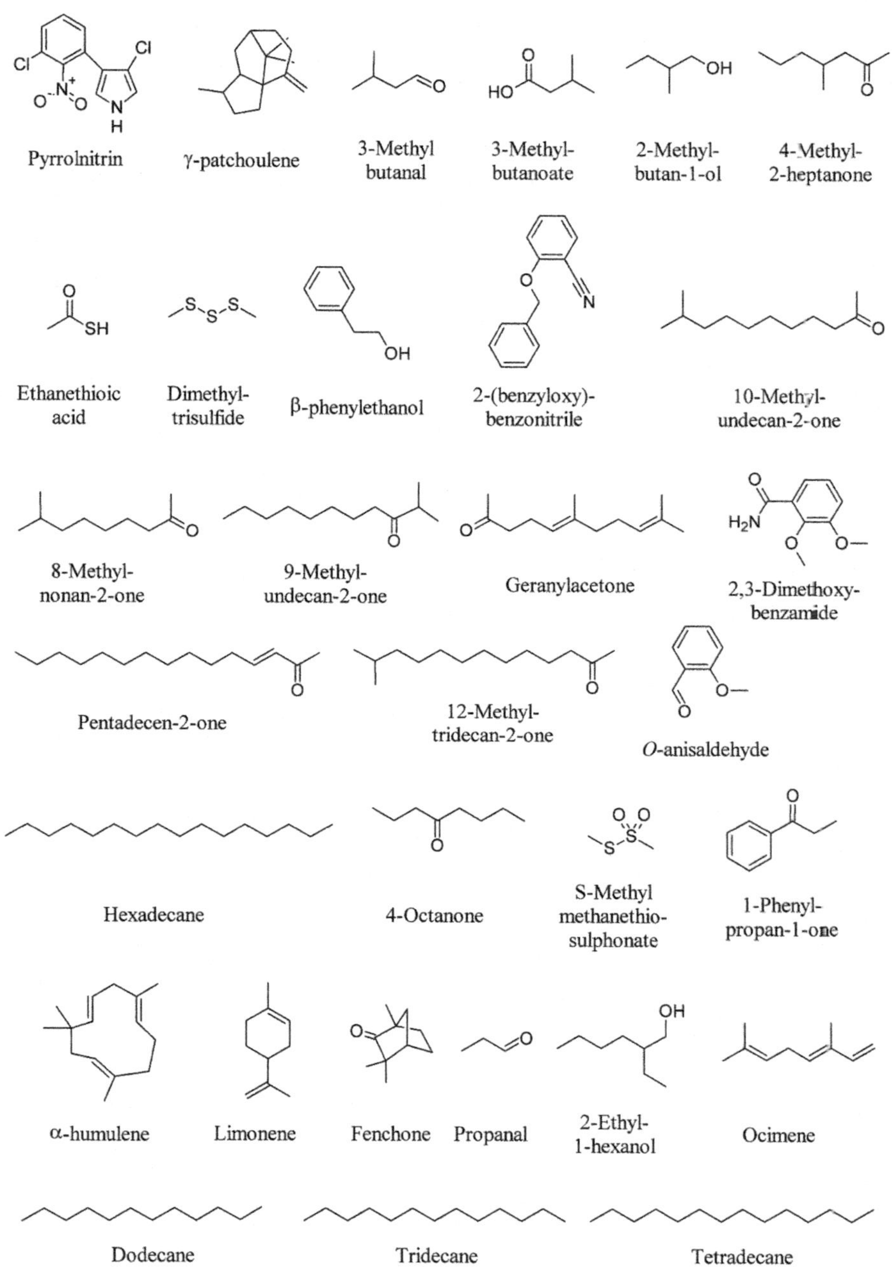

Fig. 8.9 Example of some bacterial volatile that interfere with fungi (Adapted from Kanchiswamy et al. 2015)

resulting in a lifelong asymptomatic infection. This becomes of massive importance for the diagnosis of respiratory disease when these bacteria instead grow to high levels in the extracellular spaces or cavities in the lung that are caused by a delayed-type hypersensitivity (DTH) immune response. Reported cases for some species which cause community-acquired infections such as *Burkholderia cepacia* and *Pseudomonas aeruginosa* are of high interest in terms of the settlement in the lung causing cystic fibrosis disease which is considered the most common genetically inherited disease in most countries. *Streptococcus pneumoniae* is the most common and important cause of significant lower respiratory tract diseases among healthy individuals. A recent meta-analysis of patient data points out that it is most likely that a recent unusual drop in the reported frequency as the causative agent of community-acquired pneumonia (CAP) is due to a dramatic defeat for identification of causative agents causing respiratory disease (Bartlett 2011). It appears that none have yet analyzed the potential contributions of this diverse microbe to human breath (Hiller et al. 2007). The previous studies showed that successful patient breath testing for bacterial respiratory infections to date come from the efforts in the diagnosis of tuberculosis. Diagnosis of *M. tuberculosis* by breath testing is of great specific interest as it is only in active disease when bacteria grow in extracellular spaces within the lung that tuberculosis patients are infectious (Grosset 2003). If these cases can be efficiently recognized in this way, new cases could be dramatically reduced in the remaining high prevalence regions. To our knowledge, bacterial volatiles and VOCs are limited (Kai et al. 2009). Cultures of *Pseudomonas aeruginosa* often have a distinctive "grape-like" odor due to secreting 2-amino-acetophenone which has demonstrated good potential as a volatile breath marker of respiratory infection (Scott-Thomas et al. 2010). It is unlikely that breathe volatile makers originating from bacteria and other causative agents causing respiratory infection will become mainstream, and reliable indicators of disease until their metabolic origins and chemical structures are fully defined using isotopes and laboratory cultures. We also may not find those of actual utility unless we also look simultaneously at microbes growing in their normal ecological niches. For the bacteria of high relevance to respiratory infection, these are of course the host cells and tissues in which they normally reside.

8.11.2 Fungal Volatiles in Respiratory Diagnosis (Lung Infections)

Aspergillus fumigatus causes lung infection through inhaling the spores. It causes allergic diseases and severe health diseases especially in immunocompromised patients. Ventilator-associated pneumonia (VAP) is one of the important concerns in a hospital which can develop in more than 48 h after endotracheal incubation procedure (Chastre and Fagon 2002). It is reasonable that the route of the infection and also the rate of morbidity and mortality are not constant and are always changeable among hospitals and patients. It is reasonable that the route of the infection may be complicated due to the VAP and increase the risk of death in critically ill patients (Rello et al. 2002). *A. fumigatus* and *Candida* sp. may cause invasive infections especially in immunocompromised patients. All microbes release characteristic volatiles because of their metabolic interactions (Wady et al. 2003; Garner et al. 2007). Volatile markers can functionally be used to detect and discriminate different pathogenic species being exposed or unexposed to different concentrations of antifungal treatments potential for using the qualitative volatile markers excreted by dermatophytic fungi to distinguish among antifungal treatments when using antifungal and alternative antioxidants (Naraghi et al. 2010). This clarifies that volatile marker can be used in estimating the performance of antifungal compounds. The production patterns of volatile can be used to differentiate human pathogenic bacteria (Turner and Magan 2004). To our knowledge, till now, there are not enough studies to detect if volatile-producing microbes can be used to detect antimicrobials or novel compounds which could be used as a substitute treatment (Naraghi et al. 2010), while it is possible to distinguish between effective and noneffective concentrations of volatile compound from *A. fumigatus* and their antifungal activity. Volatile compounds with fingerprints appear to be an auspicious indicator to distinguish among treatments. No more studies have been discussed which biomarkers can be found in fungal cultures when treated with antifungal volatile compounds. From previous studies, the metabolism of *A. fumigatus* in low-oxygen conditions in the tissue of the lung can be studied. They could detect the production of ethanol when glucose was the sole carbon source indicating that the cultures of *A. fumigatus* fungus which is an obligate aerobic has the ability to switch to alcohol fermentation pathways in the absence of oxygen. Collectively, *A. fumigatus* can use different fermentation pathways to grow in hypoxic environment (Willger et al. 2008). Selected ion flow-tube mass spectrometry (SIFT-MS) techniques can be applied as biomarkers to get more detailed information about the volatile compounds released from each microbe in normal conditions and under different stress conditions during drug treatments. This could have an important role in distinguishing different clinical samples by using these data sets to establish models with breath samples from different patients that should be taken into consideration. To sum up, volatile fingerprints secreted by human pathogenic disease could be used as a potential tool to determine whether infections in the lung are caused by bacteria or certain fungi; this will help us in correct diagnosis of disease and consequently prescription of the right drug.

8.11.3 The Role of Gut-Derived Volatile Organic Compounds in the Pathogenesis of Nonalcoholic Steatohepatitis

Microbial volatile organic compounds are secreted in the gut during microbial metabolism and may get inside the portal circulation and cause toxicity of the liver which can be one of the reasons for the development of nonalcoholic steatohepatitis. Recently, nonalcoholic fatty liver disease (NAFLD) is the most prevalent malady causing severe problems of the liver especially in North America. Also, inflammation and fibrosis are signs of the occurrence of the disease called nonalcoholic steatohepatitis (NASH), which may be on the long run the reason for liver cirrhosis and hepatocellular carcinoma, which is the most common type of liver cancer (Duwaerts and Maher 2014). Fatty liver disease can evolve into an inflammatory circumstance and then nonalcoholic steatohepatitis. This process is so complicated and poorly understood. However, some studies hypothesized that genetic factors may be the cause of NASH. Moreover, other cellular factors such as the formation of endoplasmic reticulum stress and reactive oxygen species (ROS) may help in the advancement of NASH (Tilg and Moschen 2010). The imbalance in the proportion of normal flora in the gut referred as dysbiosis may be the reason for the development of nonalcoholic fatty liver disease (NAFLD) and further enhancing nonalcoholic steatohepatitis (NASH). Previous studies confirmed that fecal VOC profiles can be used to discriminate among cases with nonalcoholic fatty liver disease NAFLD from healthy ones (Henao-Mejia et al. 2012; Raman et al. 2013). VOCs such as toxic molecules and ethanol are absorbed from the intestine; this may have offensive influence on the liver cells. On the other hand, the emission and uptake of VOCs from microbes in the gut may help in the activation of numerous liver enzymes and affirm the importance of VOC analysis for evaluating the enzyme functions (Mochalski et al. 2013). Also, analyses of the VOCs from the portal vein in a mouse model of nonalcoholic steatohepatitis (NASH) represent a novel agent for the identification of the roles of VOCs in liver disease and subsequently focus on the influence of metabolites on liver functions during disease outgrowth.

8.12 Conclusion

Microorganisms are one of the most valuable creatures on our planet and play indispensable roles in all daily life activities. They share with us the same community, but their own lifestyle may be more sophisticated. Thus, the behavior studies of these intelligent organisms are interesting and come back with a lot of benefit for the humanity. One of the most interesting microbiological aspects is the microbial volatile organic compounds. MVOCs are fascinatingly dynamic and complexed metabolites produced by different microbes in response to different stimuli, to fulfill different environmental functions. MVOCs are released during both the primary and secondary metabolisms. The understanding of biological, ecological, and evolutionary theories of MVOCs is the main key to maintain the environmental sustainability. While the microbial volatiles exemplify a new frontier in bio-process and in bio-prospecting, within this review, we described the possible potential roles of MVOCs in defense and as unique cellular signals. Also, the MVOCs can be considered as suitable tools for better productivity and protection of plants, to ensure the ecological sustainability. In plant defense, MVOCs play a vital role in managing plant disease with controlling either microbial or insect pathogens as an alternative way to substitute chemical pesticides which have a remarkable side effect on health and on environment. Thus, MVOCs can be considered as a promising tool for biological control and for disease management. Besides, the fungal VOCs are the key factor in food and flavor industry. On the other hand, bacteria release highly diverse organic and inorganic volatile metabolites, which can be used in different bio-processes. However, more effort for the investigation of MVOCs is needed, and this will assure the discovery of novel natural products for human exploitation and will highlight on new theories in fundamental biology.

Acknowledgment Sincere thankful for Katerina Peros for her continuous support. Additionally R. Selim would like to acknowledge her kids for their backing.

References

Ahamed A, Ahring BK (2011) Production of hydrocarbon compounds by endophytic fungi *Gliocladium species* grown on cellulose. Bioresour Technol 102(20):9718–9722

Aochi YO, Farmer WJ (2005) Impact of soil microstructure on the molecular transport dynamics of 1, 2-dichloroethane. J Geod 127:137–153

Arnold AE, Maynard Z, Gilbert G (2001) Fungal endophytes in dicotyledonous neotropical trees: patterns of abundance and diversity. Mycol Res 105:1502–1507

Atkinson R, Tuazon EC, Aschmann SM (2000) Atmospheric chemistry of 2-pentanone and 2-heptanone. Environ Sci Technol 34(4):623–631

Audrain B, Farag MA, Ryu CM, Ghigo JM (2015) Role of bacterial volatile compounds in bacterial biology. FEMS Microbiol Rev 39:222–233

Bacon CW, White JW (2000) Microbial endophytes. Marcel Dekker, New York

Bailly A, Weisskopf L (2012) The modulating effect of bacterial volatiles on plant growth current knowledge and future challenges. Plant Signal Behav 7:79–85

Banerjee D, Strobel G, Geary B, Sears J, Ezra D, Liarzi O, Coombs J (2010) Muscodor albus strain GBA, an endophytic fungus of Ginkgo biloba from United States of America, produces volatile antimicrobials. Mycology 1:179

Barr JG (1976) Effects of volatile bacterial metabolites on the growth, sporulation and mycotoxin production of fungi. J Sci Food Agric 27(4):324–330

Bartlett JG (2011) Diagnostic tests for agents of community-acquired *pneumonia*. Clin Infect Dis J 52:S296–S304

Batterman SA (1995) Sampling and analysis of biological volatile organic compounds. In: Burge HA (ed) Bioaerosols. CRC Press, Boca Raton, pp 249–268

Bayer CW, Crow S (1994) Odorous volatile emissions from fungal contamination. In: Teichman KY (ed) Proceedings from IAQ'93: operating and maintaining buildings for health, comfort and productivity. American Society of Heating, Refrigerating and Air-Conditioning Engineers, Atlanta, pp 165–170

Bentley R, Bennett JW (1988) Biosynthesis of secondary metabolites. In: Berry DR (ed) Physiology of industrial fungi. Blackwell Scientific Publications, Oxford, pp 161–183

Berglund B, Johansson I (1996) Health effects of volatile organic compounds in indoor air. Archives of the Center for Sensory Research 3:67

Bernier SP, Letoffe S, Delepierre M (2011) Biogenic ammonia modifies antibiotic resistance at a distance in physically separated bacteria. Mol Microbiol 81:705–716

Blom D, Fabbri C, Connor EC (2011) Production of plant growth modulating volatiles is widespread among rhizosphere bacteria and strongly depends on culture conditions. Environ Microbiol 13:3047–3058

Börjesson T, Stöllman U, Adamek P, Kaspersson A (1989) Analysis of volatile compounds for detection of molds in stored cereals. Cereal Chem 66(4):300–304

Börjesson T, Stöllman U, Schnürer J (1992) Volatile metabolites produced by six fungal species compared with other indicators of fungal growth on cereal grains. Appl Environ Microbiol 58:2599–2605

Bornehag CG, Blomquist G, Gyntelberg F, Järvholm B, Malmberg P, Nordvall L, Nielsen A, Pershagen G, Sundell J (2001) Dampness in buildings and health. Indoor Air 11:72–86

Bornehag CG, Sundell J, Bonini S, Custovic A, Malmberg P, Skerfving S, Sigsgaard T, Verhoeff A (2004) Dampness in buildings as a risk factor for health effects, EUROEXPO: a multidisciplinary review of the literature (1998–2000) on dampness and mite exposure in buildings and health effects. Indoor Air 14:243–257

Bos LDJ, Sterk PJ, Schultz MJ (2013) Volatile metabolites of pathogens: a systematic review. PLoS Pathog J 9(5):e1003311

Brodhagen M, Tsitsigiannis DI, Hornung E, Goebel C, Feussner I, Keller NP (2008) Reciprocal oxylipin-mediated cross-talk in the *Aspergillus* – seed pathosystem. Mol Microbiol 67:378–391

Brown SK (1999) Occurrence of volatile organic compounds in indoor air. In: Salthammer T (ed) Organic indoor air pollutants – occurrence, measurement, evaluation, pp 171–184

Champe SP, Rao P, Chang A (1987) An endogenous inducer of sexual development in *Aspergillus nidulans*. J Gen Microbiol 133:1383–1388

Chastre J, Fagon J (2002) Ventilator-associated pneumonia. Am J Respir Crit Care Med 165:867–903

Chiron N, Michelot D (2005) Odeurs de champignons: chimie et rôle dans les interactions biotiquesdune revue. Cryptogam Mycol 26:299–364

Cho IH, Namgung HJ, Choi HK, Kim YS (2008a) Volatiles and key odorants in the *pileus* and stipe of pine-mushroom (*Tricholoma matsutake Sing*). Food Chem 106:71–76

Cho SM, Kang BR, Han SH, Anderson AJ, Park JY, Lee YH, Cho BH, Yang KY, Ryu CM, Kim YC (2008b) 2R, 3R-butanediol, a bacterial volatile produced by *Pseudomonas chlororaphis* O6, is involved in induction of systemic tolerance to drought in *Arabidopsis thaliana*. Mol Plant Microbe Interact J 21:1067–1075

Christensen S, Kolomiets M (2011) The lipid language of plant-fungal interactions. Fungal Genet Biol 48:4–14

Claeson AS (2006) Volatile organic compounds from microorganisms identification and health effects. In: Ume°a University medical dissertations, New Series No. 1052. The National Institute for Working Life, Sweden, pp 1–53

Claeson AS, Sunesson AL (2005) Identification using versatile sampling and analytical methods of volatile compounds from *Streptomyces albidoflavus* grown on four humid building materials and one synthetic medium. Indoor Air 15(9):41–47

Claeson AS, Levin JO, Blomquist G, Sunesson AL (2002a) Volatile metabolites from microorganisms grown on building materials. Proc Indoor Air 1:437–442

Claeson AS, Levin JO, Blomquist G, Sunesson AL (2002b) Volatile metabolites from microorganisms grown on humid building materials and synthetic media. J Environ Monit 4:667–672

Contreras-Cornejo HA, Macías-Rodríguez L, Cortés-Penagos C, López-Bucio J (2014) The 4-phosphopantetheinyl transferase of *Trichoderma virens* plays a role in plant protection against *Botrytis cinerea* through volatile organic compound emission. Plant Soil J 379:261–274

Cortes-Barco A, Goodwin P, Hsiang T (2010a) Comparison of induced resistance activated by benzothiadiazole,(2R,3R)-butanediol and anisoparaffin mixture against anthracnose of *Nicotiana benthamiana*. Plant Pathol 59:643–653

Cortes-Barco A, Hsiang T, Goodwin P (2010b) Induced systemic resistance against three foliar diseases of *Agrostis stolonifera* by (2R,3R)-butanediol or anisoparaffin mixture. Ann Appl Biol 157:179–189

Costelloe C, Metcalfe C, Lovering A, Mant D, Hay AD (2010) Effect of antibiotic prescribing in primary care on antimicrobial resistance in individual patients: systematic review and meta-analysis. BMJ 18:340–c2096

Dainty RH, Edwards RA, Hibbard CM (1984) Volatile compounds associated with the aerobic growth of some *Pseudomonas* species on beef. J Appl Bacteriol 57(1):75–81

Dainty RH, Edwards RA, Hibbard CM, Marnewick JJ (1989) Volatile compounds associated with microbial growth on normal and high pH beef stored at chill temperatures. J Appl Bacteriol 66(4):281–289

Demyttenaere JC, Morina RM, Sandra P (2003) Monitoring and fast detection of mycotoxin-producing fungi based on headspace solid-phase microextraction and headspace sorptive extraction of the volatile metabolites. J Chromatogr A 985(1–2):127–135

Demyttenaere JC, Morina RM, DeKimpe N, Sandra P (2004) Use of headspace solid-phase microextraction and headspace sorptive extraction for the detection of the volatile metabolites produced by toxigenic *Fusarium* species. J Chromatogr A 1027(1–2):147–154

Duwaerts CC, Maher JJ (2014) Mechanisms of liver injury in non-alcoholic steatohepatitis. Curr Hepatol Rep 13(2):119–129

Effmert U, Kalderas J, Warnke R, Piechulla B (2012) Volatile mediated interactions between bacteria and fungi in the soil. J Chem Ecol 38(6):665–703

Ezquer I, Li J, Ovecka M, Baroja-Fernández E, Muñoz FJ, Montero M, Cerio JD, Hidalgo M, Sesma MT, Bahaji A, Etxeberria Ed, Pozueta-Romero J (2010) Microbial volatile emissions promote accumulation of exceptionally high levels of starch in leaves in mono- and dicotyledonous plants. Plant Cell Physiol J 51:1674–1693

Farag MA, Zhang H, Ryu CM (2013) Dynamic chemical communication between plants and bacteria through airborne signals: induced resistance by bacterial volatiles. J Chem Ecol 39(7):1007–1018

Fiddaman DJ, Rossall S (1993) The production of antifungal volatiles by Bacillus subtilis. J Appl Bacteriol 74:119–126

Fiedler K, Schutz E, Geh S (2001) Detection of microbial volatile organic compounds (MVOCs) produced by moulds on various materials. Int J Hyg Environ Health 204:111–121

Fiedler N, Laumbach R, Kelly-Mcneil K, Lioy P, Fan ZH, Zhang J, Ottenweller J, Ohman-Strickland P, Kipen H (2005) Health effects of a mixture of indoor air volatile organics, their ozone oxidation products, and stress. Environ Health Perspect 113:1542–1548

Fischer G, Müller T, Schwalbe R, Ostrowski R, Dott W (2000) Exposure to airborne fungi, MVOC and mycotoxins in bio -waste-handling facilities. Int J Hyg Environ Health 203:97–104

Fraatz MA, Zorn H (2010) Fungal flavours. In: Hofrichter M (ed) The mycota X: industrial applications, 2nd edn. Springer-Verlag, Berlin, pp 249–264

Garbeva P, Voesenek K, van Elsas JD (2004) Quantitative detection and diversity of the pyrrolnitrin biosynthetic locus in soil under different treatments. Soil Biol Biochem 36:1453–1463

Garbeva P, Hordijk C, Gerards S, deBoer W (2014a) Volatile-mediated volatile-mediated interactions between phylogenetically different soil bacteria. Front Microbiol 5:289

Garbeva P, Hordijk C, Gerards S, de Boer W (2014b) Volatiles produced by the mycophagous soil bacterium Collimonas. FEMS Microbiol Ecol 87:639–649

Garner CE, Smith S, Costello B, White P, Spencer R, Probert CSJ, Ratcliffe NM (2007) Volatile organic compounds from feces and their potential for diagnosis of gastrointestinal disease. FASEB J 21:1675–1688

Glare T, Caradus J, Gelernter W, Jackson T, Keyhani N, Kohl J (2012) Have biopesticides come of age? Trends Biotechnol 30:250–258

Gorny RL, Reponen T, Willeke K, Schmechel D, Robine E, Boissier M, Grinshpun SA (2002) Fungal fragments as indoor air biocontaminants. Appl Environ Microbiol 68(7):3522–3531

Grosset J (2003) *Mycobacterium tuberculosis* in the extracellular compartment: an underestimated adversary. Antimicrob Agents Chemother J 47:833–836

Hamberg M, Su C, Oliw E (1998) Manganese lipoxygenase. Discovery of a bisallylic hydroperoxide as product and intermediate in a lipoxygenase reaction. J Biol Chem 273:13080–13088

Hamilton-Kemp T, Newman M, Collins R, Elgaali H, Keshun Yu, Archbold D (2005) Production of the long-chain alcohols octanol, decanol and dodecanol by *Escherichia coli*. Curr Microbiol 51(2):82–86

Han H, Lee S, Moon J, Park K, Yang K, Kim B, Kim Y, Lee M, Anderson A, Kim Y (2006) GacS-dependent production of 2R,3R-butanediol by Pseudomonas chlororaphis O6 is a major determinant for eliciting systemic resistance against Erwinia carotovora but not against Pseudomonas syringae pv. tabaci in tobacco. Mol Plant Microbe Interact J 19:924–930

Henao-Mejia J, Elinav E, Jin C (2012) Inflammasome-mediated dysbiosis regulates progression of NAFLD and obesity. Nature 482(7384):179–185

Hiller NL, Janto B, Hogg JS, Boissy R, Yu S, Owell (2007) Comparative genomic analyses of seventeen *Streptococcus pneumoniae* strains: insights into the pneumococcal supra genome. Bacteriol J 189:8186–8195

Hörnsten L, Su C, Osbourn AE, Garosi P, Hellman U, Wernstedt C, Oliw EH (1999) Cloning of linoleate diol synthase reveals homology with prostaglandin H synthases. J Biol Chem 274:28219–28224

Howe LR (2007) Inflammation and breast cancer: cyclooxygenase/prostaglandin signaling and breast cancer. Breast Cancer Res 9:210

Hung R, Lee S, Bennett JW (2013) *Arabidopsis thaliana* as a model system for testing the effect of *Trichoderma* volatile organic compounds. Fungal Ecol J 6:19–26

Hyvärinen A, Meklin T, Vepsäläinen A, Nevalainen A (2002) Fungi and actinobacteria in moisture-damaged building materials – concentrations and diversity. Int Biodeteroration Biodegrad 49:27–37

Insam H, Seewald MS (2010) Volatile organic compounds (VOCs) in soils. Biol Fertil Soils 46:199–213

Jelen H, Wasowicz E (1998) Volatile fungal metabolites and their relation to the spoilage of agricultural commodities. Food Rev Int 14(4):391–426

Johansson N, Kalin M, Tiveljung-Lindell A, Giske CG, Hedlund J (2009) Etiology of community-acquired *pneumonia:* increased microbiological yield with new diagnostic methods. Clin Infect Dis J 50:202–209

Jones AP (1999) Indoor air quality and health. Atmos Environ 33:4535–4564

Juenger M, Vautz W, Kuhns M, Hofmann L, Ulbricht S, Baumbach JI (2012) Ion mobility spectrometry for microbial volatile organic compounds: a new identification tool for human pathogenic bacteria. Appl Microbiol Biotechnol 93:2603–2614

Junker RR, Tholl D (2013) Volatile organic compound mediated interactions at the plant-microbe interface. J Chem Ecol 39:810–825

Kachroo A, Kachroo P (2009) Fatty acid-derived signals in plant defense. Ann Rev Phytopathol 47:153–176

Kai M, Haustein M, Molina F, Petri A, Scholz B, Piechulla B (2009) Bacterial volatiles and their action potential. Appl Microbiol Biotechnol J 81:1001–1012

Kai M, Crespo E, Cristescu SM, Harren FJ, Francke W, Piechulla B (2010) Serratia odorifera: analysis of volatile emission and biological impact of volatile compounds on *Arabidopsis thaliana*. Appl Microbiol Biotechnol 88:965–976. doi:10.1007/s00253-010-2810-1

Kanchiswamy CN, Malnoy M, Maffei ME (2015) Bioprospecting bacterial and fungal volatiles for sustainable agriculture. Trends Plant Sci 20(4):206–211

Karlshøj K, Larsen TO (2005) Differentiation of species from the *Penicillium roqueforti* group by volatile metabolite profiling. J Agric Food Chem 53(3):708–715

Kishimoto K, Matsui K, Ozawa R, Takabayashi J (2006b) Analysis of defensive responses activated by volatile alloocimene treatment in *Arabidopsis thaliana*. Phytochem J 67:1520–1529

Kishimoto K, Matsui K, Ozawa R, Takabayashi J (2006a) Components of C6-aldehyde-induced resistance in Arabidopsis thaliana against a necrotrophic fungal pathogen, Botrytis Cinerea. Plant Sci J 170:715–723

Kishimoto K, Matsui K, Ozawa R, Takabayashi J (2007) Volatile 1-octen-3-ol induces a defensive response in *Arabidopsis thaliana*. Gen Plant Pathol J 73:35–37

Korpi A (2001) Faculty of natural and environmental sciences C129. Fungal volatile metabolites and biological responses to fungal exposure [thesis]. Kuopio University, Finland, pp 1–77

Korpi A, Jarnberg J, Pasanen AL (2009) Microbial volatile organic compounds. Crit Rev Toxicol 39:139–193

Kwon YS, Ryu CM, Lee S, Park HB, Han KS, Lee JH, Lee K, Chung WS, Jeong MJ, Kim HK, Bae DW (2010) Proteome analysis of *Arabidopsis* seedlings exposed to *bacterial* volatiles. Planta J 232:1355–1370

La Camera S, Gouzerh G, Dhondt S, Hoffmann L, Fritig B, Legrand M, Heitz T (2004) Metabolic reprogramming in plant innate immunity: the contributions of phenylpropanoid and oxylipin pathways. Immunol Rev 198:267–284

Ladygina N, Dedyukhina EG, Vainshtein MB (2006) A review on microbial synthesis of hydrocarbons. Process Biochem 41(5):1001–1014

Lee JH, Lee J (2010) Indole as an intercellular signal in microbial communities. FEMS Microbiol Rev 34:426–444

Lee SO, Kim HY, Choi GJ, Lee HB, Jang KS, Choi YH, Kim JC (2009) Mycofumigation with *Oxyporus latemarginatus* EF069 for control of post-harvest apple decay and *Rhizoctonia* root rot on moth orchid. Appl Microbiol J 106:1213–1219

Lemfack MC, Nickel J, Dunkel M, Preissner R, Piechulla B (2014) MVOC: a database of microbial volatiles. Nucleic Acids Res 42:D744–D748

Létoffé S, Audrain B, Bernier SP, Delepierre M, Ghigo JM (2014) Aerial exposure to the bacterial volatile compound trimethylamine modifies antibiotic resistance of physically separated bacteria by raising culture medium pH. MBio 5(1):e00944–e00913

Liang H, Zhang X, Jun RAO, Huanwen CHEN (2008) Microbial volatile organic compounds: generation pathway sand mass spectrometric detection. J Chin Biotechnol 28:124–133

Lorenz MC, Fink GR (2002) Life and death in a macrophage: role of the glyoxylate cycle in virulence. Eukaryot Cell 1(5):657–662

Lutz MP, Wenger S, Maurhofer M, Defago G, Duffy B (2004) Signaling between bacterial and fungal biocontrol agents in a strain mixture. FEMS Microbiol Ecol J 48:447–455

Ma Q, Fonseca A, Liu W, Fields AT, Pimsler ML, Spindola AF, Tarone AM, Crippen TL, Tomberlin JK, Wood TK (2012) *Proteus mirabilis* inter kingdom swarming signals attract blow flies. ISME J 6:1356–1366

Macias-Rubalcava ML, Hernandez-Bautista BE, Oropeza F, Duarte G, Gonzalez MC, Glenn AE, Hanlin RT, Anaya AL (2010) Allelochemical effects of volatile compounds and organic extracts from *Muscodor yucatanensis*, a tropical endophytic fungus from *Bursera simaruba*. Chem Ecol J 36:1122–1131

Maffei ME, Gertsch J, Appendino G (2011) Plant volatiles: production, function and pharmacology. Nat Prod Rep 28:1359–1380

Marilley L, Casey MG (2004) Flavors of cheese products: metabolic pathways, analytical tools and identification of producing strains. Int J Food Microbiol 90:139–159

Mattila JT, Thomas AC (2014) Nitric oxide synthase: non-canonical expression patterns. Front Immunol 5:478

Mercier J, Jimenez JI (2004) Control of fungal decay of apples and peaches by the biofumigant fungus *Muscodor albus*. Postharvest Biol Technol J 31:1–8

Mercier J, Jimenez JI (2007) Potential of the volatile-producing fungus *Muscodor albus* for control of building molds. Can J Microbiol 53:404–410

Mercier J, Manker D (2005) Biocontrol of soil-borne diseases and plant growth enhancement in green house soilless mix by the volatile-producing fungus *Muscodor albus*. Crop Prot J 24(4):355–362

Meruva NK, Penn JM, Farthing DE (2004) Rapid identification of microbial VOCs from tobacco molds using closed-loop stripping and gas chromatography/time-of-flight mass spectrometry. J Ind Microbiol Biotechnol 31(10):482–488

Minerdi D, Bossi S, Gullino ML, Garibaldi A (2009) Volatile organic compounds: a potential direct long distance mechanism for antagonistic action of *Fusarium oxysporum* strain MSA 35. Environ Microbiol 11:844–854

Minerdi D, Bossi S, Maffei ME, Gullino ML, Garibaldi A (2011) *Fusarium oxysporum* and its bacterial consortium promote *lettuce* growth and expansion A5 gene expression through microbial volatile organic compound (MVOC) emission. FEMS Microbiol Ecol J 76:342–351

Mitchell AM, Strobel GA, Moore E, Robison R, Sears J (2010) Volatile antimicrobials from *Muscodor crispans*, a novel endophytic fungus. Microbiol J 156:270–277

Mochalski P, King J, Klieber M, Unterkofler K, Hinterhuber H, Baumann M, Amann A (2013) Blood and breath levels of selected volatile organic compounds in healthy volunteers. Analyst 138:2134–2145

Mølhave L, Clausen G, Berglund B, Ceaurriz JD, Kettrup A, Lindvall T, Maroni M, Pickering AC, Risse U, Rothweiler H, Seifert B, Younes M (1997) Total volatile organic compounds (TVOC) in indoor air quality investigations. Indoor Air 7:225–240

Morey P, Wortham A, Weber A, Horner E, Black M, Muller W (1997) Microbial VOCs in moisture damaged buildings. Healthy Build 1:245–250

Muckschel B, Simon O, Klebensberger J, Graf N, Rosche B, Altenbuchner J, Pfannstiel J, Huber A, Hauer B (2012) Ethylene glycol metabolism by *Pseudomonas putida*. Appl Environ Microbiol 78(24):8531–8539

Naraghi K, Sahgal N, Adriaans B, Barr H, Magan N (2010) Use of volatile fingerprints for rapid screening of antifungal agents for efficacy against dermatophyte *Trichophyton species*. Sens Actuat B Chem J 146:521–526

Nawrath T, Mgode GF, Weetjens B, Kaufmann SH, Schulz S (2012) The volatiles of pathogenic and nonpathogenic mycobacteria and related bacteria. Beilstein J Org Chem 8:290–299

Nevalainen A, Seuri M (2005) Of microbes and men. Indoor Air 15:58–64

Nilsson A, Kihlström E, Lagesson HV, Wessén B, Szponar B, Larsson L, Tagesson C (2004) Microorganisms and volatile organic compounds in airborne dust from damp residences. Indoor Air 14:73–144

Noverr MC, Erb-Downward JR, Huffnagle GB (2003) Production of eicosanoids and other oxylipins by pathogenic eukaryotic microbes. Clin Microbiol Rev 16:517–533

Ortiz-Castro R, Contreras-Cornejo H, Macias-Rodriguez L, Lopez-Bucio J (2009) The role of microbial signals in plant growth and development. Plant Signal Behav 4:701–712

Pagans E, Font X, Sanchez A (2006) Emission of volatile organic compounds from composting of different solid wastes: abatement by bio filtration. J Hazard Mater 131:179–186

Park MS, Ahn J, Choi GJ, Choi YH, Jang KS, Kim JC (2010) Potential of the volatile-producing fungus nodulisporium sp cf016 for the control of postharvest diseases of apple. Plant Pathol J 26:253–259. doi:10.5423/PPJ.2010.26.3.253

Park HB, Lee B, Kloepper JW, Ryu CM (2013) One shot-two pathogens blocked: exposure of Arabidopsis to hexadecane, a long chain volatile organic compound, confers induced resistance against both *Pectobacterium carotovorum* and *Pseudomonas syringae*. Plant Signal Behav J 8:24619

Pasanen AL, Heinonen-Tanski H, Kalliokoski P, Jantunen MJ (1992) Fungal microcolonies on indoor surfaces an explanation for the base-level fungal spore counts in indoor air. Atmos Environ 26B(1):121–124

Peñuelas J, Asensio D, Holl D, Wenke K, Rosenkranz M, Piechulla B, Schnitzler JP (2014) Biogenic volatile emissions from the soil. Plant Cell Environ J 37:1866–1891

Perl T, Juenger M, Vautz W, Nolte J, Kuhns M, Zepelin MB (2011) Detection of characteristic metabolites of *Aspergillus fumigatus* and *Candida species* using ion mobility spectrometry-metabolic profiling by volatile organic compounds. Mycoses 54:E828–E837

Petrini O (1991) Fungal endophytes of tree leaves. In: Andrews J, Hirano SS (eds) Microbial ecology of leaves. Springer, New York, pp 179–197

Phillips M, Basa-Dalay V, Blais J, Bothamley G, Chaturvedi A, Modi KD (2012) Point-of-care breath test for biomarkers of active pulmonary tuberculosis. Tuberculosis (Edinburgh) 92:314–320

Piechulla P, Degenhardt J (2014) The emerging importance of microbial volatile organic compounds. Plant Cell Environ 37:811–812

Que YA, Hazan R, Strobel BA, Maura D, He J, Kesarwani M, Panopoulos P, Tsurumi A, Giddey M, Wilhelmy J, Mindrinos MN, Rahme LG (2013) Quorum sensing small volatile molecule promotes antibiotic tolerance in bacteria. PLoS One 8(12):e80140

Raman I, Ahmed, Gillevet PM (2013) Fecal microbiome and volatile organic compound metabolome in obese humans with nonalcoholic fatty liver disease. Clin Gastroenterol Hepatol J 11(7):868–875. e1–3

Ramirez JA, Anzueto AR (2011) Changing needs of community-acquired *pneumonia*. Antimicrob Chemother J 66:3–9

Redlich CA, Sparer J, Cullen MR (1997) Sick-building syndrome. Lancet 349:1013–1016

Rello J, Ollendorf DA, Oster G, Vera-Llonch M, Bellm L, Redman R, Kollef MH (2002) Epidemiology and outcomes of ventilator-associated *pneumonia* in a large US database. Chest J 122(6):2115–2121

Riga E, Lacey LA, Guerra N (2008) Muscodor albus, a potential biocontrol agent against plant-parasitic nematodes of economically important vegetable crops in Washington State, USA. Biol Control., Orlando 45:380–385

Rodriguez RJ, White JF, Arnold AE, Redman RS (2009) Fungal endophytes: diversity and functional roles. New Phytol 182:314–330

Romoli R, Papaleo MC, De Pascale D, Tutino ML, Michaud L, LoGiudice (2014) GC-MS volatolomic approach to study the antimicrobial activity of the antarctic bacterium *Pseudoalteromonas sp* TB41. Metabolomics 10:42–51

Ryu CM, Farag MA, Hu CH, Reddy MS, Wei HX, Paré PW, Kloepper JW (2003) Bacterial volatiles promote growth in *Arabidopsis*. P Natl Acad Sci 100(8):4927–4932

Sakuradani E, Ando A, Ogawa J, Shimizu S (2009) Improved production of various polyunsaturated fatty acids through filamentous fungus *Mortierella alpina* breeding. Appl Microbiol Biotechnol 84:1–10

Samson RA (1985) Occurrence of moulds in modern living and working environments. Eur J Epidemiol 1:54–61

Sauer F, Schäfer C, Neeb P, Horie O, Moortgat GK (1999) Formation of hydrogen peroxide in the ozonolysis of isoprene and simple alkenes under humid conditions. Atmos Environ 33(2):229–241

Schleibinger H, Brattig C, Mangler M, Samwer H, Laußmann D, Eis D, Braun P, Marchl D, Nickelmann A, Rueden H (2002) Microbial volatile organic compounds (MVOC) as indicators for fungal damage. Proc. Indoor Air 4:707–712

Schleibinger H, Brattig C, Mangler M, Laußmann D, Eis D, Braun P, Marchl D, Nickelmann A, Rueden H (2003) Are microbial volatile organic compounds (MVOC) useful predictors for hidden mould damage? In: Proceedings of healthy buildings, 2003, pp 706–710. ISIAQ, Singapore

Schmidt R, Etalo DW, de Jager VGS, Zweers H, de Boer W, Garbeva P (2016) Microbial small talk: volatiles in fungal–bacterial interactions. Front Microbiol 6:1495

Schulz S, Dickschat JS (2007) Bacterial volatiles: the smell of small organisms. Nat Prod Rep 24:814–842

Scott-Thomas AJ, Syhre S, Pattemore PK, Epton M, Laing R, Pearson J, Chambers ST (2010) 2-Aminoacetophenone as a potential breath biomarker for Pseudomonas Aeruginosa in the cystic fibrosis lung. BMC Pulm Med 10:56–10

Shea JM, Del Poeta M (2006) Lipid signaling in pathogenic fungi. Curr Opin Microbiol 9:352–358

Shim WB, Dunkle LD (2002) Identification of genes expressed during cercosporin biosynthesis in *Cercospora zeae-maydis*. Physiol Mol Plant Pathol 61:237–248

Singh SK, Strobel GA, Knighton B, Geary B, Sears J, Ezra D (2011) An endophytic *Phomopsis sp.* possessing bioactivity and fuel potential with its volatile organic compounds. Microb Ecol 61:729–739

Simon S, Petrášek J (2011) Why plants need more than one type of auxin. Plant Sci 180(3):454–460

Song GC, Ryu CM (2013) Two volatile organic compounds trigger plant self-defense against abacterial pathogen and asucking insect in cucumber under open field conditions. Int J Mol Sci 14:9803–9819

Splivallo R, Bossi S, Maffei M, Bonfante P (2007) Discrimination of truffle fruiting body versus mycelia aromas by stir bar sorptive extraction. Phytochem J 68:2584–2598

Splivallo R, Ottonello S, Mello A, Karlovsky P (2011) Truffle volatiles: from chemical ecology to aroma biosynthesis. New Phytol J 189:688–699

Stefels J (2000) Physiological aspects of the production and conversion of DMSP in marine algae and higher plants. J Sea Res 43:183–197

Stetzenbach LD (1998) Microorganisms and indoor air quality. Clin Microbiol Newsl 20:157–161

Stinson M, Ezra D, Hess WM, Sears J, Strobel G (2003) An endophytic *Gliocladium sp.* of *Eucryphia cordifolia* producing selective volatile antimicrobial compounds. Plant Sci J 165:913–922

Stone JK, Bacon CW, White JF Jr (2000) An over view of endophytic microbes: endophytism defined. In: Bacon CW, White JF Jr (eds) Microbial endophytes. Marcel Dekker, New York, pp 3–29

Strobel GA (2006) Harnessing endophytes for industrial microbiology. Curr Opin Microbiol 9:240–244

Strobel GA, Dirksen E, Sears J, Markworth C (2001) Volatile antimicrobials from *Muscodor albus* a novel endophytic fungus. Microbiol J 147:2943–2950

Strobel G, Singh SK, Riyaz-Ul-Hassan S, Mitchell AM, Geary B, Sears J (2011) An endophytic/pathogenic *Phoma sp.* from creosote bush producing biologically active volatile compounds having fuel potential. FEMS Microbiol Lett J 320:87–94

Styger G, Prior B, Bauer F (2011) Wine flavor and aroma. J Ind Microbiol Biotechnol 38:1145–1159

Su C, Sahlin M, Oliw EH (1998) Kinetics of manganese lipoxygenase with a catalytic mononuclear redox center. J Biol Chem 273:20744–20751

Sundberg C, Yu D, Franke-Whittle I, Kauppi S, Smars S, Insam H (2013) Effects of pH and microbial composition on odour in food waste composting. Waste Manag 33:204–211

Sunesson AL (1995) Volatile metabolites from microorganisms in indoor environments-sampling, analysis and identification [thesis]. Umea University and National Institute for Working Life, Sweden, pp 1–88

Sunesson AL, Nilsson CA, Andersson B, Blomquist G (1996) Volatile metabolites produced by two fungal species cultivated on building materials. Ann Occup Hyg 40:397–410

Suutari M, Rönkä E, Lignell U, Rintala H, Nevalainen A (2002) Characterisation of *Streptomyces spp.* isolated from water-damaged buildings. FEMS Microbiol Ecol 39:77–84

Tellez MR, Schrader KK, Kobaisy M (2001) Volatile components of the *cyanobacterium Oscillatoria perornata* (Skuja). J Agric Food Chem 49(12):5989–5992

Ten Brinke J, Selvin S, Hodgson A. Fisk W, Mendell M, Koshland C, Daisey J (1998) Development of new volatile organic compound (VOC) exposure metrics and their relationship to "sick building syndrome" symptoms. Indoor Air:140–152

Thorn R, Greenman J (2012) Microbial volatile compounds in health and disease conditions. J Breath Res 6(2):024001. ISSN 1752-7155

Tilg H, Moschen AR (2010) Evolution of inflammation in nonalcoholic fatty liver disease: the multiple parallel hits hypothesis. Hepatol J 52(5):1836–1846

Ting ASY, Mah SW, Tee CS (2010) Identification of volatile metabolites from fungal endophytes with biocontrol potential towards Fusarium oxysporum F.sp.cubenseRace4. Am J Agric Biol Sci 5:177–182

Tirranen LS, Gitelson II (2006) The role of volatile metabolites in microbial communities of the LSS higher plant link. Adv Space Res 38:1227–1232

Tomsheck AR, Strobel GA, Booth E, Geary B, Spakowicz D, Knighton B, Floerchinger C, Sears J, Liarzi O, Ezra D (2010) *Hypoxylon sp.*, an endophyte of *Persea indica*, producing 1, 8- cineole and other bioactive volatiles with fuel potential. Microb Ecol 60:903–914

Tonkin-Crine S, Yardley L, Little P (2011) Antibiotic prescribing for acute respiratory tract infections in primary care: a systematic review and meta-ethnography. J Antimicrob Chemother 66:2215–2223

Trivedi P, He Z, Van Nostrand JD, Albrigo G, Zhou J, Wang N (2011) Huanglongbing alters the structure and functionaldiversity of microbial communities associated with citrus rhizosphere. ISME J 6:363–383

Tsitsigiannis DI, Keller NP (2007) Oxylipins as developmental and host–fungal communication signals. Trends Microbiol 15:109–118

Turner AP, Magan N (2004) Electronic noses and disease diagnostics. Nat Rev Microbiol 2(2):161–166

Verginer M., Leitner E., Berg G (2010). Production of volatile metabolites by grape-associated microorganisms. J Agric Food Chem 58:8344–8350 10.1021/jf100393w

Verma M, Brar SK, Tyagi RD, Surampalli RY, Valero JR (2007) Antagonistic fungi, *Trichoderma spp.:* panoply of biological control. Biochem Eng J 37:1–20

Vespermann A, Kai M, Piechulla B (2007) Rhizobacterial volatiles affect the growth of fungi and Arabidopsis thaliana. Appl Environ Microbiol 73:5639–5641. doi:10.1128/AEM.01078-07

Vining LC (1990) Functions of secondary metabolites. Annu Rev Microbiol 44:395–427

Voigt CA, Schafer W, Salomon S (2005) A secreted lipase of *Fusarium graminearum* is a virulence factor required for infection of cereals. Plant J 42:364–375

von Reuss S, Kai M, Piechulla B, Francke W (2010) Octamethylbicyclo[3.2.1] octadienes from *Serratia odorifera*. Angew Chem 122:2053–2054

Wady L, Bunte A, Pehrson C, Larsson L (2003) Use of gas chromatography -mass spectrometry/solid phase micro extraction for the identification of MVOCs from moldy building materials. J Microbiol Methods 52:325–332

Waller F, Achatz B, Baltruschat H, Fodor J, Becker K, Fischer M, Heier T, Huckelhoven R, Neumann C, Wettstein D, Franken P, Kogel KH (2005) The endophytic fungus *Piriformospora indica* reprograms barley to salt-stress tolerance, disease resistance, and higher yield. Proc Natl Acad Sci U S A 102:13386–13391

Wani MA, Sanjana K, Kumar DM, Lal DK (2010) GC-MS analysis reveals production of 2-phenylethanol from *Aspergillus niger* endophytic in rose. J Basic Microbiol 50:110–114

Watson AG, Ford EJ (1972) Soil fungistasis a reappraisal. Annu Rev Phytopathol 10:327

Weise T, Kai M, Gummesson A, Troeger A, vonReuss S, Piepenborn S (2012) Volatile organic compounds produced by the phytopathogenic bacterium *Xanthomonas campestris* pv. vesicatoria 85–10. Beilstein J Org Chem 8:579–596

Weisskopf L (2013) The potential of bacterial volatiles for crop protection against phytophathogenic fungi. In Méndez-Vilas A (ed) Microbial pathogens and strategies for combating them: science, technology and education. Formatex Research Center, Badajoz, pp 1352–1363

Weschler CJ (2000) Ozone in indoor environments: concentration and chemistry. Indoor Air 10(4):269–288

Wheatley RE (2002) The consequences of volatile organic compound mediated bacterial and fungal interactions. Antonie Van Leeuwenhoek 81:357–364

Whillans FD, Lamont GS (1995) Fungal volatile metabolites released into indoor air environments: variation with fungal species and growth media. In: Morawska L, Bofinger ND, Maroni M (eds) Proceedings of the international workshop indoor air—an integrated approach, gold coast Australia, 1994. Elsevier Science and Technology Books, Oxford, pp 47–50

Whipps JM (2001) Microbial interactions and biocontrol in the rhizosphere. J Exp Bot 52:487–511

Whiteson KL, Meinardi S, Lim YW, Schmieder R, Maughan H, Quinn R, Blake RD, Conrad D, Rohwer F (2014) Breath gas metabolites and bacterial metagenomes from cystic fibrosis airways indicate active pH neutral 2, 3-butanedione fermentation. ISME J 8:1247–1258

Wilkins K, Larsen K (1995) Variation of volatile organic compounds patterns of mold species from damp buildings. Chemosphere 31(5):3225–3236

Wilkins K, Scholl S (1989) Volatile metabolites of some barley storage molds. Int J Food Microbiol 8(1):11–17

Wilkins K, Nielsen EM, Wolkoff P (1997) Patterns in volatile organic compounds in dust from moldy buildings. Indoor Air 7(2):128–134

Wilkins K, Larsen K, Simkus M (2000) Volatile metabolites from mold growth on building materials and synthetic media. Chemosphere 41:437–446

Wilkins K, Larsen K, Simkus M (2003) Volatile metabolites from indoor molds grown on media containing wood constituents. Environ Sci Pollut Res Int 10(4):206–208

Willger SD, Grahl N, Cramer RA (2008) *Aspergillus fumigatus* metabolism: clues to mechanisms of in vivo fungal growth and virulence. Med Mycol J 47:S72–S79

Wilson D (1995) Endophyte the evolution of a term, and clarification of its use and definition. Oikos 73:274–276

Wolkoff P, Clausen PA, Wilkins CK, Hougaard KS, Nielsen GD (1999) Formation of strong airway irritants in a model mixture of (+)-alpha-pinene/ozone. Atmos Environ 33(5):693–698

Wolkoff P, Clausen PA, Wilkins CK, Nielsen GD (2000) Formation of strong airway irritants in terpene/ozone mixtures. Indoor Air 10(2):82–91

Worapong J, Strobel GA, Ford EJ, Li JY, Baird G, Hess WM (2001) *Muscodor albus anam. nov., an* endophyte from *Cinnamomum zeylanicum*. Mycotaxon 79:67–79

World Health Organization (WHO) (1986) Indoor air quality and research. Euro Reports and Studies 103

Xu C, Mo M, Zhang L, Zhang K (2004) Soil volatile fungistasis and volatile fungistatic compounds. Soil Biol Biochem 36:1997–2004

Yuan Zl, Zhang Cl, Lin FC (2010) Role of diverse non-systemic fungal endophytes in plant performance and response to stress: progress and approaches. J Plant Growth Regul 29:116–126

Zhang H, Xie X, Kim M, Kornyeyev M, Holaday S, Pare PW (2008) Soil bacteria augment *Arabidopsis* photosynthesis by decreasing glucose sensing and abscisic acid levels in planta. Plant J 56:264–273

Zhang CL, Wang GP, Mao LJ, Komon-Zelazowska M, Yuan ZL, Lin FC, Druzhinina IS, Kubicek CP (2010) Muscodor fengyangensis sp. nov. from southeast China: morphology, physiology and production of volatile compounds. Fungal Biol J 114:797–808

Zhang J, Fang A, Wang B, Kim SH, Bogdanov B, Zhou Z (2011) I match: a retention index tool for analysis of gas chromatography-mass spectrometry data. J Chromatogr A 1218:6522–6530

Zou CS, Mo MH, Gu YQ, Zhou JP, Zhang KQ (2007) Possible contributions of volatile-producing bacteria to soil fungistasis. Soil Biol Biochem 39:2371–2379. doi:10.1016/j.soilbio.2007.04.009

Plant Essential Oils for Controlling Plant Pathogenic Fungi

9

Riad Sedki Riad El-Mohamedy

Abstract

Phytopathogenic fungi consider the causal agents of plant diseases than any other group of plant pathogens, about 8000 species shown to cause many plant diseases on many crops. Some of the world's great famines and human suffering can be blamed on plant pathogenic fungi. The most important natural products derived from plants are the essential oils (EOs), as it has various biological properties and medicinal and nutritional uses. Excessive usage of synthetic chemical fungicides in the environment has increased risks to humans, flora, and fauna in recent years. Moreover, the development of resistance in species of pathogenic fungi due to the indiscriminate use of chemical fungicides for controlling plant pathogenic fungi. Most plant essential oils (EOs) are relatively nontoxic to laboratory animals and fish; they meet the criteria for "reduced risk" pesticides. Many EOs show a broad spectrum of activity against plant pathogenic fungi and have a long tradition application in the protection of stored products. Moreover, many EOs and their constituents are widely used as flavoring agents in foods and beverages. The special regulatory status, combined with the wide availability of EOs from the flavor and fragrance industries, has made it possible to fast-track commercialization of EO-based pesticides as green pesticides. Such green pesticides have been received by consumers and companies to be applied against home and garden field pests; these can also prove effective in agricultural situations, particularly for organic food production. This chapter provides a review of several different aspects relating to plant essential oils (EOs) including a historical perspective, main sources and antifungal activity, bioactive single constituents of EOs and their modes of action, and, moreover, the chemical measures necessary for controlling plant pathogens and their negative impact on human health and/

R.S.R. El-Mohamedy (✉)
Plant Pathology Department, Agriculture and Biology Division, National Research Center, Dokki, Cairo, Egypt
e-mail: riadelmohamedy@yahoo.com

D.K. Choudhary et al. (eds.), *Volatiles and Food Security*,
DOI 10.1007/978-981-10-5553-9_9

or the environment. This chapter also reviews the different sources of plant essential oils. The antimicrobial activity of EOs, with particular emphasis on the antifungal properties exhibited against some serious pathogenic fungi and postharvest disease, is reviewed. Moreover, various antimicrobial tests and techniques are included.

Keywords
Essential oil • Antifungal activities • Fungal diseases • Control

9.1 Introduction

The plant pathogens such as fungi, nematodes, bacteria, and viruses cause many diseases and damages in economic crops and plants. Fungi considered the main pathogens and caused yield losses in numerous economically important crops. During the recent years, the environmental problems caused by overuse of fungicides have been the matter of concern for both scientists and public in many countries. Pesticides including fungicides used on crops each year for controlling plant have been estimated about 2.5 million tons. The worldwide damage caused by pesticides reaches $100 billion annually. The reasons for these damages are due to the high toxicity, no biodegradable properties of pesticides and the residues in soil, water resources, and crops that affect public health. Thus, scientists search the new highly selective and biodegradable pesticides, study the environmentally friendly pesticides, and develop techniques that can be used to reduce pesticide usage while maintaining crop yields. Many natural product derivatives from organisms of such plants are excellent alternatives and substitutes to synthetic pesticides as safe means to reduce negative impacts to human health and the environment.

Plant essential oils (EOs) of many plants showed a wide spectrum of antifungal against many plant pathogenic fungi. EOs are typically liquid at room temperature and get easily transformed from a liquid to a gaseous state at room or slightly higher temperature without undergoing decomposition. The amount of essential oil found in most plants ranges from 1.0 to 10.0. In certain plants, one main essential oil constituent may predominate, while in others, there are many terpenes. The basil plants (*Ocimum basilicum*) contain methyl chavicol up to 75%, and β-asarone amounts up to 70–80% of the oil. Meanwhile, the rhizomes of *Acorus calamus* plants contain linalool in the range of 50–60% (Lawrence and Reynolds 2001). Most plant essential oils (EOs) comprise of monoterpenes – compounds that contain 10 carbon atoms often arranged in a ring or in acyclic form – as well as sesquiterpenes which are hydrocarbons comprising of 15 carbon atoms. The higher terpenes in EOs may also be present as minor constituents. The most predominant groups of terpenes are cyclic compounds with saturated or unsaturated hexacyclic or an aromatic system. The another components of essential oils are bicyclic (1,8-cineole) and acyclic (linalool, citronellal) (Koul et al. 2008).

Plant essential oils (EOs) have several applications in the manufacture of perfumes, cosmetics, and soaps as well as food industry. Application of many plant essential oils as alternative antimicrobial agents for replacing chemical treatments has attracted considerable interest from the postharvest scientists especially in recent years, as many of EOs and their single constituents have been reported to inhibit postharvest pathogenic fungi. Bioactivity in the vapor phase of essential oils was recognized as a characteristic that makes them attractive for use as possible fumigants for the protection of stored product (Crespo et al. 1990; Adam et al. 1998).The fungitoxic activity of essential oils may be due to synergism among their components, since most of this activity has been reported to be enhanced when combined (Adam et al. 1998; Tullio et al. 2007).

9.2 Chemical Control of Plant Pathogenic Fungi

The chemical disease control employs the use of many commercial chemicals agents such as fungicides and antibiotics that are generally toxic to pathogens and characterized by their specific effect. These chemical agents must be effective and act safely, without harming any live organism and must also have minimal or no effect on the environment, microflora, and soil. Generally, commercial chemical fungicides are used in the control of fungal diseases on many crops. Recently, growing public concern over the health and environmental hazards associated with the increased levels of chemical pesticides has led to the development of safe, alternative, and natural methods of disease control (Lopez-Reyes et al. 2013). Throughout the 1980s and 1990s, public concern about the use of agricultural chemical pesticides grew as a result of the high risk of poisoning to all living organisms, beneficial microflora and microfauna, and the contamination of food products. The risk from chemical pesticides is due to the possibility that they may be absorbed through the skin and inhaled or ingested through consumption of contaminated plants. To overcome the risks of chemical pesticides, some roles should be taken with respect to the handling of the chemicals and the management of agriculture practices. Many countries over the world have already started educational programs for farmers to reduce the overuse of chemical pesticides and hence increase the productivity of several economic crops. Up to date, there has been a great interest in the use of essential oils and plant extracts as possible natural substitutes for conventional synthetic pesticides.

9.3 The Plant Essential Oils

The essential oils (EOs) are defined as any volatile oil(s) that have strong aromatic components and that give distinctive odor, flavor, or scent to a plant. EOs are the by-products of plant metabolism and are commonly referred to as volatile plant secondary metabolites. The EOs are found in glandular hairs or secretory cavities of plant cell wall and are present as droplets of fluid in the leaves, stems, bark, flowers, roots, and/or fruits in different plants. The aromatic characteristics of essential oils provide various functions for the plants including (1) attracting or repelling insects, (2) protecting themselves from heat or cold, and (3) utilizing chemical constituents in the oil as defense materials. Most plants contain essential oils ranging from 1 to 2%, but some plants contain amounts ranging from 0.01 to 10%. In *Ocimum basilicum* (basil), for example, methyl chavicol makes up 75% of the oil, β-asarone amounts to 70–80%, and *Acorus calamus* rhizomes contain linalool range of 50–60% (Lawrence and Reynolds 2001). Most plant essential oils comprise of monoterpenes – compounds that contain 10 carbon atoms often arranged in a ring or in acyclic form – as well as sesquiterpenes which are hydrocarbons comprising of 15 carbon atoms. In some EOs, higher terpenes may also be present as minor constituents. The most predominant terpenes are cyclic compounds with saturated or unsaturated hexacyclic. Also, bicyclic (1,8-cineole) and acyclic (linalool, citronellal) examples also make the components of essential oils. Most EOs play a role in plant defense mechanisms acting against phytopathogenic microorganisms (Pandey et al. 1982; El-Mohamedy et al. 2002; El-Mohamedy 2003; Chuang et al. 2007; Ping-Hsien et al. 2007).

9.4 Sources of Plant Essential Oils (EOs) Have an Antifungal Activity

Many scientific investigations have highlighted the importance and the contribution of many plant families, such as Asteraceae, Liliaceae, Apocynaceae, Solanaceae, Caesalpiniaceae, Rutaceae, Piperaceae, Sapotaceae, etc., that are used as medicinal plants. The antifungal activity of EOs can be attributed to the presence of some components such as carvacrol, α-terpinyl acetate, cymene, thymol, pinene, and linalool which are already known to exhibit antimicrobial activity against many microorganisms (Knobloch et al. 1985; Cimanga et al. 2002; El-Mohamedy 2003). Several in vitro studies have been published confirming the effect of essential oil and their major compounds on plant and human pathogenic fungi. The antifungal activity of essential oil derivatives from the plants of such families is summarized below.

9.4.1 Family Lamiaceae

Sage and oregano are considered to be the main genus among the Lamiaceae family. The antifungal and antibacterial activity of oregano essential oil against a number of plant pathogens such as *Aspergillus niger* v. Tieghem, *A. flavus* Link, *A. ochraceus* Wilhelm, *Fusarium oxysporum* Snyder and Hansen, *Penicillium* sp. L., *Pseudomonas aeruginosa* Schroter, *Staphylococcus aureus*, *F. solani* var. *coeruleum* (Martius) Saccardo, *Clavibacter michiganensis* S., *Phytophtora infestans* Mont., *Sclerotinia sclerotiorum* Lib., and *Xanthomonas vesicatoria* has been reported by Juven et al. (1994). Thyme (*T. capitatus* Hoffm. and Link) belongs to the genus *Thymus* of the mint family (Lamiaceae) and is a relative of the oregano genus. Thyme essential oil displayed antifungal activity in stored foods and inhibited the growth of both *B. cinerea* and *Monilinia fructicola*; also EO of *T. vulgaris* showed antifungal activity against some postharvest pathogens such as *B. cinerea*, *P. italicum*, *P. citrophthora*, *Rhizopus stolonifer*, *Fusarium oxysporum radicis-lycopersici*, *F. oxysporum lycopersici*, *F. solani*, *Rhizoctonia solani*, *Sclerotium rolfsii*, *Macrophomina phaseolina*, *Pythium* sp., and *Phytophthora* sp. (El-Mohamedy et al. 2013). Essential oils of *Majorana hortensis* (family Lamiaceae) showed antifungal activity against *P. citrophthora*, *R. stolonifera*, *C. acutatum*, and *B. cinerea* and antibacterial activity against two strains of G+ (*Bacillus megaterium* and *C. michiganensis*) and five strains of G− (*Escherichia coli*, *X. campestris*, *B. mojavensis*, *P. savastanoi*, and *P. syringae* pv. *phaseolicola*).

9.4.2 Family Magnoliaceae

Magnolia is a large genus of about 210 flowering plant species in the family Magnoliaceae. Under field conditions, the essential oil of *M. liliiflora* showed a potential antifungal effect against *P. capsici*, and this activity could be attributed to its constituents: α-terpineol, α-bourbonene, β-caryophyllene, 2-β-pinene, α-humulene, farnesene, and caryophyllene oxide components (Cimanga et al. 2002). Also, *M. liliiflora* essential oil showed potential in vitro and in vivo antifungal effects against *B. cinerea*, *Colletotrichum capsici* (Syd.) E.J. Butler and Bisby, *F. oxysporum* Snyder and Hansen, *Fusarium solani*, *P. capsici* Leonian, *Rhizoctonia solani* (Cooke) Wint., and *S. sclerotiorum* (Cimanga et al. 2002).

9.4.3 Family Asteraceae

The plants in Asteraceae are medically important in areas that don't have access to Western medicine. The antifungal activity and chemical composition of the EOs from the most aromatic plants belonging to Asteraceae family have been investigated (Odom et al. 2000). The essential oil of flower heads of garland chrysanthemum (*Chrysanthemum coronarium* L.) clearly showed the strongest inhibition against the growth of 12 agricultural plant pathogens (Alvarez-Castellanos et al. 2001). The main compounds of such oil were camphor, α- and β-pinene, and lyratyl acetate. In addition, the essential oils (EOs) from three *Artemisia* species, i.e., *Artemisia absinthium* L., *A. santonicum* L., and *A. spicigera* C. Koch, had antifungal activity potent against many plant pathogenic fungi (Govinden-Soulange et al. 2004; Kordali et al. 2005). Romagnoli et al. (2005) noted that the essential oil of *Tagetes patula* L. exerted antifungal activity against phytopathogenic fungi, *Botrytis cinerea* and *Penicillium digitatum*, providing complete growth inhibition. *Chrysactinia mexicana* Grag essential oil caused completely inhibited growth of *Aspergillus flavus* (Cardenas et al. 2005). Meanwhile EO from *Helichrysum italicum* (Roth) Don plants was active against *Pythium ultimum* (Tundis et al. 2005).

Essential oil derivatives from *Arnica longifolia*, *Aster hesperius*, and *Chrysothamnus nauseosus* showed antifungal activity against three *Colletotrichum* species (Demirci et al. 2006; Altintas et al. 2006; Nurhayat et al. 2007).

9.4.4 Family Rutaceae

Rutaceae is commonly known as the rue or citrus family. The essential oil from the epicarp of *Citrus sinensis* (L.) Osbeck exhibited absolute fungitoxicity against many phytopathogenic fungi such as *Curvularia lunata*, *Fusarium oxysporum*, *Botrytis cinerea*, *Penicillium digitatum*, and *P. itilicum* (Al-Burtamani et al. 2005; El-Mohamedy 2003). The most abundant citrus oil components were α- and β-phellandrene, limonene, β-ocimene, β-caryophyllene, and myrcene. EOs from *Citrus limon* (L.), *C. paradise*, and *C. sinensis* showed antifungal activities against five phytopathogenic fungi (Saikia et al. 2001; Sun et al. 2007). In this respect, Pitipong et al. (2009) noted that bergamot (*Citrus hystrix* DC.). EO shows the antagonistic activity against seven species of economically important rice pathogenic fungi *Alternaria brassicicola*, *Aspergillus flavus*, *Bipolaris oryzae*, *Fusarium moniliforme*, *F. proliferatum*, *Pyricularia arisea*, and *Rhizoctonia solani*.

9.4.5 Family Liliaceae

Many plants in the Liliaceae are important ornamental plants, are widely grown for their attractive flowers, and are native primarily to temperate and subtropical regions. The essential oils (EOs) from *Allium fistulosum* L., *A. sativum* L., and *A. cepa* L. showed strongly antifungal activities against three *Trichophyton* species responsible for severe mycoses in humans and many phytopathogenic fungi on may economic crops. *A. sativum* oil exhibited the strongest inhibition of growth in *T. rubrum* with an IC50 value of 61 μg/ml, while *A. cepa* and *A. fistulosum* were relatively mild (Singh and Rai 2000).

9.4.6 Family Lamiaceae

The most members of this family are useful economically for medicinal, culinary, ornamental, and various commercial utilizations. In the previous studies on the essential oils of many Lamiaceae, Baratta et al. (1998) recorded that the EOs from plants of family Lamiaceae have a broad range of biological activities, notably their antimicrobial potency; this activity is generally correlated to the chemical composition of the oil. The biological activities difference between EOs can be attributed to the variation in their chemical composition. The antifungal activity in oil of *Ocimum sanctum* L. plants against pathogenic fungi was recoded; also, EO of *Ocimum basilicum* L. is known for its wound healing property and, hence, is used in the treatment of fungal infections. The major component of this oil was linalool (Sunita and

Mahendra 2008). Essential oil of *Ocimum gratissimum* L. showed fungicidal activity against many *Candida* species such as *Candida albicans*, *C. krusei*, *C. parapsilosis*, and *C. tropicalis* (Nakamura et al. 2004; Hili et al. 1997). Lis-Balchin and Deans (1997) found that the essential oils from many species of the *Ocimum* genus, such as basil (*Ocimum micranthum* Willd.), showed antifungal activity against many pathogenic and food spoiling yeasts.

Essential oils of *Nepeta crispa* Willd. exhibited a noticeable antifungal activity against many pathogenic fungi. The main constituents were 1,8-cineole and 4α,7α-abetanepetalactone (Sonboli et al. 2004). Salgueiro et al. (2004) noted that the essential oil of *Thymbra capitata* (L.) showed antifungal activity against *Candida*, *Aspergillus*, and dermatophyte strains. Meanwhile, essential oil of *Thymus daenensis* Celak., *Zataria multiflora* Boiss., and *Thymbra spicata* L. had antifungal activities against *A. flavus* and *A. parasiticus*. These oils exhibited antifungal activity against all the strains tested, particularly for dermatophytes, with IC50 values ranging from 0.08 to 0.32 μg/ml. These oils contain carvacrol type with a high content of carvacrol and its biogenetic precursors, γ-terpinene and p-cymene. Thus, the plant essential oils are potentially safe source of antifungal agents. The essential oils of some *Thymus* species showed antimicrobial activity against viruses, bacteria, food-derived microbial strains, and fungi; these EOs possess large quantities of phenolic monoterpenes (Rasooli et al. 2006).

Sage (*Salvia officinalis* L.) and lavender (*Lavandula angustifolia* Mill.) have always been used to treat various skin diseases and as cosmetic products for skin care (Lecha et al. 2005; Andrews and Burns 2008). The efficacy of essential oils of lavender and sage can be explained by the high content of 1,8-cineole, which is capable of changing the structure and moisture of mucous membranes of fungal cells, interfering with the respiratory processes, and therefore comes to the elimination of pathogens (Andriole 1994; Gupta et al. 2003). Gupta et al. (1999) recorded that the presence of limonene and 1,8-cineole, which can influence on the changing structure of the lipid layer, the stratum corneum, increased permeability of the epidermis, and stratum corneum human, also may be one of the reasons for the good of the essential activities in vivo. The major compounds in the essential oil were α-pinene, borneol, camphene, camphor, verbenone, and bornyl acetate.

In addition, Angioni et al. (2004, 2006) noted that essential oils from the stems, leaves, and flowers of *Lavandula stoechas* L. were effective on the inactivation of *Rhizoctonia solani* and *Fusarium oxysporum* and less effective against *Aspergillus flavus*. The essential oil of rosemary (*Rosmarinus officinalis* L) showed an inhibitory effect toward *F. graminearum* and exhibited an intermediate antifungal activity (MIC = 1.10 mg/mL) against *C. albicans* (Dalleau et al. 2007).

Essential oils from several species of the Lamiaceae family, *Satureja montana* L., *Lavandula angustifolia*, *L. hybrida* Reverchon, *Origanum vulgare* L., *Rosmarinus officinalis* L., and six chemotypes of *Thymus vulgaris* L showed antifungal effect on *Candida albicans* growth. The greatest efficiency was obtained with the essential oil from the *T. vulgaris* (IC50 of 0.016 μg/ml). Lavender oil (*L. angustifolia*) and its main components, linalool and linalyl acetate, showed antifungal activity against 50 clinical isolates of *Candida albicans* (oropharyngeal and vaginal strains) (D'Auria et al. 2005). Also, *O. vulgare* oil alone inhibited many phytopathogenic fungi. The main constituents of *such* oil were carvacrol, ρ-cymene, and thymol; these compounds had high-potential activities against many pathogens, with carvacrol and thymol as strong inhibitors of *B. cinerea* in vitro (Bouchra et al. 2003; Holly and Patel 2005; Bozin et al. 2006). Skovic et al. (2002) reported that the highest and broadest activity was shown by the carvacrol containing oils such as *O. onites* and *S. thymbra*. Pina-Vaz et al. (2004) found that ρ-cymene, a constituent of *O. vulgare* oil, showed synergistic activity with thymol against fungi.

The essential oils (EOs) from *Origanum vulgare* L., *Thymus serpyllum* L., *Thymus vulgaris*, *Lavandula latifolia* Medik., *L. angustifolia., and T. vulgaris* showed antifungal activities against *T. mentagrophytes* using vapor phase test (Inouye et al. 2006). The effective tested oils inhibited the fungal growth due to the presence of phenolic compounds, namely, thymol and carvacrol (Tullio et al. 2007). The mint essential oil was found to have strong antimycotic activity against *Candida albicans* and dermatophytes (Kishore et al. 1993). Essential oils (EOs) of *Mentha arvensis* L, *Mentha piperita* L., and *T. vulgaris* showed strongly antifungal activity against mycotoxin producers *Aspergillus flavus* and *A. parasiticus*. Menthol and thymol are the major components of such essential oils (Farag et al. 1989; Rath et al. 2001).

Essential oil from the aerial parts of *Salvia mirzayanii* Rech had antifungal against *Fusarium solani* and *Candida albicans*. The inhibitory effect of the oil increased when the concentration was increased. The major components were linalool, linalyl acetate, α-terpinyl acetate, 1,8-cineole, α-cadinol, and δ-cadinene. The antimicrobial activity of such oil against pathogens could be attributed to the high amount of major components such as 5-neocedranol, α-terpinyl acetate, 1,8-cineol, bicyclogermacrene, and α-cadinene and other valuable components in oil (Patker et al. 1993; Sinha et al. 1993; Pattnaik et al. 1996; Paster et al. 1995; Chao and Young 2000; Portillo et al. 2005).

9.4.7 Family Verbenaceae

Verbenaceae is a family of mainly tropical flowering plants. Some of the medicinal and aromatic plants have been reported to be anti-infectious agents. Viljoen et al. (2005) noted that the essential oil from oregano (*Lippia berlandieri* Shauer) study showed antifungal activity of versus food-contaminant fungi. Also, Tzasna et al. (2008) reported the antimicrobial activity of *Lippia javanica* (Burm. f.) leaf essential oil against some pathogenic fungi. In addition, the essential oils (EOs) of *Lantana achyranthifolia* Desfaerial parts and *Lippia graveolens* Kunth showed highly inhibitory effect on the growth of *Fusarium sporotrichum, Aspergillus niger, Trichophyton mentagrophytes, and Fusarium moniliforme.* Meanwhile, *L. graveolens* presented higher antifungal activity than *L. achyranthifolia* (Simic et al. 2004).

9.4.8 Family Lauraceae

The Lauraceae are the laurel family of flowering plants some of which are valued for spices and perfumes. The most plants of Lauraceae contain high concentrations of essential oils. Essential oils of several plants from the Lauraceae family such as *Aniba rosaeodora* Ducke, *Laurus nobilis* L., *Sassafras albidum* (Nutt.) Nees, and *Cinnamomum zeylanicum* Blume had antifungal activity against 17 micromycetes, i.e., food poisoning and food spoilage fungi, plant pathogenic fungi, and animal pathogens (Wang et al. 2005). The essential oil of *C. zeylanicum* showed the strongest antifungal activity. Another antifungal *Cinnamomum* species is *Cinnamomum osmophloeum* Kaneh which has significant antifungal activity against wood decay fungi (Jantan et al. 2008).

Sirirat et al. (2009) and Sessou et al. (2012) demonstrated that *C. zeylanicum* bark oil has fungitoxic properties against fungi involved in respiratory tract mycoses, such as *A. niger*, *A. fumigatus*, *A. nidulans*, and *A. flavus* (Singh et al. 1995). Cinnamon oil exhibited strong inhibitory effects on *Botrytis cinerea.* It completely suppressed the mycelial growth of fungus. Essential oil of *Ocimum gratissimum* (L.) had significant fungistatic activity against all the species investigated (Prajapati 2012). *Penicillium* species and *Fusarium poae* were the most sensible to this essential oil. Furthermore, the biological activity of this oil is probably due to its prominent concentration in thymol, which is a phenolic compound. Generally, the essential oils possessing the strongest antimicrobial properties against food-borne pathogens contain a high percentage of phenolic compounds such as carvacrol, eugenol (2-methoxy-4-(2-propenyl) phenol), and thymol (Lambert et al. 2001; Ahonkhai et al. 2009). An important characteristic of thymol is its hydrophobicity, which enables it to partition in the lipids of the fungal cell membrane, disturbing the structures and rendering it more permeable so leakage of ions and other cell contents can then occur (Lambert et al. 2001; Cheng et al. 2004). Hong et al. (2004) assessed antifungal activities of essential oil of *O. gratissimum* against different food-borne pathogens, and they reported that volatile oil of *O. gratissimum* had significant antimicrobial effects against all tested fungi and bacteria. The main component in the essential oil of *A. rosaeodora* was linalool, while 1,8-cineole was dominant in *L. nobilis*. The major component in *S. albidum* essential oil was safrole, but the trans-cinnamaldehyde is the main component of the oil of *C. zeylanicum.*

9.4.9 Family Cupressaceae

Many plants that belong to Cupressaceae family are recognized for its natural resistance to decay due to high content of natural essential oils. *Calocedrus formosana* florin leaf essential oil CLE constituents displayed activity against many decay fungi, namely, *Lenzites betulina*, *Pycnoporus coccineus*, *Trametes versicolor*, and *Laetiporus sulphureus*. Two compounds, α-cadinol and murolol, exhibited the strongest antifungal activity (Cavaleiro et al. 2006). The antifungal activity of essential oils from *Chamaecyparis obtusa* (Siebold and Zucc.) and *Juniperus communis*

L. was achieved against dermatophyte, *Aspergillus* and *Candida* strains. The main component was bornyl acetate (Singh 2001; Singh et al. 2004).

9.4.10 Family Apiaceae

Many plants of family Apiaceae (parsley family) have been used as a folk medicine (Nigam and Rao 1977). Essential oil of ajwain *Trachyspermum ammi* (L.) exhibited a broad spectrum of against many fungitoxic fungi such as *Aspergillus niger*, *Fusarium moniliforme*, and *Curvularia lunata*. The zone inhibition was obtained at a 6 μl dose of the such oil. The antifungal of such oil attributed to the major component, thymol, p-cymene, γ-terpinene, β-pinene, and terpinen-4-ol. The essential oil of *Cuminum cyminum* L. showed high antifungal activity against *Aspergillus* (Iacobellis et al. 2005; Sunita and Mahendra 2008). β-Pinene, γ-terpinene, and cuminaldehyde were the main constituents; both β-pinene and γ-terpinene clearly showed antifungal activity against many pathogenic fungi (Patra et al. 2002; Kawther 2007). The antimicrobial (antimycobacterial and anticandidal) properties of essential oils from dill plants (*Anethum graveolens* L.) and fennel (*Foeniculum vulgare* Mill) against many pathogens such as *Aspergillus niger*, *A. flavus*, *A. fumigatus*, *A. ustus*, *Candida albicans*, *Epidermophyton floccosum*, *Microsporum canis*, *M. audouinii*, *M. nanum*, *M. gypseum*, *Rhizopus nigricans*, *Trichophyton tonsurans*, *and T. violaceum* have been reported by many investigators (Shabnam et al. 2012).

The fennel oil causes higher inhibition against many phytopathogenic such *Alternaria alternata*, *Fusarium oxysporum*, *Aspergillus flavus*, and *A. parasiticus* (Aggarwa et al. 2000). The antifungal effects are directly related to its main component trans-anethole of essential oil of fennel. Shabnam et al. (2012) noted that the essential oils of fennel plant roots, stem, leaves, and seeds clearly showed anticandidal activity against *C. albicans* and *C. tropicalis*. The main compounds of essential oil from aerial parts of *Bupleurum gibraltaricum* Lamarck were sabinene, α-pinene, and 2,3,4-trimethyl benzaldehyde; such compounds or crude oil had strongly antifungal active toward *Plasmopara halstedii* (Komiya et al. 2006).

9.4.11 Family Poaceae

The family Poaceae still constitutes the most economically important plant oils. These plant oils are important source of fungitoxic compounds, and they provide a renewable source of useful fungicides that can be utilized in antimycotic drugs against *Aspergillus fumigatus* and *A. niger.* There are different degrees of antifungal activity between *Cymbopogon martini* (Roxb.) Wats. and *Cymbopogon citratus* (DC.), the maximum antimycotic activity shown by *C. martini* (Sunita and Mahendra 2008; Ewansiha et al. 2012).

Essential oils from *C. citratus* had antifungal activities against some phytopathogenic fungi (*Cladosporium* sp., *Aspergillus niger*, and *Mucor*) at lower concentrations (Sun et al. 2007). EO from *C. martini* was more effective against *Candida* sp.,

Aspergillus fumigatus, and *Trichophyton rubrum* compared with the EO of *C. citrates*. Essential oil derivatives from *C. citratus* confirm the activity against many microorganisms (Akiyama et al. 2001). Lemongrass essential oil was found to be active against human dermatophyte strains such as *C. albicans* (Fernanda et al. 2012; Singh et al. 2005b).

9.4.12 Moringaceae

Moringa that is the genus in Moringaceae contains 13 species from tropical and subtropical climates that range in size from tiny herbs to massive trees. Several in vitro studies have been published confirming the effect of essential oil and their major compounds on dermatophytic fungi as well as phytopathogenic fungi. The ethanol extracts of many plants of Moringaceae showed antifungal activities in vitro against many dermatophytes such as *Trichophyton rubrum*, *T. mentagrophytes*, *Epidermophyton floccosum*, and *Microsporum canis*. *Moringa* seed oil has antifungal activities against many phytopathogenic fungi.

In this respect, El-Mohamedy and Aboelfetoh (2014) noted that seed oil and seed extract of *Moringa oleifera* showed antifungal activities against seven plant pathogenic fungi, i.e., *Fusarium oxysporum*, *Fusarium solani*, *Alternaria solani*, *Alternaria alternata*, *Rhizoctonia solani*, *Sclerotium rolfsii, and Macrophomina phaseolina* in vitro. They also recorded that seed extracts and oil of *Moringa oleifera* significantly reduced linear growth, spore germination, and dry growth weight of such pathogens. Oil and seed extract of *M. oleifera* had different degrees of inhibition against growth rate and spore germination of such pathogens. *Moringa* oil at a concentration of 2.0% showed the highest records of reduction on both growth and spore germination. *M. oleifera* oil and seed extract caused highly reduction in growth of *F. oxysporum*, *F. solani* and *A. solani*, *A. alternata* than *R. solani*, *S. rolfsii, and M. phaseolina*. El-Mohamedy and Aboelfetoh (2014) noted that seed extract and *Moringa* oil may be recommended as a bio-fungicide against many fungal pathogens.

9.4.13 Family Zingiberaceae

This family consists of flowering plants and aromatic perennial herbs. There are many species of Zingiberaceae which are important ornamental or medicinal plants. The essential oils of ginger (*Zingiber officinale* Roscoe) showed antifungal activity against many mycotoxin producers such as *Aspergillus flavus* and *A. parasiticus*. Zingiberene is the major component in ginger oil and showed low antifungal activity on the mycelial growth of *A. flavus* and *F. oxysporum* (Farag et al. 1989; Pandey et al. 2010). The essential oil of *Curcuma longa* L. and their major components such as terpinolene, α-phellandrene, and terpinene-4-ol have significant antifungal activity against wood decay fungi (Apisariyakul et al. 1995; Jantan et al. 2008; Saha and Bhupendar 2011). Bark oil *C. zeylanicum* has fungitoxic properties against fungi involved in respiratory tract mycoses, such as *A. niger*, *A. fumigatus*, *A. nidulans*, and *A. flavus*. The antifungal activity of the bark and leaf oils of *C. zeylanicum* was correlated with the high levels of cinnamaldehyde (44.2%) and eugenol (90.2%). Meanwhile, the high amounts of benzyl benzoate (>50%) in the leaf oils of *Curcuma rhynchophyllum*, *C. microphyllum*, *C. pubescens*, *C. impressicostatum*, and *C. mollissimum* were responsible for selective toxicity against many dermatophytes (Singh et al. 1995). Montes-Belmont and Carvajal (1998) and Lee et al. (2007) demonstrated that cinnamon oils has inhibitory effect against the growth and aflatoxin production by *A. flavus*. Also, Sirirat et al. (2009) and Sessou et al. (2012) noted that cinnamon oil exhibited strong inhibitory effects against *Botrytis cinerea*. Essential oil of *Ocimum gratissimum* (L.) had significant fungistatic activity against *Penicillium* species and *Fusarium poae*; the fungicidal activity of this oil is attributed to the prominent concentration of thymol, which is a phenolic compound (Prajapati 2012). The essential oils possessing the strongest antimicrobial activities of many against food-borne pathogens contain a high percentage of phenolic compounds such as carvacrol, eugenol (2-methoxy-4-(2-propenyl, phenol), and thymol (Lambert et al. 2001; Ahonkhai et al. 2009). Thymol is hydrophobic, which enables it to partition in the lipids of the pathogenic fungal cell membrane, disturbing the structures and rendering it more permeable so leakage of ions and other cell contents can then occur (Lambert et al. 2001; Cheng et al. 2004). In addition, essential oil of *O. gratissimum* had significant antimicrobial effects against many pathogenic food-borne pathogens (fungi and bacteria) (Hong et al. 2004).

9.4.14 Family Ranunculaceae

Many plants of Ranunculaceae are used as herbal medicines because these plants contain high amount of alkaloids and glycosides. For example, the oil extract from *Nigella sativa* L. plants showed antimicrobial effect against *Candida albicans* under in vitro and in vivo conditions (Aggrawal et al. 1979; Hanafy and Hatem 1991; Khan et al. 2003; Mashhadian and Rakhshandeh 2005). Also, Amrouche et al. (2011) noted that such volatile oil had the strongest antifungal activity against *Aspergillus* species. Singh et al. (2005a, b) and Al-Jabre et al. (2005) found that seed oil of *N. sativa* prevents the mycelial growth of *A. flavus* as 6.0 μl/ml of such oil exhibited 73.3% growth reduction of the pathogen. Essential oil of *N. sativa* contains high amount of thymoquinone that is believed to be the responsible constituent to antifungal activity as 2 μg/ml concentration of thymoquinone resulted in 100% growth inhibition in *A. niger* (Zhang et al. 2006).

9.4.15 Family Theaceae

The family Theaceae contains many flowering plants, composed of shrubs and trees. The essential oils from *Camellia sinensis* (L.) Kuntze showed antifungal potential against many pathogenic fungi. Many workers showed that extracts and essential oils of the tea tree plant have the antimicrobial activity against many pathogenic fungi (Hili et al. 1997; Lis-Balchin and Deans 1997). The essential oils of tea tree exhibited an intermediate antifungal activity (MIC = 2.25 mg/mL) against *C. albicans* (Singh et al. 1980; Dalleau et al. 2007). Also, the oils of Melaleuca *alternifolia plants* had antagonistic activity against many species of economically important rice pathogenic fungi, namely, *Rhizoctonia solani*, *Aspergillus flavus*, *Bipolaris oryzae*, *Fusarium moniliforme*, *F. proliferatum*, *Pyricularia arisea*, and *Alternaria brassicicola* (Saikia et al. 2001; Pitipong et al. 2009).

9.4.16 Family Meliaceae

The family Meliaceae includes many flowering plants, mostly trees and shrubs with an economically importance, most of them used for vegetable oil, soap making, and insecticides. The plant oils from many plants of this family Meliaceae are important sources for fungitoxic compounds, and they may provide a renewable source of useful fungicides; these compounds also can be utilized in antimycotic drugs against *Aspergillus fumigatus* and *A. niger*. The most common tree of this family is neem tree. In this respect, Kazmi et al. (1995), Locke (1995), Sunita and Mahendra (2008) and Sukatta et al. (2008) demonstrated that neem *Azadirachta indica* (L.) Adelb oil at 2–10% completely suppressed the pathogenic fungi, i.e., *Alternaria alternata*, *Aspergillus niger*, and *Fusarium oxysporum*.

9.4.17 Family Myrtaceae

The family Myrtaceae or (myrtle family) includes many species of woody plants with essential oils. The oils from *Syzygium aromaticum* (L.) Merrill and Perry exhibited strong inhibitory effects with complete inhibition of mycelia growth of *Botrytis cinerea* (Sirirat et al. 2009; Sessou et al. 2012). Martini et al. (1996) showed that mixing of clove oil and cinnamon oils at appropriate ratios results in an improvement of the efficacy against *Phomopsis viticola*, *Aspergillus niger*, *Alternaria alternata*, *Colletotrichum gloeosporioides*, *Lasiodiplodia theobromae*, and *Rhizopus stolonifera* the major postharvest decaying fungi of grapes. Also, Azzouz and Bullerman (1982) and Eugenia et al. (2009) found that the clove oil and its main component eugenol had antifungal activity against *Candida*, *Aspergillus*, and dermatophyte.

In general, phenolic components of the essential oils are considered as responsible for the antimicrobial activity, followed by aldehydes, ketones, and alcohols (Shelef 1983; Akgul 1989; Burt 2004). Different modes of action are involved in the antimicrobial activity of essential oils. The activity of clove oil can be related to the presence of a high concentration (85.3%) of eugenol. Such activity of this oil may be due to their hydrophobicity, responsible for their partition into the lipid bilayer of the cell membrane, leading to permeability alteration and a consequent leakage of cell contents. Moreover, Lambert et al. (2001) noted that essential oils can travel through the cell wall and cytoplasmic membrane, disrupt the structure of the different layers of polysaccharides, fatty acids, and phospholipids, and permeabilize them.

Clove oil significantly suppressed the growth *A. fumigatus* and *A. aculeatus*. Meanwhile, the essential oil of *Leptospermum petersonii* Bailey showed strong antifungal activities against mycotoxins producing fungi, i.e., *Aspergillus ochraceus*, *A. flavus*, and *A. niger* (Eunae and Kwon 2012). The major constituents of clove oil are citronellal, neral, and geranial; these compounds show antifungal activity against various dermatophytes, such as *Microsporum canis*, *Trichophyton mentagrophytes*, and *Microsporum gypseum* (Park et al. 2007).

9.4.18 Family Piperaceae

The family Piperaceae or pepper family includes many small trees, shrubs, or herbs. Most plants are used as spices. For example, the essential oils from fruits of *Piper barberi* Gamble showed antifungal activity against *Aspergillus foetidus*, *A. fumigates*, *A. ochraceus*, *A. flavus*, and *Penicillium notatum*. Also, Raju and Maridass (2011) found that *Piper nigrum* L. volatile oils show antifungal activities against *Fusarium graminearum* (Singh et al. 2004a, 2005a).

9.5 The Antifungal Activity of Plant Essential Oil and Their Constituents

The essential oils (EOs) are complex mixtures of natural organic compounds which are predominantly composed of terpenes (hydrocarbons) such as pinene, myrcene, terpinene, limonene, p-cymene, and phellandrene; terpenoids (oxygen-containing hydrocarbons), i.e., acyclic monoterpene alcohols (geraniol, linalool) and monocyclic alcohols (menthol, 4-carvomenthenol, terpineol, carveol, borneol); aliphatic aldehydes (citral, citronellal, perillaldehyde); aromatic phenols (carvacrol, thymol, safrole, eugenol); bicyclic alcohol (verbenol); monocyclic ketones (menthone, pulegone, carvone); bicyclic monoterpenic ketones (thujone, verbenone, fenchone); and acids (citronellic acid, cinnamic acid) and esters (linalyl acetate).

Some of the essential oils (EOs) contain oxides such as (1,8-cineole), sulfur-containing constituents, methyl anthranilate, coumarins, zingiberene, curcumene, farnesol, sesquiphellandrene, turmerone, and nerolidol.

The chemical structures of some of the essential oils are shown in Fig. 9.1; many essential oils possess potent biological activity and are responsible for the bitter taste and toxic properties. The toxicity of some essential oil compounds is presented in Table 9.1.

The antifungal activities of certain essential oils and/or their constituents have been investigated and showed effective against many plant pathogenic fungi such as *Botrytis cinerea* and *Monilinia fructicola* (Alviano et al. 2005; Arshad et al. 2014); *Rhizoctonia solani*, *Fusarium moniliforme*, *Sclerotinia sclerotiorum*, and *F. oxysporum* (Taso and Zhou 2000); *Cymbopogon nardus* (Al-jafari et al. 2011);

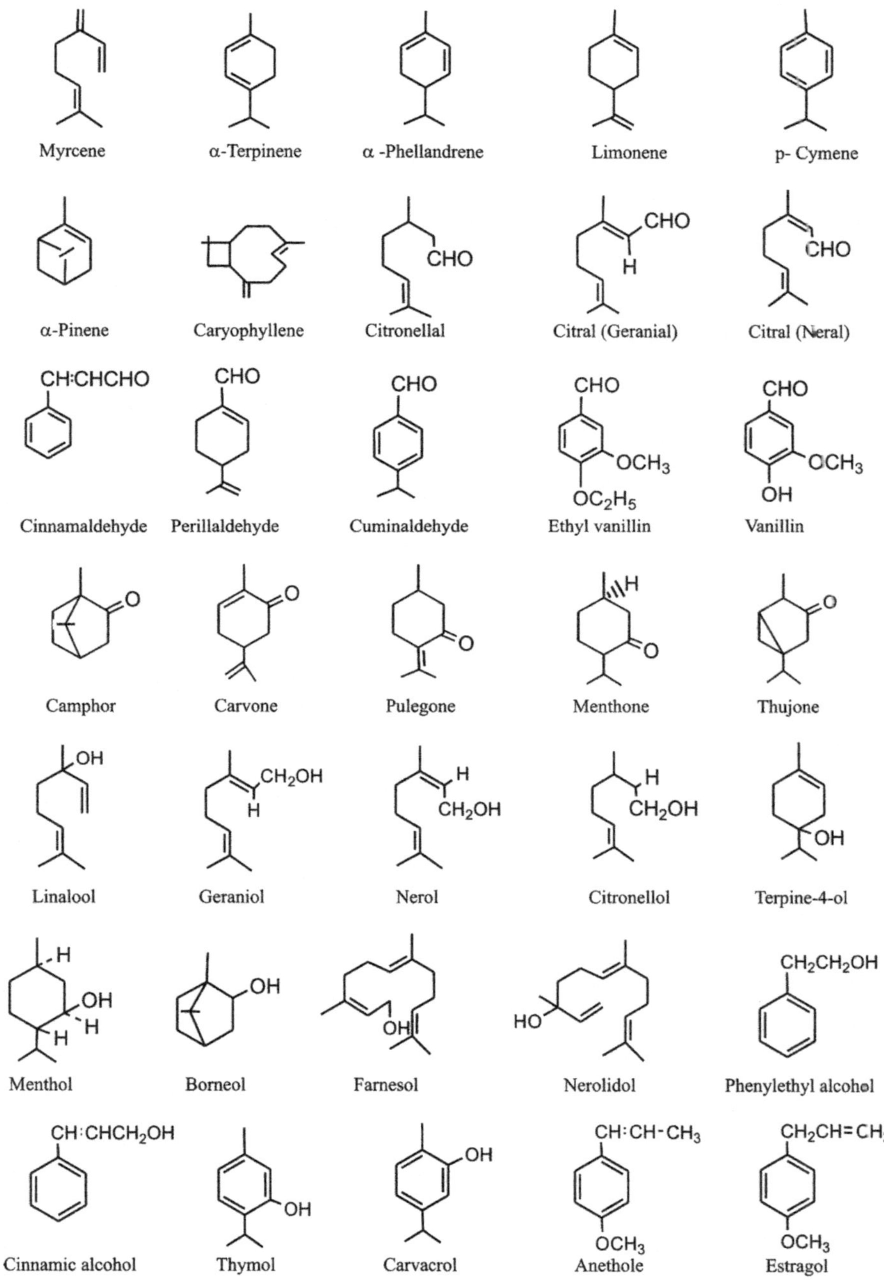

Fig. 9.1 Chemical structures of essential oil constituents

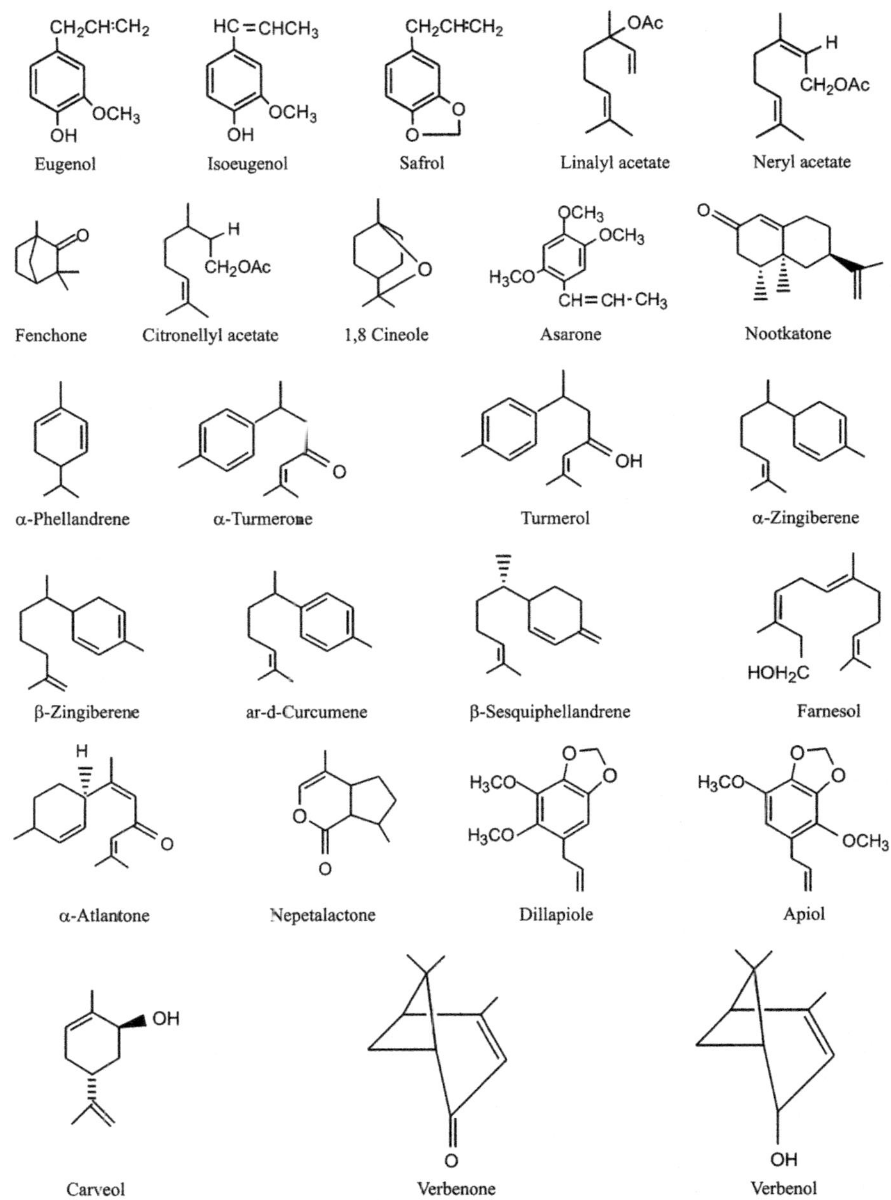

Fig. 9.1 (continued)

Table 9.1 Toxicity (mammalian) of some essential oil compounds

Compound	Animal tested	Route	LD50 (mg/kg)
2-Acetonaphthone	Mice	Oral	599
Apiol	Dogs	Intravenous	500
Anisaldehyde	Rats	Oral	1510
trans-Anethole	Rats	Oral	2090
(+) Carvone	Rats	Oral	1640
1,8-Cineole	Rats	Oral	2480
Cinnamaldehyde	Guinea pigs	Oral	1160
	Rats	Oral	2220
Citral	Rats	Oral	4960
Dillapiole	Rats	Oral	1000–1500
Eugenol	Rats	Oral	2680
3-Isothujone	Mice	Subcutaneous	442.2
d-Limonene	Rats	Oral	4600
Linalool	Rats	Oral	>1000
Maltol	Rats	Oral	2330
Menthol	Rats	Oral	3180
2-Methoxyphenol	Rats	Oral	725
Methyl chavicol	Rats	Oral	1820
Methyl eugenol	Rats	Oral	1179
Myrcene	Rats	Oral	5000
Pulegone	Mice	Intraperitoneal	150
1-Terpinene	Rats	Oral	1680
Terpinen-4-ol	Rats	Oral	4300
Thujone	Mice	Subcutaneous	87.5
Thymol	Mice	Oral	1800
	Rats	Oral	980

Sources: Dev and Koul (1997), FAO (1999), and Koul (2005)

Aspergillus niger and *A. flavus* (Paster et al. 1995; De-Billerbeck et al. 2001; Alviano et al. 2005); *Penicillium digitatum*, *F. solani*, *R. solani*, *Pythium ultimum*, and *Colletotrichum lindemuthianum* (El-Mohamedy 2003; El-Mohamedy et al. 2013, 2015); and *Alternaria padwickii* and *Bipolaris oryzae* (Nguefack et al. 2007). Also, thymol and carvacrol are active against many plant and human pathogenic fungi (El-Mohamedy 2003; Tsao and Zhou 2000). The mode of action and the mechanism of action of such EO compounds against pathogenic fungi were related to their general ability to dissolve or otherwise disrupt the integrity of cell walls and membranes (Isman and Machial 2006).

The effectiveness of some plant essential oils as soil fumigants to manage bacterial wilt disease caused by *Ralstonia solanacearum* in tomato has been investigated under greenhouse experiments As the infested with *R. solanacearum* was treated with the some plant essential oils at 400 and 700 mg per liter of soil in greenhouse, then the population densities of *R. solanacearum* were determined just before 0 and 7 days after treatment. Results indicate that populations declined to undetectable levels in thymol, palmarosa oil, and lemongrass oil treatments at both tested concentrations, whereas tea tree oil had no effect. Tomato seedlings transplanted in soil treated with 700 mg/L of thymol, 700 ml/L of palmarosa oil, and 700 ml/L of lemongrass oil were free from bacterial wilt; also, 100% of plants in thymol treatments were free of *R. solanacearum* (Pradhanang et al. 2003).

Under field experiments, many essential oils or their constituents showed effectiveness against many phytopathogenic fungi and resulted in control of many plant fungal diseases. In this respect, Yaheia et al. (2011) found that citral, methyl anthranilate, and nerol at concentration of 5.0 ml/L caused complete reduction in linear growth of *Alternaria tenuis* and *Cercospora beticola* the causal agents of *Alternaria* and *Cercospora* leaf spot diseases on sugar beet plants. EO compounds such as citral, methyl anthranilate, and nerol at 5.0 ml/L reduced the *Alternaria* leaf spot disease severity by more than 78.3 and 80.0% and *Cercospora* leaf spot severity more than 67.5% and 78.1%, respectively. El-Mohamedy et al. (2013) noted that some essential oils (EOs) such as lemongrass, thyme, citral, and nerol clearly caused strong inhibition of the growth of many root rot soilborne pathogenic fungi, i.e., *Fusarium oxysporum radicis-lycopersici*, *F. oxysporum lycopersici*, *F. solani*, *Rhizoctonia solani*, *Sclerotium rolfsii*, *Macrophomina phaseolina*, *Pythium* sp., and *Phytophthora* sp. The mycelial growth decreased significantly with the increase in concentrations of essential oils and reached minimum mycelia growth at the highest concentration used, as the complete inhibition of fungal growth was observed at 1.5% of all tested essential oils.

In addition, essential oil was applied alone or in combination with biocontrol agents and/or natural products for controlling plant pathogenic fungi. Abdel-Kader et al. (2013) tested the combination between *T. harzianum* and some essential oils in control of peanut crown rot disease under field conditions. They recorded that all concentrations of essential oils significantly reduced the growth of *A. niger*. Complete reduction was obtained with thyme and lemongrass oils at 0.5%. Also, *T. harzianum* isolates significantly reduced the growth of *A. niger*. The highest reduction was obtained with isolate no. 1 which reduced the growth by 81.1%. Meanwhile, under field conditions, all applied treatments significantly reduced the peanut crown rot disease. The highest reduction was obtained with combined treatments compost + *T. harzianum* + thyme and compost + *T. harzianum* + lemongrass which reduced the disease incidence at both pre- and postemergence growth stages.

Also, El-Mohamedy and Abd-El-latif (2015) applied humic acid and thyme essential oil alone or in combination for controlling late blight diseased of tomato plants under field conditions. Thyme oil at all tested concentrations caused complete inhibition in linear growth of *P. infestans*. The highest inhibition in linear growth was obtained with thyme at 6.0 ml/L which reduced the linear growth by 92.2%. Meanwhile, under field conditions, the authors revealed that the most effective treatments are humic acid at 6.0 or 8.0 g LG1 combined with thyme at 6.0 or 8.0 mL LG1 which reduced the late blight incidence more than 81.3 and 78.4% during the first and second growing seasons, respectively.

In another field trials, El-Mohamedy et al. (2016) noted that *Moringa oleifera* seed oil (MSO) and leaves extract (MLE) as well as chitosan (CH) alone or in integrative treatments successfully controlled early blight disease of potato plants caused by *Alternaria solani*, as chitosan at 4.0 and 6.0 g/L, MSO at 3.0%, CH + MSO at 1 g/L + 1.5% or 2 g/L+ 1.0%, and mixture of CH + MLE + MSO at 2 g/L+ 20% +1.5% resulted in complete inhibition of linear growth and sporulation of *A. solani*. Meanwhile, under field trials, integrative treatments of chitosan (CH) and *M. oleifera* leaves extract (MLE) or seed oil (MSO) successfully suppressed early blight of potato plants during 2014 and 2015 seasons. The highest records of disease reduction were obtained with the mixture of CH+MLE+MSO (1.0 g/L+20%+1.5%) followed by CH+MSO (2.0 g/L+1.5%) and CH+MLE (2.0 g/L+20%) treatments, which reduced the early blight disease severity by 94.4%, 88.2%, and 84.4% and 92.8%, 85.2%, and 80.8%, respectively, during the two growing seasons.

9.6 Plant Essential Oils (EOs) for Controlling Postharvest Pathogenic Fungi

In general, the characteristics of the fruit, desirable storage time, and decay (disease) must be based mainly for selection of the essential oil that used in a postharvest treatment. The application methods are very important in enhancing the potential of the antifungal activity of tested essential oils. The potential of using many essential oils as dipping or spraying methods to control postharvest decay has been investigated and applied in fruit and vegetables (Saikia et al. 2001; Pyun and Shin 2006; Pitipong et al. 2009). The length of storage may also negatively influence the antifungal activity of the essential oil treatments. Therefore, treatments using essential oils should only be used for short storage times, or they should be repeated after a defined time period has elapsed, depending on the fruit cultivar in question (Adam et al. 1998). The essential oils from basil (*Ocimum basilicum* L.), fennel (*Foeniculum sativum* Mill.), lavender (*Lavandula officinalis* Chaix), marjoram (*O. majorana* L.), oregano (*O. vulgare* L.), peppermint (*Mentha piperita* L.), rosemary (*Rosmarinus officinalis* L.), sage (*Salvia officinalis* L.), savory (*Satureja montana* L.), thyme (*T. vulgaris* L.), and wild mint (*Mentha arvensis* L.) showed a potentially significant antifungal activity higher than that available from chemical treatments in postharvest treatments against *B. cinerea* and *P. expansum* on apples (Adam et al. 1998).

Many plant essential oils have shown an antifungal effect against several postharvest pathogens, e.g., *B. cinerea* Pers. (Odom et al. 2000; Barkley et al. 2006; Sheetal and Singh 2008), *Aspergillus* spp. and *Penicillium* spp. (Alvarez-Castellanos et al. 2001; Govinden-Soulange et al. 2004; Kordali et al. 2005), *Fusarium* spp. (Romagnoli et al. 2005; Cardenas et al. 2005), *R. stolonifer* (Ehrenb.: Fr.) Vuill (Tundis et al. 2005; Demirci et al. 2006; Nurhayat et al. 2007), and *C. gloeosporioides* Penz. (Altintas et al. 2006).

The expression of disease will be restricted by the impeding of the initial infection and the subsequent mycelial spread beyond the infection site. So, the antifungal effect of essential oils is attributed mainly to the inhibition of both mycelial growth and spore germination (Sharma and Tripathi 2006). Carvacrol one of the major constituents of oregano essential oil is a phenol that was reported to show a high inhibition of mycelium growth in *Neofabraea alba* (E.J. Gutrie) Verkley on apples (Sun et al. 2007). Clove oil is an essential oil extracted from the clove plant, *Syzygium aromaticum* (L.) Merr. and Perry, and has been reported to act as a bioactive against *B. cinerea*, *M. fructigena* Honey, *P. expansum* Link, and *Phlyctema vagabunda* Desm (Al-Burtamani et al. 2005).

The different efficacies of essential oils are due to the differing fungitoxic properties of each single active constituent, as well as the synergic effect (Adam et al. 1998; Sun et al. 2007). This hypothesis suggests that the possible phytotoxic effects of essential oil treatments may be due to the same active components existing in each essential oil.

The combination between various postharvest treatments may improve the efficacy of controlling postharvest pathogens (Baratta et al. 1998; Sharma and Tripathi 2006; Singh and Rai 2000). El-Mohamedy et al. (2015) investigate the efficiency of chitosan essential oil amended coating application on citrus fruits to control postharvest diseases as fungicide alternatives. Different concentrations of chitosan and essential oils (EOs) such as lemongrass and citral were applied as individually or in combination treatments against *Penicillium digitatum* and *P. italicum* the main pathogens of rotting citrus fruit. They noted that, in vitro, chitosan, lemongrass EO, citral EO, and chitosan EO mixtures significantly reduced the linear growth and spore germination of *P. digitatum* and *P. italicum*. Lemongrass and citral EOs at 6 ml/L as well as chitosan + citral or lemongrass EO mixtures at 3 g/L+ 3 ml/L or 4 g/L+ 4 ml/L caused complete growth reduction of *P. digitatum* and *P. italicum*. Moreover, these treatments cause 100% protection of peel dices of orange and lime fruits artificially infected with *P. digitatum and P. italicum*. In storage trials, coated orange and lime fruits with combined treatments of chitosan + citral or lemongrass at concentrations 4.0 g/L + 4.0 ml/L or 3 g/L + 3 ml/L caused significantly protective effect against green and blue mold diseases and prevented the development of fruit decay due the mold incidence during 40 days of storage at 20 °C.

Essential oils from thyme (*T. capitatus* L.), spearmint (*M. spicata* L.), and anise (*Pimpinella anisum* L.) exhibited inhibitory effects on the development of *M. fructicola* (G. Winter) Honey (Singh and Rai 2000; Baratta et al. 1998). Rios and Recio (2005) noted that the essential oil of lemon myrtle (*Backhousia citriodora* F. Muell.) has antifungal activity against *M. fructicola* and has been shown to have a strong antimicrobial activity, mainly with respect to its potential application as a topical pharmaceutical product. The main constituent of such oil is citral that had also been reported to exhibit a fungitoxic effect against a range of postharvest pathogens. Also, El-Mohamedy et al. evaluated antifungal activity of three chemical groups of citrus essential oil constituents, i.e., alcohols, aldehydes, and esters, against linear growth and sporulation of *Penicillium digitatum*, *P. italicum*, and *Geotrichum candidum*. They found that alcohols such as nerol, alpha-terpineol, citronellol, and linalool at 4 and 6 ml/L showed 100% inhibition of linear growth and sporulation of all pathogens. Nerol showed the highest effect against *G. candidum*, where the linear growth and sporulation were completely inhibited at 2 ml/L. Concerning aldehyde compounds, only citral caused 100% inhibition of growth and sporulation of all tested fungi at 4 and 6 ml/L, but its effect on all pathogens was fungistatic. Meanwhile, citronellal had the least effect against all pathogens. Esters such as limonene showed the least effect against all pathogens at all concentrations. In vitro trials, peel discs of orange and lime fruit treated with nerol, alpha-terpineol, citronellol, or citral caused a reduction of colonization of all pathogens by 80% or more at 2 ml/L and by 67% at 1 ml/L. The authors recorded that alcohols, such as nerol, alpha-terpineol, citronellol, and aldehyde such as citral, a constituent of citrus essential oils, can be used for controlling postharvest decay of citrus fruit.

In addition, El-Mohamedy (2003) revealed that the essential oils of thyme, lemongrass, sour orange, and grapefruit at 4 and 6 ml/L concentrations showed 98–100% inhibitory effect on linear growth and sporulation of all tested pathogens. Orange and mandarin oils had inhibitory effects less than 90% at 6 ml/L and 80% at least with 4 ml/L, while lime oil gave 100% inhibitory effect at 6 ml/L and more than 90% at 4 ml/L of all tested pathogens. *Geotrichum candidum* was sensitive to all tested essential oils followed by *P. digitatum*, while *P. italicum* was more resistant. Thyme and lemongrass showed a fungitoxic effect against all pathogens at 4 and 6 ml/L, whereas sour orange, grapefruit, and lime oils had a fungicidal effect at 6 ml/L against *P. digitatum*. Orange and mandarin oils cannot completely inhibit the growth or sporulation of all pathogens at 4 or 6 ml/L. Treated peel discs of orange and lime fruits with thyme, lemongrass, sour orange, and grapefruit essential oils reduced fungal colonization of *P. digitatum* by 86–98%, *P. italicum* by 82.6–95.7%, and *G. candidum* by 90–100% at 4 ml/L of all oils. Also, Abd-alla et al. (2011) found that vapor compounds, i.e., lemongrass and thyme oils, as well as peracetic acid significantly reduced the linear growth and spore germination of *B. cinerea*. Complete inhibition of linear growth and spore germination were observed with lemongrass oil and thyme oil vapors at concentration of 100.0 μl/L of each. Meanwhile, the same effect was recorded at concentration of 0.50 ml/L. In vivo pre-inoculated strawberry fruits were submersed in a concentration of 2.0 ml/L of essential oil emulsion lemongrass and thyme oils, a method that ensured complete penetration of the oil into treated fruits resulted in about 90.0% reduction in disease incidence.

9.7 Mode of Action and Mechanism of the Essential Oil Antifungal Activity

Several investigators recorded that there is a synergetic effect between the individual oils and their chemical constituents. This synergism in the aromatic plants' component functions makes them more effective and reduces the developing resistance of any pathogenic fungi. In this respect, Juvenet al. (1994) noted that thymol, eugenol, and carvone are widely used in the control of several fungi, particularly those which contaminate various important economic crops. They also found that some single constituents of these EOs such as carvacrol, γ-terpinene, and p-cymene become more effective when they are combined together and act synergistically. Moreover, p-cymene is efficient facilitator of the transport of carvacrol across cell wall components and the cytoplasmic membrane of pathogenic fungi. Also, Nakamura et al. (2004) and Sunita and Mahendra (2008) revealed that the diameter reduction and lyses of the hypha wall of the pathogenic fungi may be attributed to the enzymatic reactions within the essential oil which act to regulate synthesis of the wall. Furthermore, the lipophilic properties of the abovementioned single components might have the ability to degrade the plasma membrane and thus to increase the permeability of the cytoplasm.

9.8 Conclusions

Plant pathogenic fungi cause serious diseases on many plants that led to losses in quantity and quality of yield. The plant essential oils (EOs) are one of the promising eco-friendly fungicide alternatives used for controlling several phytopathogenic fungi, food contaminants, and decays. In this chapter, the highlight was dropped on plant sources, antifungal properties, and the mode of action of many plant essential oil and their constituents. Most of such essential oils and their formulations reported in this chapter showed a high efficacy against many fungal pathogens either via direct contact or through incorporation into nutrient media. Many EOs such as oregano, thyme, and vervain and their active constituents such as lemongrass, nerol, carvacrol, thymol, linalool, citral, and their isomers have fungistatic and/or fungicidal activities against plant and human pathogenic fungi as well as mycotoxic producing fungi on food. In addition, recently many essential oils were applied as natural or botanical fungicides for controlling many plant pathogenic fungi under greenhouse and field conditions especially under organic forming systems. Also, a lot of compounds of EOs were applied as natural biocides for controlling postharvest and decayed fungi on fruits and vegetables. The most effective essential oils or their active ingredients and their isomers can be commercially applied effectively as seed, soil, and foliar spray treatments controlling postharvest decay fungi, pathogenic fungi on plants, and food productions.

References

Abd-AllA MA, Abd-El-Kader MM, Abd-El-Kareem F, El-Mohamedy RSR (2011) Evaluation of lemongrass, thyme and peracetic acid against gray mold of strawberry fruits. J Agric Technol 7(6):1775–1787

Abdel-Kader MM, Abdel-Kareem F, El-Mougy NS, El-Mohamady RS (2013) Integration between compost, trichoderma harzianum and essential oils for controlling peanut crown rot under field conditions. J Mycol 7:1–7

Adam K, Sivropoulu A, Kokkini S, Lanaras T, Arsenakis M (1998) Antifungal activities of Origanum vulgare subsp. hirtum, Mentha spicata, Lavandula angustifolia and Salvia fruticosa essential oils against human pathogenic fungi. J Agric Food Chem 46:1739–1745

Aggarwal KK, Ahmed A, Santha TRK, Jain N, Gupta SK, Kumar V, Khanuja SPS (2000) Antimicrobial activity spectra of Pelargonium graveolens L. and Cymbopogon winterrianus Jowitt. oil constituents and acyl derivatives. J Med Aroma Plant Sci 22:544–548

Aggrawal R, Kharya MD, Srivastava R (1979) Antimicrobial and anthelmintic activities of the essential oil of Nigella sativa Linn. Indian J Exp Biol 17(11):1264–1265

Ahonkhai I, Ayinde BA, Edogun O, Uhuwmangho MU (2009) Antimicrobial activities of the volatile oils of Ocimum bacilicum L. and Ocimum gratissimum L. (Lamiaceae) against some aerobic dental isolates. Pak J Pharm Sci 22(4):405–409

Akgul A (1989) Antimicrobial activity of black cumin (Nigella sativa L.) essential oil. J Gazi Pharmacol Fac 6:63–68

Akiyama H, Fujii K, Yamasaki O, Oono T, Iwatsuki K (2001) Antibacterial action of several tannins against Staphylococcus aureus. J Antimicrob Chemother 48(4):487–491

Al-Burtamani SKS, Fatope MO, Marwah RG, Onifade AK, Al-Saidi SH (2005) Chemical composition, antibacterial and antifungal activities of the essential oil of Haplophyllum tuberculatum from Oman. J Ethnopharmacol 96:107–112

Al-Ja'fari A-H, Vila R, Freixa B, Tomi F, Casanova J, Costa J (2011) Composition and antifungal activity of the essential oil from the rhizome and roots of Ferula hermonis. Phytochemistry 72:1406–1413

Al-Jabre S, Al-Akloby OM, Al-Qurashi AR, Akhtar N (2005) Thymoquinone, an active principle of Nigella sativa, inhibited Aspergillus niger. Pak J Med Res 42(3):102–104

Altintas A, Tabanca N, Wedge DE, Kosar M, Ozek T, Baser KHC (2006) Characterization of volatile constituents of Origanum onites and their antifungal activities against phytopathogens. Presented at the 37th International Symposium on Essential Oils (ISEO 2006) Grasse, France, on Sept 10–13

Alvarez-Castellanos PP, Bishop CD, Pascual M (2001) Antifungal activity of the essential oil of flowerheads of garland chrysanthemum (Chrysanthemum coronarium) against agricultural pathogens. J Phytochem 57(1):99–102

Alviano WS, Mendonca-Filho RR, Alviano DS, Bizzo HR, Souto-Padron T, Rodrigues ML (2005) Antimicrobial activity of Croton cajucara Benth linalool-rich essential oil on artificial biofilms and planktonic microorganisms. Oral Microbiol Immunol 20:101–105

Amrouche A, Benmehdi H, Moussaoui A, Mebarki K, Chaoufi A, Saneba A (2011) Evaluation of antifungal activity of some oils from Algerian medicinal plants against Aspergillus flavus strain produced aflatoxins. J Appl Pharm Sci 01(08):48–53

Andrews MD, Burns M (2008) Common tinea infections in children. Am Fam Physician 77:1415–1420

Andriole VT (1994) Current and future antifungal therapy: new targets for antifungal agents. J Antimicrobial Chemo 44:151–162

Angioni A, Barra A, Cereti E, Barile D, Coisson JD, Arlorio M (2004) Chemical composition, plant genetic differences, antimicrobial and antifungal activity investigation of the essential of Rosmarinus officinalis L. J Agric Food Chem 52(11):3530–3535

Angioni A, Barra A, Coroneo V, Dessi S, Cabras P (2006) Seasonal plant part chemical variability and antifungal activity investigation of Lavandula stoechas L. ssp. Stoechas essential oils. J Agric Food Chem 54:4364–4370

Apisariyakul A, Vanitanakom N, Bunddhasukh D (1995) Antifungal activity of turmeric oil extracted from Curcuma longa. J Ethnopharmacol 49:163–169

Arshad Z, Muhammad AH, Rashad WK, Muhammad MK (2014) Role of essential oils in plant diseases protection: a review. Int J Chem Biochem Sci 6:11–17

Azzouz MA, Bullerman LB (1982) Comparative antimycotic effects of selected herbs, spices, plant components and commercial antifungal agents. J Food Prot 45:1298–1301

Baratta MT, Dorman HJD, Deans SG, Figueiredo AC, Barroso JG, Ruberto G (1998) Antibacterial and anti-oxidant properties of some commercial essential oils. Flav Frag J 13:235–244

Barkley TM, Brouillet L, Strother JL (2006) In: Flora of North America Editorial Committee. (ed) 1993 Magnoliophyta: asteridae: asteraceae, part 1 vol. 19, part 2 vol. 20, and part 3 vol. 21. Flora of North America North of Mexico, New York

Bouchra C, Achouri M, Idrissi HLM, Hmamouchi M (2003) Chemical composition and antifungal activity of essential oils of seven Moroccan Labiatae against Botrytis cinerea Pers: Fr. J Ethnopharmacol 89:165–169

Bozin B, Duckic NM, Simin N, Anackov G (2006) Characterization of the volatile composition of essential oils of some Lamiaceae spice and the antimicrobial and antioxidant activities of the entire oils. J Agric Food Chem 54:1822–1828

Burt S (2004) Essential oils: their antibacterial properties and potential applications in foods – a review. Int J Food Microbiol 94:223–253

Cardenas NC, Zavala MA, Aguirre JR, Perez C, Perez S (2005) Chemical composition and antifungal activity of essential oil of Chrysactinia mexicana gray. J Agric Food Chem 53(11):4347–4349

Cavaleiro C, Pinto E, Gonçalves MJ, Salgueiro L (2006) Antifungal activity of Juniperus essential oils against dermatophyte. Aspergillus and Candida strains. J Appl Microbiol 100:1333–1338

Chao SC, Young DG (2000) Screening of inhibitory activity of essential oils on selected bacteria, fungi and viruses. J Essent Oil Res 12:630–649

Cheng SS, Wu CL, Chang HT, Kao YT, Chang ST (2004) Antitermitic and antifungal activities of essential oil of Calocedrus formosana leaf and its composition. J Chem Ecol 30(10):1957–1967

Chuang PH, Lee CW, Chou JY, Murugan M, Shieh BJ, Chen HM (2007) Antifungal activity of crude extracts and essential oil of Moringa oleifera Lam. Bioresour Technol 98:232–236

Cimanga K, Kambu K, Tona L, Apers S, De Bruyne T, Hermans N (2002) Correlation between chemical composition and antibacterial activity of essential oils of some aromatic medicinal plants growing in the Democratic Republic of Congo. J Ethnopharmacol 79:213–220

Crespo ME, Jimenez J, Gomis E, Navarro C (1990) Antimicrobial activity of essential oil of Thymus serpylloides subspecies gadorensis. Micro Bios 61:181–184

D'Auria FD, Tecca M, Strippoli V, Salvatore G, Battinelli L, Mazzanti G (2005) Antifungal activity of Lavandula angustifolia essential oil against Candida albicans yeast and mycelial form. Med Mycol 43(5):391–396

Dalleau S, Cateau E, Berges T, Berjeaud J, Limbert C (2007) In vitro activity of essential oils and their major components against Candida albicans yeasts growing planktonically and as biofilms. Int J Antimicrob Agents 29(2):147

De-Billerbeck VG, Roques CG, Bessiere JM, Fonvieille JL, Dargent R (2001) Effects of Cymbopogon nardus (L) essential oil on the growth and morphogenesis of Aspergillus niger. Can J Microbiol 47:9–17

Demirci B, Baser KHC, Tabanca N, Wedge DE (2006) Antifungal activity of Haplopappus greenei essential oil towards phytopathogenic Colletotrichum species. J Agric Food Chem 54:3146–3150

Dev S, Koul O (1997) Insecticides of natural origin. Harwood Academic Publishers, Amsterdam

El-Mohamedy RSR (2003) Effect of various essential oils against some post harvest pathogenic fungi of citrus fruits. Egypt J Appl Sci Zagazg Univ Cairo 18(2):42–56

El-Mohamedy RSR, Abd-El-latif FM (2015) Field application of Humic acid and thyme essential oil for controlling late blight disease of tomato plants under field conditions. Asian J Plant Pathol 9(4):167–174

El-Mohamedy RSR Abdallah AM (2014) Antifungal activity of Moringa oleifera oil and seed extract against some plant pathogenic fungi Middle East J Agric Res, 3(2): 242–249, 2014

El-Mohamedy RSR, Abd-El-Kareem F, Abd-Alla MA (2002) Effect of some constituents of citrus essential oil against post harvest pathogenic fungi of citrus fruits. Arab Univ Agric Sci Ain Shams Univ Cairo 10:335–350

El-Mohamedy RSR, Abdel-Kader MM, Abd-El-Kareem F, El-Mougy NS (2013) Essential oils, inorganic acids and potassium salts as control measures against the growth of tomato root rot pathogens in vitro. J Agric Technol 9(6):1507–1520

El-Mohamedy RSR, El-Gamal NG, Bakeer ART (2015) Application of chitosan and essential oils as alternatives fungicides to control green and blue moulds of citrus fruits Int. J Curr Microbiol App Sci 4(6):629–643

El-Mohamedy RSR, Abdallah AM, Ghoname AA (2016) Field application of chitosan and Moringa oleifera extracts as fungicides alternatives to control early blight and improvement growth and yield quality of potato. Plant Pathol J 15(4):135–143

Eugenia P, Vale-Silva L, Cavaleiro C, Salgueiro L (2009) Antifungal activity of the clove essential oil from Syzygium aromaticum on Candida, Aspergillus and dermatophyte species. J Med Microbiol 58(11):1454–1462

Eunae K, Il-Kwon P (2012) Fumigant antifungal activity of Myrtaceae essential oils and constituents from Leptospermum petersonii against three Aspergillus species. Molecules 17:10459–10469

Ewansiha JU, Garba SA, Mawak JD, Oyewole OA (2012) Antimicrobial activity of Cymbopogon Citratus (lemon grass) and its phytochemical properties. Front Sci 2(6):214–220

FAO (1999) The use of spices and medicinals as bioactive protectants for grains. Agric Serv Bull 137:201–213

Farag RS, Daw ZY, Abo- Raya SH. (1989) Influence of some spice essential oils on Aspergillus parasiticus growth and production of aflatoxins in synthetic medium. J Food Sci 54:74–76

Fernanda CS, Sara MC, Virginia MS, Deila MS, Botelho NL, Luis RB (2012) Evaluation of antifungal activity of essential oils against potentially mycotoxigenic Aspergillus flavus and Aspergillus parasiticus. Braz J Pharma 22(5):1002–1010

Govinden-Soulange J, Magan N, Gurib-Fakim A, Gauvin A, Smadja J, Kodja H (2004) Chemical composition and in vitro antimicrobial activities of the essential oils from endemic Psiadia species growing in mauritius. Biol Pharm Bull 27(11):1814–1818

Gupta AK, Katz HI, Shear NH (1999) Drug interactions with itraconazole, fluconazole and terbinafine and their management. J Am Acad Dermatol 41:237–249

Gupta AK, Chow M, Daniel CR, Aly R (2003) Treatment of Tinea pedis. Dermatol Clin 21:431–462

Hanafy MS, Hatem ME (1991) Studies on the antimicrobial activity of Nigella sativa seed (black cumin). J Ethnopharmacol 34:275–278

Hili P, Evans CS, Veness RG (1997) Antimicrobial action of essential oils: the effect of dimethylsulphoxide on the activity of cinnamon oil. Appl Environ Microbiol 24:269–275

Holly RA, Patel D (2005) Improvement in shelf-life and safety of perishable foods by plant essential oils and smoke antimicrobials. Food Microbiol 22:273–292

Hong EJ, Na KJ, Choi IG, Choi KC, Jeung EB (2004) Antibacterial and antifungal effects of essential oils from coniferous trees. Biol Pharm Bull 27:863–866

Iacobellis NS, Cantore PL, Capasso F, Senatore F (2005) Antibacterial activity of Cuminum cyminum L. and C. arumcarvi L. essential oils. J Agric Food Chem 53:57–61

Inouye S, Uchida K, Abe S (2006) Vapor activity of 72 essential oils against a Trichophyton mentagrophytes. J Infect Chemother 12:210–216

Isman MB, Machial CM (2006) Pesticides based on plant essential oils: from traditional practice to commercialization. In: Rai M, Carpinella MC (eds) Naturally occurring bioactive compounds. Elsevier, BV, Amsterdam, pp 29–44

Jantan IB, Moharam BAK, Santhanam J, Jamal JA (2008) Correlation between chemical composition and antifungal activity of the essential oils of eight Cinnamomum species. Pharma Biol 46:406–412

Juven BJ, Kanner J, Schved F, Weisslowicz H (1994) Factors that interact with the antibacterial action of thyme essential oil and its active constituents. J Appl Bacteriol 76:626–631

Kawther FA (2007) Antimicrobial activity of essential oils of some medicinal plants from Saudi Arabia. Saudi J Biol Sci 14(1):53–60

Kazmi SAR, Shahzad S, Niaz I (1995) Effect of neem oil on in vitro growth of root infecting fungi. Pak J Bot 27(1):217–220

Khan MA, Ashfaq MK, Zuberi HS, Mahmood MS, Gilani AH (2003) The in vivo antifungal activity of the aqueous extract from Nigella sativa seed. Phytother Res 17(2):183–186

Kishore N, Mishra AK, Chansouria JP (1993) Fungal toxicity of essential oils against dermatophytes. Mycoses 36:211–225

Knobloch L, Weigand H, Weis N, Schwarn HM, Vigenschow H (1985) Action of terpenoids on energy metabolism. In: Brunke E-J (ed) Progress in essential oil research. Walter de Gruyter, USA, pp 429–448

Komiya M, Takeuchi T, Harada E (2006) Lemon oil vapor causes an anti-stress effect via modulating the 5-HT and DA activities in mice. Behav Brain Res 172(2):240–249

Kordali S, Cakir A, Mavi A, Kilic H, Yildirim A (2005) Screening of chemical composition and antifungal and antioxidant activities of the essential oils from three Turkish artemisia species. J Agric Food Chem 53(5):1408–1416

Koul O (2005) Insect antifeedants. CRC Press, Bota Racon

Koul O, Suresh W, Dhaliwal GS (2008) Essential oils as green pesticides: potential and constraints. Biopestic Int 4(1):63–84

Lambert RJW, Skandamis PN, Coote P, Nychas GJE (2001) A study of the minimum inhibitory concentration and mode of action of oregano essential oil, thymol and carvacrol. J Appl Microbiol 9:453–462

Lawerence BM, Reynolds RJ (2001) Progress in essential oils. Perf Flavour 26:44–52

Lecha M, Effendy I, Feuilhade de CM, Di Chiac-chio N, Baran R (2005) Treatment options development of consensus guidelines. J Eur Acad Dermatol Venereol 19:25–33

Lee SO, Choi GJ, Jang KS, Lim HK, Cho KY, Kim JC (2007) Antifungal activity of five plant essential oils as fumigant against postharvest and soil borne plant pathogenic fungi. Plant Pathol J 23(2):97–102

Lis-Balchin M, Deans SG (1997) Bioactivity of selected plant essential oils against Listeria monocytogenes. J Appl Bacteriol 82:759–762

Locke JE (1995) Fungi. In: Schmutterer H (ed) The neem tree, source of unique National Products for integrated pest management, medicine, industry and other proposes. V.C.H, Weinheim, pp 118–125

Lopez-Reyes JG, Spadaro D, Prelle A, Garibaldi A, Gullino ML (2013) Efficacy of plant essential oils on postharvest control of rots caused by fungi on different stone fruits in vivo. J Food Prot 76:631–639

Martini H, Weidenborner M, Adams S, Kunz B (1996) Eugenol and carvacrol: the main fungicidal compounds in clove. Ital J Food Sci 1:63–67

Mashhadian NV, Rakhshandeh H (2005) Antibacterial and antifungal effects of Nigella sativa extracts against S. aureus, P. aeruginosa and C. albicans. Pak J Med Sci 21(1):47–52

Montes-Belmont, Carvajal (1998) Control of Aspergillus flavus in maize with plant essential oils and their components. J Food Prot 61:616–619

Nakamura CV, Ishida K, Faccin LC, Filho BP, Cortez DA, Rozental S (2004) In vitro activity of essential oil from Ocimum gratissimum L. against four Candida species. Res Microbiol 155(7):579–586

Nguefack J, Nguikwie SK, Fotio D, Dongmo B, Zollo PH, Amvam LV, Nkengfack AE, Poll L (2007) Fungicidal potential of essential oils and fractions from Cymbopogon citratus, Ocimum gratissimum and Thymus vulgaris to control Alternaria padwickii and Bipolaris oryzae, two seed-borne fungi of rice (Oryza Sativa L). J Essent Oil Res 19:581–587

Nigam SS, Rao TST (1977) Antimicrobial efficacy of some Indian essential oils. In: Kapoor LD, Ramakrishnan (eds) Advance in essential oil industry. Today's and Tomorrow's printer and publishers, New Delhi, pp 177–180

Nurhayat T, Betul D, Sara LC, Kemal HB, David EW (2007) Chemical composition and antifungal activity of Arnica longifolia, Aster hesperius, and Chrysothamnus nauseosus essential oils. J Agric Food Chem 55:8430–8435

Odom RB, James WD, Berger TG (2000) Andrews' diseases of the skin: clinical dermatology. W.B. Saunders Company, Philadelphia, p 1135

Pandey DK, Tripathi NN, Tripathi RD, Dixit SN (1982) Fungitoxic and phytotoxic properties of the essential oil Caesulia axillaris Roxb. Angew Bot 56:259–267

Pandey KP, Mishra RK, Kamran A, Mishra P, Bajaj AK, Dikshit A (2010) Studies on antidermatophytic activity of waste leaves of Curcuma longa L. Physiol Mol Biol Plants 16:177–185

Park MJ, Gwak KS, Yang I, Choi WS, Jo HJ, Chang JW (2007) Antifungal activities of the essential oils in Syzygium aromaticum (L.) Merr.Et Perry and Leptospermum petersonii Bailey and their constituents against various dermatophytes. J Microbiol 45:460–465

Paster N, Menasherov M, Ravid U, Juven BJ (1995) Antifungal activity of oregano and thyme essential oils applied as fumigant against fungi attacking stored grain. J Food Prot 58:81–85

Patker KL, Usha CM, Shetty SH, Paster N, Lacey J (1993) Effect of spice oils on growth and aflatoxin B1 production by Aspergillus flavus. Lett Appl Microbiol 17:49–51

Patra M, Shahi SK, Midgely G, Dikshit A (2002) Utilization of essential oil as natural antifungal against nail-infective fungi. Flavour Frag J 17:91–94

Pattnaik S, Subramanyam VR, Kole C (1996) Antibacterial and antifungal activity of ten essential oils in vitro. Micro Bios 86:237–246

Pina-Vaz C, Rodrigues G, Pinto E, Costa-de-Oliveira S, Tavares C, Salgueiro L (2004) Antifungal activity of Thymus oils and their major compounds. J Eur Acad Dermatol Venereol 18:73–78

Ping-Hsien C, Chi-Wei L, Jia-Ying C, Murugan M, Bor-Jinn S, Hueih-Min C (2007) Antifungal activity of crude extracts and essential oil of Moringa oleifera lam. Bioresour Technol 98:232–236

Pitipong T, Jularat U, Apinya P Phirayot K (2009) Screening for The antifungal activity of essential oils from Bergamot oil (Citrus hystrix DC.) and Tea tree oil (Melaleuca alternifolia) against economically rice pathogenic fungi: a driving force of organic rice cv. KDML 105 production. Asian J Food Agric Ind; special issue, S374–S380

Portillo MC, Viramontes S, Muñoz LN, Gastelum MG, Nevarez GV (2005) Antifungal activity of Mexican oregano (Lippia berlandieri Shauer.) J Food Prot 68(12):2713–2717

Pradhanang PM, Momol MT, Olson SM, Jones JB (2003) Effects of plant essential oils on Ralstonia solanacearum population density and bacterial wilt incidence in tomato. Plant Dis 87:423–427

Prajapati A (2012) Synthesis, antimicrobial and insecticidal activity studies of 5-nitro N′-(Arylidenhydrazidomethyl indole) 2-(substituted aryl) -3-(N′-Indolyl Acetamiddyl)-4-Oxothiazolidines. Res J Recent Sci 1:99–104

Pyun MS, Shin S (2006) Antifungal effects of the volatile oils from allium plants against Trichophyton species and synergism of the oils with ketoconazole. Phytomedicine 13(6):394–400

Raju G, Maridass M (2011) Composition, antifungal and cytotoxic activities of essential oils of Piper barberi fruits. Int J Bio Technol 2(2):100–105

Rasooli I, Rezaei MB, Allameh A (2006) Ultrastructural studies on antimicrobial efficacy of thyme essential oils on Listeria monocytogenes. Int J Infect Dis 10:236–241

Rath CC, Dash SK, Mishra RK (2001) In vitro susceptibility of Japanese mint (Mentha arvensis L.) essential oil against five human pathogens. Indian Perfume 45:57–61

Rios JL, Recio MC (2005) Medicinal plants and antimicrobial activity. J Ethnopharmacol 100:80–84

Romagnoli C, Bruni R, Andreotti E, Rai MK, Vicentini CB, Mares D (2005) Chemical characterization and antifungal activity of essential oil of capitula from wild Indian Tagetes patula L. Protoplasma 225(1–2):57–65

Saha R, Bhupendar K (2011) Pharmacognosy and pharmacology of Nigella Sativa- a review. Int Res J Pharm 2(11):36–39

Saikia D, Khanuja SPS, Kahol AP, Gupta CS, Kumar S (2001) Comparative antifungal activity of essential oils and constituents from three distinct genotypes of Cymbopogon sp. Curr Sci 86:1264–1266

Salgueiro LR, Piato E, Goncalves MJ, Pina-Vaz C, Cavaleiro C, Rodriques AG (2004) Chemical composition and antifungal activity of the essential oil of Thymbra capitata. Planta Med 70(6):572–575

Sessou P, Farougou S, Alitonou G, Djenontin TS, Yèhouénou B, Azokpota P (2012) Chemical composition and antifungal activity of essential oil of fresh leaves of Ocimum gratissimum from Benin against six Mycotoxigenic fungi isolated from traditional cheese wagashi. Int Res J Bio Sci 1(4):22–27

Shabnam J, Sobia M, Ibatsam K, Rauf A, Saleem MH (2012) Comparative antimicrobial activity of clove and fennel essential oils against food borne pathogenic fungi and food spoilage bacteria. Afr J Biotechnol 11(94):16065–16070

Sharma N, Tripathi A (2006) Fungitoxicity of the essential oil Citrus sinensis on postharvest pathogens. World J Microbiol Biotechnol 22:587–593

Sheetal V, Singh SP (2008) Current and future status of herbal medicines. Vet World 1(11):347–350

Shelef LA (1983) Antimicrobial effects of spices. J Food Saf 6:29–44

Simic A, Sokovic MD, Ristic M, Grujic S, Vukojevic J, Marin PD (2004) The chemical composition of some Lauraceae essential oils and their antifungal activities. Phytother Res 18(9):713–717

Singh N (2001) Trends in the epidemiology of opportunistic fungal infections: predisposing factors and the impact of antimicrobial use practices. Clin Inf Dis 33(10):1692–1696

Singh R, Rai B (2000) Antifungal potential of some higher plants against Fusarium causing with diseases of Cajanus cajan. Micro Bios 102:165–173

Singh VP, Singh HB, Singh RB (1980) The fungicidal effect of neem extracts on some soil borne pathogens of gram (Cicer arietinum). Mycologia 72:1077–1093

Singh HB, Srivastava M, Singh AB, Srivastava AK (1995) Cinnamon bark oil, a potent fungitoxicant against fungi causing respiratory tract mycosis. Allergy 50:995–999

Singh G, Sumitra M, Catalan C, Lampasona MP (2004) Chemical antioxidant and antifungal activities of volatile oil of black pepper and its acetone extract. J Sci Food Agric 84:1878–1884

Singh G, Marimuthu P, Heluani CS, Catalan C (2005a) Antimicrobial and antioxidant potentials of essential oil and acetone extract of Myristica Fragrans Houtt. (Aril part). J Food Sci 70:141–148

Singh G, Sumitra M, Catalan C, Lampasona MP (2005b) Chemical, antifungal, antioxidant and sprout suppressant studies on ginger essential oil and its oleoresin. Flav Frag J 20:1–6

Sinha KK, Sinha AK, Prasad G (1993) The effect of clove and cinnamon oils on growth of and aflatoxin production by Aspergillus flavus. Lett Appl Microbiol 16:114–117

Sirirat S, Wimolpun R Sanit S (2009) Antifungal activity of essential oils derived from some medicinal plants against grey mould (Botrytis cinerea). Asian J Food Ag-Ind.; Special Issue, S229–S233

Skovic M, Tzakaou O, Pitarokili D, Couladis M (2002) Antifungal activity of selected aromatic plants growing wild in Greece. Nahrung/Food 46:317–320

Sonboli A, Salehi P, Yousefzadi M, Shahid B (2004) Antimicrobial activity and chemical composition of the essential oil of Nepeta crispa Willd. from Iran. Z Naturforsch 59(9–10):653–656

Sukatta U, Haruthaithanasan V, Chantarapanont W, Dilokkunanant U, Suppakul P (2008) Antifungal activity of clove and cinnamon oil and their synergistic against postharvest decay fungi of grape in vitro. Kasetsart J (Nat Sci) 42:169–174

Sun OL, Gyung JC, Kyoung SJ, He KL, Kwang YC, Jin-Cheol K (2007) Antifungal activity of five plant essential oils as fumigant against postharvest and soil borne plant pathogenic fungi. Plant Pathol J 23(2):97–102

Sunita B, Mahendra R (2008) Antifungal activity of essential oils from Indian medicinal plants against human pathogenic Aspergillus fumigatus and A. niger. World J Med Sci 3(2):81–88

Tsao R, Zhou T (2000) Antifungal activity of monoterpenoids against postharvest pathogens Botrytis cinerea and Monilinia fructicola. J Essent Oil Res 12:113–121

Tullio V, Nostro A, Mandras N, Dugo P, Banche G, Cannatelli MA (2007) Antifungal activity of essential oils against filamentous fungi determined by broth microdilution and vapour contact methods. J Appl Microbiol 102:1544–1550

Tundis R, Statti GA, Conforti F, Bianchi A, Agrimonti C, Sachetti G (2005) Influence of environmental factors on composition of volatile constituents and biological activity of Helichrysum italicum (Roth) Don (Asteraceae). Nat Prod Res 19(4):379–387

Tzasna H, Margarita C, Ana Maria G, Angel D, Sameul M, Patricia DJ (2008) Antifungal activity of the essential oils of two verbenaceae: Lantana achyranthifolia and Lippia graveolens of Zapotitlan de las Salinas, Puebla (Mexico). Bol Latinoam Caribe Plantas Medicinales Aromaticas 7(4):202–206

Viljoen AM, Subramoney S, Van Vuuren SF, Baser KHC, Demicri B (2005) The composition, geographical variation and antimicrobial activity of Lippia javanica. Verbenaceae leaf essential oils. J Ethnopharmacol 96:271–277

Wang SY, Chen PF, Chang ST (2005) Antifungal activities of essential oils and their constituents from indigenous cinnamon (Cinnamomum osmophloeum) leaves against wood decay fungi. Bioresour Technol 96(7):813–818

Yaheia OF, Abd-El-Kareem F, Abd-El-Latif FM, El-Mohammedy (2011) Effects of citrus essential oil compounds on management leaf spot disease on sugar beet plant under field condition. J Agric Technol 7(3):869–877

Zhang ZZ, Li YB, Qi L, Wan XC (2006) Antifungal activities of major tea leaf volatile constituents toward Colletotrichum camelliae Massea. J Agric Food Chem 54(11):3936–3940

Bacterial Volatiles: Potential Applications in Plant Growth and Health

10

Mohd Musheer Altaf, Mohd Sajjad Ahmad Khan, and Iqbal Ahmad

Abstract

Microorganisms including bacteria produce several kinds of volatile organic compounds (VOCs). In the last decade, an increasing research confirmed the role of VOCs as environmental friendly and potential growth inducers. These volatile compounds can help in mitigating dependence on synthetic chemicals and recombinant DNA technology. Recent experiments carried out under field conditions successfully established the role of bacterial volatiles in increasing food production. However, the information on the contribution of bacterial volatiles in plant growth and development is scattered. In this chapter, we have discussed the role of volatile organic compounds in microbe-microbe and microbe-plant interactions. Effect of VOCs as inducers for enhancing crop productivity is reviewed. Problems associated with field applications are also highlighted.

Keywords

Bacterial volatile organic compounds • Plant growth • Microbe-microbe interaction • Bacteria-plant interaction • Rhizosphere

M.M. Altaf (✉) • I. Ahmad
Department of Agricultural Microbiology, Faculty of Agricultural Sciences, Aligarh Muslim University, Aligarh 202002, India
e-mail: mohdmusheer@rediffmail.com

M.S.A. Khan
Department of Biology, College of Medicine, Imam Abdulrahman Al Faisal University, Dammam 31451KSA, Saudi Arabia

D.K. Choudhary et al. (eds.), *Volatiles and Food Security*,
DOI 10.1007/978-981-10-5553-9_10

10.1 Introduction

Agriculture in the twenty-first century faces challenges, associated with loss of soil fertility, increased use of synthetic fertilizers and pesticides, drought, salinity, fluctuating climatic conditions, and growing attack on crops by pathogens and pest. The global need to enhance agricultural production from rapidly shrinking land resources had placed considerable strain on the fragile agroecosystem (Tilak et al. 2005; Gomiero et al. 2011). It is the fact that intense external input-oriented agriculture, which was part of our Green Revolution Strategy, has depleted soil fertility considerably in our major agricultural systems. This is reflected in the declining trend of total production from the same land, diminishing response of food grain increase to applied fertilizer nutrient reduction in the organic matter content of the fertile soil and increase in the productivity, and constraints to land such as groundwater depletion, salinity, acidity, and alkalinity (Raman 2005). Worldwide the remarkable enhancement in agricultural and industrial production causes severe environmental degradation. Modern agriculture severely modified and polluted the natural environment due to the wide application of chemical fertilizers, herbicides, and pesticides. These fertilizers, herbicides, and pesticides are known to be mutagenic and carcinogenic (Abboud 2014). Agricultural sustainability, environmental safety, and increased crop production can be achieved by employing eco-friendly methods (Bhardwaj et al. 2014).

Despite this, there is an immediate requirement to accelerate this revolution as there is continuous increase in world population, which will be increased to 9.7 billion by 2050 (UN-DESA 2015), and this whole population will satisfy their need of food, fuel, feed, and fibers through agriculture. Further, the present global population is relying strongly on fossil oil for their energy needs and various other chemicals. Therefore, this fossil oil is exhausting at an alarming rate, and substitutes for fossil oil are required to be discovered (Jansen et al. 2011).

Volatiles are compounds with low molecular masses (<300 Da), low polarity, and high vapor pressure (0.01 kPa or higher at 20 °C). Volatile organic compounds are crucial information compounds which are directly involved in necessary conversation processes among all kingdoms of life (Hare 2011; Dweck et al. 2015; Kai et al. 2016). These compounds are universally found on planet earth owing to their physicochemical characteristics (high vapor pressure, low boiling point, and low molecular mass). Collectively, these properties assist vaporization. Distinctive volatiles are monoterpenes, aromatic compounds, and fatty acid by-products. It is clear that volatiles are components of the atmosphere. However, they also emerge in porous materials, such as plants, rocks, and soil pores, and get mixed in aqueous media. They are from both anthropogenic and biogenic origin. Although animal and plant volatile emissions have been largely evaluated in the past, the volatiles of microorganisms especially bacteria have been typically ignored. Currently the richness of bacterial volatile organic compounds (VOCs) has been identified. It is anticipated that while about 10,000 microbial species are presently characterized, no less than a million of unknown species are present on planet earth. In this situation, amazingly very little population of bacteria has been scrutinized so far for their VOC system. The additional microbes we explore in the future, the more volatiles with novel composition are likely to be revealed as this new research field develops (von Reuss et al. 2010; Lemfack et al. 2014).

VOCs occur in the environment and work over long range. In addition to airborne volatile-related interactions, furthermore the belowground volatile exchanges are also examined. Currently it has been found that plants can also detect these volatile compounds and respond to these vapors by boosting or diminishing their growth and development. The earliest recognized evidence of bacterial volatile-related influence on plants was reported by Cook and Stall (1969). It took over 30 years until 2003 when Ryu and coworkers (Ryu et al. 2003) detained the subject once more and unlock the access to a novel area of research designated as the volatile-related effects of bacteria on plants. After that, different effects have been accounted, varying from large plant biomass increase to plant death (Bailly and Weisskopf 2012; Kanchiswamy et al. 2015a; Kai et al. 2016). In spite of the strong effects noticed, the explanation of the dependable compounds is not as sophisticated as anticipated. A handful of bacterial volatile organic compounds (VOCs) have been clearly recognized as responsible for the detected harmful or beneficial effects on plant growth and development. This is related to the elevated intricacy of the volatile mixture secreted by bacteria. So far ~1,000 bacterial volatile organic compounds (VOCs) have been characterized. The variety of bacterial ecological habitats advocated that this might be a total distortion of the diversity of volatile organic and inorganic molecules secreted by bacteria (Schulz and Dickschat 2007; Kai et al. 2009; Lemfack et al. 2014).

There are no less than two causes to consider that volatiles might play a key communication role among bacteria and plants under normal circumstances. The primary cause is that volatiles are recognized to be employed by plants as caution or attraction signal in plant-plant (Heil and Karban 2010) and plant-insect interactions (Hare 2011), signifying that detection methods survive which allow plants to identify volatiles. The involved process is explained by concerning the rhizosphere environment. This appears encouraging for volatile-related messaging, because the associates are spatially close to each other and volatiles are further expected to build up and attain their action threshold in the rhizosphere than in the airborne atmosphere. Twelve years following the initial description of plant growth promotion by bacterial volatiles, the phenomenon has been established by independent experiments employing diverse strains and plant model systems (Kai et al. 2016). But still the compounds related to the recognized impacts and other metabolic changes taking place in plants upon encountering of bacterial volatiles are weakly inferred.

The aim of this chapter is to summarize the current knowledge that how the bacterial volatiles mediate plant growth and development and to mitigate the problems associated with field application. We have also reviewed suggestion and future directions of research that would help us to better understand this fascinating new feature of bacteria-plant interaction.

10.2 Methods for Identification and Characterization of Bacterial Volatile Compounds

Bacteria secrete a plethora of volatiles, and in the last few years, an ever-increasing understanding regarding the release of a large quantity of bacterial volatiles has been documented. The current examinations have undoubtedly established that bacteria use their volatiles during communications with new organisms in order to manipulate populations and communities (Kai et al. 2009; Romoli et al. 2014). A broad range of molecules has been characterized in bacterial discharge of a large number of chemicals. Bacteria are recognized to either positively or negatively influence other organism's health.

Bacterial volatiles characteristically comprise alkenes, alcohols, ketones, terpenes, benzenoids, pyrazines, acids, and esters (Piechulla and Degenhardt 2014). Most microbial volatiles are believed as by-products of primary and secondary metabolism. They are produced primarily by oxidation of glucose from different intermediates (Korpi et al. 2009; Schmidt et al. 2015). Even though there are common volatiles secreted by diverse, over and over again, dissimilar microorganisms, other volatiles are unique for certain strains (Schulz and Dickschat 2007; Garbeva et al. 2014a, b). The quantity and configuration of volatiles secreted by microbes can differ depending on culturing conditions (Blom et al. 2011; Garbeva et al. 2014a, b). Other key factors impacting the secretion of volatiles are the physiological condition of the producing microorganism, oxygen accessibility, moisture, temperature, and pH (Insam and Seewald 2010; Romoli et al. 2014).

Investigation of VOC is a complicated job owing to the broad array of volatiles' large quantity and the intricacy of mixtures and/or matrixes in which they are generally present and source of secretion. The primary method for vigorous and reproducible VOC investigation thus includes three key features: (1) VOC extraction, (2) analysis, and (3) recognition (Audrain et al. 2015). A diversity of methods has been established to trap volatiles secreted by microbes. The amount of measurable volatiles normally enhanced with the diversity and sensitivity of the extraction systems employed and the different trapping equipments (Wenke et al. 2012). It is established that the closed-loop stripping apparatus (CLSA) is extensively employed to pre-concentrate volatiles from aqueous samples (Meruva et al. 2004). Finally, direct and real-time gas sampling of bacterial headspace can be performed using secondary electrospray ionization-mass spectrometry (SESI-MS), which has several advantages, such as a sensitive detection limit and the potential for high-throughput sample analysis (Zhu and Hill 2013).

Typical methods to examine VOC profiles depend on gas chromatography together with mass spectrometry (GC-MS), characterized by a powerful separation capacity and highly sensitive recognition performance. Certainly, current GC-MS software advances the detection limit by adjacent peak deconvolution and background subtraction (Farag et al. 2006; Farag and Wessjohann 2012) strategies. Optimal GC-MS analysis comprises the use of the time-of-flight (TOF)-MS detector, particularly when it is attached with 2D column gas chromatography (GC × GC). The high degree of separation afforded by the two columns and the high sensitivity of TOF-MS are appropriate explanation for analysis of trace volatile components (Meruva et al. 2004; Salvador et al. 2013). On the other hand, the GC-MS-based technique frequently involves pre-concentration and does not permit rapid in situ analysis. Selected-ion flow tube-mass spectrometry (SIFT-MS), ion-mobility spectrometer (IMS), and electronic noses (eNoses) are thus regularly preferred for real-time analysis of volatiles (Dolch et al. 2012). SIFT-MS and IMS are compatible with in situ real-time measurement of VOCs (Bos et al. 2013). IMS and SIFT-MS are very promising analytical methods, with short measurement times performed in ambient conditions and high sensitivity (Tiebe et al. 2009; Kunze et al. 2013).

Characterization of molecules found in a volatile sample can be realized by comparing mass spectra with spectra from different databases like the Wiley or NIST libraries. Still, these libraries have a tendency to deceive the inexperienced user, since the neighboring hit inside the library might uncritically be taken as a positive detection. These databases list almost 1000 microbial volatile compounds and include information about their discharging microbes, the biochemical pathways, and their biological impacts (Lemfack et al. 2014). In case of unidentified volatiles, separation steps are needed and followed by spectral analysis employing MS and nuclear magnetic resonance (NMR) spectroscopy. Multivariate statistical processes, largely grouped into two categories as unsupervised and supervised, are required to estimate possible dissimilarity between VOCs secreted from different species, particularly when evaluating numerous samples. Employing of supervised methods such as partial least squares (PLS) or PLS discriminant analysis allows comparing samples analyzed with different machines and not in the same time, though great care must be taken in introducing these models, since such analyses often tend to overfit data, leading to abnormal outcome (Farag 2014). Lastly, a correlation-based multiblock hierarchical principal component analysis (MB-PCA) and MB-PLS enable the comparison of data sets derived from different analytical approaches, i.e., NMR versus MS.

Identified VOC descriptions are often relatively intricate, and it is thus essential to run blank controls to establish which signals correspond to volatile molecules secreted by bacteria or by the culture medium itself (Tait et al. 2014). An additional main challenge in VOC analysis lies in repeated result inconsistency due to genomic disparity among bacterial species subtypes, changing the metabolic pathway leading to volatile molecules (Tait et al. 2014). In addition to medium composition, dissimilarity in metabolite accessibility according to growth phase, oxygen concentration, or initial inoculum also influences the production and emission of VOCs (Bos et al. 2013; Broekaert et al. 2013; Kunze et al. 2013). Examination of in vitro and in vivo VOC descriptions has exposed that the host response can also affect volatile emission (Bos et al. 2013; Audrain et al. 2015).

Bacterial volatiles can be trapped in open or closed airflow systems. The volatiles of this active headspace are captured on polymeric adsorption matrices (SuperQ, Tenax, Lewatit, and activated charcoal). In open volatile collection methods (Ryu et al. 2003; Kai et al. 2007, 2010), purified and sterile air enters the test vessel. Half of the influx air is sucked out and is transported to an adsorption trap; as a result, a definite quantity of excess air librated. As a result, outside gaseous molecules and bacterial contaminations can be mitigated. In closed systems, the total headspace air is analyzed since the airflow circulates continuously through the bacterial culture and through the trap (Dickschat et al. 2004; Schulz et al. 2004). This closed-loop stripping apparatus (CLSA) was established by Boland et al. (1984). Molecules trapped in open or closed systems are either eluted with a solvent (methanol, dichloromethane, and pentane) and analyzed using gas chromatography/mass spectrometry (GC-MS) or directly thermally desorbed.

An additional opportunity to take out bacterial volatiles encounters the static headspace of bacterial cultures using solid-phase microextraction (SPME). This method was introduced by Arthur and Pawliszyn (1990). A thin film of an extracting phase immobilized over the surface of a fused silica fiber makes possible the adsorption of compounds situated in the headspace. According to the properties of predictable volatiles, special coatings are available for extraction, e.g., polydimethylsiloxan, carboxen, and divinylbenzene or mixture of these adsorbents. The SPME technique provides advantages over other techniques such as the process is solventless, is simple in situ sampling, and has a short analytical time. Thus far, numerous bacterial headspace-SPME investigations have been performed (Schulz et al. 2004; Farag et al. 2006). Other static approaches (diffusive sampling) were also recognized (Larsen and Frisvad 1994) using polymeric substances (Carbon black, Tenax). They were filled into stainless steel tubes and straightforwardly placed into the Petri dishes to trap volatiles from the headspace of diverse microbial cultures (Schöller et al. 1997), or activated charcoal was placed in the lid of the Petri dishes (Gust et al. 2003).

Every volatile compilation technique pointed out above was merged with GC-MS techniques. As a replacement for GC-MS, the compilation methods can also be connected to proton transfer reaction/mass spectrometer (PTR/MS) (Mayr et al. 2003; Bunge et al. 2008; Kai et al. 2010) or selected-ion flow tube-mass spectrometer (SIFT-MS) (Allardyce et al. 2006; Thorn et al. 2010). While GC-MS depicts volatile outline that is dependent on the analyses of defined retention times, PTR/MS and SIFT-MS allow uninterrupted supervision of volatile emission. Another substantial benefit of PTR/MS is that prior to analysis, no pre-concentration step or chromatography is required. PTR/MS decides the m/z ratio of a compound and no fragmentation pattern; therefore, the application of natural isotopic ratios and literature exploration are essential to make a sophisticated guess to recognize the molecules. To mitigate this drawback, a substitute system can be used to identify and differentiate volatiles like secondary electron spray ionization-mass spectrometry (SESI-MS) (Zhu et al. 2010). It has to be understood that all precise methods described now simply permit the discovery and determination of a definite spectrum of volatiles secreted from the bacteria. To obtain a complete collection of volatiles, it is inevitable to mix the diverse volatile compilation techniques (Wenke et al. 2012).

10.3 Bacterial Volatiles in Microbial Interactions

Bacterial volatiles are capable of affecting the metabolism of each other, but the function of volatiles in communications among bacterial species has been poorly investigated. In view of the physically divided distribution of bacterial populations (microcolonies) in the leaky soil matrix, it has recently been recommended that VOCs play important functions in interspecies bacterial cooperation (Garbeva et al. 2014a). It is anticipated that rhizosphere-colonizing microbes might devote a large fraction of the energy achieved from metabolizing root secretions to manufacture bioactive VOCs. Volatiles secreted by *Collimonas pratensis* and *Serratia plymuthica* encouraged the development of *Pseudomonas fluorescens*, while volatiles produced by *Paenibacillus* sp., *Pedobacter* sp., and the combination of all these four bacteria did not affect *P. fluorescens* development. The highest numbers of distinctive volatile molecules are narrated to be secreted by *C. pratensis* and *S. plymuthica*, together with *S*-methyl thioacetate, methyl thiocyanate, benzonitrile, and dimethyl disulfide (DMDS). Definite VOCs secreted by *C. pratensis* included 3-hexanone, 2-methyl propanal, ethenylacetate, 3-methyl-2-pentanone, methyl 2-methylbutanoate, methyl 3-methylbutanoate, 4-methyl-3-penten-2-one, 3-methyl-2-heptanone, myrcene, terpinene, and methyl salicylate. Specific VOCs produced by *S. plymuthica* include 2-pentadecanone, 1H-pyrrole, ethyl butanoate, chlorobenzene, dimethylsulfone, 2-octanone, and 5-dodecanone (Garbeva et al. 2014a).

In recent times it has been confirmed that bacteria are capable to reduce the development of *Burkholderia cepacia* complex (Bcc) strains through the production of VOCs (Papaleo et al. 2012, 2013; Orlandini et al. 2014). Previous recorded data confirmed that two *Pseudoalteromonas* strains were competent enough to totally reduce the growth of most Bcc strains (Romoli et al. 2011). Methyl-2,3,3,4-tetrahydroxytetrahydrofuran, indole and its derivatives, quinolones, and (*S*)-3 hydroxytridecan-4-one as signals have also been illustrated (Diggle et al. 2006; Ryan and Dow 2008).

An AHL-based interaction among individual bacterial cells has been noticed in the rhizosphere, and rhizobacteria have been found to sustain and generate VOCs within the plant. Bacterial volatile organic compounds secreted by some *P. fluorescens* and *S. plymuthica* strains suppressed the growth of *Agrobacterium tumefaciens* and *A. vitis* strains under in vitro *conditions*. DMDS was the major volatile secreted by *S. plymuthica*; it powerfully inhibited *Agrobacterium* development in vitro and was produced by tomato plants treated with *S. plymuthica*. 1-Undecene was the chief volatile produced by the *P. fluorescens* strain. It was accomplished that VOCs, and in particular DMDS, may be implicated in the inhibition of oncogenicity in plants (Dandurishvili et al. 2011). The most important VOCs secreted by the *P. chlororaphis* strain 449 were 1-undecene, 2-nonanone, and 2-heptanone along with 2-undecanone and lower amounts of DMDS. The profile of VOCs secreted by the *S. proteamaculans* 94 strain is considerably different from that secreted by *P. chlororaphis* strain 449, with DMDS being the main headspace VOC emitted by the former (Popova et al. 2014). When these VOCs were examined on the *A. tumefaciens* strain C58 and the cyanobacterium *Synechococcus* sp. strain PCC 7942, a powerful

A. tumefaciens bacteriostatic result of DMDS was established and totally inhibited the growth and development of the cyanobacterium strain *Synechococcus* sp. 2-Nonanone and 2-heptanone were effective on both microorganisms, whereas 2-undecanone entirely suppressed the growth of *Synechococcus*; however, it does not significantly influence *A. tumefaciens*. 1-Undecene does not considerably influence the growth and development of any of the two microbes examined (Popova et al. 2014).

10.4 Bacterial Volatiles in Microbe-Plant Interactions

Hundreds of diverse bacterial VOCs have been recognized, consisting alkanes, alkenes, alcohols, esters, ketones, sulfur compounds, and terpenoids. The appearance of a distinctive volatile outline or molecules is the result of precise metabolism or metabolic pathway(s) operating in the bacteria (Kai et al. 2009). These volatiles are perfect communicating compounds since they are present in the biosphere in ample amount and can work over long range (Wheatley 2002). Several bacteria selectively inhabit soil strongly connected with the plant roots, utilizing nutrient-rich root discharges. These bacteria are jointly described as rhizobacteria and a lot of them encourage plant growth (Bhattacharyya and Jha 2012), while the root atmosphere they inhabit is known as the rhizosphere (Mendes et al. 2013). Within rhizobacteria, pseudomonads have been regarded as a significant group of rhizosphere microbes (Goswami et al. 2013). VOCs secreted by rhizospheric bacteria are implicated in their communication with phytopathogenic microbes and colonizing plants and demonstrate antimicrobial and plant growth-promoting traits (Vespermann et al. 2007; Kanchiswamy et al. 2015a). Fluorescent *Pseudomonas* strains assist in preservation of soil fitness and defend the crops from phytopathogens. Rhizospheric bacterial strains can change both plant growth development and root-system structure by secreting different VOCs (Gutierrez-Luna et al. 2010).

Universally recognized bacterial VOCs such as 2-pentanone, 4-heptanone, 2-heptanol, 2-undecanone, 2-tridecanone, and 2-pentadecanone (Schulz and Dickschat 2007; Weise et al. 2014) and well-recognized molecules like sodorifen, a bicyclic oligomethyloctadiene secreted by *Serratia odorifera* (Kai et al. 2010), are capable of meddling with plants. In an experiment performed by Blom et al. (2011), 42 soilborne bacterial strains were evaluated for their volatile-dependent influence on 6-day-old seedlings of *Arabidopsis thaliana*. Thirty-six molecules of bacterial source were chosen for additional examination and within these, 1-hexanol, indole, and pentadecane encouraged plant growth. Cocultivation of *A. thaliana* using *S. odorifera* in bipartite Petri dishes, which only permitted volatiles to disperse from one side to the other, results in remarkable growth suppression of plants (Vespermann et al. 2007). DMDS and ammonia are among the most bioactive molecules (Kai et al. 2010). Huang et al. (2012) monitored that use of DMDS secreted by a *Bacillus cereus* strain has considerably shielded tobacco (*Nicotiana tabacum*) and corn (*Zea mays*) plants against *Botrytis cinerea* and *Cochliobolus heterostrophus*, respectively. Also, DMDS supplementation has been observed to extensively decrease the articulation of *Nicotiana attenuata* sulfur-assimilation genes, as well as methionine production and reprocessing (Meldau et al. 2013).

Two compounds, 3-hydroxy-2-butanone also branded as acetoin and 2,3-butanediol (2,3-BD), were secreted constantly from strains of *Bacillus subtilis* and *B. amyloliquefaciens* and were establish to extensively improve total leaf surface region and induced systemic resistance (ISR) of *A. thaliana* (Ryu et al. 2003; Rudrappa et al. 2010). 2,3-BD was also one of the main VOCs created by *Enterobacter aerogenes*, an endophytic bacterium that lives in corn plants. The secretion of 2,3-BD by *E. aerogenes* provides corn plants greater resistant in opposition to the northern corn leaf blight fungus *Setosphaeria turcica* (D'Alessandro et al. 2014). The uneven production of acetophenone, tridecanal, tetradecanal, 6,10,14-trimethyl-2-pentadecanone, and benzaldehyde secreted by diverse lemon rhizobacteria demonstrated that the outcome reported in *Arabidopsis* roots is relative to the kind and quantity of molecules formed by the bacteria (Gutierrez-Luna et al. 2010). 3-Hexanone secreted by strains of *Burkholderia ambifaria* considerably augmented *Arabidopsis* biomass, as obtained by acetophenone and DMDS (Groenhagen et al. 2013). Effect of volatile organic compounds (VOCs) released by rhizobacteria on plant cell physiology and how they influence growth and productivity is briefly represented in Fig. 10.1.

10.5 Potential of Bacterial Volatiles in Plant Growth and Health

Significant advancement has been made in comprehending the vital function of VOCs. Bacterial VOCs adjust plant development and protection and interspecies communication among plant, bacteria, fungi, and nematodes. It also works as a magnet to pull bitter enemies, as biocontrol instrument, and discover appropriate use as pest/insect/herbivore supervision (Leroy et al. 2011; Davis et al. 2013; Weise et al. 2013; D'Alessandro et al. 2014). These progressive data on VOCs demonstrate their serious function in multitrophic communications and their significance in both the environment and sustainable agriculture structure. Throughout the diverse phases of plant growth, widespread interaction takes place among soil microbes and plants in which indicator compounds from the two associates take part in vital functions. VOCs implicated in complicated inter- and intraspecific communications, above- and belowground, ensuing in genetic, phenotypic, and morphologic modification of the cooperating microorganisms (Effmert et al. 2012; Piechulla and Degenhardt 2014; Penuelas et al. 2014). Diverse bacterial species are capable to sense the plant host and start their inhabitation approaches in the rhizosphere by manufacturing canonical plant growth modifiable compounds like auxins and/or cytokinins. Extra indicator molecules from bacteria participate in key functions related to plant morphogenetic processes. These molecules facilitate bacterial cells to control gene outcome relying on inhabitants' density (Ortiz-Castro et al. 2009). It was establish that AHLs can be sensed by plants, modify gene expression in roots and shoots, and adjust protection and cell development reactions (Ortiz-Castro et al.

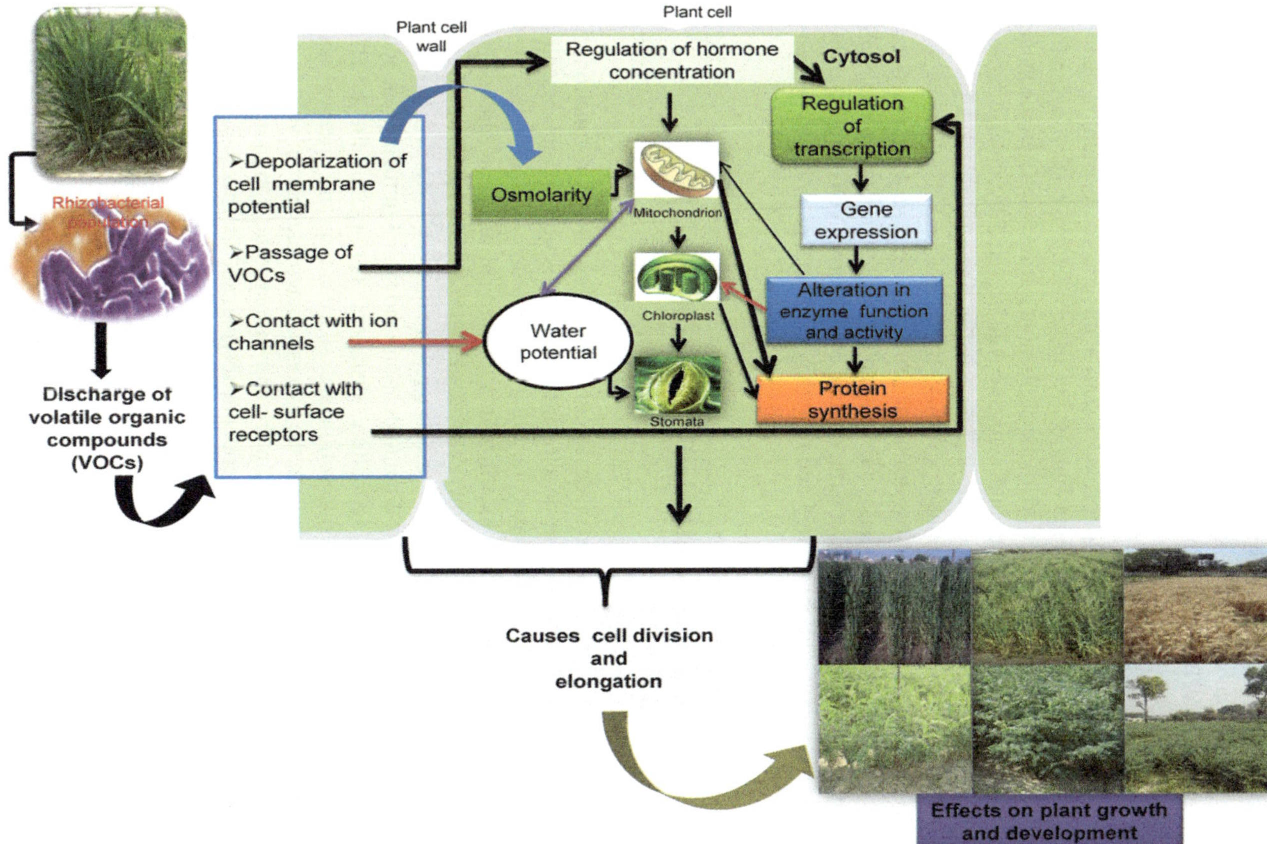

Fig. 10.1 Modulating effect of volatile organic compounds (VOCs) released by rhizobacteria on plant growth and productivity (Reproduced from Kanchiswamy et al. 2015a)

2008; Audrain et al. 2015). In particular, medium (C8–C14)-chained AHL molecules (*N*-hexanoylhomoserine lactone, *N*-3-oxo-hexanoyl-homoserine lactone, *N*-octanoyl-homoserine lactone, *N*-decanoyl-homoserine lactone, *N*-dodecanoyl-homoserine lactone, and *N*-tetradecanoylhomoserine lactone) demonstrated a dose-related impact on root structure, changing primary root development, lateral root arrangement, and root hair growth of *Arabidopsis* (Ortiz-Castro et al. 2008; Kanchiswamy et al. 2015a).

Large numbers of bacteria that trigger induced systemic resistance (ISR) demonstrate to do so via salicylic acid (SA) independent pathway linking jasmonate (JA) and ethylene signals. VOCs from *Bacillus amyloliquefaciens* strain IN937a activate ISR via ethylene-independent communication pathway, while VOCs from *Bacillus subtilis* strain GB03 found to operate through an ethylene-subjected pathway, although autonomous of the SA or JA communicating pathways (Ryu et al. 2004). This discovery has offered new approaches into the function of VOCs as originator of protection reactions in plants. Primarily, in the course of developing an examination method to review the development encouragement capability of rhizobacteria in vitro, Ryu et al. (2003, 2004) establish that bacterial volatiles are implicated in plant growth and development. An evaluation of growth encouragement stimulated by bacterial volatiles in *Arabidopsis* exposed that inoculation with the above cited GB03 or IN937a strains considerably enhanced the growth of *Arabidopsis*, in contrast to water control with the *Escherichia coli* strain DH5α. The two main molecules secreted from cultures of strains GB03 and IN937a, although not from cultures of the other strains, were recognized as 2,3-BD and its originator acetoin. The qualitative and quantitative structure of volatile mixture secreted by the growth-promoting strains fluctuate considerably from those of the null growth-promoting *E. coli* strain DH5α. External use of marketable acetoin and 2,3-BD causes the dose-reliant increase of plant growth, which enhances the impacts of the volatile mixture secreted by the abovementioned two *Bacillus* sp. (Kanchiswamy et al. 2015a).

Within plant growth-promoting rhizobacteria (PGPR) preparation of plant resistance, introduction of the primed condition is considered to result in an enhancement in the level of activity of cellular parts that participate in significant roles in protection signaling, while this method is not linked with direct modification in gene outcome in leaves. The priming action of 2,3-B, which results in decreasing plant vulnerability to infection, has been established in numerous studies (Park et al. 2013; D'Alessandro et al. 2014).

In other experiment, the use of 2,3-BD was found unsuccessful to trigger ISR against *Pseudomonas syringae* pv. *tabaci* but did stimulate the ISR response against *P. carotovora* subsp. *carotovora*, signifying that diverse protection reactions were generated in reply to different pathogens. The originator acetoin was in contrast revealed to generate ISR against *P. syringae* in *Arabidopsis* (Rudrappa et al. 2010). Bacterial volatile organic compounds of *Serratia plymuthica* and *Stenotrophomonas maltophilia* considerably repressed development and trigger H_2O_2 generation in *Arabidopsis* in double culture. Expression experiments carried out with dissimilar time exposed distorted transcript intensity for 889 genes and 655 genes in reaction to *S. plymuthica* or *S. maltophilia* volatiles, respectively. Additionally, especially volatile-receptive genes were extensively related with those influenced by abiotic stress, and genes responsive to both treatments were enriched for W-box motifs in their promoters and transcription factors (ERF2, ZAT10, MYB73, and WRKY18). Fascinatingly, the vulnerability of *wrky18* mutant lines to volatiles was extensively deferred, signifying a crucial function of WRKY18 in bacterial volatile reactions (Wenke et al. 2012). Volatiles secreted from several microbial species varying from Gram-negative to Gram-positive bacteria put forth an impact on leaf starch metabolism. Amazingly, the entire bacterial species examined secreted VOCs which powerfully enhance starch buildup in leaves of both mono- and dicotyledonous plants (Ezquer et al. 2010; Kanchiswamy et al. 2015a). Table 10.1 presents a summary of bacterial volatile organic compounds and their effects on plant growth and development.

Table 10.1 Summary of bacterial volatile organic compounds and their effects on plant growth and development

Bacterial species and strains	Plant species involved	Volatile organic compounds	Effect on plant	References
Arthrobacter agilis UMCV2	*Medicago sativa*	N,N-Dimethyl-hexadecanamine	Growth promotion	Velazquez-Becerra et al. (2011)
Bacillus amyloliquefaciens IN937a	*Arabidopsis thaliana*	2,3-Butanediol acetoin	Growth promotion and ISR	Ryu et al. (2003, 2004)
Bacillus amyloliquefaciens SQR-9	*Lycopersicon esculentum*	Combination of VOCs	Control of plant pathogens	Raza et al. (2016a)
Bacillus amyloliquefaciens T-5	-do-	Combination of volatile compounds including benzenes, ketones, aldehydes, alkanes, acids	Plant growth promotion	Raza et al. (2016b)
Serratia plymuthica IC1270 and *Pseudomonas fluorescens* B-4117	-do-	Dimethyl disulfide and other volatiles	Plant growth promotion by inhibition of *Agrobacterium tumefaciens* and *Agrobacterium vitis*	Dandurishvili et al.(2011)
Bacillus megaterium XTBG34	*Arabidopsis thaliana*	2-Penthylfuran	Growth promotion	Zou et al. (2010)
Bacillus subtilis GBO3	*Arabidopsis thaliana*	2,3-Butanediol acetoin	Growth promotion and ISR	Ryu et al. (2003, 2004)

(continued)

Table 10.1 (continued)

Bacterial species and strains	Plant species involved	Volatile organic compounds	Effect on plant	References
Pseudomonas chlororaphis O6	*Nicotiana tabacum* and *Arabidopsis thaliana*	2,3-Butanediol	Growth promotion, ISR, and desiccation tolerance	Han et al. (2006) and Cho et al. (2008)
Pseudomonas fluorescens SS101	*Nicotiana tabacum*	13-Tetradecadien-1-ol, 2-butanone, and 2-methyl-n-1-tridecene	Plant growth promotion	Park et al. (2015)
Rhizosphere strains (isolated from rhizosphere of *Citrus aurantifolia*) L263, L266, L272a, L254, L265a, and L265b	*Arabidopsis thaliana*	Combination of VOCs	Growth promoting and changes in root-system architecture	Gutierrez-Luna et al. (2010)
Different species of rhizobacteria	*Brachypodium distachyon*	-ND-	Biomass production and changes in root-system architecture	Delaplace et al. (2015)
Pseudomonas palleroniana R43631, *Bacillus* sp. R47065, R47131, *Paenibacillus* sp. B3a R49541, and *Bacillus simplex* M3–4 R49538	*Solanum tuberosum* subsp. *andigenum*	Volatile mixture	Increased yield	Velivelli et al. (2015)
Different species of *Bacillus*, *Serratia*, and *Staphylococcus*	*Lactuca sativa*	Acetoin and other volatiles	Growth promotion	Fincheira et al. (2016)

Partially adapted from Kanchiswamy et al. (2015b)
ND not detected

10.6 Challenges for the Deployment of Bacterial Volatiles Under Field Conditions

The exploration, extraction, and commercialization of new compounds from nature with biotechnological utilization are known as biodiversity prospecting. For most part of the last century, microorganisms were mainly studied for exploring conventional organic compounds possessing therapeutic characteristics with novel functions. An intensive investigation for novel compounds within VOCs will involve a standard change in scientist's ideology (Shannon et al. 2012; Kanchiswamy et al. 2015b). Bacterial VOCs present a new unexplored area in biodiversity prospecting. Though significant development has been achieved in comprehending VOCs utility in plant growth under laboratory environments, still we lack behind in employing this information under field environments. Comparatively, new experiments carried out on volatile uses under open-field environment advocates that VOCs can be useful in activating resistance against both phytopathogens and herbivores (Song and Ryu 2013; Sharifi and Ryu 2016). We further require improving in conditions for the efficient utilization of volatiles under field conditions. But, there are lots of restrictions related to the field use of bacterial volatiles, for instance, (a) detection of biologically active volatiles, (b) upgrading the level of precise volatiles or mixture of volatile molecules, and (c) utilization at field stages. One of the most intricate parts of bacterial volatiles is their physicochemical characteristics. For example, 2,3-butanediol field applications of tobacco result in considerable suppression of disease sign, but no important influence was reported when cucumber plants was nursed to battle with phytopathogen *P. syringae* (Cortes-Barco et al. 2010; Song and Ryu 2013). On the other hand, a synthetic volatile blend depending on the configuration of the VOCs (mostly alcohols and esters) impersonated the suppressing results of the natural VOCs secreted by *Saccharomyces cerevisiae* on citrus black spot originated due to the fungus *Guignardia citricarpa* postharvest. Therefore, volatiles secreted by yeast of synthetic blend can work as a potential process to manage citrus black spot or other postharvest diseases (Fialho et al. 2010, 2011). Presently, bacterial volatiles are effectively employed at field stage as foliar spray and by soil drenching. However, there are no parallel reports that present an improved perception of efficient and advanced techniques for field utilization.

10.7 Conclusion and Future Prospects

Latest information related to bacterial volatile organic compounds increased our comprehension that these small molecules can significantly assist in the growth and development of crop plants. But, still additional field experiments and data are required to improve the efficacy and importance of these molecules. We all are well versed with the harmful impacts of synthetic chemicals on human and crop health. Moreover, genetically modified crops can serve as a substitute, but majority of the world is still not accepting these crops due to various constraints. Thus, there is a need to adopt substitute to synthetic chemicals, presented in the form of bacterial volatiles as they are novel, cost effective, competent, and environmental friendly, which can be used for crop welfare.

References

Abboud MAA (2014) Bioimpact of application of pesticides with plant growth hormone (gibberellic acid) on target and non-target microorganisms. J Saudi Chem Soc 18:1005–1010

Allardyce RA, Langford VS, Hill AL, Murdoch DR (2006) Detection of volatile metabolites produced by bacterial growth in blood culture media by selected ion flow tube mass spectrometry (SIFT-MS). J Microbiol Methods 65:361–365

Arthur CL, Pawliszyn J (1990) Solid phase microextraction with thermal desorption using fused silica optical fibers. Anal Chem 62:2145–2148

Audrain B, Farag MA, Ryu CM, Ghigo JM (2015) Role of bacterial volatile compounds in bacterial biology. FEMS Microbiol Rev 39:222–233

Bailly A, Weisskopf L (2012) The modulating effect of bacterial volatiles on plant growth: current knowledge and future challenges. Plant Signal Behav 7:79–85

Bhardwaj D, Ansari MW, Sahoo RK, Tuteja N (2014) Biofertilizers function as key player in sustainable agriculture by improving soil fertility, plant tolerance and crop productivity. Microb Cell Factories 13:66

Bhattacharyya P, Jha D (2012) Plant growth-promoting rhizobacteria (PGPR): emergence in agriculture. World J Microbiol Biotechnol 28:1327–1350

Blom D, Fabbri C, Connor E, Schiestl F, Klauser D, Boller T, Eberl L, Weisskopf L (2011) Production of plant growth modulating volatiles is widespread among rhizosphere bacteria and strongly depends on culture conditions. Environ Microbiol 13:3047–3058

Boland W, Ney P, Jaenicke L, Gassmann G (1984) A "closed-loop-stripping" technique as a versatile tool for metabolic studies of volatiles. In: Schreier P (ed) Analysis of volatiles. Walter De Gruyter & Co, Berlin, pp 371–380

Bos LD, Sterk PJ, Schultz MJ (2013) Volatile metabolites of pathogens: a systematic review. PLoS Pathog 9:e1003311

Broekaert K, Noseda B, Heyndrickx M, Vlaemynck G, Devlieghere F (2013) Volatile compounds associated with *Psychrobacter* spp. and *Pseudoalteromonas* spp., the dominant microbiota of brown shrimp (*Crangon crangon*) during aerobic storage. Int J Food Microbiol 166:487–493

Bunge M, Araghipour N, Mikoviny T, Dunkl J, Schnitzhofer R, Hansel A, Schinner F, Wisthaler A, Margesin R, Mörk TD (2008) On-line monitoring of microbial volatile metabolites by proton transfer reaction-mass spectrometry. Appl Environ Microbiol 74:2179–2186

Cho SM, Kang BR, Han SH, Anderson AJ, Park JY, Lee YH, Cho BH, Yang KY, Ryu CM, Kim YC (2008) 2R, 3R-butanediol, a bacterial volatile produced by *Pseudomonas chlororaphis* O6, is involved in induction of systemic tolerance to drought in *Arabidopsis thaliana*. Mol Plant Microb Interact 21:1067–1075

Cook AA, Stall RE (1969) Necrosis in leaves induced by volatile materials produced in vitro by bacteria. Phytopathol Notes 59:259–260

Cortes-Barco AM, Goodwin PH, Hsiang T (2010) Comparison of induced resistance activated by benzothiadiazole, (2R, 3R)-butanediol and an isoparaffin mixture against anthracnose of *Nicotiana benthamiana*. Plant Pathol 59:643–653

D'Alessandro M, Erb M, Ton J, Brandenburg A, Karlen D, Zopfi J, Turlings TC (2014) Volatiles produced by soil-borne endophytic bacteria increase plant pathogen resistance and affect tritrophic interactions. Plant Cell Environ 37:813–826

Dandurishvili N, Toklikishvili N, Ovadis M, Eliashvili P, Giorgobiani N, Keshelava R, Tediashvili M, Vainstein A, Khmel I, Szegedi E, Chernin L (2011) Broad-range antagonistic rhizobacteria *Pseudomonas fluorescens* and *Serratia plymuthica* suppress *Agrobacterium crown* gall tumours on tomato plants. J App Microbiol 110:341–352

Davis TS, Crippen TL, Hofstetter RW, Tomberlin JK (2013) Microbial volatile emissions as insect semiochemicals. J Chem Eco 39:840–859

Delaplace P, Delory BM, Baudson C, Mendaluk-Saunier de Cazenave M, Spaepen S, Varin S, Brostaux Y, du Jardin P (2015) Influence of rhizobacterial volatiles on the root system architecture and the production and allocation of biomass in the model grass *Brachypodium distachyon* (L.) P. Beauv. BMC Plant Biol 15:1–15

Dickschat JS, Wenzel SC, Bode HB, Müller R, Schulz S (2004) Biosynthesis of volatiles by the myxobacterium *Myxococcus xanthus*. Chembiochem 5:778–787

Diggle SP, Cornelis P, Williams P, Camara M (2006) 4-quinolone signalling in *Pseudomonas aeruginosa*: old molecules, new perspectives. Int J Med Microbiol 296:83–91

Dolch ME, Hornuss C, Klocke C, Praun S, Villinger J, Denzer W, Schelling G, Schubert S (2012) Volatile compound profiling for the identification of gram-negative bacteria by ion-molecule reaction-mass spectrometry. J Appl Microbiol 113:1097–1105

Dweck HKM, Ebrahim SAM, Thoma M, Mohamed AA, Keesey IW, Trona F, Lavista-Llanos S, Svatoš A, Sachse S, Knaden M, Hansson BS (2015) Pheromones mediating copulation and attraction in *Drosophila*. Proc Natl Acad Sci U S A:E2829–E2835

Effmert U, Kalderas J, Warnke R, Piechulla B (2012) Volatile mediated interactions between bacteria and fungi in the soil. J Chem Ecol 38:665–703

Ezquer I, Li J, Ovecka M, Baroja-Fernandez E, Jose Munoz F, Montero M, Díaz de Cerio J, Hidalgo M, Sesma MT, Bahaji A, Etxeberria E, Pozueta-Romero J (2010) Microbial volatile emissions promote accumulation of exceptionally high levels of starch in leaves in mono- and dicotyledonous plants. Plant Cell Physiol 51:1674–1693

Farag MA (2014) Comparative mass spectrometry and nuclear magnetic resonance metabolomic approaches for nutraceuticals quality control analysis: a brief review. Recent Pat Biotechnol 8:17–24

Farag MA, Wessjohann LA (2012) Volatiles profiling in medicinal licorice roots using steam distillation and solid-phase microextraction (SPME) coupled to chemometrics. J Food Sci 77:C1179–C1184

Farag MA, Ryu CM, Summer LW, Pare PW (2006) GC-MS SPME profiling of rhizobacterial volatiles reveals prospective inducers of growth promotion and induced systemic resistance in plants. Phytochemistry 67:2262–2268

Fialho MB, Toffano L, Pedroso MP, Augusto F, Pascholati SF (2010) Volatile organic compounds produced by *Saccharomyces cerevisiae* inhibit the in vitro development of *Guignardia citricarpa*, the causal agent of citrus black spot. World J Microbiol Biotechnol 26:925–932

Fialho MB, Ferreira LFR, Monteiro RTR, Pascholati SF (2011) Antimicrobial volatile organic compounds affect morphogenesis-related enzymes in *Guignardia citricarpa*, causal agent of citrus black spot. Biocontrol Sci Tech 21:797–807

Fincheira P, Venthur H, Mutis A, Parada M, Quiroz A (2016) Growth promotion of Lactuca sativa in response to volatile organic compounds emitted from diverse bacterial species. Microbiol Res 93:39–47

Garbeva P, Hordijk C, Gerards S, De Boer W (2014a) Volatile-mediated interactions between phylogenetically different soil bacteria. Front Microbiol 5:285–290

Garbeva P, Hordijk C, Gerards S, de Boer W (2014b) Volatiles produced by the mycophagous soil bacterium *Collimonas*. FEMS Microbiol Ecol 87:639–649

Gomiero T, Pimentel D, Paoletti MG (2011) Environmental impact of different agricultural management practices: conventional vs. organic agriculture. Crit Rev Plant Sci 30:95–124

Goswami D, Vaghela H, Parmar S, Dhandhukia P, Thakker JN (2013) Plant growth promoting potentials of *Pseudomonas* spp. strain OG isolated from marine water. J Plant Interact 8:281–290

Groenhagen U, Baumgartner R, Bailly A, Gardiner A, Eberl L, Schulz S, Weisskopf L (2013) Production of bioactive volatiles by different *Burkholderia ambifaria* strains. J Chem Ecol 39:892–906

Gust B, Challis GL, Fowler K, Kieser T, Chater KF (2003) PCR-targeted Streptomyces gene replacement identifies a protein domain needed for biosynthesis of the sesquiterpene soil odor geosmin. Proc Natl Acad Sci U S A 100:1541–1546

Gutierrez-Luna FM, Lopez-Bucio J, Tamirano-Hernandez J, Valencia-Cantero E, de la Cruz HR, Ias-Rodriguez L (2010) Plant growth-promoting rhizobacteria modulate root-system architecture in *Arabidopsis thaliana* through volatile organic compound emission. Symbiosis 51:75–83

Han SH, Lee SJ, Moon JH, Park KH, Yang KY, Cho BH, Kim KY, Kim YW, Lee MC, Anderson AJ, Kim YC (2006) GacS-dependent production of 2R, 3R-butanediol by *Pseudomonas chlororaphis* O6 is a major determinant for eliciting systemic resistance against *Erwinia carotovora* but not against *Pseudomonas syringae* pv. tabaci in tobacco. Mol. Plant Microbe Interact 19:924–930

Hare JD (2011) Ecological role of volatiles produced by plants in response to damage by herbivorous insects. Annu Rev Entomol 56:161–180

Heil M, Karban R (2010) Explaining evolution of plant communication by airborne signals. Trends Ecol Evol 25:137–144

Huang CJ, Tsay JF, Chang SY, Yang HP, Wu WS, Chen CY (2012) Dimethyl disulfide is an induced systemic resistance elicitor produced by *Bacillus cereus* C1L. Pest Manag Sci 68:1306–1310

Insam H, Seewald MSA (2010) Volatile organic compounds (VOCs) in soils. Biol Fertil Soils 46:199–213

Jansen RM, Wildt J, Kappers IF, Bouwmeester HJ, Hofstee JW, van Henten EJ (2011) Detection of diseased plants by analysis of volatile organic compound emission. Annu Rev Phytopathol 49:157–174

Kai M, Effmert U, Berg G, Piechulla B (2007) Volatiles of bacterial antagonists inhibit mycelial growth of the plant pathogen *Rhizoctonia solani*. Arch Microbiol 187:351–360

Kai M, Haustein M, Molina F, Petri A, Scholz B, Piechulla B (2009) Bacterial volatiles and their action potential. Appl Microbiol Biotechnol 81:1001–1012

Kai M, Crespo E, Cristescu SM, Harren FJM, Piechulla B (2010) *Serratia odorifera*: analysis of volatile emission and biological impact of volatile compounds on *Arabidopsis thaliana*. Appl Microbiol Biotechnol 88:965–976

Kai M, Effmert U, Piechulla B (2016) Bacterial-plant-interactions: approaches to unravel the biological function of bacterial volatiles in the rhizosphere. Front 7:108

Kanchiswamy CN, Malnoy M, Maffei ME (2015a) Chemical diversity of microbial volatiles and their potential for plant growth and productivity. Front Plant Sci 6:151

Kanchiswamy CN, Malnoy M, Maffei ME (2015b) Bioprospecting bacterial and fungal volatiles for sustainable agriculture. Trends Plant Sci 20:206–211

Korpi A, Järnberg J, Pasanen AL (2009) Microbial volatile organic compounds. Crit Rev Toxicol 39:139–193

Kunze N, Göpel J, Kuhns M, Jünger M, Quintel M, Perl T (2013) Detection and validation of volatile metabolic patterns over different strains of two human pathogenic bacteria during their growth in a complex medium using multi-capillary column-ion mobility spectrometry (MCC-IMS). Appl Microbiol Biotechnol 97:3665–3676

Larsen TO, Frisvad JC (1994) A simple method for collection of volatile metabolites from fungi based on diffusive sampling from Petri dishes. J Microbiol Methods 19:297–305

Lemfack MC, Nickel J, Dunkel M, Preissner R, Piechulla B (2014) mVOC: a database of microbial volatiles. Nucleic Acids Res 42:D744–D748

Leroy PD, Sabri A, Heuskin S, Thonart P, Lognay G, Verheggen FJ, Francis F, Brostaux Y, Felton GW, Haubruge E (2011) Microorganisms from aphid honeydew attract and enhance the efficacy of natural enemies. Nat Commun 2:348

Mayr D, Margesin R, Klingsbichel E, Hartungen E, Jenewein D, Mark TD, Schinner F. und(2003) Rapid detection of meat spoilage by measuring volatile organic compounds by using proton transfer reaction mass spectrometry. Appl Environ Microbiol 69:4697–4705

Meldau DG, Meldau S, Hoang LH, Underberg S, Wünsche H, Baldwin IT (2013) Dimethyl disulfide produced by the naturally associated bacterium *Bacillus* sp. B55 promotes *Nicotiana attenuata* growth by enhancing sulfur nutrition. Plant Cell 25:2731–2747

Mendes R, Garbeva P, Raaijmakers JM (2013) The rhizosphere microbiome: significance of plant beneficial, plant pathogenic, and human pathogenic microorganisms. FEMS Microbiol Rev 37:634–663

Meruva NK, Penn JM, Farthing DE (2004) Rapid identification of microbial VOCs from tobacco molds using closed-loop stripping and gas chromatography/time-of-flight mass spectrometry. J Ind Microbiol Biot 31:482–488

Orlandini V, Maida I, Fondi M, Perrin E, Papaleo MC, Bosi E, de Pascale D, Tutino ML, Michaud L, Lo Giudice A, Fani R (2014) Genomic analysis of three sponge-associated *Arthrobacter* Antarctic strains, inhibiting the growth of *Burkholderia cepacia* complex bacteria by synthesizing volatile organic compounds. Microbiol Res 169:593–601

Ortíz-Castro R, Martínez-Trujillo M, López-Bucio J (2008) N-acyl-Lhomoserine lactones: a class of bacterial quorum-sensing signals alter postembryonic root development in *Arabidopsis thaliana*. Plant Cell Environ 31:1497–1509

Ortíz-Castro R, Contreras-Cornejo HA, Macías-Rodríguez L, López-Bucio J (2009) The role of microbial signals in plant growth and development. Plant Sign Behav 4:701–712

Papaleo MC, Fondi M, Maida I, Perrin E, Lo Giudice A, Michaud L, Mangano S, Bartolucci G, Romoli R, Fani R (2012) Sponge-associated microbial Antarctic communities exhibiting antimicrobial activity against *Burkholderia cepacia* complex bacteria. Biotechnol Adv 30:272–293

Papaleo MC, Romoli R, Bartolucci G, Maida I, Perrin R, Fondi M, Orlandini V, Mengoni A, Emiliani G, Tutino ML, Parrilli E, de Pascale D, Michaud L, Lo Giudice A, Fani R (2013) Bioactive volatile organic compounds from Antarctic (sponges) bacteria. New Biotechnol 30:824–838

Park HB, Lee B, Kloepper JW, Ryu CM (2013) One shot-two pathogens blocked: exposure of *Arabidopsis* to hexadecane, a long chain volatile organic compound, confers induced resistance against both *Pectobacterium carotovorum* and *Pseudomonas syringae*. Plant Sign Behav 8:e24619

Park YS, Dutta S, Ann M, Raaijmakers JM, Park K (2015) Promotion of plant growth by *Pseudomonas fluorescens* strain SS101 via novel volatile organic compounds. Biochem Biophys Res Commun 461:361–365

Peñuelas J, Asensio D, Tholl D, Wenke K, Rosenkranz M, Piechulla B, Schnitzler JP (2014) Biogenic volatile emissions from the soil. Plant Cell Environ 37:1866–1891

Piechulla B, Degenhardt J (2014) The emerging importance of microbial volatile organic compounds. Plant Cell Environ 37:811–812

Popova AA, Koksharova OA, Lipasova VA, Zaitseva JV, Katkova-Zhukotskaya OA, Eremina SI, Mironov AS, Chernin LS, Khmel IA (2014) Inhibitory and toxic effects of volatiles emitted by strains of *Pseudomonas* and *Serratia* on growth and survival of selected microorganisms, *Caenorhabditis elegans*, and *Drosophila melanogaster*. Biomed Res Int 2014:125704

Raman KV (2005) Sustaining soil fertility. The Hindu Survey of Ind Agric 165–167

Raza W, Ling N, Yang L, Huang Q, Shen Q (2016a) Response of tomato wilt pathogen *Ralstonia solanacearum* to the volatile organic compounds produced by a biocontrol strain *Bacillus amyloliquefaciens* SQR-9. Sci Rep 6:24856

Raza W, Wang J, Wu Y, Ling N, Wei Z, Huang Q, Shen Q (2016b) Effects of volatile organic compounds produced by *Bacillus amyloliquefaciens* on the growth and virulence traits of tomato bacterial wilt pathogen *Ralstonia solanacearum*. Appl Microbiol Biotechnol 100:7639–7650

Romoli R, Papaleo MC, de Pascale D, Tutino ML, Michaud L, LoGiudice A, Fani R, Bartolucci G (2011) Characterization of the volatile profile of Antarctic bacteria by using solid-phase microextraction-gas chromatography-mass spectrometry. J Mass Spectro 46:1051–1059

Romoli R, Papaleo MC, de Pascale D, Tutino ML, Michaud L, LoGiudice A, Fani R, Bartolucci G (2014) GC-MS volatolomic approach to study the antimicrobial activity of the antarctic bacterium *Pseudoalteromonas* sp TB41. Metabolomics 10:42–51

Rudrappa T, Biedrzycki ML, Kunjeti SG, Donofrio NM, Czymmek KJ, Paré PW, Bais HP (2010) The rhizobacterial elicitor acetoin induces systemic resistance in *Arabidopsis thaliana*. Commun Integr Biol 3:130–138

Ryan RP, Dow J (2008) Diffusible signals and interspecies communication in bacteria. Microbiology 154:1845–1858

Ryu CM, Farag MA, Hu CH, Reddy MS, Wei HX, Paré PW, Kloepper JW (2003) Bacterial volatiles promote growth in Arabidopsis. Proc Natl Acad Sci U S A 100:4927–4932

Ryu C-M, Farag MA, Hu C-H, Reddy MS, Kloepper JW, Paré PW (2004) Bacterial volatiles induce systemic resistance in Arabidopsis. Plant Physiol 134:1017–1026

Salvador AC, Baptista I, Barros AS, Gomes NCM, Cunha A, Almeida A, Rocha SM (2013) Can volatile organic metabolites be used to simultaneously assess microbial and mite contamination level in cereal grains and coffee beans? PLoS One 8:e59338

Schmidt R, Cordovez V, de Boer W, Raaijmakers J, Garbeva P (2015) Volatile affairs in microbial interactions. ISME J 9:2329–2335

Schöller CEG, Molin S, Wilkins K (1997) Volatile metabolites from some gram-negative bacteria. Chemosphere 35:1487–1495

Schulz S, Dickschat JS (2007) Bacterial volatiles: the smell of small organisms. Nat Prod Rep 24:814–842

Schulz S, Fuhlendorff J, Reichenbach H (2004) Identification and synthesis of volatiles released by the myxobacterium *Chondromyces crocatus*. Tetrahedron 60:3863–3872

Shannon U, Hung R, Bennett JW (2012) Fungal volatile organic compounds: a review with emphasis on their biotechnological potential. Fungal Biol Rev 26:73–83

Sharifi R, Ryu CM (2016) Making healthier or killing enemies? Bacterial volatile-elicited plant immunity plays major role upon protection of Arabidopsis than the direct pathogen inhibition. Commun Integr Biol 9:4–e1197445

Song GC, Ryu CM (2013) Two volatile organic compounds trigger plant self-defense against a bacterial pathogen and a sucking insect in cucumber under open field conditions. Int J Mol Sci 14:9803–9819

Tait E, Perry JD, Stanforth SP, Dean JR (2014) Identification of volatile organic compounds produced by bacteria using HS-SPME-GCMS. J Chromatogr Sci 52:363–373

Thorn RMS, Reynolds DM, Greenman J (2010) Multivariate analysis of bacterial volatile compound profiles for discrimination between selected species and strains in vitro. J Microbiol Methods 84:258–264

Tiebe C, Miessner H, Koch B, Hübert T (2009) Detection of microbial volatile organic compounds (MVOCs) by ion-mobility spectrometry. Anal Bioanal Chem 395:2313–2323

Tilak KVBR, Ranganayaki N, Pal KK, De R, Saxena AK, Nautiyal CS, Mittal S, Tripathi AK, Johri BN (2005) Diversity of plant growth and soil health supporting bacteria. Curr Sci 89:136–150

United Nations Department of Economic and Social Affairs (UN-DESA) (2015) United Nations Population Division. http://www.un.org/en/development /desa /population/ theme/trends/ index.shtml. Accessed 12 Feb 2017

Velazquez-Becerra C, Macias-Rodriguez LI, Lopez-Bucio J, Altamirano-Hernandez J, Flores-Cortez I, Valencia-Cantero E (2011) A volatile organic compound analysis from *Arthrobacter agilis* identifies dimethylhexadecylamine, an amino-containing lipid modulating bacterial growth and *Medicago sativa* morphogenesis *in vitro*. Plant Soil 339:329–340

Velivelli SLS, Kromann P, Loján P, Rojas M, Franco J, Suarez JP, Prestwich BD (2015) Identification of mVOCs from Andean rhizobacteria and field evaluation of bacterial and mycorrhizal inoculants on growth of potato in its center of origin. Microb Ecol 69:652–667

Vespermann A, Kai M, Piechulla B (2007) Rhizobacterial volatiles affect the growth of fungi and *Arabidopsis thaliana*. Appl Environ Microbiol 73:5639–5641

von Reuß SH, Kai M, Piechulla B, Francke W (2010) Octamethylbicyclo[3.2.1]octadienes from the Rhizobacterium *Serratia odorifera*. Angew Chem Int Ed Engl 49:2009–2010

Weise T, Kai M, Piechulla B (2013) Bacterial ammonia causes significant plant growth inhibition. PLoS One 8:e63538

Weise T, Thuermer A, Brady S, Kai M, Daniel R, Gottschalk G, Piechulla B (2014) VOC emission of various *Serratia* species and isolates and genome analysis of *Serratia plymuthica* 4Rx13. FEMS Microbiol Lett 352:45–53

Wenke K, Weise T, Warnke R, Valverde C, Wanke D, Kai M, Piechulla B (2012) Bacterial volatiles mediating information between bacteria and plants. In: Witzany G (ed) Biocommunication of plants. Springer Verlag, Heidelberg, pp 327–348

Wheatley RE (2002) The consequences of volatile organic compound mediated bacterial and fungal interactions. Antonie Van Leeuwenhoek 81:357–364

Zhu J, Hill JE (2013) Detection of *Escherichia coli* via VOC profiling using secondary electrospray ionization-mass spectrometry (SESIMS). Food Microbiol 34:412–417

Zhu J, Bean HD, Kuo YM, Hill JE (2010) Fast detection of volatile organic compounds from bacterial cultures by secondary electrospray ionization–mass spectrometry. J Clin Microbiol 48:4426–4431

Zou C, Li Z, Yu D (2010) *Bacillus megaterium* strain XTBG34 promotes plant growth by producing 2-pentylfuran. J Microbiol 48:460–466

Bacterial Volatile in Rhizosphere

11

Ömür Baysal and R. Soner Silme

Abstract

Investigation on antagonism, to follow competition between target pathogen and biocontrol agent is the main step to select the effective candidate in vitro and in vivo conditions. Rhizosphere is a harbour of bacteria and fungi. Their interactions and ongoing communication network are in balance and continued with prominent molecules mediating this signal traffic 'volatile organic compounds' (VOCs). Volatile compounds emitted by different soil bacteria affect the growth of plants and gene expression. Bacterial volatiles can have an important role in communication and antagonistic interactions within the soil bacterial community. In this complicated microflora, to decipher the chemical language through microbial interactions will be beneficial for sustainable control measurements. Here we evaluated volatile-mediated interactions of a microbial community mimicking the natural conditions in the heterogeneous soil environment along the rhizosphere. This chapter summarizes biological and ecological significance of VOC-mediated resistance and briefly provides updated knowledge on this concept.

Keywords

Bacteria • Signalling • VOCs mediated resistance • Induction of resistance • Microbial community • Secreted compounds

---

Ö. Baysal (✉)
Faculty of Science, Department of Molecular Biology and Genetics,
Mugla Sıtkı Koçman University, 48000 Kötekli/Muğla, Turkey
e-mail: omurbaysal@mu.edu.tr

R.S. Silme
Scientific Research Projects, Istanbul University,
34452, Main Campus, Beyazit/Fatih-Istanbul, Turkey
e-mail: rsoner@istanbul.edu.tr

D.K. Choudhary et al. (eds.), *Volatiles and Food Security*,
DOI 10.1007/978-981-10-5553-9_11

11.1 Introduction

Soil microflora is a very complex fauna sheltering different microorganisms and bacteria which is one of the major inhabitants of soil rhizosphere that is a part of under the influence of the plant roots. This microflora directly and indirectly has an influence on both plant growth and resistance to pathogens. Bacteria form colonization using root exudates and their secretions to create antagonistic microflora for inhibiting pathogenic microorganism causing plant disease and promoting plant growth. Therefore, they have been used as a bio-fertilizer to enhance nitrogen fixation and solubilization of required elements for plants besides phytostimulator and biocontrol agents (Correa et al. 2009; Sumayo et al. 2013).

Moreover, volatiles emitted from rhizobacteria have plant growth promotion effect by regulation of secreted hormones depending on signalling cascades (Bhattacharyya et al. 2015a, b). As in practice these properties are beneficial for seed treatment and soil drenching (Yu and Lee 2013). Numerous dynamic interactions are occurring in rhizosphere that changes nutrient transformation and exchange of genetic material within microorganisms. These cases directly have an impact on microbiome and characteristic property of rhizosphere. In some special circumstances, microbiome balance is disrupted by unconscious treatments on soil that impairs dynamism of microflora (Chowdhury et al. 2013). The beneficial microorganisms are efficient tool to rearrangement of this impairment on microbial community. Some commercial biocontrol agents are available in the purpose of remediation of microbiome. On the other hand, shifting of bacterial community has importance in view of structural changing in rhizosphere of certain cultivated plants (Kozdroj 2008).

Different inhibitory mechanisms of beneficial bacteria such as iron depletion, respiratory poison, antibiotic secretation, emited volatiles reduce virulence and help to plants for combating with disease caused by plant pathogenic fungi. As another concept, induced systemic resistance using non-pathogenic bacteria stimulates plant defence and activation of signalling pathways related to responsive genes *in planta* (Baysal et al. 2008). Our recent study showed that decomposition of essential oils by Enterobacter sp. results in production of indole-3-acetic acid with chitinase that inhibits growth of pathogens (Silme and Baysal 2016).

This chapter is particularly focused on recent findings on volatile compounds of bacteria considering their effects on plant pathogenic fungi and its possible use in practice.

11.2 Volatile Compounds of Bacteria and Its Effect on Plant Pathogenic Fungi

Plant pathogenic fungi cause economic severe losses on crops and threats to production yield in the world. Therefore, environmentally friendly control methods are necessary to minimize the problem in infected fields. Although chemicals are preferred to effectively control of disease, pathogen develops resistance to these compounds, which globally affects food security. Volatile compounds emitted by non-cyanogenic bacteria would be used in practice if it had no toxicity (HCN production) and lack of specificity. These disadvantages result in limitation in practice. Integrated pest management strategies involve with less toxic application and methods based on biocontrol approaches. Some major pathogens are not cultivated in in vitro condition, because they have obligate parasites and biotrophic life manner. Within control strategies, antagonistic property depending on cell wall-degrading enzymes of microorganism is considered to be an efficient weapon for selected biocontrol agents; until now their volatile-mediated antagonistic effect has been ignored. Whereas this effect can be suggested in order to improve effective strategies to obligate pathogens. For instance, a common pathogen caused by *Rhizoctonia solani* infecting fields of cultivation can be controlled by volatile-mediated antagonistic bacteria (Kai et al. 2007).

In sight of these knowledge, updated research results are summarized in Table 11.1. The bacteria genus belonging *Pseudomonas* and *Bacillus* are relatively more efficient rather than other species for biocontrol purposes.

When recent literatures are taken into account, pH level has no enhancer or diminishing effect on volatile-mediated antagonistic effect; also these similar results on growth-promoting volatiles in plants have been reported (Bailly and Weisskof 2012).

Our recent findings showed that the production of 3-hydroxy-2-butanone (acetoin) by EU07 and FZB24 is a prominent property that can be correlated with bacterial VOCs (acetoin or 3-hydroxy-2-butanone) that can serve as agents for triggering growth promotion as observed in *Arabidopsis thaliana* (Baysal et al. 2013). Previously, we reported similar observations in which EU07 treatment of plants resulted in increased plant height in comparison to that observed with QST713 (Baysal et al. 2008). External applications of commercial acetoin and 2,3-butanediol obtained by two *Bacillus* species resulted in a dose-dependent stimulation of plant growth (Chen et al. 2008). Moreover, the production of acetoin and 2,3-butanediol by plant growth-promoting bacteria (PGPR) has been reported to increase systemic disease resistance and drought tolerance (Wan et al. 2008; Huang 2012). Therefore, these properties can be associated with VOC's stimulation effect and indirectly induction of resistance.

Interestingly our MALDI-TOF MS analyses indicated that genes related to the GacA system may also be involved in the *Bacillus*-pathogen interactions. GacA has been previously suggested to be an essential system for the synthesis of extracellular protease and secondary metabolites (Kobaisy et al. 2001). The GacS/GacA system was also shown to be associated with the production of antibiotic compounds (Ann 2012) that display a possible correspondence with VOC's production.

Methylation of small molecules and macromolecules has importance to be crucial in metabolism, cell signalling and epigenetic programming. It can be conferred by SAM-dependent methyltransferases (Vespermann et al. 2007; Alimi et al. 2012). Actual studies showed SAM synthetase and other methyltransferases were identified to play a role in phenylpropanoid biosynthesis and in signal transduction on another soilborne pathogen (Ting et al. 2011; Elshafie et al. 2012).

Table 11.1 Overview of literature reporting inhibition of major plant pathogenic fungi by bacterial volatiles. Ref., literature reference

Fungi	Effects	Bacteria	Volatiles	Refs.
Alternaria alternata	Mycelial growth inhibition and pigmentation	*Burkholderia ambifaria*	Dimethyl trisulphide, 2-nonanone, 2-undecanone	Groenhagen et al. (2013)
Alternaria brassicae	Mycelial growth inhibition	*Paenibacillus polymyxa* BMP-11	1-Octen-3-ol, benzothiazole, citronellol	Zhao et al. (2011)
Botrytis cinerea	Mycelial growth inhibition	*Paenibacillus polymyxa* BMP-11	1-Octen-3-ol, benzothiazole	Zhao et al. (2011)
	Reduced disease on tobacco plants (greenhouse)	*Bacillus cereus*	Dimethyl disulphide	Huang (2012)
Colletotrichum capsici	Inhibition of mycelial growth and of pigmentation	*Paenibacillus polymyxa* BMP-11	1-Octen-3-ol, benzothiazole, citronellol	Zhao et al. (2011)
Fusarium oxysporum	Inhibition of mycelial growth, of spore germination and alteration of morphology	*Bacillus subtilis (Eu07)*, *Bacillus amyloliquefaciens*	Benzenes, ketones and aldehydes	Baysal et al. (2013)
	Mycelial growth inhibition	*Paenibacillus polymyxa* BMP-11	1-Octen-3-ol, benzothiazole, citronellol	Zhao et al. (2011)
	Reduced pigmentation	*Paenibacillus polymyxa* BMP-11	Citronellol	Zhao et al. (2011)

Sclerotinia sclerotiorum	Inhibition of mycelial growth and of germination from ascospores and sclerotia	*Pseudomonas* sp.	Benzothiazole, cyclohexanol, n-decanal, dimethyl trisulphide, 2-ethyl 1-hexanol and nonanal	Fernando et al. (2005)
Rhizoctonia solani	Mycelial growth inhibition	*Paenibacillus polymyxa* BMP-11	1-Octen-3-ol, benzothiazole, citronellol	Zhao et al. (2011)
	Mycelial growth inhibition	*Xanthomonas campestris* pv. *vesicatoria* 85–10	Decan-2-one	Weise et al. (2012)
	Mycelial growth inhibition	*Burkholderia ambifaria*	Dimethyl disulphide, dimethyl trisulphide, 4-octanone, S-methyl methanethiosulphonate, acetophenone, phenylpropan-1-one, 1-phenyl-1,2-propanedione, 2-undecanone	Groenhagen et al. (2013)
Phytophthora capsici	Mycelial growth inhibition	*Paenibacillus polymyxa* BMP-11	1-Octen-3-ol, benzothiazole, citronellol	Zhao et al. (2011)

It is known that 1-aminocyclopropane-1-carboxylic acid (ACC) synthase converts S-adenosylmethionine (AdoMet) into ACC that changes to ethylene by ACC oxidase. PGPR may enhance plant growth by lowering a plant's ethylene levels. In these cases, the immediate precursor of ethylene is ACC. This compound is hydrolysed by bacteria-expressing ACC-deaminase activity (Ting et al. 2011; Elshafie et al. 2012). Although the details of the molecular interactions remain to be unclarified, the ethylene and cytokinin signalling pathways appear to be involved in growth promotion induced by VOCs (Yuan et al. 2012). The volatiles of *Bacillus subtilis* GB03 and *B. amyloliquefaciens* IN937a have been ascribed as 2R-R butanediol, and acetoin is also a precursor of this molecule (Syu 2001). Volatile-mediated signalling is a complicated case and transduction mechanism related to hormone regulation in which cytokinin, ethylene, auxin, salicylic acid and jasmonic acid pathways play different roles (Ryu et al. 2005; Cho et al. 2008). Indole-3-acetic acid (IAA, 3-IAA) is a natural plant hormone of the auxin class (Simon and Petrazek 2011). Our recent study showed to degradation of essential oils by *Enterobacter* sp. that causes production of indole-3-acetic acid with chitinase and inhibition on pathogen growth (Silme and Baysal 2016). From this aspect, salicylic acid signalling pathways may be considered to be a precursor of PR-related protein induction; therefore, it can be hypothesized that VOC-mediated resistance can be related to auxin metabolism and activation of PR proteins through salicylic acid signalling.

Our detected upregulated protein yuaE (gi|154687215) (function unknown and unique) obtained from EU07 showed similarity to SAM-dependent methyltransferases; the same mechanism in the inhibition of soilborne pathogens can be suggested. There are also upregulated two-response regulator protein (gi|76667926) and two-component response regulator related to signal transduction, which shows that signalling is active (Baysal et al. 2013). SAM-dependent methyltransferases are highly likely to be responsible for pathogen inhibition in the early phase of recognition in addition to the known cell wall-degrading enzymes, acetoin and other components. Therefore, these cases can be a precursor of initiation as resistance response to pathogen secreted by VOC-mediated antifungal activity of a biocontrol agent that results in suppression on soilborne pathogen. VOC-mediated antifungal activities of bacterial strains are occurring through inorganic volatiles such as HCN or ammonia (NH_3) (Kai et al. 2010; Trivedi et al. 2011), which inhibits the growth of fungi and secretion of inhibiting compounds. Dimethyl di-trisulphide and other sulphur compounds, benzaldehydes and long-chain ketones like 2-undecanone or 2-nonanone have been identified as antifungal compounds that their effects can change minor level depending on pH (Fiddaman and Rossall 1993).

11.3 Open Questions and Future Challenges

Recent studies based on volatile-mediated antifungal activity of bacteria are popular subject that attracts researcher interests for sustainable agriculture and eco-friendly approach to limit chemical use as alternatives in organic farming. Rhizobacteria using volatiles to inhibit fungal disease will be a new control strategy that would be a practical application. Further studies will contribute to understand volatile-mediated mechanism more in detail that new compounds can be elucidated, which can be used in practice for agricultural purposes. Further studies are required to detect and identify new bio-agent candidates with high potential in view of VOC-mediated resistance activity to control pathogenic fungi that cause yield and quality losses in many parts of the world.

In this context, new studies should be carried out using transcriptomics and proteomics to understand the behaviour of biocontrol agents and plant interaction under different stress conditions (abiotic-biotic) that will help in improving maintainable and sustainable approach and deciphering mysteries of microbial communication that is indispensable knowledge for integrated pest management strategies.

References

Alimi M, Soleimani MJ, Darzi MT (2012) Characterization and application of microbial antagonists for control of Fusarium head blight of wheat caused by *Fusarium graminearum* using single and mixture strain of antagonistic bacteria on resistance and susceptible cultivars. Afr J Microbiol Res 6:326–334

Ann YC (2012) Rhizobacteria of pepper (*Piper nigrum*) and their antifungal activities. Afr J Microbiol Res 6:4185–4193

Bailly A, Weisskopf L (2012) The modulating effect of bacterial volatiles on plant growth: current knowledge and future challenges. Plant Signal Behav 7:79–85

Baysal O, Calışkan M, Yeşilova O (2008) An inhibitory effect of a new *Bacillus subtilis* strain (EU07) against *Fusarium oxysporum* f. sp. *radicis lycopersici*. Physiol Mol Plant Pathol 73:25–32

Baysal O, Lai D, Xu HH, Siragusa M, Calışkan M, Carimi F, Teixeira da Silva JA, Tor M (2013) A proteomic approach provides new insights into the control of soil-borne plant pathogens by *Bacillus* species. PLoS One 8:e53182

Bhattacharyya D, Garladinne M, Lee YH (2015a) Volatile indole produced by rhizobacterium *Proteus vulgaris* JBLS202 stimulates growth of *Arabidopsis thaliana* through auxin, cytokinin, and brassinosteroid pathways. J Plant Growth Regul 34:158–168

Bhattacharyya D, Yu SM, Lee YH (2015b) Volatile compounds from *Alcaligenes faecalis* JBCS1294 confer salt tolerance in Arabidopsis thaliana through the auxin and gibberellin pathways and differential modulation of gene expression in root and shoot tissues. Plant Growth Regul 75:297–306

Chen H, Xiao X, Wang J, Wu LJ, Zheng ZM, Yu ZL (2008) Antagonistic effects of volatiles generated by *Bacillus subtilis* on spore germination and hyphal growth of the plant pathogen, *Botrytis cinerea*. Biotechnol Lett 30:919–923

Cho S, Kang B, Han S, Anderson A, Park J, Lee Y et al (2008) 2R,3R-butanediol, a bacterial volatile produced by *Pseudomonas chlororaphis* O6, is involved in induction of systemic tolerance to drought in *Arabidopsis thaliana*. Mol Plant-Microbe Interact 21:1067–1075

Chowdhury SP, Dietel K, Rändler M, Schmid M, Junge H, Borriss R, Grosch R (2013) Effects of *Bacillus amyloliquefaciens* FZB42 on lettuce growth and health under pathogen pressure and its impact on the rhizosphere bacterial community. PLoS One 8:e68818

Correa OS, Montecchia MS, Berti MF, Fernández Ferrari MC, Pucheu NL, Kerber NL, Garcia AF (2009) *Bacillus amyloliquefaciens* BNM122 a potential microbial biocontrol agent applied on soybean seeds, causes a minor impact on rhizosphere and soil microbial communities. Appl Soil Ecol 41:185–194

Elshafie HS, Camele I, Racioppi R, Scrano L, Iacobellis NS, Bufo SA (2012) In vitro antifungal activity of *Burkholderia gladioli* pv. *agaricicola* against some phytopathogenic fungi. Int J Mol Sci 13:16291–16302

Fernando WGD, Ramarathnam R, Krishnamoorthy AS, Savchuk SC (2005) Identification and use of potential bacterial organic antifungal volatiles in biocontrol. Soil Biol Biochem 37:955–964

Fiddaman DJ, Rossall S (1993) The production of antifungal volatiles by Bacillus subtilis. J Appl Bacteriol 74:119–126

Fiddaman PJS (1994) Rossall effect of substrate on the production of antifungal volatiles from *Bacillus subtilis*. J Appl Bacteriol 76:395–405

Groenhagen U, Baumgartner R, Bailly A, Gardiner A, Eberl L, Schulz S, Weisskopf L (2013) Production of bioactive volatiles by *Burkholderia ambifaria* strains. J Chem Ecol 39:892–906

Huang CJ (2012) Dimethyl disulfide is an induced systemic resistance-elicitor produced by *Bacillus cereus* C1L. Pest Manag Sci 68:1306

Kai M, Effmert U, Berg G, Piechulla B (2007) Volatiles of bacterial antagonists inhibit mycelial growth of the plant pathogen *Rhizoctonia solani*. Arch Microbiol 187:351–360

Kai M, Crespo E, Cristescu S, Harren F, Francke W, Piechulla B (2010) *Serratia odorifera*: analysis of volatile emission and biological impact of volatile compounds on *Arabidopsis thaliana*. Appl Microbiol Biotechnol 88:965–976

Kobaisy M, Tellez MR, Webber CL, Dayan FE, Schrader KK, Wedge DE (2001) Phytotoxic and fungitoxic activities of the essential oil of kenaf (*Hibiscus cannabinus* L.) leaves and its composition. J Agric Food Chem 49:3768–3771

Kozdrój J (2008) Microbial community in the rhizosphere of young maize seedlings is susceptible to the impact of introduced pseudomonads as indicated by FAME analysis. J Gen Appl Microbiol 54:205–210

Ryu CM, Hu CH, Locy RD, Kloepper JW (2005) Study of mechanisms for plant growth promotion elicited by rhizobacteria in *Arabidopsis thaliana*. Plant Soil 268:285–292

Silme RS, Baysal Ö (2016) Biodegradation of essential oil by *Enterobacter* sp. Induces plant growth and inhibits damping-off pathogens. Fresenius Environ Bull 25:3493–3506

Simon S, Petrášek J (2011) Why plants need more than one type of auxin. Plant Sci 180(3):454–460

Sumayo M, Hahm MS, Ghim SY (2013) Determinants of plant growth promoting *Ochrobactrum lupini* KUDC1013 involved in induction of systemic resistance against *Pectobacterium carotovorum* subsp *carotovorum* in tobacco leaves. Plant Pathol J 29:174–181

Syu M (2001) Biological production of 2,3-butanediol. Appl Microbiol Biotechnol 55:10–18

Ting ASY, Mah SW, Tee CS (2011) Detection of potential volatile inhibitory compounds produced by endobacteria with biocontrol properties towards *Fusarium oxysporum* f. sp *cubense* race 4. World J Microbiol Biotechnol 27:229–235

Trivedi P, He Z, Van Nostrand JD, Albrigo G, Zhou J, Wang N (2011) Huanglongbing alters the structure and functionaldiversity of microbial communities associated with citrus rhizosphere. ISME J 6:363–383

Vespermann A, Kai M, Piechulla B (2007) Rhizobacterial volatiles affect the growth of fungi and *Arabidopsis thaliana*. Appl Environ Microbiol 73:5639–5641

Wan MG, Li GQ, Zhang JB, Jiang DH, Huang HC (2008) Effect of volatile substances of *Streptomyces platensis* F-1 on control of plant fungal diseases. Biol Control 46:552–559

Weise T, Kai M, Gummesson A, Troeger A, von Reuss S, Piepenborn S, Kosterka F, Sklorz M, Zimmermann R, Francke W, Piechulla B (2012) Volatile organic compounds produced by the phytopathogenic bacterium *Xanthomonas campestris* pv. *vesicatoria* 85-10. Beilstein J Org Chem 8:579–596

Yu SM, Lee YH (2013) Plant growth promoting rhizobacterium *Proteus vulgaris* JBLS202 stimulates the seedling growth of Chinese cabbage through indole emission. Plant Soil 370:485–495

Yuan J, Raza W, Shen QR, Huang QW (2012) Antifungal activity of *Bacillus amyloliquefaciens* NJN-6 volatile compounds against *Fusarium oxysporum* f. sp *cubense*. Appl Environ Microbiol 78:5942–5944

Zhao LJ, Yang XN, Li XY, Mu W, Liu F (2011) Antifungal, insecticidal and herbicidal properties of volatile components from *Paenibacillus polymyxa* strain BMP-11. Agric Sci China 10:728–736

Microbial Volatiles as Chemical Weapons Against Pathogenic Fungi

12

Enespa and Prem Chandra

Abstract

The soil is one of the main habitations of fungi and bacteria. Their interactions are part of a communication web that keeps microhabitats in balance in this amphitheater. Protuberant negotiator molecules of these inter- and intra-organismic relationships are inorganic and organic microbial volatile compounds (MVOCs). Various mixtures of gas-phase and carbon-based compounds are called volatile VOCs produced by microbes and have the capability to diffuse through the atmosphere and soils due to their small size. The volatile organic compounds benzothiazole, cyclohexanol, n-decanal, dimethyl trisulfide, 2-ethyl-1-hexanol, and nonanal are emitted by microorganisms. VOCs have possible potential as an alternative to harmful pesticides, fungicides, and bactericides as well as genetic modification. They play an important role in the inhibition of sclerotial activity, limiting ascospore production, and reducing disease levels in plant pathogenic fungi. Their role as below- and aboveground signals has been established decadely. Recently it is suggested evidently that they might have an important role in below- and aboveground level and involved in microbial-root interactions. Similarly, microbial VOCs appear to be involved in antagonism, mutualism, intra- and interspecies regulation of cellular and developmental processes, and modification of their surrounding environments. Various researchers specify that the MVOCs might provide an alternative to the use of chemicals to protect plants from pathogens and provide a setting for better crop welfare. It is well known that MVOCs can modify the plant physiology and microorganisms. In this assessment, we suggest that MVOCs can be exploited as an eco-friendly,

Enespa
Department of Botany, Dayanand P.G. College, Bachhrawan, Raebareli 229301, U.P., India

P. Chandra (✉)
Department of Environmental Microbiology, School for Environmental Sciences, Babasaheb Bhimrao Ambedkar (A Central) University, Lucknow 226025, U.P., India
e-mail: p.chandrabbau@gmail.com

D.K. Choudhary et al. (eds.), *Volatiles and Food Security*,
DOI 10.1007/978-981-10-5553-9_12

cost-effective, disease-resistant, and sustainable strategy for agricultural practices. Our effort is making a comprehensive chapter of below- and aboveground interactions of microbial volatile diversity and their role against pathogenic fungi.

Keywords
Bacteria • Ecological agriculture • Fungi • Microbial volatile compounds

12.1 Introduction to Microbial Volatile Organic Compounds (MVOCs)

In earthly environments microbes such as fungi and bacteria live in the complex multi-species linkages (Frey-Klett et al. 2011; Hung et al. 2015; Schmidt et al. 2016). In these complexes, both fungi and bacteria formed secondary metabolites of various biochemical classes in excess quantity (Schulz and Dickschat 2007; Morath et al. 2012; Muller et al. 2013). Antibiotics and antibiotic like substances or signaling molecules and diffusible molecules of various secondary metabolites are important in interactions between fungi and bacteria (Frey-Klett et al. 2011; Haq et al. 2014). The mixture of volatile organic compounds (VOCs) such as alkenes, alcohols, ketones, terpenes, benzenoids, pyrazines, acids, and esters mainly is released by bacteria, whereas alcohols, benzenoids, aldehydes, alkenes, acids, esters, ketones and their derivatives of benzene, and cyclohexanes mainly are released by the fungal microbes (Chiron and Michelot 2005; Korpi et al. 2009; Ortiz-Castro et al. 2009; Piechulla and Degenhardt 2014).

These carbon-based solid and liquid volatile compounds are released by soil microbes, which freely enter in the gas phase by evaporating at 0.01 kPa approximately at temperature 20 °C (Pagans et al. 2006). Mostly these volatile organic compounds are small compounds (up to C20) with low molecular mass (<400 Da) and have high vapor pressure, low boiling point, low polarity, and a lipophilic moiety (Vespermann et al. 2007). The lipophilic solutions of volatile compounds are freely crossing the membranes (Pichersky et al. 2006; Wheatley 2002). These physico-chemical properties accelerate the diffusion and vaporization in soil as well as zone of the rhizosphere through gas filled pores and water (Schmidt et al. 2015) and stimulate the plant interaction and communication with the surrounding ecology (Dudareva et al. 2006). Emissions of volatile by microbes such as bacteria and fungi have been studied by several novelists for a long time (Stotzky and Schenck 1976; Kai et al. 2009). The production of volatile is species-specific and work as infochemicals for inter- and intra-organismic communications, cell-to-cell statement signals, a potential carbon discharge valve, or growth-inhibiting agents (Kai et al. 2009). The mode of action of inhibition, when it includes one phytopathogen microbe among interrelating organisms, will be of attention to investigators in regions such as phytopathology, soil destruction, and biocontrol. In agriculture, microbial VOCs have been used as part of biological control strategies to prevent the growth of plant pathogens.

These volatiles played important role in the food chain of microbes because they are assimilated into organic matter; stimulate the biological machinery such as laccase activity, nitrification, nitrogen mineralization, etc.; worked as electron acceptors or donors for support of metabolic reactions; played an important role in quorum sensing/quenching; and are actively used in defense mechanism and communication signals (Effmert et al. 2012). The efflux of volatile organic compounds (VOCs) examined by enormous readings from terrestrial ecosystems has focused on the production of such substances by plants (Kesselmeier and Staudt 1999).They are formed mainly by oxidation of glucose from various intermediates (Korpi et al. 2009). The underlying biosynthetic pathways are aerobic, heterotrophic carbon metabolism, fermentation, amino acid catabolism, terpenoid biosynthesis, fatty acid degradation, and sulfur reduction (Penuelas et al. 2014). The VOCs like alkanes, alkenes, alcohols, carbonyls, esters, isoprenes, terpenes, and acids can influence the communication between microorganisms and plants (Kesselmeier and Staudt 1999; Ryu et al. 2005; Kai et al. 2006). Volatile and nonvolatile organic compounds produced by microbes have received much attention and demonstrate strong inhibition against fungal phytopathogen (Xu et al. 2004; Fernando et al. 2005). These antagonists can be bacteria, like *Pseudomonas* spp., and other beneficiary fungi, such as *Trichoderma* spp. (Minerdi et al. 2009), serve as signaling molecules mediating plant-microbe interaction. Microbial volatile organic compounds released by microbes like bacteria and fungi could inhibit the germination of spore and growth of phytopathogenic fungi (Fiddaman and Rossall 1994); it is suggested that VOCs released by bacteria and other microbes against some soilborne fungal diseases could be a mechanism of biocontrol (McKee and Robinson 1988), or these can act as antibiotics and cause growth inhibition and have more injurious effects on pathogenic microorganisms. The release of VOCs by microbes of soil has been reported to stimulate the growth of the plant (Ryu et al. 2003) and encourage complete resistance in crops. Currently, some microbial volatile compounds are accessible for field applications (Farag et al. 2013; Chung et al. 2016). Rhizobacterial isolates reported previously, such as *Serratia plymuthica*, *Serratia odorifera*, *Stenotrophomonas maltophilia*, *Stenotrophomonas rhizophila*, *Pseudomonas fluorescens*, and *Pseudomonas trivialis,* secrete intricate mixtures of volatiles that constrain the growth of *Rhizoctonia solani* (Kai et al. 2006). In this chapter, we can contribute our scientific literature and current knowledge about microbial VOCs (Fig. 12.1), which emphasize on current improvements in our capability to isolate and recognize the VOCs and their potential for exploitation in biotechnological prospects, as well as on stimulating novel outcomes against biocontrol of fungal phytopathogens.

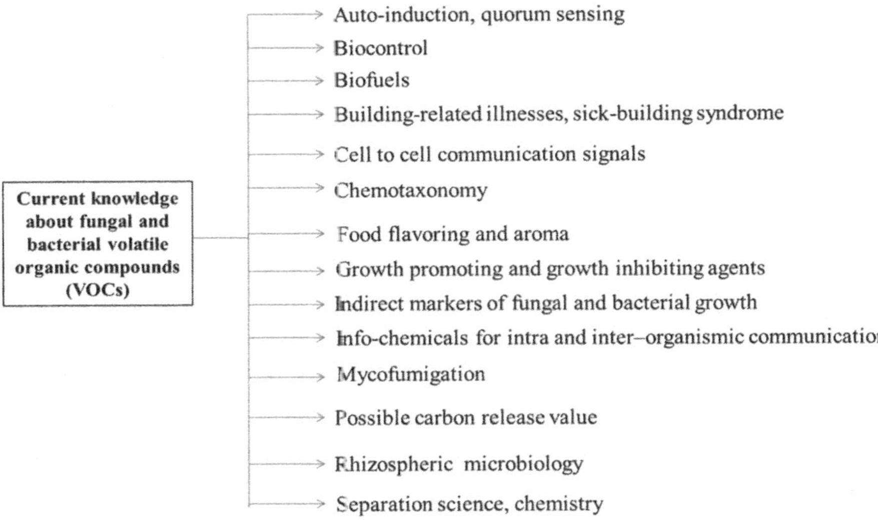

Fig. 12.1 The microbial volatile organic compounds that contribute our knowledge in various subdisciplines

12.2 Microbial VOC Collection and Detection Methods

Due to technological and methodological constraints, the study of fungal volatiles has protected behind the study of other fungal metabolites. Furthermore, the production of VOCs is biologically active. The profile of VOCs will differ, and it completely depends on the substrate, incubation period, nutrients media, temperature, and various environmental factors of given strains or species (Pasanen et al. 1997; Nilsson et al. 2004; Fiedler et al. 2005). There has been significant progress in the separation of various compounds over the last half century. The volatile organic compounds released by fungi are detected by gas chromatography-mass spectrometry (GC-MS) and high-performance liquid chromatography (HPLC) which are used recently due to its powerful separation and highly sensitive detection abilities (Matysik et al. 2009).The solid adsorbent such as Tenax can be used for the concentration of culture headspace, followed by thermal desorption into the GC-MS. After that, the volatile compounds are identified by using a database or library of mass spectra or by comparative study of known standards of retention times and spectrum. Solid-phase microextraction (SPME) is another method of adsorbing and desorbing VOCs in culture headspace. This method has become popular in current years because it decreases the time of preparation by combining extraction, concentration, and introduction into one step while increasing sensitivity over other extraction methods (Zhang and Li 2010). Shortest profiling of living fungal cultures can be mechanized for headspace-SPME-GC-MS (Stoppacher et al. 2010). The GC-MS cannot determine novel volatile novel compounds; this is the drawback. The traditional method for vapor distillation and solvent extraction simultaneous distillation

extraction (SDE) is used to the determination of VOCs from *Penicillium roqueforti* and compares to SPME method (Jelen 2003). Selected-ion flow tube-mass spectrometry (SIFT-MS) provides trace VOCs in moderately complex gas mixtures rapidly. The VOC production from *Aspergillus*, *Candida*, *Mucor*, *Fusarium*, and *Cryptococcus* species also is detected by this technique (Scotter et al. 2005). Secondary electrospray ionization-mass spectrometry (SESI-MS) is a real-time clinical diagnostic tool for bacterial infection. It is known that volatile organic compounds (VOCs), produced in different combinations and quantities by bacteria as metabolites, generate characteristic odors for certain bacteria. Secondary electrospray ionization-mass spectrometry (SESI-MS) is a powerful VOC determination tool for bacteria. The instrument proton transfer reaction-mass spectrometry (PTR-MS) and GC/MS determine the profile of volatiles emitted by *Xanthomonas* c. pv. *vesicatoria* (Weise et al. 2012). PTR-MS can be used to quantify fungal VOCs since it has a fine detection capability and a fine scale time response (Ezra et al. 2004a). Additionally, analyses can be run in real time without sample preparation, derivatization, or concentration with the advantage of having sensitivities is comparable to GC-MS. This technique has been used to quantify the VOCs of *Muscodor albus* (Ezra et al. 2004a). Carbotrap A and B (Supelco) and bentonite-shale materials were used for trapping and collection fungal VOCs. These compounds were placed in a stainless steel column and were recovered; further separation and analysis of potential are determined by nuclear magnetic resonance (NMR) spectroscopy to identify innovative compounds produced by fungi (Booth et al. 2011). An advance technique used for the detection of volatile compounds is called electronic nose. A multi-sensor array has an information processing unit with pattern recognition software and reference library and combines in e-nose system (Wilson and Baietto 2011). The VOC production studies investigated from various microbes and mixed microbial communities of soil by various techniques are listed in Table 12.1.

12.3 Plant Microbe Interactions and Biocontrol

The microbial VOCs have the potential for biotechnological applications in agriculture, pharmaceutical, and medicine industry. In the field of agriculture, the microbial VOCs have their potential in biological control (biocontrol) agents to control the fungal pest to employ pest management strategy by reducing chemical fungicide use on crop plants. Volatiles are important in the functioning of both atmospheric ("aboveground") and lithospheric ecosystems ("belowground"). The inhibition of pathogenic fungi at below- and aboveground by the emission of MVOCs is given in Fig. 12.2.

12.3.1 Belowground Effects on Aboveground Diversity

The resources of high diversity and microbial species in soil could feed back to a high diversity above ground, where various species or functional groups of volatile

Table 12.1 Method applied for the detection of VOCs from different bacterial and fungal species

Methods	Organisms investigated	Habitat/cultivation media	VOCs found	References
SPME/GC-MS	*Bacillus amyloliquefaciens* FZB42 and *Bacillus atrophaeus* LSSC22	Tobacco plant	2-Pentylfuran, 13-tetradecadien-1-ol, 2-butanone, and 2-methyl-n-1-tridecene	Tahir et al. (2017)
SPME/GC-MS	*Pseudomonas aeruginosa* PAO1, *A. fumigatus*	Modified minimal medium	Dimethyl sulfide (DMS), dimethyl disulfide (DMDS), 2,5-dimethylpyrazine (2,5-DMP), 1-undecene, 2-nonanone, 2-undecanone, and 2-aminoacctophcnonc (2-AAP)	Briard et al. (2016)
SPME/GC-MS	*B. amyloliquefaciens* SQR-9	*Lycopersicon esculentum*, cv. Jiangshu	2-Undecanone, undecanal, 2-tridecanone	Raza et al. (2016)
PTR-TOF MS, GC-MS, electronic nose (e-nose) analysis	*Erwinia amylovora*, *Pseudomonas syringae* pv. *syringae*	Rooted plantlets, Murashige and Skoog (MS) medium	2-ethoxy-2-methyl propane, 2,4,4-trimethyl-1-pentene, and 2-methyl-furan	Cellini et al. (2016)
PTR-MS/ PTR-TOF-MS	Microbial community	Temperate soil under different compost load	Diverse VOCs (seven tentatively identified)	Seewald et al. (2010)
GC-MS	Microbial community	Hyperthermic, hypersaline soils	Diverse VOCs (72 identified)	McNeal and Herbert (2009)
GC-MS	*Aspergillus* spp., *Cladosporium cladosporioides*, *Penicillium* spp.,	Dichloran glycerol agar	Diverse VOCs	Matysik et al. (2009)
PTR-MS	*Escherichia coli*, *Shigella flexneri*, *Salmonella enterica*, *Candida tropicalis*	Complex media	Diverse VOCs, several unidentified and some identified compounds of low molecular weight <150 u	Bunge et al. (2008)
GC-MS/growth inhibition of bacterial cultures	*Fusarium oxysporum* strain MSA 35	Agar (as described in experimental procedures)	Diverse VOCs	Minerdi et al. (2009)

SPME-GC-MS	*Bacillus subtitles* G8	Soil isolate, cultivated on agar plates	Diverse VOCs, alkyls, alcohols, esters, ketones, acid, amines, oxides, phenols, and heterocyclic compounds	Liu et al. (2008)
GC-MSD (mass selective detector)	Microbial community	Orange waste	Monoterpenes, isoprene, other VOCs	Wang and Wu (2008)
GC-MS	*Muscodor albus* E-6 Endophytic fungus of *Guazuma ulmifolia*	Cultivated on PDA	Diverse VOCs	Strobel et al. (2007)
HS-SPME-GC-MS	*Hypholoma fasciculare Resinicium bicolor* Wood-decaying fungi	Cultivated on malt broth	Diverse VOCs	Hynes et al. (2007)
GC-MS	Alcaligenaceae, Bacillales, Micrococcaceae, Rhizobiaceae, Xanthomonadaceae fungi	Soil bacterial isolates	Diverse VOCs	Zou et al. (2007)
HS-SBSE GC-MS	Truffle (3 *Tuber* spp.)	Fruiting bodies (119 identified)	Diverse VOCs	Spilvallo et al. (2007b)
PTR-MS	Microbial community	Organic waste	various VOCs	Mayrhofer et al. (2006)
GC-MS	Microbial community	Different Mediterranean soils	Diverse VOCs (25 identified)	Serrano and Gallego (2006)
GC-MS, SPME	*Bacillus amyloliquefaciens* IN937a, *E. coli* DH5alpha, *B. subtilis GB03, Pseudomonas fluorescens* 89B61	Rhizosphere	Diverse VOCs, 10 identified, 28 not characterized, branched chain alcohols	

(continued)

Table 12.1 (continued)

Methods	Organisms investigated	Habitat/cultivation media	VOCs found	References
GC-MS	*Stenotrophomonas maltophilia* R3089, *Serratia plymuthica* HRO-C48, *Stenotrophomonas rhizophila* P69, *Serratia odorifera* 4Rx13, *Pseudomonas trivialis* 3Re2-7, *S. plymuthica* 3Re4-18, *Bacillus subtilis* B2g, *P. fluorescens* L13-6-12	Rhizosphere isolates, cultivated on NBII	Diverse unidentified VOCs	Kai et al. (2006)
CLSA, GC-MS	*Cytophaga-Flavobacterium-Bacterioides* group	Maritime arctic strains cultured on agar	Diverse VOCs, mainly methyl ketones	Dickschat et al. (2005a)
GC-MS	*Streptomyces* GWS-BW-H5	North seawater, cultured on agar, liquid culture	Diverse VOCs	Dickschat et al. (2005b)
GC-MS	*Pseudomonas fluorescens*, *P. corrugata*, *P. chlororaphis*, *P. aurantiaca*	Isolated from canola stem, soybean plants, cultivated on Luria-Bertani broth (LBB)	Fungistatic VOCs	Fernando et al. (2005)
GC-MS	Benthic cyanobacteria (*Calothrix, Plectonema*)	Cyanobacterial medium	Limonene, cyclohexanone, straight-chain aldehydes	Höckelmann and Jüttner (2004)
GC-MS/PTR-MS	*Muscodor albus*	Soil, grown on potato dextrose agar (PDA)	Diverse VOCs	Ezra et al. (2004a)
SPME/HSSE GC-MS	*Fusarium* spp.	MEA and PDA	Sesquiterpenes, mainly trichodiene	Demyttenaere et al. (2004)
PTR-MS	*Pseudomonas* spp., Enterobacteriaceae, lactic bacteria, *Enterococcus* spp.	Air and vacuum packed meat (beef and pork)	Diverse VOCs, several unidentified and some identified compounds of low molecular weight <150 u	Mayr et al. (2003a)

GC-FID, GC-MS	*Streptomyces* spp.	Yeast starch agar	Diverse VOCs (120 characterized), mainly terpenoids	Schöller et al. (2002)
SPME-GC-MS	*Sclerotinia minor*, *S. sclerotium*, *S. rolfsii*	Lettuce and bean isolates, cultivated on PDA	Diverse VOCs	Fravel et al. (2002)
GC-MS	*Muscodor albus*	Endophytic fungus of *Cinnamomum*, cultivated on PDA	Diverse VOCs	Strobel et al. (2001)
GC-FID	*Rhizopus* spp.	Cassava bagasse, apple pomace, soybean, amaranth grain, and soil bean oil	Diverse VOCs	Christen et al. (2000)

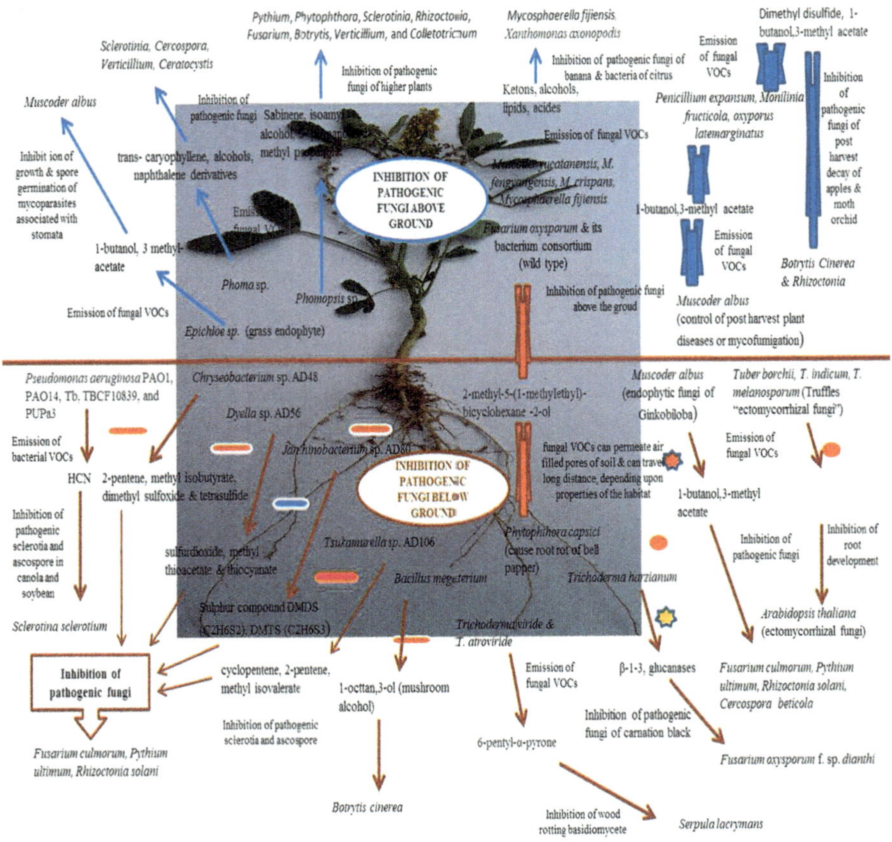

Fig. 12.2 Inhibition of pathogenic fungi at below and above ground by the emission of MVOCs

compounds are diligently connected to belowground groups (Van der Heijden et al. 1998). A positive correlation was found between the plant diversity and species of endomycorrhizal fungi potentially since the various fungal species contaminate the different plant's species, though another justification exists for the arrangements they detected (Morath et al. 2012). However, this type of design does not embrace at the measure of well-designed types of mycorrhizae: The high diversity of plants can be associated with low-diversity arbuscular mycorrhizal communities and low diversity of plants to high-diversity ectomycorrhizal communities (Allen et al. 1995). The belowground populations of microbes also distress the nutrient availability and litter collection. The response through these mechanisms could stimulate the diversity of aboveground communities, while the indication for such response is mixed.

12.3.2 Interaction of Belowground Microbes

The failure of propagules of fungi to propagate or the inhibition of fungal hyphae growth in favorable atmospheric conditions by soil fungistasis (Watson and Ford 1972) has been postulated to happen because of the release of inhibitory compounds or competition for nutrients in the soil. The growth of three fungal species is inhibited by these volatile compounds, *Paecilomyces lilacinus*, *Pochonia chlamydosporia*, and *Clonostachys rosea*, suggesting that the occurrence of volatile fungistasis may not need the direct competition between soil microorganisms and may have a different mechanism than direct fungistasis (Xu et al. 2004). VOCs can infiltrate air-filled pores of soil and cover long distances; this situation completely depends on the properties of the habitat (Aochi and Farmer 2005). For example, without any direct contact between the strains, *Fusarium oxysporum* is the wild-type antagonistic released volatiles and the fungal growth of a plant pathogenic strain of *F. oxysporum* inhibited by its bacterial consortium (Minerdi et al. 2009). This type of properties may make fungal VOCs as a valuable addition to biocontrol strategies. The soilborne diseases can be controlled by VOCs released from endophyte *M. albus*. When it is mixed in soil, it controls disease of damping-off of broccoli of *Rhizoctonia solani* and root rot of bell pepper disease caused by *Phytophthora capsici* (Mercier and Manker 2005).

12.3.3 Interaction of Aboveground Microbes

The novel species of endophytic fungi are isolated from beneath the epidermal cell layer tissues of plants. There is no harm caused by endophytic fungus that lives within their host plant (Bacon and White 2000). The endophytic fungus releases a mixture of VOCs which provide additional lines of resistance against phytopathogens; it may benefit to host plant (Macias-Rubalcava et al. 2010). For example, the VOCs produced by *Muscodor albus* resist the growth and eradicate plant pathogenic microbes such as fungi and bacteria (Strobel et al. 2001). Moreover, the other isolate like *Muscodor yucatanensis and Muscodor fengyangensis* also produced the VOCs and inhibited the growth of various pathogenic species of bacteria, fungi, and oomycota (Atmosukarto et al. 2005; Macias-Rubalcava et al. 2010; Zhang et al. 2010). Conclusively, the mixture of VOCs produced by *Muscodor crispans* repressed the growth of a wide range of phytopathogens such as *Mycosphaerella fijiensis* fungi (the black sigatoka pathogen of bananas) and the severe bacterial pathogen of citrus, *Xanthomonas axonopodis* pv. *citri* (Mitchell et al. 2010). The VOCs released by Muscodor sp. (Strobel et al. 2001) have esters, alcohols, lipids and ketones, while a unique mixture of VOCs such as sabinene (a monoterpene with a peppery odor) formed in a *Phomopsis* species and a mixture of other compounds such as 1-butanol, 3-methyl, benzeneethanol, 1-propanol, 2-methyl, and 2-propanone are also determined in this species (Singh et al. 2011). Mixtures of commercially available compounds, such as valencene and bulnesene, and chemically synthesized mixtures of VOCs, such as 3-methylbutyl ester, 2-methyl, propanoic acid, and 1-butanol,

3-methyl-, acetate, were tested against the phytopathogens and generated related inhibitory effects to the *M. albus*-manufactured VOCs (Strobel et al. 2001). However, when the mixture of VOCs was split down into numerous classes of complexes, the same inhibitory properties were not attained, suggesting that it is the collection of VOCs that contributes to the antifungal movement (Strobel et al. 2011).

12.4 Inhibition of Germination and Mycelial Growth of Phytopathogenic Fungus by Bacterial Volatiles

Dobbs and Hinson (1953) illustrated the phenomenon of fungistasis, which can be on the growth and germination of soilborne fungi due to the negative influence of bacterial volatiles. The volatiles manufactured by *Streptomyces griseus* encouraged early sclerotia construction in *Sclerotium cepivorum* and *Rhizoctonia solani* and reduced sporulation in *Gloeosporium aridum* (Morath et al. 2012). The volatile but-3-en-2-one fashioned by *Streptomyces griseoruber* caused the strong inhibition of spore germination in *Cladosporium cladosporioides* (Herrington et al. 1987). Approximately 1080 isolates of bacteria screened out for fungistatic action by Zou et al. (2007), in which the 328 isolates belong to the family of Xanthomonadaceae, Rhizobiaceae, Alcaligenaceae, and Micrococcaceae and to the order of Bacillales that were recognized as declining the mycelial growth and the germination of *Paecilomyces lilacinus* and *Pochonia chlamydosporia*. For the inhibition of spore germination in both fungi, the soil direct fungistasis and soil volatile fungistasis were responsible. Simultaneously both properties of fungi carefully diminished the spore germination and vanished after autoclaving. Numerous volatiles were recognized, but the trimethylamine, benzaldehyde, and N,N-dimethyl octylamine disclosed the strong antifungal activity (Chuankun et al. 2004). The investigational operations confirm that merely volatile metabolites would stimulate fungal growth. The comprehensive studies are directed using various bacterial species like *Bacillus subtilis*, *Pseudomonas fluorescens*, *Pseudomonas trivialis*, *Burkholderia cepacia*, *Staphylococcus epidermidis*, *Stenotrophomonas maltophilia*, *Stenotrophomonas rhizophila*, *Serratia odorifera*, and *Serratia plymuthica*; all rhizospheric bacteria repressed the mycelial growth of phytopathogenic fungi, such as *Fusarium culmorum*, *Aspergillus niger*, *Fusarium solani*, *Paecilomyces carneus*, *Microdochium bolleyi*, *Phoma eupyrena*, *Penicillium waksmanii*, *Phoma betae*, *Sclerotinia sclerotiorum*, *Rhizoctonia solani*, *Trichoderma strictipile*, and *Verticillium dahliae* (Vespermann et al. 2007; Kai et al. 2008). The individual bacteria-fungus combination depends upon the amount of inhibition. Remarkably, the *Fusarium solani* turned out to be resilient against the bacterial volatiles. Many unknown bacterial volatile compounds are also manufactured, but some compounds such as dodecanal, dimethyl disulfide (DMDS), dimethyl trisulfide (DMTS), 2-phenylethanol, and 1-undecene could be identified (Kai et al. 2006). In dual-culture experiments, dimethyl disulfide and 1-undecene compound are applied particularly which inhibits the growth of *F. culmorum* definitely (Kai et al. 2009). Another report also deep-rooted the antifungal act of volatiles manufactured by hostile rhizospheric bacteria.

The growth of fungus *Gaeumannomyces graminis* var. tritici is completely inhibited by *Pseudomonas fluorescens* and *Pseudomonas pumila* bacteria which cause take-all disease (TAD) in *Triticum aestivum* crop (Yang et al. 2011), and the growth of *Fusarium oxysporum* is decreased by *Gluconacetobacter diazotrophicus* (Logeshwarn et al. 2011), *Bacillus subtilis*, *B. pumilus*, and *B. cereus* slowed down the development of *Botrytis mali* fungus (Jamalizadeh et al. 2010). The bacterium *B. subtilis* exhibited antifungal bustle against *Aspergillus alternate*, *Cladosporium oxysporum*, *Fusarium oxysporum*, *Paecilomyces lilacinus*, *Paecilomyces variotii*, and *Pythium afertile* (Chaurasia et al. 2005), and *Rhizoctonia solani* and *Pythium ultimum* (Fiddaman and Rossall 1993), *Phytophthora sojae* caused the damping-off disease in soybean diminished the growth by *Bacillus* spp. (Tehrani et al. 2002). *Paenibacillus polymyxa* produced 1-octen-3-ol, benzothiazole, and citronellol volatile, which strongly inhibited the mycelial growth and reduced propagation of eight mycological pathogens, including *Botrytis cinerea* (Zhao et al. 2011). *Pseudomonas* sp. released volatiles, which inhibited up to 90% ascospore germination, isolated from canola and soybean plants (Fernando et al. 2005). The growth inhibition exposed in various fungal pathogens directly by volatile compounds of rhizospheric bacteria *Bacillus pumilus* and mycorrhizal fungus *Glomus mosseae* in mandarin plants and improves the growth of it (Chakraborty et al. 2011). The antifungal derivative compounds 2-phenylethanol and phellandrene were identified from *Streptomyces platensis*, which were responsible for clamping down the growth of mycelium in *Rhizoctonia solani*, *Sclerotinia sclerotiorum*, and *Botrytis cinerea* phytopathogenic fungus (Wan et al. 2008). Various *Pseudomonas* spp. recognized for HCN production as active antifungal compounds (Voisard et al. 1989; Haas and Défago 2005), though the HCN production could be associated with fungistasis and its antifungal effect and verified only in in vitro conditions. In dual-culture tests, the rhizobacterial isolates were screened out for HCN production and diffusible antifungal metabolites, and it is tested against *Verticillium dahliae* and *Rhizoctonia solani* in greenhouse experiments (Tehrani et al. 2001; Afsharmanesh et al. 2006). Extraordinarily, the HCN manufacturers revealed utmost productivity when functional to the topsoil, whereas non-manufacturers were more competent when applied to seeds. The conidial germination of *Botrytis cinerea* and *Penicillium nigricans* fungus is inhibited by ammonium sulfate solution and ammonia discharge from soil. But there were no any effects of ammonia seen on the fungi *Fusarium culmorum* and *Verticillium nigrescens*. Volatiles released by bacteria against phytopathogenic fungi with a focus on major crop pathogens are shown in Table 12.2.

12.5 Effect of Bacterial Volatiles on Fungal Morphology

Morphological changes in fungi have been seen after the treatment of bacterial volatiles in various experiments. Volatiles secreted by *Bacillus subtilis* bacterium caused the abnormality in hyphae with deformations and enhanced the vacuolation in *Rhizoctonia solani* and *Pythium ultimum* (Fiddaman and Rosall 1993). The septa of fungi totally vanished and conidia converted thick-walled and asymmetrical in

Table 12.2 Bacterial volatile compounds used as an inhibitor against phytopathogenic fungal growth

Bacteria	Volatiles	Effects	Fungi	Reference
Streptomyces sp. CEN26	2,5-Bis (hydroxymethyl) furan monoacetate (BHMF-OAc)	Affected conidial germination	*Alternaria brassicicola*	Phuakjaiphaeo et al. (2016)
Bacillus subtilis	2,3-Butanediol	Growth inhibited	*Trichoderma* sp., *Ralstonia solanacearum*	Ye et al. (2016)
Dyella sp. AD56	Sulfur dioxide, methyl thioacetate, dimethyl sulfoxide, dimethyl trisulfide, dimethyl disulfide	Growth inhibition	*F. culmorum P. ultimum R. solani*	Tyc et al. (2015) and Hunziker et al. (2015)
Janthinobacterium sp. AD80	Dimethyl disulfide, thiocyanates, thioesters, and sulfides	Growth inhibition	*F. culmorum P. ultimum R. solani*	
Pseudomonas moraviensis, *Pseudomonas veronii*, *P. chlororaphis*, and *P. fluorescens*	1-Undecene, dimethyl disulfide, dimethyl trisulfide, 2-acetylfuran, 1-decene, 1-undecene, 1-dodecene, and an undecadiene	Mycelial growth, sporangium formation, germination, changes its sporulation behavior	*Phytophthora infestans*	
Tsukamurella sp. AD106	Dimethyl trisulfide, dimethyl disulfide, thiocyanates, thioesters, and sulfides	Growth inhibition	*F. culmorum P. ultimum R. solani*	
Paenibacillus ehimensis KWN38	Butanol	Fungal growth inhibitory	*R. solani* AG1-IA, *R. cerealis*, *Phytophthora capsici*, *F. oxysporum* f. sp. *lycopersici*, *Pythium aphanidermatum*, *Colletotrichum caudatum*	Naing et al. (2014)
Streptomyces alboflavus	2-Methylisoborneol, 1,4-dimethyl adamantane, and 1,2,3-trimethyl benzene	Inhibit storage fungi	*F. moniliforme*, *Aspergillus flavus*, *Aspergillus ochraceus*, *Aspergillus niger*, and *Penicillium citrinum*	Kanchiswamy et al. (2015)

Burkholderia tropica MTo431	α-Pinene, DMDS, ocimene, limonene, and fenchone	Inhibited the mycelial growth	*Colletotrichum gloeosporioides*, *Fusarium culmorum*, *F. oxysporum*, *Sclerotium rolfsii*	Kanchiswamy et al. (2015) and Tenorio Salgado et al. (2013)
Burkholderia ambifaria	Dimethyl trisulfide, 2-nonanone, 2-undecanone, 4-octanone, S-methylmethane thiosulfonate, acetophenone, phenylpropan-1-one, 1-phenyl-1,2 propanedione	Mycelial growth inhibition	*Alternaria alternate*, *Rhizoctonia solani*	Groenhagen et al. (2013)
Pseudomonas sp.	Benzothiazole, cyclohexanol, decanal, dimethyl trisulfide, 2-ethyl-1-hexanol, and nonanal	Inhibition of mycelial growth and of germination from ascospores and sclerotia	*Sclerotinia sclerotiorum*, *Rhizoctonia solani*	Fernando et al. (2005) and Bitas et al. (2013)
Xanthomonas campestris pv. *vesicatoria* 85–10	Decan-2-one	Mycelial growth inhibition	*Rhizoctonia solani*	Weise et al. (2012)
Bacillus cereus	Dimethyl disulfide	Reduced disease on tobacco plants (greenhouse)	*Botrytis cinerea*	Huang (2012)
Bacillus amyloliquefaciens (NJN-6)	Benzenes, ketones, and aldehydes	Inhibition of mycelial growth and of spore germination	*F. oxysporum* f. sp. *cubense*	Yuan et al. (2012)
Burkholderia gladioli pv. *agaricola*	Limonene	Inhibited fungal growth and reduced the growth rate	*F. oxysporum* and *Rhizoctonia solani*	Elshafie et al. (2012)

(continued)

Table 12.2 (continued)

Bacteria	Volatiles	Effects	Fungi	Reference
Paenibacillus polymyxa BMP-11	1-Octen-3-ol, benzothiazole, citronellol	Mycelial growth inhibition and of pigmentation	*Alternaria brassicae* and *Phytophthora capsici*, *Rhizoctonia solani*, *Fusarium oxysporum*, *Botrytis cinerea*	Zhao et al. (2011)
Sporosarcina ginsengisoli	Dimethyl disulfide, 1-undecene	Inhibition of mycelial growth	*F. culmorum* PR19-12-11, *R. solani* AG3, *S. sclerotiorum*, *V. dahlia* V25	Kai et al. (2009)
S. maltophilia, *Alcaligenes faecalis*, *Arthrobacter nitroguajacolius*	Acetamide, methanamine, 1-butanamine, benzaldehyde, phenylacetaldehyde, 1-decene, benzothiazole	Inhibition of mycelial growth	*Paecilomyces lilacinus*, *Pochonia chlamydosporia*	Zou et al. (2007)

shape in *Aspergillus alternaria* due to bacterial volatiles produced *Bacillus subtilis.* It also instigated hyphae and conidial deformations in *Cladosporium oxysporum, Fusarium oxysporum, Paecilomyces lilacinus, Paecilomyces variotii, and Pythium afertile* (Chaurasia et al. 2005). Volatiles emitted from *Staphylococcus pasteuri* showed swollen and contorted patterns in fungal mycelium after treatment with 1-octen-3-ol (Barbieri et al. 2005). The germination of fungal spores and growth of hyphae can be inhibited by bacterial volatiles (Herrington et al. 1987). Moreover, the bacterial volatile exposure has been reported to change the enzyme activity, fungal morphology, and gene expression (Vespermann et al. 2007; Minerdi et al. 2009; Kai et al. 2009; Garbeva et al. 2014b). For instance, the activity of laccases and tyrosinases can be intensely affected by bacterial volatiles (Wheatley 2002). The amount of prevention depends on the specific fungus-bacteria or fungus-fungus interaction, and sensitivity to volatiles can differ between various fungal species (Vespermann et al. 2007; Garbeva et al. 2014b). There is no effect on *F. solani* by bacterial volatiles, while the *Pythium* species (oomycetes) are extremely sensitive to bacterial volatiles reported in various experimental studies (Effmert et al. 2012; Garbeva et al. 2014a). The fungus *O. latemarginatus* produced volatile which is slightly resistant against *F. oxysporum*, whereas *Magnaporthe grisea* was sensitive. Recently, it is observed that the late blight oomycete pathogen *Phytophthora infestans* was highly sensitive to bacterial volatiles. The hydrogen cyanide and 1-undecene volatile compounds were responsible for the growth resistant of fungi (Hunziker et al. 2015).

12.6 Influence of Bacterial Volatiles on Mycorrhizal Fungi

In plant roots, a complex symbiotic community of mycorrhiza, which is associated with fungi and bacteria, was found (Jung et al. 2012). Mycorrhiza also reduced the activity of phytopathogenic fungi directly due to physical contact and also releases the volatile compounds playing a monitoring character in a mycorrhizal machinery establishment (Bonfante and Anca 2009). The mycorrhiza helper bacteria (MHB) associated with rhizobacteria released volatile compounds, which have toxicity against the phytopathogenic fungi (Miransari 2011). Various fungal volatiles also showed the growth inhibition and spore germination to other fungal species. The endophytic fungi such as *Muscodor albus* and *Oxyporus latemarginatus* intensely exhibited the inhibitory effect on the growth and development of numerous phytopathogenic fungi, i.e., *Botrytis cinerea* and *Rhizoctonia solani* (Strobel et al. 2001). It is also reported that the volatiles are released from *Muscodor albus* to kill the fungus of human pathogens like *Aspergillus fumigatus* and *Candida albicans* (Strobel et al. 2001). However, the fungi have symbiotic relationship with bacteria. Hyphae of *Fusarium oxysporum* associated with bacteria release the volatile sesquiterpene caryophyllene, which suppressed the expression of two virulence genes. And when it is cured from the bacterial symbionts, caryophyllene was not detected, and *F. oxysporum* became pathogenic (Minerdi et al. 2009).

12.7 Microbial VOCs as Chemical Weapons Against Other Microbes

Several microbes like bacteria and fungus that are well furnished with this type of weaponry have been used as biocontrol agents to manage phytopathogens (Pal and McSpadden Gardener 2007). Several bacteria that released the volatile organic compounds have been seen to inhibit mycelial growth and spore germination or caused the abnormal morphological changes in fungi (Effmert et al. 2012). In the in vitro and field experiments, it has been observed that some *Pseudomonas* spp., which are isolated from soybean and canola plants regularly, produced the volatile organic compounds (VOCs) that inhibit the mycelial growth and germination of sclerotia and ascospores of *Sclerotinia sclerotiorum* phytopathogen (Fernando et al. 2005). Some volatiles from these isolates such as cyclohexanol, benzothiazole, n-decanal, dimethyl trisulfide, 2-ethyl-1-hexanol, and nonanal (Fig. 12.2) totally inhibited the sclerotial germination and mycelial growth. VOCs like acetamide, benzothiazole, 1-butanamine, benzaldehyde, methanamine, phenylacetaldehyde, and 1-decene appear to play characters in fungistasis.

Some biocontrol fungi appear to service VOC to control pathogenic fungi (Bruce et al. 2003; Humphris et al. 2002; Hynes et al. 2007). Various species of *Trichoderma* have been shown to efficiently parasitize or constrain a wide range of soilborne fungal pathogens by playing various mechanisms, like mycoparasitism, nutrient competition, and secretion of inhibitory compounds and hydrolytic enzymes (Harman 2011; Lorito et al. 2010). The VOC secreted from *Trichoderma viride* and *T. aureoviride* that inhibits the growth and production of proteins from a wood-rotting basidiomycete *Serpula lacrymans*. But there is no effect of *T. pseudokoningii* on any of the *Serpula* isolate tests (Humphris et al. 2002). *F. oxysporum* f. sp. *ciceris*, a soilborne fungal pathogen that causes chickpea wilt disease, showed the degree of growth inhibition by VOC secreted by numerous isolates of three *Trichoderma* spp. The growth of lettuce plants improved via VOC secreted from *F. oxysporum* strain MSA35 (Minerdi et al. 2011) and also released the VOC in the presence of ectosymbiotic bacteria that inhibit the growth of pathogenic strains of *F. oxysporum* (Bitas et al. 2013). Microbial volatile organic compounds secreted by fungi against phytopathogenic fungi are given in Table 12.3.

12.8 Conclusion

We have arranged the data which collected from disparate scientific fields with the determined of highlighting. Microbial VOCs are found abundantly with biotechnological utility and they facilitate plentiful interactions between organisms through different ecological niches. The study of nonvolatile natural products and their chemical interactions are the best-known example. However, we have the assurance that the advances in analytical and genomics techniques will produce much associated multidisciplinary cooperation, which focused on gas-phase organic molecules, the research for microbial species anticipates in finding novel biotechnological

Table 12.3 Fungal volatile organic compounds secreted by fungi against phytopathogenic fungi

Fungal antagonists	Volatiles	Effects	Pathogenic fungi	References
H. anthochroum Blaci	2-Methyl-5-(1-methylethyl)-bicyclohexan-2-ol and 2, 6-dimethyl-2,4,6-octatriene	Inhibiting effect on growth of oomycetes	*Pythium ultimum*, *Phytophthora capsici*, *Alternaria solani*, and *Fusarium oxysporum*	Ulloa-Benıtez. (2016)
Hypsizygus marmoreus	2-Methylpropanoic acid 2,2-dimethyl-1-(2-hydroxy-1-methylethyl) propyl ester	Inhibitory effect against conidial germination	*A. brassicicola (O-264)*	Oka et al. (2015)
Trichoderma virens	β-Caryophyllene; β-elemene; germacreneD; δ-cadinene	Growth inhibition	*Botrytis cinerea*	Contreras-Cornejo H A et al. (2014)
Ampelomyces sp.	*m*-Cresol	Inhibition of mycelial growth	*Pseudomonas syringae* pv.	Naznin et al. (2014)
Cladosporium cladosporioides CL-1	α-Pinene, β-caryophyllene, tetrahydro-2,2,5,5 tetramethyl furan, dehydroaromadendrene, and sativene	Growth inhibition of mycelium	*Pseudomonas syringae*	Paul and Park (2013)
Mycoleptodonoides aitchisonii	1-Phenyl-3-pentanone	Strongly inhibited the mycelial growth, spore germination	*Alternaria alternata*, *A. brassicicola*, *A. brassicae*, *Colletotrichum orbiculare*, and *Corynespora cassiicola*	Nishino et al. (2013)
Epichloe typhina	Sesquiterpenes, chokols A–G	Fungitoxic	*Cladosporium phlei*	Kumar and Kaushik (2012)
Phaeosphaeria nodorum	Ethyl acetate, 3-methylbutan-1-ol, acetic acid, 2-propyl-1-ol, and 2-propenenitrile	Volatiles inhibited growth and reduced the width of the hyphae	*M. fructicola*	Pimenta et al. (2012)

(continued)

Table 12.3 (continued)

Fungal antagonists	Volatiles	Effects	Pathogenic fungi	References
Saccharomyces cerevisiae CR-1	3-Methylbutan-1-ol, 2-methylbutan-1-ol, 2-phenylethanol, ethyl acetate, and ethyl octanoate	Inhibit the vegetative development	*Guignardia citricarpa*	Fialho et al. (2011)
Phoma sp.	Series of sesquiterpenoids, some alcohols, and several reduced naphthalene derivatives	Antifungal and fuel properties; some of the test organisms with the greatest sensitivity	*Verticillium*, *Ceratocystis*, *Cercospora*, and *Sclerotinia*	Strobel et al. (2011)
Phomopsis sp.	Sabinene; isoamyl alcohol; 2-methyl propanol; 2-propanone	Worked as antibiotic effects	*Pythium*, *Phytophthora*, *Sclerotinia*, *Rhizoctonia*, *Fusarium*, *Botrytis*, *Verticillium*, *Colletotrichum*	Singh et al. (2011)
Saccharomyces cerevisiae	Ethyl acetate, 2-methylbutan-1-ol, 3-methylbutan-1-ol, 2-phenylethanol, and ethyl octanoate	Growth inhibition	*G. citricarpa*	Fialho et al. (2010)
Trichoderma viride, *Trichoderma harzianum*	6-Pentyl-α-pyrone, β-1-3-glucanases	Phytotoxicity during seedling formation, seedling blight suppression	*Fusarium oxysporum*, *Rhizoctonia solani* (Israel), *Pythium ultimum* (USA)	El-Hasan and Buchenauer (2009)
Kyu-W63	5-Pentyl-2-furaldehyde and 5-(4-pentenyl)-2-furaldehyde	Strongly inhibited growth of mycelia	*Fusarium graminearum*, *Aspergillus restrictus*, *A. fumigatus*, *Penicillium islandicum*	Koitabashi et al. (2004)

Irpex lacteus	5-Pentyl-2-furaldehyde	Suppressed the growth	*F. oxysporum* f. sp. *lycopersici*, *Blumeria graminis*, *Fusarium oxysporum*, *Colletotrichum fragariae*, *Botrytis cinerea*	Koitabashi and Tsushima (2007)
Candida albicans	Farnesol	Inhibition of mycelial development, Apoptosis in altered morphology and reduced fitness	*Aspergillus nidulans*, *Fusarium graminearum*	Semighini et al. (2008, 2006) and Hornby et al. (2001)
Muscodor albus	Ethyl acetate, propanoic acid, 2-methyl-, methyl ester, ethanol, acetic acid, 2-methylpropyl ester, propanoic acid, 2-methyl-, butyl ester, 1-butanol, 2-methyl, acetate, naphthalene, azulene, bulnecene	Inhibited the growth of fungi	*Pythium ultimum*, *Phytophthora cinnamomi*, *Rhizoctonia solani*, *Ustilago hordei*, *Stagonospora nodorum*, *Sclerotinia sclerotiorum*, *Aspergillus fumigatus*, *Verticillium dahliae*, *Tapesia vallundae*, *Candida albicans*, *F. solani*, *Cercospora beticola*, *Xylaria* sp.	Strobel et al. (2001) and Ezra et al. (2004a)

products of volatile compounds. Usually, this chapter focused on the discovery of secondary metabolites of industrially or other commercial values of VOCs, especially from soil microorganisms. These microbes may increase the wealth and production of crops.

References

Afsharmanesh H, Ahmadzadeh M, Sharifi-Tehrani A (2006) Biocontrol of *Rhizoctonia solani*, the causal agent of bean damping-off by *fluorescent pseudomonads*. Commun Agric Appl Biol Sci 71:1021–1029

Allen LA, Zhao XJ, Caughey W, Poyton RO (1995) Isoforms of yeast cytochrome c oxidase subunit V affect the binuclear reaction center and alter the kinetics of interaction with the isoforms of yeast cytochrome c. J Biol Chem 270(1):110–118

Aochi YO, Farmer WJ (2005) Impact of soil microstructure on the molecular transport dynamics of 1, 2-dichloroethane. Geoderma 127:137–153

Atmosukarto I, Castillo U, Hess W, Sears J, Strobel G (2005) Isolation and characterization of *Muscodor albus* I-41.3s, a volatile antibiotic producing fungus. Plant Sci 169:854–861

Babaeipoor E, Mirzaei S, Danesh YR, Arjmandian A, Chaichi M (2011) Evaluation of some antagonistic bacteria in biological control of *Gaeumannomyces graminis* var tritici causal agent of wheat take-all disease in Iran. Afr J Microbiol Res 5:5165–5173

Bacon CW, White JF (2000) Microbial endophytes. Marcel Dekker, New York, pp 341–388

Barbieri E, Gioacchini AM, Zambonelli A, Bertini L, Stocchi V (2005) Determination of microbial volatile organic compounds from *Staphylococcus pasteuri* against tuber borchii using solid-phase microextraction and gas chromatography/ion trap mass spectrometry. Rapid Commun Mass Sp 19:3411–3415

Barco-Cortes AM, Goodwin PH, Hsiang T (2010a) Comparison of induced resistance activated by benzothiadiazole, (2R, 3R)- butanediol and an isoparaffin mixture against anthracnose of *Nicotiana benthamiana*. Plant Pathol 59:643–653. doi:10.1111/j.1365-3059.2010.02283

Barco-Cortes AM, Hsiang T, Goodwin PH (2010b) Induced systemic resistance against three foliar diseases of *Agrostis stolonifera* by (2R, 3R) – butanediol or an isoparaffin mixture. Ann Appl Biol 157:179–189. doi:10.1111/j.1744-7348.2010.00417

Bitas V, Kim HS, Bennett JW, Kang S (2013) Sniffing on microbes: Diverse roles of microbial volatile organic compounds in plant health. Am Phytopathol Soci 26(8):835–843. doi.org/10.1094/MPMI-10-12-0249-CR

Bonfante P, Anca I (2009) Plants, mycorrhizal fungi, and bacteria: a network of interaction. Annu Rev Microbiol 63:363–383

Booth E, Strobel G, Knighton B, Sears J, Geary B, Avci R (2011) A rapid column technique for trapping and collecting of volatile fungal hydrocarbons and hydrocarbon derivatives. Biotechnol Lett 33:1963–1197

Briard B, Heddergott C, Latgé J (2016) Volatile compounds emitted by *Pseudomonas aeruginosa* stimulate the growth of the fungal pathogen *Aspergillus fumigatus*. MBio 7(2):e00219-16. doi:10.1128/mBio.00219-16

Bruce A, Stewart D, Verrall S, Wheatley RE (2003) Effect of volatiles from bacteria and yeast on the growth and pigmentation of sapstain fungi. Int Biodeterior Biodegrad 51:101–108

Bunge M, Araghipour N, Mikoviny T, Dunkl J, Schnitzhofer R, Hansel A, Schinner F, Wisthaler A, Margesin R, Märk TD (2008) On-line monitoring of microbial volatile metabolites by proton transfer reaction–mass spectrometry. Appl Environ Microb 74:2179–2186

Cellini A, Biondi E, Blasioli S, Rocchi L, Farneti B, Braschi I, Savioli S, Rodriguez-Estrada MT, Biasioli F, Spinelli F (2016) Early detection of bacterial diseases in apple plants by analysis of volatile organic compounds profiles and use of electronic nose. Ann Appl Biol 168:409–420

Chakraborty U, Chakraborty BN, Allay S, De U, Chakraborty AP (2011) Dual application of *Bacillus pumilus* and *Glomus mosseae* for improvement of health status of mandarin plants. Acta Hortic 892:215–230

Chaurasia B, Pandey A, Palni LMS, Trivedi P, Kumar B, Colvin N (2005) Diffusible and volatile compounds produced by an antagonistic *Bacillus subtilis* strain cause structural deformations in pathogenic fungi in vitro. Microbiol Res 160:75–81

Chiron N, Michelot D (2005) Odeurs de champignons: chimie et role dans les interactions biotiques – une revue. Cryptogam Mycol 26:299–364

Christen P, Bramorski A, Revah S, Soccol CR (2000) Characterization of volatile compounds produced by *Rhizopus* strains grown on agro-industrial solid wastes. Bioresour Technol 71:211–215

Chuankun X, Minghe M, Leming Z, Keqin Z (2004) Soil volatile fungistasis and volatile fungistatic compounds. Soil Biol Biochem 36:1997–2004

Chung JH, Song GC, Ryu CM (2016) Sweet scents from good bacteria: case studies on bacterial volatile compounds for plant growth and immunity. Plant Mol Biol 90:677–687. doi:10.1007/s11103-015-0344-8

Contreras-Cornejo HA, Macías-Rodríguez L, Alfaro-Cuevas R, López-Bucio J (2014) *Trichoderma* spp. improve growth of *Arabidopsis* seedlings under salt stress through enhanced root development, Osmolite production, and Na+ elimination through root exudates. MPMI 27(6):503–514. doi.org/10.1094/MPMI-09-13-0265-R

Demyttenaere JCR, Moriña RM, DeKimpe N, Sandra P (2004) Use of headspace solid-phase microextraction and headspace sorptive extraction for the detection of the volatile metabolites produced by toxigenic *Fusarium* species. J Chromatogr A 1027:147–154

Dickschat JS, Helmke E, Schulz S (2005a) Volatile organic compounds from arctic bacteria of the Cytophaga–Flavobacterium–Bacteroides group: a retrobiosynthetic approach in chemotaxonomic investigations. Chem Biodivers 2:318–353

Dickschat JS, Martens T, Brinkhoff T, Simon M, Schulz S (2005b) Volatiles released by a Streptomyces species isolated from the North Sea. Chem Biodivers 2:837–865

Dobbs CG, Hinson WH (1953) A widespread fungistasis in the soil. Nat Lond 172:197–199

Dudareva N, Negre F, Nagegowda DA, Orlova I (2006) Plant volatiles: recent advances and future perspectives. Crit Rev Plant Sci Phila 25:417–440

Effmert U, Kalderas J, Warnke R, Piechulla B (2012) Volatile mediated interactions between bacteria and fungi in the soil. J Chem Ecol 38:665–703

El-Hasan A, Buchenauer H (2009) Actions of 6-Pentyl-alpha-pyrone in controlling seedling blight incited by *Fusarium moniliforme* and inducing Defence responses in maize. J Phytopathol 157(11–12):697–707

Elshafie HS, Camele I, Racioppi R, Scrano L, Iacobellis NS, Bufo SA (2012) In Vitro antifungal activity of *Burkholderia gladioli* pv agarici- cola against some phytopathogenic fungi. Int J Mol Sci 13:16291–16302. doi:10.3390/ijms131216291

Ezra D, Hess WM, Gary A (2004a) Strobel new endophytic isolates of *Muscodor albus*, a volatile-antibiotic-producing fungus. Microbiology 150:4023–4031. doi:10.1099/mic.0.27334-0

Farag MA, Zhang HM, Ryu CM (2013) Dynamic chemical communication between plants and bacteria through airborne signals: induced resistance by bacterial volatiles. J Chem Ecol 39:1007–1018. doi:10.1007/s10886-013-0317-9

Fernando WGD, Ramarathnam R, Krishnamoorthy AS, Savchuk SC (2005) Identification and use of potential bacterial organic antifungal volatiles in biocontrol. Soil Biol Biochem 37:955–964

Fialho MB, Toffano L, Pedroso MP, Augusto F, Pascholati SF (2010) Volatile organic compounds produced by *Saccharomyces cerevisiae* inhibit the in vitro development of *Guignardia citricarpa*, the causal agent of citrus black spot. World J Microbiol Biotechnol 26:925–932. doi:10.1007/s11274-009- 0255-4

Fialho MB, Romanholo Ferreira LF, Rosim Monteiro RT, Pascholati SF (2011) Antimicrobial volatile organic compounds affect morphogenesis-related enzymes in *Guignardia citricarpa*, the causal agent of citrus black spot. Biocontr Sci Technol 21:797–807. doi:10.1080/09583157.2011.580837

Fiddaman PJ, Rossal S (1994) Effect of substrate on the production of antifungal volatiles from *Bacillus subtilis*. J Appl Bacteriol Oxford 76:395–405
Fiddaman PJ, Rossall S (1993) The production of antifungal volatiles by *Bacillus subtilis*. J Appl Bacteriol 74:119–126
Fiedler N, Laumbach R, Kelly-Mcneil K, Lioy P, Fan ZH, Zhang J, Ottenweller J, Ohman-Strickland P, Kipen H (2005) Health effects of a mixture of indoor air volatile organics, their ozone oxidation products, and stress. Environ Health Perspect 113:1542–1548
Fravel DR, Connick WJ, Grimm CC, Lloyd SW (2002) Volatile compounds emitted by sclerotio of *Sclerotinia minor*, *Sclerotinia sclerotiorum*, and *Sclerotium rolfsii*. J Agr Food Chem 50:3761–3764
Frey-Klett P, Burlinson P, Deveau A, Barret M, Tarkka M, Sarniguet A (2011) Bacterial-fungal interactions: hyphens between agricultural, clinical, environmental, and food microbiologists. Microbiol Mol Biol Rev 75:583–609. doi:10.1128/MMBR.00020-11
Garbeva P, Hordijk C, Gerards S, De Boer W (2014a) Volatile-mediated interactions between phylogenetically different soil bacteria. Front Microbiol 5:285–290
Garbeva P, Hordijk C, Gerards S, de Boer W (2014b) Volatiles produced by the mycophagous soil bacterium Collimonas. FEMS Microbiol Ecol 87:639–649
Groenhagen U, Baumgartner R, Bailly A, Gardiner A, Eberl L, Schulz S, Weisskopf L (2013) Production of bioactive volatiles by *Burkholderia ambifaria* strains. J Chem Ecol 39(7):892–906
Haas D, Défago G (2005) Biological control of soil-borne pathogens by *fluorescent pseudomonads*. Nat Rev Microbiol 3:307–319
Haq IU, Zhang M, Yang P, VanElsas JD (2014) The interactions of bacteria with fungi in soil: emerging concepts. Adv Appl Microbiol 89:185–215. doi:10.1016/B978-0-12-800259-9.00005-6
Harman G (2011) *Trichoderma* not just for biocontrol anymore. Phytoparasitica 39:103–108
Herrington PR, Craig JT, Sheridan JE (1987) Methyl vinyl ketone – a volatile fungistatic inhibitor from *Streptomyces griseoruber*. Soil Biol Biochem 19:509–512
Höckelmann C, Jüttner F (2004) Volatile organic compound (VOC) analysis and sources of limonene, cyclohexanone and straight chain aldehydes in axenic cultures of Calothrix and Plectonema. Water Sci Technol 49:47–54
Hornby JM, Jensen EC, Lisec AD (2001) Quorum sensing in the dimorphic fungus *Candida albicans* is mediated by farnesol. Appl Environ Microb 67:2982–2992
Huang CJ (2012) Dimethyl disulfide is an induced systemic resistance-elicitor produced by *Bacillus cereus* C1L. Pest Manag Sci 68(9):1306
Humphris SN, Bruce A, Buultjens E, Wheatley RE (2002) The effects of volatile microbial secondary metabolites on protein synthesis in *Serpula lacrymans*. FEMS (Fed Eur Microbiol Soc) Microbiol Lett 210:215–219
Hung R, Lee S, Bennett JW (2015) Fungal volatile organic compounds and their role in ecosystems. Appl Microbiol Biotechnol 99:3395–3405. doi:10.1007/s00253-015-6494-4
Hunziker L, Boenisch D, Groenhagen U, Bailly A, Schulz S, Weisskopf L (2015) Pseudomonas strains naturally associated with potato plants produce volatiles with high potential for inhibition of *Phytophthora infestans*. Appl Environ Microbiol 81:821–830. doi:10.1128/AEM.02999-14
Hynes J, Müller C, Jones T, Boddy L (2007) Changes in volatile production during the course of fungal mycelial interactions between *Hypholoma fasciculare* and *Resinicium bicolor*. J Chem Ecol 33:43–57
Jamalizadeh M, Etebarian HR, Aminian H, Alizadeh A (2010) Biological control of Botrytis Mali on apple fruit by use of *Bacillus* bacteria, isolated from the rhizosphere of wheat. Arch Phytopathol Pl 43:1836–1845
Jelen HH (2003) Use of solid phase microextraction (SPME) for profiling fungal volatile metabolites. Lett Appl Microbiol 36:263–267
Jung RE, Zembic A, Pjetursson BE, Zwahlen M, Thoma DS (2012) Systematic review of the survival rate and the incidence of biological, technical, and aesthetic complications of single crowns on implants reported in longitudinal studies with a mean follow-up of 5 years. Clin Oral Implants Res 6:2–21. doi:10.1111/j.1600-0501.2012.02547.x

Kai M, Effmert U, Berg G, Piechulla B (2006) Volatiles of bacterial antagonists inhibits mycelial growth of the plant pathogen *Rhizoctonia solani*. Arch Microbiol 187:351–360

Kai M, Vespermann A, Piechulla B (2008) The growth of fungi and Arabidopsis thaliana is influenced by bacterial volatiles, Plant Signal Behav 3(7):482–484. doi: 10.4161/psb.3.7.5681

Kai M, Haustein M, Molina F, Petri A, Scholz B, Piechulla B (2009) Bacterial volatiles and their action potential. Appl Microbiol Biotechnol 81:1001–1013

Kanchiswamy CN, Malnoy M, Maffei ME (2015) Chemical diversity of microbial volatiles and their potential for plant growth and productivity. Front Plant Sci 6:151. doi:10.3389/fpls.2015.00151

Kesselmeier J, Staudt M (1999) Biogenic volatile organic compounds (VOC): an overview on emission, physiology, and ecology. J of Atmospheric Chem 33(1):23–88

Koitabashi M, Tsushima S, (2007) Studies on biocontrol of air-borne plant disease by a filamentous fungus producing antifungal volatiles. JARQ 41(4):261–265

Koitabashi M, Kajitani Y, Hirashima K (2004) Antifungal substances produced by fungal strain Kyu-W63 from wheat leaf and its taxonomic position. J Gen Plant Pathol 70:124–130. doi:10.1007/s10327-003-0095-2

Korpi A, Jarnberg J, Pasanen AL (2009) Microbial volatile organic compounds. Crit Rev Toxicol 39:139–193. doi:10.1080/10408440802291497

Kumar S, Kaushik N (2012) Metabolites of endophytic fungi as a novel source of bio fungicide: a review. Phytochem Rev 11:507–522. doi:10.1007/s11101- 013-9271-y

Liu W, Mu W, Zhu B, Liu F (2008) Antifungal activities and components of VOCs produced by *Bacillus subtilis* G8. Curr Res Bacteriol 1:28–34

Logeshwarn P, Thangaraju M, Rajasundari K (2011) Antagonistic potential of *Gluconacetobacter diazotrophicus* against *Fusarium oxysporum* in sweet potato (*Ipomea batatas*). Arch Phytopathol Pl 44:216–223

Lorito M, Woo SL, Harman GE, Monte E (2010) Translational research on *Trichoderma*: from 'omics to the field. Annu Rev Phytopathol 48:395–417

Macias-Rubalcava ML, Hernandez-Bautista BE, Oropeza F, Duarte G, Gonzalez MC, Glenn AE, Hanlin RT, Anaya AL (2010) Allelochemical effects of volatile compounds and organic extracts from *Muscodor yucatanensis*, a tropical endophytic fungus from *Bursera simaruba*. J Chem Ecol 36:1122–1131

Matysik S, Herbarth O, Mueller A (2009) Determination of microbial volatile organic compounds (MVOCs) by passive sampling onto charcoal sorbents. Chemosphere 76:114–119

Mayr D, Margesin R, Klingsbichel E, Hartungen E, Denewein D, Schinner F, Märk TD (2003) Rapid detection of meat spoilage by measuring volatile organic compounds by using proton transfer reaction mass spectrometry. Appl Environ Microbiol 69:4697–4705

Mayrhofer S, Mikoviny T, Waldhuber S, Wagner AO, Innerebner G, Franke-Whittle IH, Märk TD, Hansel A, Insam H (2006) Microbial community related to volatile organic compound (VOC) emission in household biowaste. Environ Microbiol 8:1960–1974

McKee ND, Robinson PM (1988) Production of volatile inhibitors of germination and hyphal extension by *Geotrichum candidum*. Trans Br Mycol Soc 91:157–190

McNeal KS, Herbert BE (2009) Volatile organic metabolites as indicators of soil microbial activity and community composition shifts. Soil Sci Soc Am J 73:579–588

Mercier J, Manker D (2005) Biocontrol of soil-borne diseases and plant growth enhancement in greenhouse soilless mix by the volatile-producing fungus *Muscodor albus*. Crop Prot 24:355–362

Minerdi D, Bossi S, Gullino ML, Garibaldi A (2009) Volatile organic compounds: a potential direct long-distance mechanism for antagonistic action of *Fusarium oxysporum* strain MSA 35. Environ Microbiol 11:844–854

Minerdi D, Bossi S, Maffei ME, Gullino ML, Garibaldi A (2011) *Fusarium oxysporum* and its bacterial consortium promote lettuce growth and expansion A5 gene expression through microbial volatile organic compound (MVOC) emission. FEMS (Fed Eur Microbiol Soc) Microbiol Ecol 76:342–351

Miransari M (2011) Interactions between arbuscular mycorrhizal fungi and soil bacteria. Appl Microbiol Biotechnol 89(4):917–930. doi:10.1007/s00253-010-3004-6

Mitchell AM, Strobel GA, Moore E, Robison R, Sears J (2010) Volatile antimicrobials from *Muscodor crispans*, a novel endophytic fungus. Microbiology 156:270–277

Morath SU, Hung R, Bennett JW (2012) Fungal volatile organic compounds: are views with an emphasis on their biotechnological potential. Fungal Biol Rev 26:73–83. doi:10.1016/j.fbr.2012.07.001

Muller A, Faubert P, Hagen M, ZuCastell W, Polle A, Schnitzler JP (2013) Volatile profiles of fungi–chemotyping of species and ecological functions. Fungal Genet Biol 54:25–33. doi:10.1016/j.fgb.2013.02.005

Naing KW, Anees M, Kim SJ, Nam Y, Kim YC, Kim KY (2014) Characterization of antifungal activity of *Paenibacillus ehimensis* KWN38 against soilborne phytopathogenic fungi belonging to various taxonomic groups. Ann Microbiol 64:55–63

Naznin HA, Kiyohara D, Kimura M, Miyazawa M, Shimizu M, Hyakumachi M (2014) Systemic resistance induced by volatile organic compounds emitted by plant growth-promoting fungi in *Arabidopsis thaliana*. PLoS One 9:868–882. doi:10.1371/journal.pone.0086882

Nilsson A, Kihlstrom E, Lagesson V, Wessen B, Szponar B, Larsson L, Tagesson C (2004) Microorganisms and volatile organic compounds in airborne dust from damp residences. Indoor Air 14:74–82

Nishino S, Parada RY, Ichiyanagi T, Maekawa N, Shimomura N, Otani H (2013) 1-phenyl-3-pentanone, a volatile compound from the edible mushroom *Mycoleptodonoides aitchisonii* active against some Phytopathogenic fungi. J Phytopathol 161:515–521

Oka K, Ishihara A, Sakaguchi N, Nishino S, Parada RY, Nakagiri A, Otani H (2015) Antifungal activity of volatile compounds produced by an edible mushroom *Hypsizygus marmoreus* against Phytopathogenic fungi. J Phytopathol 163:987–996

Ortiz-Castro R, Contreras-Cornejo H, Macias-Rodriguez L, Lopez-Bucio J (2009) The role of microbial signals in plant growth and development. Plant Signal Behav 4:701–712

Pagans E, Font X, Sanchez A (2006) Emission of volatile organic compounds from composting of different solid wastes: abatement by biofiltration. J Hazard Mater 131:179–186

Pal KK, McSpadden Gardener B (2007) Biological control of plant pathogens. Plant Health Instr. doi:10.1094/PHI-A-2006-1117-02

Pasanen P, Korpi A, Kalliokosi P, Pasanen AL (1997) Growth and volatile metabolite production of *Aspergillus versicolor* in house dust. Environ Int 23:425–432

Paul D, Park KS (2013) Identification of volatiles produced by *Cladosporium cladosporioides* CL-1, a fungal biocontrol agent that promotes plant growth. Sensors 13:13969–13977. doi:10.3390/s131013969

Peñuelas J, Asensio D, Tholl D, Wenke K, Rosenkranz M, Piechulla B (2014) Biogenic volatile emissions from the soil. Plant Cell Environ 37:1866–1891

Phuakjaiphaeo C, Chang CI, Ruangwong O, Kunasakdakul K (2016) Isolation and identification of an antifungal compound from endophytic *Streptomyces* sp. CEN26 active against *Alternaria brassicicola*. Lett Appl Microbiol 63:38–44

Pichersky E, Noel JP, Dudareva N (2006) Biosynthesis of plant volatiles: nature's diversity and ingenuity. Sci Wash 311:808–811

Piechulla B, Degenhardt J (2014) The emerging importance of microbial volatile organic compounds. Plant Cell Environ 37:811–812

Pimenta RS, Moreira da Silva JF, Buyer JS, Janisiewicz WJ (2012) Endophytic fungi from plums (*Prunus domestica*) and their antifungal activity against *Monilinia fructicola*. J Food Protect 75:1883–1889. doi:10.4315/0362- 028X.JFP-12-156

Raza W, Ling N, Yang L, Huang SQ (2016) Response of tomato wilt pathogen *Ralstonia solanacearum* to the volatile organic compounds produced by a biocontrol strain *Bacillus amyloliquefaciens* SQR-9. Sci Rep 6:248–256. doi:10.1038/srep24856

Ryan RP, Dow JM (2008) Diffusible signals and interspecies communication in bacteria. Microbiology 154:1845–1858. doi:10.1099/mic.0.2008/017871-0

Ryu CM, Farag MA, Hu CH, Reddy MS, Wei HX, Pare PW, Kloepper JW (2003) Bacterial volatiles promote growth in *Arabidopsis*. Proc Natl Acad Sci U S A 100:4927–4932

Ryu CM, Kim J, Choi O, Park SY, Park SH, Park CS (2005) Nature of a root-associated *Paenibacillus polymyxa* from field-grown winter barley in Korea. J Microbiol Biotechnol 15:984–991

Schmidt R, Cordovez V, DeBoer W, Raaijmakers J, Garbeva P (2015) Volatile affairs in microbial interactions. ISMEJ:2329–2335. doi:10.1038/ismej.2015.42

Schmidt R, Etalo DW, de Jager V, Gerards S, Zweers H, de Boer W, Garbeva P (2016) Microbial small talk: volatiles in fungal–bacterial interactions. Front Microbiol 6:1495. doi:10.3389/fmicb.2015.01495

Schöller CEG, Gürtler H, Pedersen R, Molin S, Wilkins K (2002) Volatile metabolites from Actinomycetes. J Agric Food Chem 50:2615–2621

Schulz S, Dickschat JS (2007) Bacterial volatiles: the smell of small organisms. Nat Prod Rep 24(4):814–842

Scotter JM, Langford VS, Wilson PF, Mcewan MJ, Chambers ST (2005) Real-time detection of common microbial volatile organic compounds from medically important fungi by selected ion flow tube-mass spectrometry (SIFT-MS). J Microbiol Methods 63:127–134

Seewald MSA, Bonfanti M, Singer W, Knapp BA, Hansel A, Franke-Whittle I, Insam H (2010) Substrate-induced VOC emissions from compost amended soils under aerobic and anaerobic incubation. Biol Fertil Soils 46:371–382

Semighini CP, Hornby JM, Dumitru R, Nickerson KW, Harris SD (2006) Farnesol-induced apoptosis in *Aspergillus nidulans* reveals a possible mechanism for antagonistic interactions between fungi. Mol Microbiol 59:753–764

Semighini CP, Murray N, Harris SD (2008) Inhibition of *Fusarium graminearum* growth and development by farnesol. FEMS Microbiol Lett 279:259–264

Serrano A, Gallego M (2006) Sorption study of 25 volatile organic compounds in several Mediterranean soils using headspace–gas chromatography–mass spectrometry. J Chromatogr 1118:261–270

Singh SK, Strobel GA, Knighton B, Geary B, Sears J, Ezra D (2011) An endophytic *Phomopsis* sp. possessing bioactivity and fuel potential with its volatile organic compounds. Microb Ecol 61:729–739

Spilvallo R, Novero M, Bertea CM, Bossi S, Bonfante P (2007) Truffle volatiles inhibit growth and induce an oxidative burst in *Arabidopsis thaliana*. New Phytol 175:417–424

Stoppacher N, Kluger B, Zeilinger S, Krska R, Schuhmacher R (2010) Identification and profiling of volatile metabolites of the biocontrol fungus *Trichoderma atroviride* by HS-SPME-GC-MS. J Microbiol Methods 81:187–193

Stotzky G, Schenck S (1976) Volatile organic compounds and microorganisms. Crit Rev Microbiol 4:333–382

Strobel GA, Dirkse E, Sears J, Markworth C (2001) Volatile antimicrobials from *Muscodor albus*, a novel endophytic fungus. Microbiology 147:2943–2950

Strobel GA, Kluck K, Hess WM, Sears J, Ezra D, Vargas PN (2007) *Muscodor albus* E-6, an endophyte of *Guazuma ulmifolia* making volatile antibiotics: isolation, characterization and experimental establishment in the host plant. Microbiology 153:2613–2620

Strobel G, Singh SK, Riyaz-Ul-Hassan S, Mitchell AM, Geary B, Sears J (2011) An endophytic/pathogenic *Phoma* sp. from creosote bush producing biologically active volatile compounds having fuel potential. FEMS Microbiol Lett 320:87–94

Tahir HAS, Gu Q, Wu H, Niu Y, Huo R, Gao X (2017) Bacillus volatiles adversely affect the physiology and ultrastructure of *Ralstonia solanacearum* and induce systemic resistance in tobacco against bacterial wilt. Sci Rep 7:40481. doi:10.1038/srep40481

Tehrani AS, Disfani FA, Hedjaroud GA, Mohammadi M (2001) Antagonistic effects of several bacteria on *Verticillium dahliae* the causal agent of cotton wilt. Meded Rijksuniv Gent Fak Landbouwkd Toegep Biol Wet 66:95–101

Tehrani AS, Zebarjad A, Hedjaroud GA, Mohammadi M (2002) Biological control of soybean damping-off by antagonistic rhizobacteria. Meded Rijksuniv Gent Fak Landbouwkd Toegep Biol Wet 67:377–380

Tenorio-Salgado S, Tinoco R, Vazquez-Duhalt R, Caballero-Mellado J, Perez-Rueda E (2013) Identification of volatile compounds produced by the bacterium *Burkholderia tropica* that inhibit the growth of fungal pathogens. Bioengineered 4:236–243. doi:10.4161/bioe.23808

Tyc O, Zweers H, de Boer W, Garbeva P (2015) Volatiles in inter-specific bacterial interactions. Front Microbiol 6:1412. doi:10.3389/fmicb.2015.01412

Ulloa-Benıtez A, Medina-Romero YM, Sanchez-Fernandez RE, Lappe-Oliveras P, Roque-Flores G, Duarte Lisci G, Herrera Suarez T, Macıas-Rubalcava ML (2016) Phytotoxic and antimicrobial activity of volatile and semi-volatile organic compounds from the endophyte *Hypoxylon anthochroum* strain Blaci isolated from *Bursera lancifolia* (Burseraceae). J Appl Microbiol 121:380–400

Van der Heijden MGA, Klironomos JN, Ursic M, Moutoglis P, Streitwolf-Engel R, Boller T, Wiemken A, Sanders IR (1998) Mycorrhizal fungal diversity determines plant biodiversity, ecosystem variability, and productivity. Nature 396:69–72. doi:10.1038/23932

Vespermann A, Kai M, Piechulla B (2007) Rhizobacterial volatiles affect the growth of fungi and *Arabidopsis thaliana*. Appl Environ Microbiol 73(17):5639–5641

Voisard C, Keel C, Haas D, Défago G (1989) Cyanide production by *Pseudomonas fluorescens* helps to suppress black root of tobacco under gnotobiotic conditions. EMBO J 8:351–358

Wan M, Li G, Zhang J, Jiang D, Huang HC (2008) Effect of volatile substances of *Streptomyces platensis* F-1 on control of plant fungal diseases. Biol Control 46:552–559

Wang X, Wu T (2008) Release of isoprene and monoterpenes during the aerobic decomposition of orange wastes from laboratory incubation experiments. Environ Sci Technol 42:3265–3270

Watson AG, Ford EJ (1972) Soil fungistasis – a reappraisal. Annu Rev Phytopathol 10:327–348

Weise T, Kai M, Gummesson A, Troeger A, von Reuss S, Piepenborn S, Kosterka F, Sklorz M, Zimmermann R, Francke W, Piechulla B (2012) Volatile organic compounds produced by the phytopathogenic bacterium *Xanthomonas campestris* pv *Vesicatoria* 85-10. Beilstein J Organic Chem 8:579–596

Wheatley RE (2002) The consequences of volatile organic compound mediated bacterial and fungal interactions. Antonie Van Leeuwenhoek Amst 81:357–364

Wilson AD, Baietto M (2011) Advances in electronic-nose technologies developed for biomedical applications. Sensors 11:1105–1176

Xu C, Mo M, Zhang L, Zhang K (2004) Soil volatile fungistasis and volatile fungistatic compounds. Soil Biol Biochem 36:1997–2004

Ye HS, Ahn YR, Song GC, Ghim SY, Lee S, Lee G, Ryu CM (2016) Impact of a bacterial volatile 2,3-Butanediol on *Bacillus subtilis* rhizosphere robustness. Front Microbiol 7:993. doi:10.3389/fmicb.2016.00993

Yuan J, Raza W, Shen Q, Huang Q (2012) Antifungal activity of *Bacillus amyloliquefaciens* NJN-6 volatile compounds against *Fusarium oxysporum* f. sp. *cubense*. Appl Environ Microbiol 78(16):5942–5944

Zhang Z, Li G (2010) A review of advances and new developments in the analysis of biological volatile organic compounds. Microchem J 95:127–139

Zhang CL, Wang GP, Mao LJ, Komon-Zelazowska M, Yuan ZL, Lin FC, Druzhinina IS, Kubicek CP (2010) *Muscodor fengyangensis* sp. nov. from southeast China: morphology, physiology, and production of volatile compounds. Fungal Biol 114:797–808

Zhao LJ, Yang XN, Li XY, Mu W, Liu F (2011) Antifungal, insecticidal and herbicidal properties of volatile components from *Paenibacillus polymyxa* strain BMP-11. Agric Sci China 10(5):728–736

Zou C-S, Mo M-H, Gu Y-Q, Zhou J-P, Zhang K-Q (2007) Possible contributions of volatile-producing bacteria to soil fungistasis. Soil Biol Biochem 39:2371–2379

Race of Arms: Herbivore-Induced Volatiles and Their Co-evolution

13

Rabindranath Bhattacharyya and Raktim Bhattacharya

Abstract

Plants' defence against herbivory is a long studied one. And among different modes of defences, the use of volatiles by the plants is unique. The main idea behind this use of volatiles is to call the potential predators for the herbivores. Plants employ a strange strategy here, and this helps them to survive through any major herbivore attack. They can even warn the neighbouring plants about the coming of a possible attack. All these happen due to some volatile chemical compounds emitted by the herbivore-infested plants. The goal of this chapter is to study in detail the intricate relationship of plant, herbivore and carnivore in this tritrophic ecosystem and to determine whether there is any role of evolution and to calculate the overall ecological cost to govern this complex machinery of life.

Keywords

Herbivory • Plant defence • Volatil chemicals • Tritrophic

13.1 Introduction

The relationship between herbivores and plants is a long studied one (Odum and Barret 2005); all the relationships under the banner of plant-herbivore relationship are not harmful. Some are antagonistic and some are mutualistic (Chapman and Reiss 1999). Mutualistic relation between plants and herbivores benefits both the groups. But the examples of this type of reaction are quite rare, apart from

R. Bhattacharyya (✉) • R. Bhattacharya
Department of Life Sciences, Presidency University,
86/1, College Street, Kolkata 700073, West Bengal, India
e-mail: rabindranathbpc@yahoo.co.in; rabindranathbpc@gmail.com

D.K. Choudhary et al. (eds.), *Volatiles and Food Security*,
DOI 10.1007/978-981-10-5553-9_13

pollination and seed dispersal. But in this field, almost most of the relations are antagonistic in nature, i.e. only one group of organism is benefitted and in most of the cases the benefit goes to the herbivore (Molles 1999). The plants gain nothing except losing biomass. So to deter herbivores over the time, plants devised many defensive methods, such as chemical defences, mechanical defences, mimicry, camouflage, etc. (Kardong 2005). But among all these defensive methods, none is as sophisticated as the mode of chemical defences. In this chapter we are going to discuss a specific type of chemical defence used by the plants to protect themselves against the herbivores, i.e. by using volatile chemicals.

One of the best studied plant defences against herbivores is the chemical defence. Most of the plants produce a number of secondary metabolites, which are most of the time derivatives of the primary compounds (Kardong 2005). These secondary metabolites are stored in the leaf, root, bark, etc., and they provide the plants with the weapons to deal with the herbivores. Some secondary metabolites taste bitter, some pungent and some are even poisonous, so that in general herbivores used to avoid these plants.

But among this quite large list of secondary metabolites, some chemicals are there which are not directly involved in plant defences but used as volatile chemical compounds, which are being released from the plant leaves when they are under attack from the herbivores or any plant-feeding insects. If we want to state it in a most general way, then it is; plants use these volatile chemicals as the mode of communication between themselves (Taiz and Zeiger 2003). So when a particular plant is under attack, it releases some of these chemicals so plants of same species in the vicinity will get some sort of warning that a possible attack is coming. But this type of communication mostly happens in case of pathogen attacks. Plants normally respond to pathogen attacks by secreting two volatile compounds, viz. salicylic acid and jasmonic acid (Ozawa et al. 2000). By using these two most common volatile compounds which also act like plant growth regulators, plant communicate within them sending warning about an increasing pathogen attack (Taiz and Zeiger 2003).

The goal of this chapter is to explore the different volatile chemical compounds released by the plants which are also known as herbivore-induced volatile compounds (HIVC) and their effects on herbivores as well as their overall impact on the ecosystem.

13.2 Types of Herbivore-Induced Volatiles

The herbivore-induced volatile compounds (HIVC) are broadly classified in two different groups as follows (Dicke 1999):

(a) Specific Volatiles: These types of HIVC in general are emitted by the plants when they are damaged by some mechanical means. In most of the cases, the amount of these volatiles is very trace, and most often they resulted from the mechanical injuries and in some cases intact plant parts (Dicke et al. 1990a, b; Turlings et al. 1990a, b). Since this is specific for this very reason, the composition of the compounds depends on the specific plants (Takabayashi et al. 1991; Turlings et al. 1993). This has been found in different plants, viz. corn, lima bean, cucumber, etc. (Dicke 1994; Takabayashi et al. 1994a, b; Turlings et al. 1995).

(b) Non-specific Volatiles: These second types of volatiles are produced in case of a large-scale herbivore attack. That leads to the production of these volatile compounds which are qualitatively similar to the compounds produced at the time of specific response, and like those, these compounds are also produced from mechanically wounded plant parts or sometimes undamaged portions. But the main difference of non-specific volatiles with from specific volatiles is that in this case, the amount released from the herbivore-infested plants is much larger than those mechanically wounded plants. Some of the examples of this category are potato, cabbage, tomato, etc. (Blaakmeer et al. 1994; Mattiacci et al. 1994; Agelopoulos and Keller 1994; Bolter et al. 1997; Dicke et al. 1998).

Though these two mechanisms are well developed and different from each other, they have one thing in common, i.e. this large-scale release of volatile compounds irrespective of their specificity can attract the carnivores, and it has been found that carnivores can discriminate between the volatiles released from mechanically damaged plants and plants attacked by herbivores (Dicke 1999).

13.3 Volatiles to Attract Carnivores

It has been found that carnivores can differentiate between volatiles' odours from unaffected plants and plants that are being harvested by the herbivores (Roth et al. 1982; Sabelis et al. 1984). Sabelis and Dicke first suggested in 1985 that the odours which are calling the carnivores are originally produced by the affected plants (Sabelis and Dicke 1985). To achieve this deduction, they used an experimental set-up of a tritrophic ecosystem which consists of plants (lima bean), herbivore spider mites (*Tetranychus urticae*) and carnivore predatory mites (*Phytoseiulus persimilis*) (Dicke 1994). After the experiment the chemical analysis revealed a diverse array of chemicals, and all of them are plant-related volatiles such as fatty acid derivatives, terpenoids and phenolic methyl salicylate. Researchers have identified three major terpenoids which were the main factors to attract the carnivorous predatory mites. These terpenoids are (E)-β-ocimene; 4,8-dimethyl-1,3(E),7-nonatriene; and linalool [Dicke and Sabelis 1988; Dicke et al. 1990a).

Since this pioneering study, many researchers carried out same type of study with different model systems of ecosystems such as plants infested by caterpillars of beet army worm (*Spodoptera exigua*) and parasitoid *Cotesia marginiventris* (Turlings et al. 1990a), cabbage plants infested by cabbage white caterpillars (*Pieris brassicae* and *Pieris rapae*) and the parasitoids *Cotesia glomerata* and *Cotesia rubecula* (Wiskerke and Vet 1991), but one of the best studied model of ecosystem in this field apart from the pioneering study on lima bean is the relation between corn plants, caterpillars of *Spodoptera exigua* and the parasitoid *Cotesia marginiventris* (Turlings and Tumlinson 1992). The most surprising fact is that though all these ecosystems vary enormously in their structure and design, the basic findings are very much similar to that of lima bean ecosystem.

Several chemical analyses were done on the volatiles isolated from these different experiments, and in all the cases, terpenes were recorded as the major volatile emitted from the plants attacked by the herbivores, though there is a lack of data when it comes to terpenoid emission by unaffected plants of the same cultivar. And in one case, it has been found that there is no emission of terpenoids when soybean plants are infested by *Spodoptera exigua* (Turlings et al. 1993).

One main reason for herbivore-induced chemicals is that herbivory induces cell damage which in turn releases the stored materials within the cell; among them alcohols, aldehydes, esters, etc. are predominant, and they are mostly derived from lipoxygenase pathway, and as time passes specific terpenoids which are volatile in nature are produced from these compounds in response to mechanical damage induced by herbivory (Dicke 1994). The plant biomass when concerned is much larger than the predating herbivores. In comparison to that, the amount of volatile materials emitted by the plants is minute. But plants emit the volatiles at a rate of per unit biomass. This makes the plant volatiles quite detectable by the carnivores. So these odour sources make the carnivores to be attracted towards the herbivore-infested plants (Geervliet et al. 1994).

As already discussed earlier in Sect. 8.2, mechanical damage and herbivore-induced damage evoke different responses in plants. This differs both in volatile composition and quantity. This differential response is actually mediated by the secretions from the herbivores. If the oral secretion of an herbivore is applied in a mechanical wound, then that will result in the emission of volatiles which are characteristically similar with the volatiles emitted during damage by an herbivore (Mattiacci et al. 1994). However it has been found that the response of the plant is systematic. The odour source is much larger than the damage-inflicting herbivores, and that is not restricted to few plant parts. The main role in these cases is played by an elicitor compound, which travels from a damaged leaf to an undamaged one and induces the response, and this odour source also comprises the neighbouring plants (Dicke et al. 1993).

13.4 Varieties of Carnivore-Attracting Volatiles

Different types of plants deal differently with different types of herbivores. This may be due to the differences in plant genotypes (Dicke 1999). This difference can be detected in the chemicals released as volatiles when a plant is attacked by herbivores (Loughrin et al. 1995). Not only this but also the relative contribution of each chemical compound in the total mixture of the emitted volatiles varies. This variation can lead to attraction of carnivores, and by this carnivores can differentiate between the herbivore infestations (Dicke 1988). It is considered that natural selection plays a great role in provoking these differences.

It has been found that even within the same plant, these differences prevail. One example of this is cucumber plant. The spider mite infestation of cucumber triggers the volatile emission. But surprisingly the amount of volatile emitted from the old leaves is far below than that of young leaves (Takabayashi et al. 1994a, b). This evidence more strongly suggests the role of natural selection because by this technique, the plant is directing the predatory mites towards its new leaves which are in greater need for protection than the older ones. So a plant can discriminate between which parts are essential for its growth and can direct the predatory carnivore in that direction; this also helps the plants in competition with the neighbouring plants of the same species. Plant responses can also vary depending on the herbivore species that are attacking the plants. Such as carnivorous mites can differentiate between different herbivorous mites that are attacking a plant. Carnivores can even estimate quantitatively the number of different herbivore species that are attacking the same plant by the presence of a particular volatile compound in the total volatile mixture emitted by the plant (Geervliet et al. 1997). Apart from this, it has also been found that different instars are resulting in different types of plant responses; one such example is found in corn plants when infested by young caterpillars (*Pseudaletia separata*); the novel volatile compounds produced by the plant bring parasitic wasps (*Cotesia kariyai*). But if the plants are infested by old caterpillars, the rate of production of volatiles becomes much less than in the previous case (Takabayashi et al. 1995). This is probably because the rate of food consumption in case of old caterpillars is very low compared to the young ones (Takabayashi et al. 1998).

The variation found mostly when a same herbivore species attracts different plant species (Loughrin et al. 1995). That leads to the production of numerous novel volatile compounds which enable the carnivore not only quantitatively but also qualitatively (Geervliet et al. 1997). This variation is then used by the carnivores to forage on the herbivores, and carnivores can also differentiate between different plant species from a mixture of volatile compounds [Vet and Dicke 1992; Geervliet et al. 1998). Besides these variations in chemical nature of volatiles, the carnivores are also influenced by some physical factors like day length, intensity of the light, water availability, temperature, humidity, etc. (Takabayashi et al. 1994a, b).

13.5 Controlling Carnivore Behaviour

The idea that carnivores can differentiate between different plant volatiles and can even estimate the number of herbivore species attacking the plants is quite tantalizing. But through different experiments, it has been found that actually, the volatiles emitted by the plants are controlling the behaviour of the carnivores. This may be due to the fact that carnivores can temporarily specialize on a topic by using their learning behaviour (Vet and Dicke 1992). It is quite common occurrence among the insect parasitoids that they can learn different plant stimuli (Vet et al. 1990); this same phenomenon has also been reported in case of arthropod population (Dicke et al. 1990a). Adaptive solution for prey catching depends on the degree of variation in the carnivore population through the evolution of the species. This is true in both the herbivore and the plant level. If the carnivore becomes too specialize, then there will be less chance that it can learn through this type of relations. This effect is very much abundant for the carnivores which are more or less generalized when it comes to prey catching, and it has been observed there is a substantial positive effect on the learning behaviour of the carnivores imposed by the plant volatiles (Vet and Dicke 1992). When the carnivore started to learn by experience, they by time will start to less differentiate between the volatiles even when there are significant amount of differences in the volatile composition (Geervliet et al. 1998). So this in turn can affect the foraging capacity (Papaj and Vet 1990).

13.6 Relation Between Plant and Carnivore

It is quite clear that the calling of the carnivores (induced attraction) befits the plants in many ways. Besides these HIPV can have some other major benefits. Some examples are: these volatiles are most often rich in fatty acid derivatives, and they are toxic to some plant pathogens especially against some phytopathogenic fungi (Zeringue and McCormick 1989; Croft et al. 1993) and sometimes toxic to herbivores (Pallini et al. 1997). This phenomenon gives rise to the doubt whether these HIPV are acting like a direct defence system by plants or carnivores are simply exploiting this to locate prey (Turlings and Tumlinson 1992). This question is more or less unanswered, but it has been postulated since carnivores are the major factor that control the herbivore population (Cappuccino and Price 1995); it may be possible that carnivores have synchronized their responses in terms of plants. Emission of these carnivore-attracting volatiles can reduce the herbivore population surrounding the plants (Vet and Dicke 1992). As mentioned earlier, plants emit same type of volatiles in response to herbivory as well as mechanical damage. It has also been mentioned that carnivores can differentiate between the volatiles emitted by the plants to judge whether they are emitting from mechanical wound or there is any herbivore attack (Geervliet et al. 1994). It has also observed that by judging the amount of volatiles, carnivores can even differentiate between particular herbivore species (Sabelis and Baan 1983) and even different herbivore stages (Takabayashi et al. 1998). All these indicate to one thing that specificity is very robust and there is a clear distinction between two basic types of induced defensive responses, i.e. induced direct defence and induced indirect defence (Tallamy and Raupp 1991). Carnivores achieve this due to their specificity which in turn governs their foraging behaviour, and this is the basis for their ability to differentiate between different odour blends. Sometimes these volatiles attract the ineffective natural enemies of the herbivores, and this is one of the disadvantages of this system; this is because these ineffective natural enemies often outcompete the effective natural enemies of the herbivores (Bruin et al. 1995). Predation by the carnivores depends on the food availability, and this predation pressure is effective in those ecosystems where the herbivore and the plant population have a stronghold (Sabelis and Baan 1983; Dicke and Sabelis 1988). It has already been mentioned that the carnivore-attracting volatiles are produced systematically (Dicke et al. 1990a). Carnivores normally use the concentration gradient of the volatiles to locate the herbivore-infested plant parts. And these systemic volatiles make the plant much more detectable to the herbivores (Bolter et al. 1997).

13.7 Effect on Population Dynamics

There are quite a large number of researches on induced indirect defence response in context with herbivory and herbivore-induced plant volatiles. But most of these researches are focused on population dynamics. As already described in the above sections, the model ecosystems consisting of spider mites and predatory mites are also considered as a model for experiments in population dynamics (Dicke 1999). Spider mite's presence depends on several factors and among which predator and prey dispersion plays a major role. A stochastic system was developed by Sabelis in 1981 to deal with this type of population dynamics problem (Sabelis 1981). This model includes different life history parameters such as predator-prey abundance, rate of predation and predator behaviour, and all these parameters can be quantified by using this model. Predator behaviour is the most important factor when it comes to the predator-prey relations; this is because predator behaviour is largely controlled by the HIPV. Since the above-mentioned incidence is a widespread, more emphasis should be given on this phenomenon when it comes to estimating the population dynamics. But in some cases, it has been found that carnivore-attracting volatiles are not playing any major role. Migrations of predators and prey can also influence the population dynamics, and this is very evident in case of metapopulations (Sabelis and Diekmann 1988). In these cases the dispersal rate and moment are very much essential, and that is normally greatly influenced by the plants (Dicke 1999). So in overall sense, plants are producing a top-down and bottom-up effect which in turn influences a wide range of factors controlling the population dynamics of both herbivores and carnivores (Anholt 1997).

13.8 Responses of Herbivores

In earlier section it has been stated that plant volatiles generally attract the carnivores and carnivores can also detect the plant volatiles. The same scenario happens in case of herbivores also. Herbivores which are very specific in their feeding habit normally search for plants by detecting these plant volatiles (Visser and Thiery 1986), though from a healthy plant the amount of emitted volatile is very low compared to infested plants. But herbivores can detect that (Turlings et al. 1990a, b). To the herbivores these plant volatiles are the indication that some more herbivores are present in the vicinity. But one thing is certain that this volatile signal also indicates that the area may be dense with enemies, i.e. the carnivores. So to the herbivores which are foraging, this information is very complex. They may be attracted towards the plant or they may be repelled from the plant. And both these cases have been found. Herbivorous scarabaeid and chrysomelid beetles normally attracted to the plants by these volatile signals (Loughrin et al. 1994; Bolter et al. 1997); same phenomenon of attraction has also been noticed in case of moths [Landolt 1993) and spider mites (Dicke 1986; Pallini 1998). Moths (Landolt 1993) and spider mites (Dicke 1986) have also been recorded for repellence along with aphids (Bernasconi et al. 1998). The overall behavioural pattern is very complex as this whole relation consists of basically three organisms, plant species, herbivore species and carnivore species. In one study it has been found that chrysanthemum plants when infested with the larvae of *Mamestra brassicae* moths were not attractive to conspecific moths (Rojas 1999). Spider mites generally show interest in response to the volatiles which are being released from spider mite-infested plants but from the organs where no infestation has occurred. This obviously indicates that spider mites are directed towards a place where less spider mite density will be found and thus less competition for food. This phenomenon is completely in contrast with the following experiment in which volatiles from only spider mite-infested leaves were offered. In this case there is repellence among the spider mites (Dicke 1986). Not only that spider mites though slightly attracted towards cucumber plants which are also infested with spider mites, but they are strongly repelled if the same cucumber plant is infested by thrips (*Frankliniella occidentalis*) (Pallini et al. 1997).

Herbivores are only attracted to a plant by the emitted volatiles when the attracted herbivores and the herbivores which are already in there are conspecific (Dicke 1986; Landolt 1993). But in some cases same phenomenon was also noticed even when there are heterospecific herbivores (Wiskerke and Vet 1991). One explanation for this lies in the detectability of the volatiles. Volatiles emitted from the herbivore-infested plants are much more detectable than those emitting from unwounded plants (Vet and Dicke 1992). Sometimes this can lead to competition among the herbivores. And competition may result in low amount of food. But competition is somewhat better than predation. Because in case of predation, there is every possibility that the herbivore may be killed, in case of a potentially enemy dense area. So when herbivores are being attracted towards the plants, they have to consider all these facts.

If the viewpoint of the plants is concerned, plants generally do not produce volatile when there is almost no chance to call the carnivores, because that will be an utter waste of energy, or in case if the volatiles are no longer attracting the carnivores. One example of this situation was found in maize when maize plants are attacked by L5 instar caterpillars of *Pseudaletia separate* which do not attract the parasitoid *Cotesia kariyai*. But if the same plants are attacked by L1–L3 instar caterpillars of the same species, then that will attract the parasitoids. So it seems that plants can somehow make sense of the fact that most damage are done by the young caterpillars so they need to be controlled (Takabayashi et al. 1995). Sometimes carnivores use plants for their own purposes like nesting, territory markings, etc., and it has been found that volatiles emitted from these cases do not attract herbivores. They used to avoid these plants used by the carnivores. This has been found in case of spider mites. They used to prefer those plants which are habituated by the conspecific spider mites and used to avoid plants where both the conspecific spider mites and their predators are present (Pallini 1998). One explanation for this may be due to a pheromone released by the spider mites that are exposed to the predators (Janssen et al. 1997). Besides this as a precautionary measure, spider mites avoid the nonvolatile cues that are left on the plant by their predators (Kriesch and Dicke 1997; Grostal and Dicke 1999).

So in an overall view, it can be said that whether a volatile will call the herbivore or the carnivore depends on the detectability of the volatile. If herbivores gathered the knowledge that the volatiles emitting from the plant are indicating that there are predators around, then generally they avoid that plant. Herbivores normally prefer competition rather than predation. But this can be driven by the internal state of the herbivore like hunger, experience, etc. (Dicke 2000).

13.9 Plant Communication

Effect of plant volatiles on the defence of their neighbouring plants was first discovered in the 1980s (Baldwin and Schultz 1983; Rhoades 1985). Since then various physiological and biochemical studies have been undertaken to explore this highly promising area, and it has been demonstrated that plants can produce induced defences after getting exposed to volatiles from other plants (Arimura et al. 2000). Among these volatiles one is most notable and that volatile chemical is methyl jasmonate (Farmer and Ryan 1990) or methyl salicylate (Shulaev et al. 1997). Several behavioural studies were also undertaken at that time which indicated that plants that have been exposed to volatiles from the spider mite-infested plants started to emit volatiles that will attract the predatory mites (Bruin et al. 1992; Oudejans and Bruin 1994). This groundbreaking discovery indicates that like animals, plants also can warn their surrounding plants about a possible attack. It has also been found that plants which are unharmed can emit volatiles after getting exposed to the volatiles from unaffected plants, but in that case volatiles emitting from the unharmed plants are of lower amount. Methyl jasmonate, a volatile hormone, can induce the bean plants to attract predatory mites (Dicke et al. 1999). This directly proves that plants can produce carnivore-attracting volatiles after getting exposed to volatiles from affected plants. Jasmonic acid induces a special type of blend in bean plants which is very similar to the blend induced in spider mite-infested plants (Hopke et al. 1994), and this blend can attract the predatory mites though this data is not confirmatory as it has been found that methyl jasmonate is not emitted by the herbivore-infested plants (Turlings et al. 1990a, b). To solve this dilemma, nowadays different molecular biological approaches have been taken. By these methods now the genes responsible for herbivore-induced volatile production can be studied by overexpressing the gene under the influence of volatiles from neighbouring herbivore-infested plants. And from these gene expression analysis, it has been found that plants can induce different genes such as proteinase inhibitor genes, pathogenesis-related genes or lipoxygenase, phenylalanine ammonia lyase or farnesyl pyrophosphate synthetase, etc. when these plants were subjected to volatiles from neighbouring infested plants (Farmer and Ryan 1990; Arimura et al. 2000). Sometimes volatiles from affected plants fail to induce any defences in the surrounding plants, but even in that case, these volatiles can affect the plant-plant communications. This induced defence may also affect the competition among the neighbouring plants (Augner 1995; Dicke and Vet 1999). Total energy expended for the biosynthesis of these herbivore-induced volatiles is very low, but the ecological costs are the most important in this case (Dicke and Sabelis 1989). The total cost till date may have been underestimated; this is due to the fact that the ecological cost has been studied in conditions considering no competition among the plants (Van Dam and Baldwin 1998). So if the competition among the plants is considered, then this ecological cost will climb higher, and on this issue further research is needed on the interactions among the affected plants and their neighbours.

13.10 The Ultimate Cost

Resource partitioning plays a major role in the growth rate of both wild and cultivated plants. This growth is very much dependent on the production of new leaf area. Due to herbivory plants generally increase their structural defences which in turn reduce the resource available for growth and divert resources from the production of new leaf area (Chapin 1989).

The relative growth rate of a plant can be calculated by multiplying net assimilation rate with leaf area ratio (Hunt 1979; Lambers 1987). It can be expressed mathematically as follows:

$$\mathrm{RGR} = \mathrm{NAR} \times \mathrm{LAR}$$

where

RGR = Relative growth rate
NAR = Net assimilation rate
LAR = Leaf area ratio

Net assimilation rate can be defined as net carbon assimilation per unit time, and leaf area ratio can be defined as the ratio of total leaf area to total plant dry mass.

It has been proposed that the differential assimilation of photoassimilates in the new leaf area is mainly responsible for the phenotypic as well as genotypic differences in the growth rate [78]. So in this case for the plant defence, there is an opportunity cost. This theory of plant defence was later more modified to extend it to the evolutionary theory of plant defence (Coley et al. 1985) and can be expressed by the following equation:

$$\frac{\mathrm{d}C}{\mathrm{d}t} = G \times C \times \left(1 - \mathrm{kD}^{\alpha}\right) - \left(H - \mathrm{mD}^{\beta}\right)$$

where

$\frac{\mathrm{d}C}{\mathrm{d}t}$ = Realized growth

G = Maximum inherited growth rate of the plant permitted by the environment (in $g\ g^{-1}d^{-1}$)
C = Initial biomass of the plant (in g)
$(1 - kD^{\alpha})$ = Proportional reduction in potential growth rate due to resources invested in defence (in $g\ g^{-1}$)
$(H - mD^{\beta})$ = Reduction in realized growth due to herbivory (in $g\ d^{-1}$)
H = Herbivore pressure on the habitat assuming no defence
mD^{β} = Herbivory decreased by defence

The two parameters k and α denote constants that relate an investment in defence for the reduction in growth, and the other two parameters m and β are constants that denote shape of the defence effectiveness curve (Coley et al. 1985).

This model combines the evolved defence with herbivory and used to calculate the realized growth in ecological time scale. This model is considered as the ultimate model to determine the plant-herbivore relationships and is very much effective in dealing with the plant volatiles induced by the herbivores.

13.11 Conclusion

From the discussions throughout this chapter, it can be concluded that plants respond to herbivory in a variety of ways. And among this one method is to emit some volatile chemical compounds which have a multipurpose ability, like calling the predators of the attacking herbivores, warning neighbouring plants, etc. But surprisingly plants always do not generate these chemicals. It seems that they can differentiate between herbivores, different instars of larvae, for example. The emission of volatiles is very low when a plant is attacked by L5 instar caterpillars, but when the same plant is attacked by L1–L3 instars of the same caterpillars, they emit a strong signal of volatiles which can call the carnivores. Plants can emit volatiles even when they are not affected by the herbivores such as when a wound is caused by mechanical means, but the carnivores can differentiate between the volatiles and can detect the presence of herbivores by judging the amount of different volatile chemicals in the blend. This surprising ability of the carnivores makes this whole system a tritrophic one, where plant, herbivore and carnivores all are playing their specific parts. Herbivores also can detect these volatiles and can discriminate between plants where carnivores are present or where not. Many external factors also play great roles in this complex ecological as well as behavioural pattern of the animals. When an affected plant emits signals in the form of volatiles, the neighbouring plants become warned and started to emit carnivore-attracting volatiles beforehand. This in an overall sense reduces the impact of herbivore attack. But all these complex processes come at a heavy price, the ecological cost of these is quite high, and it can be calculated by using differential equations considering defence of plants against herbivory through evolution in an ecological time scale. This ultimate model can help us to judge the impact of herbivory on plants as well as to judge the effect of herbivore-induced plant volatiles on the herbivores.

References

Agelopoulos NG, Keller MA (1994) Plant-natural enemy association in the tritrophic system *Cotesia rubecula-Pieris rapae*-brassicaceae (cruciferae): II. Preference of *C. rubecula* for landing and searching. J Chem Ecol 20(7):1735–1748

Anholt BR (1997) How should we test for the role of behaviour in population dynamics? Evol Ecol 11(6):633–640

Arimura GI, Ozawa R, Shimoda T, Nishioka T, Boland W, Takabayashi J (2000) Herbivory-induced volatiles elicit defence genes in lima bean leaves. Nature 406(6795):512–515

Augner M (1995) Low nutritive quality as a plant defence: effects of herbivore-mediated interactions. Evol Ecol 9(6):605–616

Baldwin I, Schultz JC (1983) Talking trees. Science 221(4607):277–279

Bernasconi ML, Turlings TC, Ambrosetti L, Bassetti P, Dorn S (1998) Herbivore-induced emissions of maize volatiles repel the corn leaf aphid, *Rhopalosiphum maidis*. Entomol Exp Appl 87(2):133–142

Blaakmeer A, Geervliet JB, Loon JV, Posthumus MA, Beek TV, Groot A (1994) Comparative headspace analysis of cabbage plants damaged by two species of Pieris caterpillars: consequences for in-flight host location by Cotesia parasitoids. Entomol Exp Appl 73(2):175–182

Bolter CJ, Dicke M, Van Loon JJ, Visser JH, Posthumus MA (1997) Attraction of Colorado potato beetle to herbivore-damaged plants during herbivory and after its termination. J Chem Ecol 23(4):1003–1023

Bruin J, Dicke M, Sabelis MW (1992) Plants are better protected against spider-mites after exposure to volatiles from infested conspecifics. Experientia 48(5):525–529

Bruin J, Sabelis MW, Dicke M (1995) Do plants tap SOS signals from their infested neighbours? Trends Ecol Evol 10(4):167–170

Cappuccino N, Price PW (1995) Population dynamics: new approaches and synthesis. Academic Press, San Diego

Chapin FS III (1989) The cost of tundra plant structures: evaluation of concepts and currencies. Am Nat 133:1–19

Chapman JL, Reiss MJ (1999) Ecology: principles and applications, 2nd edn. Cambridge University Press, Cambridge, UK

Coley PD, Bryant JP, Chapin FS (1985) Resource availability and plant antiherbivore defense. Science 230(4728):895–899

Croft KP, Juttner F, Slusarenko AJ (1993) Volatile products of the lipoxygenase pathway evolved from *Phaseolus vulgaris* (L.) leaves inoculated with *Pseudomonas syringae* pv *phaseolicola.* Plant Physiol 101(1):13–24

Dicke M (1986) Volatile spider-mite pheromone and host-plant kairomone, involved in spaced-out gregariousness in the spider mite Tetranychus urticae. Physiol Entomol 11(3):251–262

Dicke M (1988) Prey preference of the phytoseiid mite Typhlodromus pyri: response to volatile kairomones. Exp Appl Acarol 4:1–13

Dicke M (1994) Local and systemic production of volatile herbivore-induced terpenoids: their role in plant-carnivore mutualism. J Plant Physiol 143(4):465–472

Dicke M (1999) Evolution of induced indirect defense of plants. The ecology and evolution of inducible defenses. Princeton University Press, Princeton, pp 62–88

Dicke M (2000) Chemical ecology of host-plant selection by herbivorous arthropods: a multitrophic perspective. Biochem Syst Ecol 28(7):601–617

Dicke M, Gols R, Ludeking D, Posthumus MA (1999) Jasmonic acid and herbivory differentially induce carnivore-attracting plant volatiles in lima bean plants. J Chem Ecol 25(8):1907–1922

Dicke M, Sabelis MW (1988) How plants obtain predatory mites as bodyguards. Neth J Zool 38:148–165

Dicke MA, Sabelis MW (1989) Does it pay plants to advertize for bodyguards. Towards a cost-benefit analysis of induced synomone production. In: Lambers H, Cambridge ML, Konings H,

Pons TL (eds) Causes & consequences of variation in growth rate and productivity of higher plants. SPB Academic Publishing, The Hague, pp 341–358

Dicke M, Sabelis MW, Takabayashi J, Bruin J, Posthumus MA (1990a) Plant strategies of manipulating predator prey interactions through allelochemicals: prospects for application in pest control. J Chem Ecol 16(11):3091–3118

Dicke M, van Beek TA, Posthumus MA, Ben Dom N, van Bockhoven H, De Groot AE (1990b) Isolation and identification of volatile kairomone that affects acarine predator-prey interaction: involvement of host plant in its production. J Chem Ecol 16:381–396

Dicke M, Van Baarlen P, Wessels R, Dijkman H (1993) Herbivory induces systemic production of plant volatiles that attract predators of the herbivore: extraction of endogenous elicitor. J Chem Ecol 19(3):581–599

Dicke M, van Beek TA, Posthumus MA, Ben Dom N, van Bockhoven H, De Groot AE (1990b) Isolation and identification of volatile kairomone that affects acarine predator-prey interaction. Involvement of host plant in its production. J Chem Ecol 16:381–396

Dicke M, Vet LE (1999) Plant-carnivore interactions: evolutionary and ecological consequences for plant, herbivore and carnivore. In: Olff H, Brown VK, Drent RH (eds) Herbivores: between plants and predators. Blackwell Science, Oxford, pp 483–520

Farmer EE, Ryan CA (1990) Interplant communication: airborne methyl jasmonate induces synthesis of proteinase inhibitors in plant leaves. Proc Natl Acad Sci 87(19):7713–7716

Geervliet JB, Posthumus MA, Vet LE, Dicke M (1997) Comparative analysis of headspace volatiles from different caterpillar-infested or uninfested food plants of *Pieris* species. J Chem Ecol 23(12):2935–2954

Geervliet JB, Vet LE, Dicke M (1994) Volatiles from damaged plants as major cues in long-range host-searching by the specialist parasitoid *Cotesia rubecula*. Entomol Exp Appl 73(3):289–297

Geervliet JB, Vreugdenhil AI, Dicke M, Vet LE (1998) Learning to discriminate between infochemicals from different plant-host complexes by the parasitoids *Cotesia glomerata* and *C. rubecula*. Entomol Exp Appl 86(3):241–252

Grostal P, Dicke M (1999) Direct and indirect cues of predation risk influence behavior and reproduction of prey: a case for acarine interactions. Behav Ecol 10(4):422–427

Hopke J, Donath J, Blechert S, Boland W (1994) Herbivore-induced volatiles: the emission of acyclic homoterpenes from leaves of *Phaseolus lunatus* and *Zea mays* can be triggered by a β-glucosidase and jasmonic acid. FEBS Lett 352(2):146–150

Hunt R (1979) Plant growth analysis: the rationale behind the use of the fitted mathematical function. Ann Bot 43(2):245–249

Janssen A, Bruin J, Jacobs G, Scraag R, Sabelis MW (1997) Predators use volatiles to avoid prey patches with conspecifics. J Anim Ecol 66:223–232

Kardong KV (2005) An introduction to biological evolution, 1st edn. McGraw-Hill, New York

Kriesch S, Dicke M (1997) Avoidance of predatory mites by the two-spotted spider mite Tetranychus urticae: the role of infochemicals. Proc Sect Exp Appl Entomol-Neth Entomol Soc 8:121–126

Lambers H (1987) Does variation in photosynthetic rate explain variation in growth rate and yield? Neth J Agric Sci 35:505–519

Landolt PJ (1993) Effects of host plant leaf damage on cabbage looper moth attraction and oviposition. Entomol Exp Appl 67(1):79–85

Loughrin JH, Manukian A, Heath RR, Tumlinson JH (1995) Volatiles emitted by different cotton varieties damaged by feeding beet armyworm larvae. J Chem Ecol 21(8):1217–1227

Loughrin JH, Manukian A, Heath RR, Turlings TC, Tumlinson JH (1994) Diurnal cycle of emission of induced volatile terpenoids by herbivore-injured cotton plant. Proc Natl Acad Sci 91(25):11836–11840

Mattiacci L, Dicke M, Posthumus MA (1994) Induction of parasitoid attracting synomone in Brussels sprouts plants by feeding of Pieris brassicae larvae: role of mechanical damage and herbivore elicitor. J Chem Ecol 20(9):2229–2247

Molles MC Jr (1999) Ecology: concepts and applications, 1st edn. McGraw-Hill, New York

Odum EP, Barret GW (2005) Fundamentals of ecology, 5th edn. Cengage Learning, New Delhi

Oudejans AM, Bruin J. (1994) Does spider-mite damage induce information transfer between plants of different species? Mededelingen-Faculteit Landbouwkundige en Toegepaste Biologische Wetenschappen Universiteit Gent (Belgium)

Ozawa R, Arimura G, Takabayashi J et al (2000) Involvement of Jasmonate and salicylate related signalling pathways for the production of specific herbivore induced volatiles in plants. Plant Cell Physiol 41(4):391–398

Pallini A. (1998) Odour-mediated indirect interactions in an arthropod food web. PhD thesis, University of Amsterdam, Amsterdam

Pallini A, Janssen A, Sabelis MW (1997) Odour-mediated responses of phytophagous mites to conspecific and heterospecific competitors. Oecologia 110(2):179–185

Papaj DR, Vet LE (1990) Odor learning and foraging success in the parasitoid, Leptopilina heterotoma. J Chem Ecol 16(11):3137–3150

Rhoades DF (1985) Offensive-defensive interactions between herbivores and plants: their relevance in herbivore population dynamics and ecological theory. Am Nat 1:205–238

Rojas JC (1999) Electrophysiological and behavioral responses of the cabbage moth to plant volatiles. J Chem Ecol 25(8):1867–1883

Roth JP, King EJ, Hensley SD (1982) Plant, host and parasite interactions in the host selection sequence of the tachinid *Lixophaga diatraeae*. Environ Entomol 11:273–277

Sabelis MW. (1981) Biological control of two-spotted spider mites using phytoseiid predators: I. Modelling the predator-prey interaction at the individual level. Centre for agricultural publishing and documentation

Sabelis MW, Afman BP, Slim PJ (1984) Location of distant spider mite colonies by *Phytoseiulus persimilis*: localization and extraction of kairomone. Acarol VI 1:431–440

Sabelis MW, Baan HV (1983) Location of distant spider mite colonies by phytoseiid predators: demonstration of specific kairomones emitted by *Tetranychus urticae* and *Panonychus ulmi*. Entomol Exp Appl 33(3):303–314

Sabelis MW, Dicke M (1985) Long range dispersal and searching behaviour. In: Helle W, Sabelis MW (eds) Spider mites: their biology, natural enemies and control, World Crop Pests 1b. Elsevier, Amsterdam, pp 141–160

Sabelis MW, Diekmann O (1988) Overall population stability despite local extinction: the stabilizing influence of prey dispersal from predator-invaded patches. Theor Popul Biol 34(2):169–176

Shulaev V, Silverman P, Raskin I (1997) Airborne signalling by methyl salicylate in plant pathogen resistance. Nature 386(6626):738

Taiz L, Zeiger E (2003) Plant physiology, 3rd edn. Panima Publishing Corporation, New Delhi

Takabayashi J, Dicke M, Posthumus MA (1991) Variation in composition of predator-attracting allelochemicals emitted by herbivore-infested plants: relative influence of plant and herbivore. Chemoecology 2(1):1–6

Takabayashi J, Dicke M, Posthumus MA (1994a) Volatile herbivore-induced terpenoids in plant-mite interactions: variation caused by biotic and abiotic factors. J Chem Ecol 20(6):1329–1354

Takabayashi J, Dicke M, Takahashi S, Posthumus MA, Van Beek TA (1994b) Leaf age affects composition of herbivore-induced synomones and attraction of predatory mites. J Chem Ecol 20(2):373–386

Takabayashi J, Sato Y, Horikoshi M, Yamaoka R, Yano S, Ohsaki N, Dicke M (1998) Plant effects on parasitoid foraging: differences between two tritrophic systems. Biol Control 11(2):97–103

Takabayashi J, Takahashi S, Dicke M, Posthumus MA (1995) Developmental stage of herbivore *Pseudaletia separata* affects production of herbivore-induced synomone by corn plants. J Chem Ecol 21(3):273–287

Tallamy DW, Raupp MJ (1991) Phytochemical induction by herbivores. Wiley, New York

Turlings TCJ, Tumlinson JH (1992) Systemic release of chemical signals by herbivore-injured corn. Proc Natl Acad Sci U S A 89:8399–8402

Turlings TC, Loughrin JH, Mccall PJ, Röse US, Lewis WJ, Tumlinson JH (1995) How caterpillar-damaged plants protect themselves by attracting parasitic wasps. Proc Natl Acad Sci 92(10):4169–4174

Turlings TCJ, Scheepmaker JWA, Vet LEM, Tumlinson JH, Lewis WJ (1990a) How contact foraging experiences affect preferences for host related odors in the larval parasitoid Cotesia marginiventris (Cresson) (hymenoptera: Braconidae). J Chem Ecol 16:1577–1589

Turlings TC, Tumlinson JH, Lewis WJ (1990b) Exploitation of herbivore-induced plant odors by host-seeking parasitic wasps. Science 250:1251–1253

Turlings TC, Wäckers FL, Vet LE, Lewis WJ, Tumlinson JH (1993) Learning of host-finding cues by hymenopterous parasitoids. Insect learning. Chapman & Hall, New York, pp 51–78

Van Dam NM, Baldwin IT (1998) Costs of jasmonate-induced responses in plants competing for limited resources. Ecol Lett 1(1):30–33

Vet LE, Dicke M (1992) Ecology of infochemical use by natural enemies in a tritrophic context. Annu Rev Entomol 37(1):141–172

Vet LE, Wäckers FL, Dicke M (1990) How to hunt for hiding hosts: the reliability-detectability problem in foraging parasitoids. Neth J Zool 41(2):202–213

Visser JH, Thiery D (1986) Effects of feeding experience on the odour-conditioned anemotaxes of Colorado potato beetles. Entomol Exp Appl 42(2):198–200

Wiskerke JSC, Vet LEM (1991) Comparison of two *Cotesia* species foraging for solitarity and gregariously feeding *Pieris* host species. Proc Exp Appl Entomol 2:190–195

Zeringue HJ, McCormick SP (1989) Relationships between cotton leaf-derived volatiles and growth of aspergillus flavus. J Am Oil Chem Soc 66(4):581–585

Potential of Microbial Volatile Organic Compounds for Crop Protection Against Phytopathogenic Fungi

14

B.N. Reddy and A. Hindumathi

Abstract

Microbial volatile organic compounds (MVOCs) are produced by a wide range of microorganisms ranging from bacteria and fungi. Earlier experimental results indicated that MVOCs are eco-friendly and can be exploited as agents that improve plant growth, productivity, and disease resistance in agricultural practices. Studies conducted in open fields indicated that MVOCs can effectively contribute to enhanced crop production and protection against diseases. In this review chapter, we have discussed on the diversity of MVOCs and further focused on their potential in exploiting these bioactive molecules in sustainable eco-friendly agriculture for improving plant growth, production, and protection.

Keywords

Bacteria • Fungi • Microbial volatile organic compounds

Abbreviations

~	More than
%	Percentage
β	Beta
C	Carbon
˚C	Degree Celsius
BD	Butanediol

B.N. Reddy (✉) • A. Hindumathi
Mycology and Plant Pathology Laboratory, Department of Botany, Osmania University, Hyderabad, Telangana 500007, India
e-mail: reddybn1@yahoo.com

D.K. Choudhary et al. (eds.), *Volatiles and Food Security*,
DOI 10.1007/978-981-10-5553-9_14

DMDS	Dimethyl disulfide
et al.	Associates
ISR	Induced systemic resistance
kPa	Kilopascals
MVOCs	Microbial volatile organic compounds
PGPF	Plant growth-promoting fungi
PGPR	Plant growth-promoting rhizobacteria
sp.	Species (singular)
spp.	Species (plural)
VOCs	Volatile organic compounds
VPEs	Volatile-producing endophytes

14.1 Introduction

The rhizosphere of plants is the habitat comprising many different types of organisms. Microorganisms are universal in the biosphere. They are often found in large quantities and diverse compositions (microbiome). A wide range of volatile metabolites, also known as volatile organic compounds (VOCs) are produced by organisms ranging from microbes to animals to plants from diverse ecosystems that are known to affect their environments and each other (Stotzky and Schenk 1976; Alstrom 2001; Schöller et al. 2002; Dickschat et al. 2004; Fernando et al. 2005; Schulz and Dickschat 2007; Effmert et al. 2012; Kramer and Abraham 2012).

Biogenic volatiles when compared with other secondary metabolites, such as enzymes, antibiotics, and toxins, are typically small compounds (up to C20) having low molecular weight (100–500 daltons); high vapor pressure (0.01 kPa or higher at 20 °C), i.e., they can easily evaporate at room temperature; low boiling point; low polarity; and a lipophilic moiety. These properties facilitate evaporation easily at normal temperatures and pressures and diffusion through both water- and gas-filled pores in soil and rhizosphere environments. Hence, microbial volatiles play significant roles in marine as well as terrestrial environments (Romoli et al. 2014).

Major advantage for this kind of antimicrobial agents is that they can travel large distances in a structurally heterogeneous environments composed of solids, liquids, and gases (Wheatley 2002), because of the volatile nature of these compounds. These properties make VOCs useful for interspecies communication as "infochemicals" or "semio-chemicals," particularly in nonaqueous environments. There is a growing recognition of the extent of chemical communication in the biosphere and the role that volatile chemicals play in biological signaling. Chemical signaling occurs within individual organisms, between individuals of the same species, and also between different species.

VOCs specifically called as microbial volatile organic compounds (MVOCs) are produced by microorganisms such as fungi and bacteria during metabolism. Volatile organic compounds (VOCs) are a complex mixture of low-molecular-weight lipophilic compounds derived from different biosynthetic pathways. Recently, Maffei et al. (2011) proposed the term "volatilome" to describe their complexity. Microbial volatile organic compounds (MVOCs) are chemically diverse and are variations of carbon-based molecules composed of low-molecular-weight alcohols, aldehydes, amines, ketones, esters, lactones, thiols, terpenes, aromatic and chlorinated hydrocarbons, and sulfur-based compounds. They serve as chemical windows through which the fundamental information about the molecular basis of microbial activities is released (Liang et al. 2008; Korpi et al. 2009; Thorn and Greenman 2012). As on today, 100s of bacteria and fungi were described as soil MVOC producers (Effmert et al. 2012). The chemical structure of ~1000 volatiles produced by a wide range of bacterial and fungal genera and species has been described (Effmert et al. 2012; Lemfack et al. 2014). Bacterial volatiles are dominated by alkenes, alcohols, ketones, terpenes, benzenoids, pyrazines, acids, and esters, whereas fungal volatiles are dominated by alcohols, benzenoids, aldehydes, alkenes, acids, esters, and ketones (Piechulla and Degenhardt 2014).

Most microbial volatiles are side products of primary and secondary metabolism, and they are formed mainly by oxidation of glucose from various intermediates (Korpi et al. 2009). The main metabolic pathways for microbial volatiles are given in. The amount and composition of volatiles produced by microorganisms are known to vary according to culturing conditions (Claeson et al. 2007; Blom et al. 2011; Garbeva et al. 2014a, b). Physiological state of the producing microorganisms, oxygen availability, moisture, pH, and temperature are the other important factors known to influence the production of volatiles (Insam and Seewald 2010; Romoli et al. 2014).

For a long time, the importance of microbial volatiles for the ecology of microorganisms has not been recognized, probably due to the lack of appropriate detection techniques. However, in the last decade, the number of studies on microbial volatiles has increased in different research areas such as food, medical, agricultural, and environmental sciences. Studies have evidenced that microbial volatiles can act as infochemicals in interactions among microbes and between microbes and their eukaryotic hosts. Now, the biggest challenge is to unravel the biological and ecological functions of these microbial volatiles.

The growing evidence indicates that MVOCs are eco-friendly, biologically based products and can be exploited as ecologically sound approach and as a cost-effective sustainable strategy for their use in agricultural practice for enhancement of plant growth, productivity, and disease resistance. MVOCs, as naturally occurring chemicals, have potential as possible alternatives to harmful synthetic chemical additives like fertilizers, pesticides, fungicides, and bactericides. Recent studies conducted under open-field conditions demonstrated that MVOCs efficiently contribute to sustainable crop protection and production.

Soil bacteria and fungi often possess traits that enable them to act as antagonists by suppressing soilborne plant diseases, by excreting antimicrobial metabolites that directly or indirectly promote plant growth (Gupta et al. 2000; Haas and Defago 2005; Handelsman and Stabb 1996; Whipps 2001). Many of these specialized compounds, such as antibiotics, cause growth inhibition or have more deleterious effects on organisms. Microbial volatiles play significant role in antagonistic interactions between microorganisms harboring the same ecological niche.

In this review, we focused on the diversity of MVOCs by describing microbial–plants and microbial–microbial interactions. Furthermore, we also discussed exploitation of these bioactive molecules in a sustainable agricultural context for the improvement of plant growth, development, and health.

Over the past 10 years, research on MVOCs–plant interactions has led to the conceptual understanding of the complex and dynamic nature of MVOCs and their potential role in enhancing crop protection and productivity in a sustainable way. Fungal and bacterial species detect plant host and initiate their colonization strategies in the rhizosphere by producing plant growth-regulating substances such as auxins and/or cytokinins (Ortiz-Castro et al. 2009). Many well-known volatile organic compounds of microbiological origin exhibit biological activity. Both fungal and bacterial VOCs have many physiological properties, especially the ability to inhibit the growth of other microbes.

14.2 Bioactive Fungal Volatiles

Several fungi show the ability to synthesize and produce a large number of volatile organic compounds (MVOCs). The VOCs produced by fungi have received limited attention in terms of their relationship to plant pathogenesis or growth promotion. Fungi emit cocktails of dozens to hundreds of unique volatile compounds that belong to many chemical classes such as alcohols, aldehydes, acids, ethers, esters, ketones, hydrocarbons, terpenes, and sulfur compounds (Korpi et al. 2009).

Volatile compounds are known to be ideal for serving as intra- and interspecific chemical signals or "infochemicals" because they can be sensed from a long distance. These compounds can diffuse to long distances from their point of origin, and they can persist and migrate in soil environments, areas of dense vegetation, and other microhabitats that harbor interacting populations of bacteria and fungi. VOCs produced by a given fungal species can have multiple effects on other organisms such as defense, environmental monitoring, and nutrient acquisition (Minerdi et al. 2009). There are several multifunctional and multi-organismic volatile-based interactions reported to occur in ecosystems. Many fungal VOCs are known to occur as either simple hydrocarbons, heterocycles, aldehydes, ketones, alcohols, phenols, thioalcohols, thioesters, benzene derivatives, or cyclohexanes (Korpi et al. 2009; Ortiz-Castro et al. 2009). Ryu et al. (2003) first reported that the microbial VOCs 2,3-butanediol and acetoin promoted the growth of *Arabidopsis thaliana*. Yamagiwa et al. (2011) first reported that plant growth-promoting fungi (PGPF) derived VOC-mediated growth promotion in plants. VOCs of certain PGPFs including the biocontrol fungus *Trichoderma* have been shown to enhance plant growth (MacDonald et al. 1986; Yamagiwa et al. 2011).

Volatiles being small molecules can easily diffuse through porous soils and travel over a long distance via the atmosphere and could be an effective tool for rhizosphere microorganisms to influence neighboring organisms. Reino et al. (2008) reviewed the capability of *Trichoderma* spp. to produce a number of volatile secondary metabolites (e.g., pyrones, sesquiterpenes). Volatile secondary metabolites have been demonstrated to play an important role in mycoparasitism of *Trichoderma* as well as its interaction with plants (Vinale et al. 2008).

Some soil saprophytes (*Trichoderma* sp.) (Yuan et al. 2012), fungi isolated from humid building materials (Schleibinger et al. 2008; Araki et al. 2009), grass endophytic *Epichloë* species belonging to *Ascomycota* (Schiestl et al. 2006), rust fungi (*Puccinia monoica* and *Uromyces pisi* of *Basidiomycota*) (Kaiser 2006), truffles (*Tuber* sp. of *Ascomycota*) (Splivallo et al. 2011), and mushroom sporocarps (Fraatz and Zom 2010) are reported to have the potential to generate MVOCs. Hung et al. (2013) reported that most of fungal MVOCs exert either potent inhibitory or stimulatory effects on plants. Yamagiwa et al. (2011) reported the plant growth-promoting fungus (PGPF), *Talaromyces* sp., emitted β-caryophyllene that significantly enhanced the growth of *Brassica campestris* seedlings and their resistance to *Colletotrichum higginsianum*.

Several fungi have been investigated as biocontrol agents, of which members of the genus *Trichoderma*, a common soil fungus, have received the most attention. *Trichoderma* species, commonly found in soil and root ecosystems, have been extensively studied for their beneficial effects on plant growth including the production of antibiotics and for their ability to compete against other fungi and pathogenic microorganisms (Harman et al. 2004). Several *Trichoderma* species are known to parasitize plant pathogens such as *Fusarium oxysporum*, *Phytophthora capsici*, and *Rhizoctonia solani*. In addition, fungi in the genus *Trichoderma* directly influence the growth of plant structures leading to an increase in plant biomass above ground and adventitious root formation below ground. The mechanisms by which *Trichoderma* simultaneously suppresses plant pathogens and enhances plant growth encompass mycoparasitism, antibiosis, competition, solubilization and sequestration of inorganic nutrients, and induced resistance. These properties have led to its use as a biofertilizer, bioprotectant, and biocontrol agent in agriculture. Proximal association of the fungi with the crop confers protection from pathogens and improves biomass production. Many species of *Trichoderma* including *T. atroviride*, *T. hamatum*, and *T. harzianum* are known mycoparasites. It is demonstrated that mycoparasitism is facilitated through the production of chitinases, glucans, and glucosidases (Elad et al. 1982; Inbar and Chet 1995). Moreover, it has been postulated that *Trichoderma* acts as a bioprotectant through the production of metabolites that suppress or inhibit the growth of other fungi. In addition, *T. asperellum* has been shown to activate induced systemic resistance in *Cucumis sativus*, and induction of this response makes plants less susceptible to attack by common plant pathogens such as *Botrytis cinerea* (Korolev et al. 2008). These studies have confirmed that *Trichoderma* and its exudates promote plant growth. In contrast, the gaseous emanations of *Trichoderma* have received little attention. The long-distance exposure to *T. viride* VOCs resulted in growth-promoting effects such as increase in fresh weight and root mass and also increase in the concentration of chlorophyll in leaves.

Therefore, an investigation of the volatiles produced by *Trichoderma* provides fertile ground for developing a new understanding of the mechanisms involved in both the triggering of the mycoparasitism response and of the phenomenon of plant growth promotion. The traditional use of *Trichoderma* as a soil treatment increases the concentration of volatiles generated by *Trichoderma* species.

Although it is known that fungi produce a large number of volatiles, many fungal VOCs such as 1-octen-3-ol are produced in large quantities. High concentrations of VOCs can have significant effects on plants in the vicinity. 1-octen-3-ol, sometimes called "mushroom alcohol," represented 33–78% of the volatile fraction of *Agaricus bisporus*, 66% of *Cantharellus cibarius*, 49–82% in *Boletus edulis*, and 90% in *Lactarius torminosus*. In particular, the chiral compound 1-octen-3-ol is one of the most abundant VOCs produced by nearly all species of fungi tested to date. In summary, the presence of fungal VOCs can have beneficial, detrimental, or neutral outcomes on plant growth and health. A few fungal VOCs are also reported to have growth inhibitory effects on plants (Splivallo et al. 2007).

Diby and Kyung (2013) showed that *Cladosporium cladosporioides* CL-1 fungal plant growth-promoting agent obtained from the rhizosphere of red pepper produced VOCs that significantly increased growth of the tobacco seedlings and their root development.

14.3 MVOCs of Fungal Endophytes

Endophytes are an important group of plant-associated fungal symbionts, present in both belowground and aboveground tissues. Volatile-producing endophytes (VPEs) are of primary interest because they produce MVOCs that are antibiotic in nature (Khan et al. 2014). VPEs produce broad-spectrum compounds having different physicochemical and biological properties that make them useful in industry as well as agriculture (Yuan et al. 2012). In recent years, some of the VOCs from endophytes have shown to have antibiotic activity, while others have potential for possible use as fuel compounds or "biodiesel." The microorganisms that live intercellularly within plant tissues without causing any evident negative effects are called endophytes. Fungal endophytes are a source of novel secondary metabolites, and also they are known to produce bioactive VOCs. Fungal endophytes in the genus *Muscodor* emit volatile substances that have strong antibacterial effects (Strobel et al. 2001; Strobel 2006). VOCs produced by an endophyte *Muscodor albus* ("stinky white fungus") have inhibitory or lethal to a wide range of bacteria and pathogenic fungi.

GC–MS analysis showed that *M. albus* produced a mixture of volatile acids, alcohols, esters, ketones, and lipids, which individually had inhibitory but not lethal effects against *Fusarium solani*, *Pythium ultimum*, and *Rhizoctonia solani* (Strobel et al. 2001). These VOCs when applied collectively acted synergistically to kill a broad range of plant-pathogenic fungi and bacteria (Strobel et al. 2001). This selective antimicrobial effect can be used against undesirable pathogens (termed as "mycofumigation"). *M. albus* has been used for the biological control of damping off in broccoli seeds grown in greenhouse soilless mix (Mercier and Jiménez 2004). Lee et al. (2009) reported that *Oxyporus latemarginatus* (an endophyte) isolated from pepper plants showed positive mycofumigation ability against post-harvest decay organisms. VOCs from endophytes have been reported to have other biotechnological applications.

14.4 Bacterial Volatiles

Bacterial species also produce a wealth of bioactive volatiles. A survey of 26 species belonging to genus *Streptomyces* revealed mixtures of alcohols, alkanes, alkenes, ketones, terpenoids, and thiols (Schöller et al. 2002). *Myxococcus xanthus* produced 42 different volatiles (Dickschat et al. 2005b), and the volatile compound of a marine *Streptomyces* species exhibited antibiotic properties (Dickschat et al. 2005a).

During the past few years, awareness concerning the emission of an unexpected high number of bacterial volatiles is being increased, and recent studies have showed that bacteria employ their volatile compounds during interactions with other organisms (Kai et al. 2009; Romoli et al. 2014). Bacteria are known to either positively or negatively affect other organisms' fitness, and recent studies showed that bacterial volatiles play an important role in bacterial–plant, bacterial–bacterial, and bacterial–fungal interactions.

Some bacteria live closely associated with the plant roots (rhizosphere) in the soil, exploiting nutrient-rich exudates that plants deliver into the soil. These bacteria are collectively defined as rhizobacteria also referred as plant growth-promoting rhizobacteria (PGPR), and many of them promote plant growth (Kevin 2003; Bhattacharyya and Jha 2012). The PGPRs promote plant growth by producing siderophores, 1-aminocyclopropane-1-carboxylate, and plant hormones, by solubilizing nutrients, fixing nitrogen, etc. (Ahemad and Khan 2012). In addition, PGPR strains also protect plants from phytopathogens by producing antibiotics, hydrolytic enzymes, and volatile organic compounds (VOCs), competing for space and nutrients with pathogens and inducing systematic resistance in plants. Among different mechanisms of action, the production of VOCs has its own importance. The volatile organic compounds (VOCs) produced by plant growth-promoting rhizobacteria have been reported to improve plant growth, inhibit bacterial and fungal pathogens and nematodes, and induce systemic resistance in plants against different phytopathogens, which revealed the importance of VOCs in plant–microbe interactions (Raza et al. 2013).

Pseudomonad rhizobacteria have been considered to be the most important rhizosphere organisms (Goswami et al. 2013). Rhizobacteria emitted VOCs that are involved in their interaction with plant-pathogenic microorganisms and host plants, thereby showing antimicrobial and plant growth-modulating activities (Vespermann et al. 2007) and often inducing systemic resistance in plants against phytopathogens (Ryu et al. 2003, Kishimoto et al. 2007). Fluorescent *Pseudomonas* strains help in soil health maintenance and protect the crops from pathogens (Hol et al. 2013). Emission of differential VOCs by the rhizospheric bacterial strains can modulate both plant growth promotion and root-system architecture (Gutierrez-Luna et al. 2010). Recent analytical studies have provided a comprehensive profile of rhizobacterial volatiles. These MVOCs have molecular masses below 300 Da and are lipophilic with relatively low boiling points.

It has been demonstrated that certain PGPRs stimulate plant growth without being in physical contact with roots through release of volatile compounds (Ryu et al. 2003, 2004; Farag et al. 2006). Further, certain PGPR can restrict fungal growth by emission of volatile organic compounds (VOCs) (Ryu et al. 2004; Ortiz-Castro et al. 2008). Bacterial VOCs have been shown to serve various roles such as signal compounds for inter- and intraspecies as well as cell-to-cell communication, stimulate or inhibit plant growth, as well as affect phytopathogens (Wenke et al. 2010).

In particular, PGPRs are known to produce VOCs that have growth-promoting properties on *Arabidopsis* (Farag et al. 2006; Zhang et al. 2007) and also induce systemic resistance in *Arabidopsis* (Ryu et al. 2003; Farag et al. 2006). Strains of *Bacillus subtilis*, *B. amyloliquefaciens*, and *Enterobacter cloacae* by emitting several volatiles promoted plant growth, of which acetoin and 2,3-butanediol (2,3-BD) gave the highest level of growth promotion and induced systemic resistance (ISR) of *A. thaliana* (Ryu et al. 2003; Rudrappa et al. 2010). *Enterobacter aerogenes* (endophytic bacterium) that colonizes corn plants produced 2,3-BD one of the major MVOCs. 2,3-BD produced by *E. aerogenes* rendered corn plants more resistant to Northern corn leaf blight fungus *Setosphaeria turcica* (D'Alessandro et al. 2014).

SPME-GC–MS (solid-phase microextraction [SPME] coupled to gas chromatography–mass spectrometry GC–MS) analysis revealed an array of volatile compounds emitted from rhizobacteria that are involved in plant growth promotion (Farag et al. 2006).

Volatiles emitted by *B. subtilis* induced beneficial effects on plant growth like increased plant size and weight, cell numbers, and modulation of root-system architecture (Xie et al. 2009; Gutiérrez-Luna et al. 2010). VOCs also stimulated the synthesis of plant hormones such as indole-3-acetic acid (IAA), cytokinin, and gibberellins. VOCs are important components of the signaling process in the interaction of plant growth-promoting bacteria (PGPR) with plants (Vespermann et al. 2007). Several bacterial VOCs including alcohols, ammonia, HCN, and phenzine-1-caroxylic acid have antifungal properties that contribute to the biocontrol effects of PGPR (Whipps 2001; Kai et al. 2009). The VOCs of PGPR are associated with induced systemic resistance (Farag et al. 2006). *Pseudomonas fluorescence* S97 and S241 have been reported to produce VOCs like hydrogen cyanide that inhibited the growth of plants (Alstrom and Burns 1989). The rhizobacterial isolates *Serratia plymuthica*, *Serratia odorifera*, *Stenotrophomonas maltophilia*, *Stenotrophomonas rhizophila*, *Pseudomonas fluorescens*, and *Pseudomonas trivialis* were reported to emit volatile compounds that inhibit the growth of *Rhizoctonia solani (*Kai et al. 2007).

Antifungal effects of organic volatiles were previously shown to inhibit germination or mycelial growth of *S. sclerotiorum* (Fernando et al. 2005; Huang et al. 1997), and unidentified compounds from *Bacillus subtilis* caused structural deformation of pathogenic fungi. Bacterial volatiles can also promote fungal growth (Huang et al. 1997; Wheatley 2002).

Bacillus amyloliquefaciens NJN-6 was reported to produce VOCs that inhibited the growth and spore germination of *Fusarium oxysporum* f. sp. *cubense* (Yuan et al. 2012). *Veillonella* species and *Bacteroides fragilis* produced VOCs that showed antibacterial activity. A VOC tridecane produced by *Paenibacillus polymyxa* E681 induced salicylic acid and ethylene-signaling marker genes PR1 and VSP2, respectively (Lee et al. 2012). Similarly, Ryu et al. (2003) reported plant growth promotion by the VOCs produced by *B. subtilis* GB03 and *B. amyloliquefaciens* IN937a. These reported results showed the importance of VOCs in plant growth promotion and inhibition of plant pathogens.

Pseudomonas fluorescens SS101 produced three VOCs, 3-tetradecadien-1-ol, 2-butanone, and methyl-n-1-2-tridecene, that improved the growth of tobacco (Park et al. 2015). There is not much information available about the actual mechanism of VOCs to improve plant growth and other characteristics. However, Ryu et al. (2003) reported studies of partial mechanism information like cytokinin signaling pathway appeared to play some role in growth promotion with exposure to *B. subtilis* GB03 VOCs. Similarly, *Bacillus* sp. B55 produced dimethyl disulfide (DMDS) that improved plant growth of tobacco by enhancing sulfur nutrition (Meldau et al. 2013). In another report, VOCs like *m*-cresol and methyl benzoate produced by plant growth-promoting fungi, *Cladosporium* sp. and *Ampelomyces* sp., elicit induced systemic resistance (ISR) in plants against the pathogen *Pseudomonas syringae* pv. tomato DC3000, and salicylic acid and jasmonic acid/ethylene might be involved in the ISR mediated by the VOCs (Naznin et al. 2014).

The microbial strains *B. amyloliquefaciens* T-5, NJN-6, FZB42, and SQR-9; *Pseudomonas fluorescence* WR-1, PF-5, Q2-87; and *Paenibacillus polymyxa* WR-2 and C5 were reported to produce VOCs that improved plant growth. The strain SQR-9 has been reported as an excellent biocontrol agent to colonize roots and control soilborne diseases effectively (Xu et al. 2014).

Bacillus atrophaeus CAB-1 produced volatile compounds that include a range of alcohols, phenols, amines, and alkane amides and hexadecane, 2,3-dimethoxy benzamide, and *O*-anisaldehyde. *O*-Anisaldehyde was found to exert the highest inhibition on the mycelial growth of the fungal pathogen *Botrytis cinerea*.

High concentrations of DMDS, 2-undecanone, dimethyl trisulfide, 4-octanone, *S*-methyl methanethiosulfonate, and 1-phenylpropan-1-one emitted by *Burkholderia ambifaria* significantly inhibited the growth of two phytopathogenic fungi, *R. solani* and *Alternaria alternata*. The mycelial growth of plant-pathogenic fungi, *Colletotrichum gloeosporioides*, *Fusarium culmorum*, *F. oxysporum*, and *Sclerotum rolffsi*, was significantly inhibited by 15 *Burkholderia tropica* strains. The volatile profile of *B. tropica* strain MTo431 showed the presence of several MVOCs known to play an important role in the antagonistic antifungal mechanism. MVOCs produced by *Burkholderia gladioli* pv. *agaricola* strains reduced the growth rate of *F. oxysporum* and *R. solani* of which limonene was found to be the most effective compound (Elshafie et al. 2012).

Earlier studies reported that VOC 2-pentylfuran, emitted by *Bacillus megaterium* strain XTBG34, caused growth promotion in *A. thaliana*. Furan compounds have been found to exhibit plant growth regulatory properties and are therefore recommended for commercial use in agriculture to improve plant health.

Fernando et al. (2005) evidenced that bacteria isolated from roots of canola, stubble, and soybean showed antifungal activity in split-plate assays and nonanal, N-decanol, cyclohexanol, ethyl-1-hexanol, benzothiazole, and dimethyl trisulfide were identified as the inhibitory volatiles.

Kai et al. (2009) reported bacterial VOCs such as amines, benzaldehyde, benzothiazole, decanal, cyclohexanol, dimethyl trisulfide, 2-ethyl-1-hexanol, and nonanal as fungicidal agents. Earlier laboratory studies demonstrated that several strains belonging to *B. amyloliquefaciens* subsp. *plantarum* (Dunlap et al. 2015) can associate with different plants and provide disease suppression or abiotic stress tolerance (Danielsson et al. 2007; Abd El-Daim et al. 2014).

Earlier study demonstrated that application of two VOCs, 3-pentanol and 2-butanone in cucumber plants, successfully elicited induced resistance against biotrophic bacterial pathogen *P. syringae* in open-field trials.

Although several studies have shown that volatile compounds can be used as signaling molecules in soil microbial communication, so far it is not clear how volatiles are perceived as signals by the microorganisms. To date, little is still known about the regulatory pathways and genes involved in biosynthesis and production of volatiles. Future challenges are therefore to further elucidate chemical diversity of microbial volatiles, to discover regulatory pathways and genes involved in the biosynthesis of volatiles in soil bacteria and fungi, to determine biologically relevant concentrations, to resolve the importance of volatiles in ecosystem processes, and ultimately to determine the soil health status of agricultural soils.

With respect to soil microbial volatiles, most attention has been given to the suppressive effects of bacterial volatiles on soil eukaryotes that are harmful to agricultural crops, e.g., plant-pathogenic fungi (Zou et al. 2007; Garbeva et al. 2014a, b). Rhizobacterial isolates *S. plymuthica*, *S. odorifera*, *Stenotrophomonas maltophilia*, *Stenotrophomonas rhizophila*, *P. fluorescens*, and *P. trivialis* emit complex MVOCs known to inhibit growth of many phytopathogenic and non-phytopathogenic fungi (Vespermann et al. 2007).

14.5 Conclusion

Microbial volatile organic compounds form interface between plants and microorganisms above and below ground environment where most of the interactions take place. Understanding the complex nature of MVOCs is the key to use a suitable tool for crop protection and production in sustainable agricultural practice.

The importance of MVOCs for crop plants will enable to adopt in sustainable crop protection and to develop production strategies on the promotional activities of growth, development, and defense system of plants. For a better understanding of the diversity and role of MVOCs at field application level, more research studies are needed to be conducted for sustainable use of naturally produced MVOCs for crop welfare.

References

Abd El-Daim IA, Bejai S, Meijer J (2014) Improved heat stress tolerance of wheat seedlings by bacterial seed treatment. Plant Soil 379:337–350

Ahemad M, Khan MS (2012) Evaluation of plant growth promoting activities of rhizobacterium *Pseudomonas putida* under herbicide-stress. Ann Microbiol 62:1531–1540

Alstrom S (2001) Characteristics of bacteria from oil seed rape in relation to their biocontrol activity against *Verticillium dahliae*. J Phytopathol 149:57–64

Alstrom I, Burns RG (1989) Cyanide production by rhizobacteria as a possible mechanism of plant growth inhibition. Biol Fertil Soils 7:232–238

Araki A, Eitaki Y, Kawai T, Kanazawa A, Takeda M, Kishi R (2009) Diffusive sampling and measurement of microbial volatile organic compounds in indoor air. Indoor Air 19:421–432

Bhattacharyya P, Jha D (2012) Plant growth-promoting rhizobacteria (PGPR): emergence in agriculture. World J Microbiol Biotechnol 28:1327–1350

Blom D, Fabbri C, Connor EC (2011) Production of plant growth modulating volatiles is widespread among rhizosphere bacteria and strongly depends on culture conditions. Environ Microbiol 13:3047–3058

Claeson AS, Sandstrom M, Sunesson AL (2007) Volatile organic compounds (VOCs) emitted from materials collected from buildings affected by microorganisms. J Environ Monit 9:240–245

D'Alessandro M, Erb M, Ton J, Brandenburg A, Karlen D, Zopfi J (2014) Volatiles produced by soil-borne endophytic bacteria increase plant pathogen resistance and affect tritrophic interactions. Plant Cell Environ 37:813–826

Danielsson J, Reva O, Meijer J (2007) Protection of oilseed rape (*Brassica napus*) toward fungal pathogens by strains of plant associated *Bacillus amyloliquefaciens*. Microb Ecol 54:134–140

Diby P, Kyung SP (2013) Identification of volatiles produced by *Cladosporium cladosporioides* CL-1, a fungal biocontrol agent that promotes plant growth. Sensors 13:13969–13977

Dickschat JS, Wenzel SC, Bode HB, Müller R, Schulz S (2004) Biosynthesis of volatiles by the Myxobacterium *Myxococcus xanthus*. Chem Biol Chem 5:778–787

Dickschat JS, Martens R, Brinkhoff T, Simon M, Schulz S (2005) Volatiles releases by *Streptomyces* species isolated from the North Sea. Chem Biodivers 2:837–865

Dunlap CA, Kim SJ, Kwon SW (2015) Phylogenomic analysis shows that *Bacillus amyloliquefaciens* subsp. *plantarum* is a later heterotypic synonym of *Bacillus methylotrophicus*. Int J Syst Evol Microbiol 65:2104–2109

Effmert U, Kalderas J, Warnke R, Piechulla B (2012) Volatile mediated interactions between bacteria and fungi in the soil. J Chem Ecol 38:665–703

Elad Y, Chet I, Henis Y (1982) Degradation of plant pathogenic fungi by *Trichoderma harzianum*. Can J Microbiol 28:719–725

Elshafie HS, Camele I, Racioppi R, Scrano L, Iacobellis NS, Bufo SA (2012) In vitro antifungal activity of *Burkholderia gladioli* pv. *agaricicola* against some phytopathogenic fungi. Int J Mol Sci 13:16291–16302

Farag MA, Ryu CM, Summer LW, Pare PW (2006) GC-MS SPME profiling of rhizobacterial volatiles reveals prospective inducers of growth promotion and induced systemic resistance in plants. Phytochemistry 67:2262–2268

Fernando WG, Ramarathnam R, Krishnamoorthy AS, Savchuk SC (2005) Identification and use of potential bacterial organic antifungal volatiles in biocontrol. Soil Biol Biochem 37:955–964

Fraatz MA, Zom H (2010) Fungal flavors. In: Esser K, Hofrichter M (eds) The mycota X: industrial applications. Springer, Heilderberg, pp 249–268

Garbeva P, Hordijk C, Gerards S, de Boer W (2014a) Volatile-mediated interactions between phylogenetically different soil bacteria. Front Microbiol 5:289

Garbeva P, Hordijk C, Gerards S, de Boer W (2014b) Volatiles produced by the mycophagous soil bacterium *Collimonas*. FEMS Microbiol Ecol 87:639–649

Goswami D, Vaghela H, Parmar S, Dhandhukia P, Thakker JN (2013) Plant growth promoting potentials of *Pseudomonas* spp. strain OG isolated from marine water. J Plant Interact 8:281–290

Gupta AM, Gopal KVB, Tilak R (2000) Mechanism of plant growth promotion by rhizobacteria. Indian J Exp Biol 38:856–862

Gutierrez-Luna FM, Lopez-Bucio J, Tamirano-Hernandez J, Valencia-Cantero E, de la Cruz HR, Macias-Rodriguez L (2010) Plant growth-promoting rhizobacteria modulate root-system architecture in *Arabidopsis thaliana* through volatile organic compound emission. Symbiosis 51:75–83

Haas D, Defago G (2005) Biological control of soil-borne pathogens by fluorescent pseudomonads. Nat Rev Microbiol 3:307–319

Handelsman J, Stabb EV (1996) Biocontrol of soilborne plant pathogens. Plant Cell 8:1855–1869

Harman GE, Howell CR, Viterbo A, Chet I, Lorito M (2004) *Trichoderma* species – opportunistic, avirulent plant symbionts. Nat Rev Microbiol 2:43–56

Hol W, Bezemer T, Biere A (2013) Getting the ecology into interactions between plants and the plant growth-promoting bacterium *Pseudomonas fluorescens*. Front Plant Sci 4:81

Huang H, Huang JW, Saidon G, Erickson R (1997) Effect of allyl alcohol and fermented agricultural wastes on carpogenic germination of sclerotia of *Sclerotinia sclerotiorum* and colonization of *Trichoderma* spp. Can J Plant Pathol 9:43–46

Hung R, Lee S, Bennett JW (2013) *Arabidopsis thaliana* As a model system for testing the effect of *Trichoderma volatile* organic compounds. Fungal Ecol 6:19–26

Inbar J, Chet I (1995) The role of recognition in the induction of specific chitinases during mycoparasitism by *Trichoderma harzianum*. Microbiology 141:2823–2829

Insam H, Seewald MS (2010) Volatile organic compounds (VOCs) in soils. Biol Fertil Soils 46:199–213

Kai M, Effmert U, Berg G, Piechulla B (2007) Volatiles of bacterial antagonists inhibit mycelial growth of the plant pathogen *Rhizoctonia solani*. Arch Microbiol 187:351–360

Kai M, Haustein M, Molina F, Petri A, Scholz B, Piechulla B (2009) Bacterial volatiles and their action potential. Appl Microbiol Biotechnol 81:1001–1012

Kaiser R (2006) Flowers and fungi use scents to mimic each other. Science 311:806–807

Kevin VJ (2003) Plant growth promoting rhizobacteria as biofertilizers. Plant Soil 255:571–586

Khan AL, Waqas M, Hussain J, Al-Harrasi A, Al-Rawahi A, Al-Hosni K (2014) Endophytes *Aspergillus caespitosus* LK12 and *Phoma* sp. *LK*13 of *Moringa peregrina* produce gibberellins and improve rice plant growth. J Plant Interact 9:731–737

Kishimoto K, Matsui K, Ozawa R, Takabayashi J (2007) Volatile 1-octen-3-ol induces a defensive response in *Arabidopsis thaliana*. J Gen Plant Pathol 73:35–37

Korolev N, David DR, Elad Y (2008) The role of phytohormones in basal resistance and *Trichoderma* -induced systemic resistance to *Botrytis cinerea* in *Arabidopsis thaliana*. Biol Control 53:667–683

Korpi A, Jarnberg J, Pasanen AL (2009) Microbial volatile organic compounds. Crit Rev Toxicol 39:139–193

Kramer R, Abraham WR (2012) Volatile sesquiterpenes from fungi: what are they good for? Phytochem Rev 11:15–37

Lee SO, Kim HY, Choi GJ, Lee HB, Jang KS, Choi YH, Kim JC (2009) Mycofumigation with *Oxyporus latemarginatus* EF069 for control of postharvest apple decay and *Rhizoctonia* root rot on moth orchid. J Appl Microbiol 106:1213–1219

Lee B, Farag MA, Park HB, Kloepper JW, Lee SH, Ryu CM (2012) Induced resistance by a long-chain bacterial volatile: elicitation of plant systemic defense by a C13 volatile produced by *Paenibacillus polymyxa*. PLoS One 7:e48744

Lemfack MC, Nickel J, Dunkel M, Preissner R, Piechulla B (2014) mVOC: a database of microbial volatiles. Nucleic Acids Res 42:D744–D748

Liang H, Zhang X, Jun RAO, Huanwen CHEN (2008) Microbial volatile organic compounds: generation pathways and mass spectrometric detection. J Chin Biotechnol 28:124–133

MacDonald EMS, Powell GK, Regier DA, Glass NL, Roberto F, Kosuge T, Morris RO (1986) Secretion of zeatin, ribosylzeatin, and "ribosyl-1"-methylzeatin by *Pseudomonas savastanoi*. Plant Physiol 82:742–747

Maffei ME, Gertsch J, Appendino G (2011) Plant volatiles: production, function and pharmacology. Nat Prod Rep 28:1359–1380

Meldau DG, Meldau S, Hoang LH, Underberg S, Wünsche H, Baldwin IT (2013) Dimethyl disulfide produced by the naturally associated bacterium *Bacillus* sp B55 promotes Nicotiana *attenuata* growth by enhancing sulfur nutrition. Plant Cell 25:2731–2747

Mercier J, Jimenez JI (2004) Control of fungal decay of apples and peaches by the biofumigant fungus *Muscodor albus*. Postharvest Biol Technol 31:1–8

Minerdi D, Bossi S, Gullino ML, Garibaldi A (2009) Volatile organic compounds: a potential direct long-distance mechanism for antagonistic action of *Fusarium oxysporum* strain MSA 35. Environ Microbiol 11:844–854

Naznin HA, Kiyohara D, Kimura M, Miyazawa M, Shimizu M, Hyakumachi M (2014) Systemic resistance induced by volatile organic compounds emitted by plant growth-promoting fungi in *Arabidopsis thaliana*. PLoS One 9:e86882

Ortiz-Castro R, Martinez-Trujillo M, Lopez-Bucio J (2008) N-acyl-Lhomoserine lactones: a class of bacterial quorum-sensing signals alter postembryonic root development in *Arabidopsis thaliana*. Plant Cell Environ 31:1497–1509

Ortiz-Castro R, Contreras-Cornejo H, Macias-Rodriguez L, Lopez-Bucio J (2009) The role of microbial signals in plant growth and development. Plant Signal Behav 4:701–712

Park YS, Dutta S, Ann M, Raaijmakers JM, Park K (2015) Promotion of plant growth by *Pseudomonas fluorescens* strain SS101 via novel volatile organic compounds. Biochem Biophys Res Commun 461:361–365

Piechulla B, Degenhardt J (2014) The emerging importance of microbial volatile organic compounds. Plant Cell Environ 37:811–812

Raza W, Faheem M, Yousaf S, Rajer FU, Yamin M (2013) Volatile and nonvolatile antifungal compounds produced by *Trichoderma harzianum* SQR-T037 suppressed the growth of *Fusarium oxysporum* f. sp. *niveum*. Sci Lett 1:21–24

Reino JL, Guerrero RF, Hernández-Galán R, Collado IG (2008) Secondary metabolites from species of the biocontrol agent *Trichoderma*. Pytochem Rev 7:89–123

Romoli R, Papaleo M, De Pascale D, Tutino M, Michaud L, LoGiudice A (2014) GC-MS volatolomic approach to study the antimicrobial activity of the antarctic bacterium *Pseudoalteromonas* sp TB41. Metabolomics 10:42–51

Rudrappa T, Biedrzycki ML, Kunjeti SG, Donofrio NM, Czymmek KJ, Pare PW (2010) The rhizobacterial elicitor acetoin induces systemic resistance in *Arabidopsis thaliana*. Commun Integr Biol 3:130–138

Ryu CM, Farag MA, Hu CH, Reddy MS, Wei HX, Paré PW, Kloepper JW (2003) Bacterial volatiles promote growth in *Arabidopsis*. Proc Natl Acad Sci U S A 100:4927–4932

Ryu CM, FaragMA HCH, Reddy MS, Kloepper JW, Pare PW (2004) Bacterial volatiles induce systemic resistance in *Arabidopsis*. Plant Physiol 134:1017–1026

Schiestl FP, Steinebrunner F, Schulz C, von Reuss S, Francke W, Weymuth C (2006) Evolution of 'pollinator' – attracting signals in fungi. Biol Lett 2:401–404

Schleibinger H, Laussmann D, Bornehag CG, Eis D, Rueden H (2008) Microbial volatile organic compounds in the air of moldy and mold-free indoor environments. Indoor Air 18:113–124

Schulz S, Dickschat JS (2007) Bacterial volatiles: the smell of small organisms. Nat Prod Rep 24:814–842

Splivallo R, Novero M, Bertea CM, Bossi S, Bonfante P (2007) Truffle volatiles inhibit growth and induce an oxidative burst in *Arabidopsis thaliana*. New Phytol 175:417–424

Splivallo R, Ottonello S, Mello A, Karlovsky P (2011) Truffle volatiles: from chemical ecology to aroma biosynthesis. New Phytol 189:688–699

Stotzky G, Schenk S (1976) Volatile organic compounds and microorganisms. Crit Rev Microbiol 4:333–382

Strobel G (2006) Muscodor albus and its biological promise. J Ind Microbiol Biotechnol 33:514–522

Strobel GA, Dirkse E, Sears J, Markworth C (2001) Volatile antimicrobials from *Muscodor albus* a novel endophytic fungus. Microbiology 147:2943–2950

Thorn RMS, Greenman J (2012) Microbial volatile compounds in health and disease conditions. J Breath Res 6:024001

Vespermann A, Kai M, Piechulla B (2007) Rhizobacterial volatiles affect the growth of fungi and *Arabidopsis thaliana*. Appl Environ Microbiol 73:5639–5641

Vinale F, Sivasithamparam K, Ghisalberti EL, Marra R, Barbetti MJ, Li H, Woo SL, Lorito MA (2008) Novel role for *Trichoderma* secondary metabolites in the interactions with plants. Physiol Mol Plant Pathol 72:80–86

Wenke K, Kai M, Piechulla B (2010) Belowground volatiles facilitate interactions between plant roots and soil organisms. Planta 231:499–506

Wheatley RE (2002) The consequences of volatile organic compound mediated bacterial and fungal interactions. Anton Leeuwenhoek 81:357–364

Whipps JM (2001) Microbial interaction and biocontrol in the rhizosphere. J Exp Bot 52:487–511

Xie X, Zhang H, Pare PW (2009) Sustained growth promotion in *Arabidopsis* with long-term exposure to the beneficial soil bacterium *Bacillus subtilis* (GB03). Plant Signal Behav 4:948–953

Xu Z, Zhang R, Wang D, Qiu M, Feng H, Zhang N, Shen Q (2014) Enhanced control of cucumber wilt disease by *Bacillus amyloliquefaciens* SQR9 by altering the regulation of its DegU phosphorylation. Appl Environ Microbiol 80:2941–2950

Yamagiwa Y, Inagaki Y, Ichinose Y, Toyoda K, Hyakumachi M, Shiraishi T (2011) *Talaromyces wortmannii* FS2 emits β-caryophyllene, which promotes plant growth and induces resistance. J Gen Plant Pathol 77:336–341

Zhang H, Kim MS, Krishnamachari V, Payton P, Sun Y, Grimson M, Farag MA, Ryu CM, Allen R, Melo IS (2007) Rhizobacterial volatile emissions regulate auxin homeostasis and cell expansion in *Arabidopsis*. Planta 226:839–851

Zou CS, Mo MH, Gu YQ, Zhou JP, Zhang KQ (2007) Possible contributions of volatile-producing bacteria to soil fungistasis. Soil Biol Biochem 39:2371–2379

Herbivore-Induced Plant Volatiles 15

A. Sabitha Rani and G. Sulakshana

Abstract

In nature, plants are subjected to various abiotic and biotic stresses of which stress due to herbivore attack is of major concern. There are many direct and indirect mechanisms which operate within the plants to combat herbivores. The direct defence is generally of two types, i.e. physical and chemical. The physical defence includes the presence of morphological barriers such as trichomes, silica deposition and cell wall lignifications. The chemical methods involve the synthesis of many toxic compounds like terpenoids, alkaloids, phenols, etc. Indirect defence involves production of volatile compounds from the host plant which can attract natural enemies of herbivores. These compounds are known as herbivore-induced plant volatiles (HIPVs) and play an important role in communicating with the parasitoids to prey upon the attacking herbivores. HIPVs not only help the plants to attract with the natural enemies of herbivores but also warn the neighbouring plants from the upcoming danger. Thus, HIPVs provide a reliable mechanism for natural management of insect pests in plants. This chapter presents the herbivore-induced plant volatiles, types, biosyntheses, emission and their role in plant defence.

Keywords

Herbivore-induced plant volatiles • Herbivores • Biosynthesis • Emission of volatiles • Jasmonates • Salicylic acid • Plant defence

A. Sabitha Rani (✉) • G. Sulakshana
Department of Botany, Osmania University College for Women, Koti, Hyderabad 500095, Telangana, India
e-mail: sabitaammana@yahoo.com

D.K. Choudhary et al. (eds.), *Volatiles and Food Security*,
DOI 10.1007/978-981-10-5553-9_15

15.1 Introduction

Plants produce a large number of volatile organic compounds which are utilized as perfumes, flavouring agents and other industrial purposes. In nature, these chemicals help the plants in pollination and seed dispersal. Significantly, these volatile compounds protect the plants from their enemy herbivores by repulsion or intoxication and by attracting predator or parasitoid insects to prey upon herbivores (Reinhard et al. 2004; Vancanneyt et al. 2001; Mercke et al. 2004). Also, these compounds act as chemical signals to communicate with the natural enemies of herbivores and with the neighbouring plants. There are about 1,700 volatile compounds isolated from 90 different plant families. The plant volatiles constitute around 1 % of the total plant secondary metabolites (Jarmo and James 2012). The volatile compounds are released from leaves, flowers and fruits into the atmosphere or into the soil from roots, to protect the plants against herbivores and pathogens (Dudareva and Pichersky 2006).

15.2 Role of Plant Volatiles

1. *Attract pollinators and seed disseminators*: Most of the flowering plants release different kinds of volatile compounds from various plant parts to attract the insects which help in pollination and dispersal of seeds (Buchmann and Nabhan 1996). Pollinators use these volatile blends to discriminate flowers of a particular plant species that rely on them for pollination. The level of volatile emission differs within a species due to various factors such as diurnal rhythms, age of flower and status of pollination. It also depends on environmental conditions like light, temperature, moisture, etc. (Dudareva et al. 2004). For example, moth-pollinated *Clarkia breweri* flowers emit a higher level of floral scent rich in linalool and aromatic esters, whereas lower levels were reported in the flowers of *Clarkia concinna*, a bee-pollinated plant (Raguso and Pichersky 1995). The volatile compounds produced by the fruits play an important role in the attraction of animals and birds, which help in seed dispersal. Luft et al. (2003) reported that *Pteropus pumilus* and *Ptenochirus jagori*, commonly known as fruit bats, can detect fruits and state of their ripeness based on the fruit odour.
2. *As antimicrobial compounds*: The floral volatiles of some of the plants were reported to have antimicrobial properties (Friedman et al. 2002). Flowers with rich nutrients and moisture are more prone to pathogen attack. Hence, volatiles with antimicrobial activity released from floral tissues can provide protection to the flowers from the pathogens. In *Arabidopsis thaliana*, stigma emits a major floral volatile, the sesquiterpene (E)-β-caryophyllene, which was reported to possess antimicrobial properties (Chen et al. 2003). Similarly 1,8-cineole, an oxygenated monoterpene, is known for its antimicrobial nature (Hammer et al. 2003).

3. *As antioxidants*: Some of the volatile compounds serves as antioxidants and protects the plants from different types of stresses induced by various environmental factors. Isoprene emitted from plants protects them from oxidative stress induced by ozone (Loreto et al. 2001). Volatile chemicals released from various plant parts act against reactive oxygen species (ROS) generated during thermal and oxidative stress (Sharkey et al. 2005; Vickers et al. 2009).
4. *In plant defence*: In response to herbivore attack and damage, plants release a mixer of volatile compounds for their self-defence. The chemicals emitted by plants can directly affect the herbivores by its toxic, repellent and deterring properties. In indirect defence, the volatiles may attract the natural enemies like parasitic wasps, flies and predatory mites that can protect the plant from further damage by herbivores (Bernasconi et al. 1998; Vet and Dicke 1992). Sometimes, the volatile organic compounds can mediate both the direct and indirect defence mechanism at a time. The volatiles produced in *Nicotiana attenuata* were reported to deter lepidopteron oviposition and also attracted the herbivore natural enemies (Kessler and Baldwin 2001). The common volatiles involved in most of the direct and indirect defences include metabolites from lipoxygenase (LOX) and shikimic acid pathways. The products of terpenoid pathway such as monoterpenes, sesquiterpenes and homoterpenes also play a major role in defence against herbivores (Pichersky and Gershenzon 2002).
5. *In tritrophic interactions*: The interaction of plant-herbivore-carnivore is a tritrophic interaction which was first reported by Price et al. (1980). The herbivore-induced plant volatiles (HIPVs) mediate the tritrophic interactions, in which plant volatiles act as cues for natural enemies of herbivores to locate their pray or the host plant (Mumm and Dicke 2010). For example, the interactions between lima beans plants (*Phaseolus lunatus*), herbivorous spider mites (*Tetranychus urticae*) and carnivorous mites (*Phytoseiulus persimilis*) are tritrophic interaction. In this, the infestation of lima bean leaves with spider mites triggers the release of volatile compounds from plants that attract the predatory mites (carnivorous) to prey on herbivorous spider mites (Takabayashi and Dicke 1996).
6. *In plant-plant interaction*: Volatile compounds released from herbivore-attacked plants bring about priming of plant defence mechanism in nearby plants. They initiate defence gene expression and result in emission of volatiles in healthy leaves of the same plant and also in unaffected leaves of neighbouring plants. As a result of this signalling and communication mechanism, plants exhibit increased attractiveness to carnivores insects and decrease the damage caused by herbivores (Ruther and Kleier 2005; Arimura et al. 2002).

15.3 Herbivore-Induced Plant Volatiles (HIPVs)

Volatile compounds synthesized in plants due to the attack by herbivore are called as herbivore-induced plant volatiles (HIPVs). Based on the metabolic origin, herbivore-induced plant volatiles are classified into four major classes. They are terpenes, fatty acid derivatives, amino acid derivatives and phenylpropanoid or benzenoid compounds. Among these volatiles, terpenoids constitute the major group in plant defence. Besides this, genus- or species-specific volatile compounds are also produced by plants. For example, the characteristic garlic-like odour in alliaceous members is emitted due to the release of volatile selenium compounds.

15.4 Biosynthesis of HIPVs

The herbivore-induced plant volatiles are low molecular weight compounds and synthesized by various metabolic pathways. Biosynthesis of HIPVs is spatially, temporally and developmentally regulated (Dudareva et al. 1996; Pare and Tumlinson 1997).

The synthesis of volatile chemicals generally occurs in the epidermal tissues and easily diffuses into the surrounding environment (Kolosova et al. 2001) or rhizosphere (Chen et al. 2004). The mechanism involved in transportation of volatiles from the site of synthesis into the atmosphere was reported to involve four major steps, i.e. trafficking within the cell, transport across the plasma membrane and cell wall, transfer through the cuticle and evaporation at the site of the cuticle. The formation of volatile compound in all of the plant organs shows similar developmental patterns. In early stages of organ development, such as young leaves, immature fruits or flowers ready for pollination, more synthesis of volatiles was observed, whereas it remained constant or decreased with the increase in the life span of the organs (Bouwmeester et al. 1998; Dudareva et al. 2000).

15.5 Terpenoids

Terpenoids constitute one of the important volatiles in plant defence against herbivores. Terpenoids consist of hemiterpenes, monoterpenes, sesquiterpenes, homoterpenes and diterpenes (Fig. 15.1). They are derived from C_5 precursors by two important pathways. Monoterpenes and diterpenes are synthesized in plastids via methyl erythritol phosphate (MEP pathway) from geranyl diphosphate (GDP) and geranylgeranyl diphosphate (GGDP), respectively (Eisenreich et al. 1998). Sesquiterpenes are produced in cytosol through mevalonic acid pathway (MVA pathway) from farnesyl diphosphate (FDP) (Mc Caskill and Croteau 1995).

15.6 Phenylpropanoids and Benzenoids

Phenylpropanoids and benzenoids are derived from L-phenylalanine which are inverted to trans-cinnamic acids. This reaction is catalysed by an enzyme L-phenylalanine ammonia lyase. Hydroxycinnamic acid on methylation and hydroxylation forms a variety of cinnamic acids, aldehydes and alcohols. The intermediate compounds of these pathways are emitted as volatiles (Humphreys and

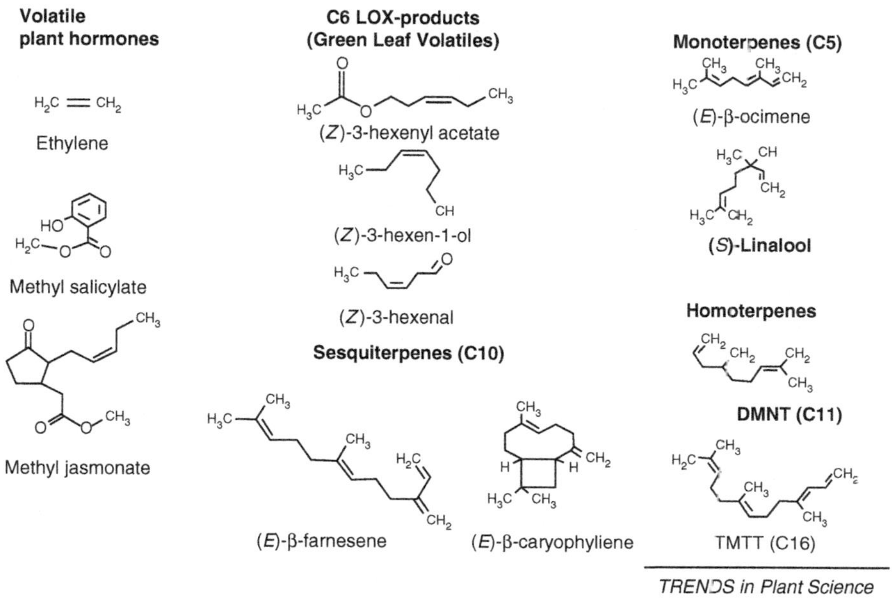

Fig. 15.1 Some of the plant volatile compounds

Chapple 2002). Benzenoid compounds are also formed from trans-cinnamic acid. During the synthesis trans-cinnamic acid, chain is shortened by a C_2 unit through CoA-dependent-β-oxidative pathway (Boatright et al. 2004).

15.7 Volatile Fatty Acid Derivatives

Volatile fatty acid derivatives are one of the common HIPVs released soon after the tissue disruption in host plants. They are derived from C_{18} unsaturated fatty acids like linoleic and linolenic acid by deoxygenation, catalysed by the enzyme lipoxygenases (Feussner and Wasternack 2002). These are also known as green leaf volatiles (GLVs). GLVs consist of C_6 aldehydes, (z)-3-hexenal n-hexenal and their derivatives (Bruinsma et al. 2009) (Fig. 15.2).

15.8 Amino Acid Volatile Derivatives

Amino acids like valine, alanine, leucine, isoleucine and methionine serve as precursors for the synthesis of plant volatiles such as aldehydes, alcohols, esters and nitrogen- and sulphur-containing volatiles. Deamination of amino acids results in α-keto acid which on decarboxylation, reductions, oxidations and esterification forms formaldehyde, alcohols and esters (Reineccius 2006). Methionine serves as a

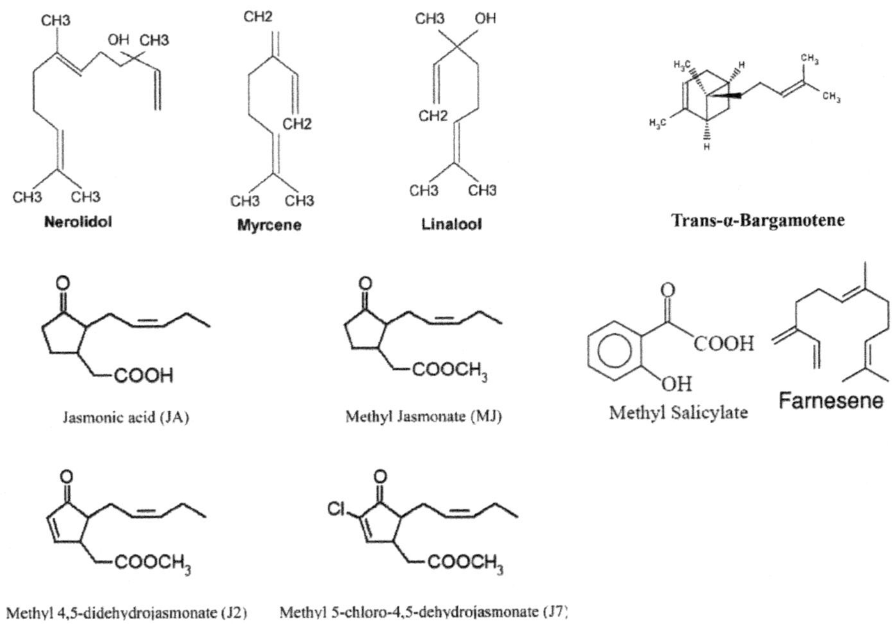

Fig. 15.2 Structure of some plant volatile organic compounds

precursor of sulphur-containing volatiles such as volatile thioesters and dimethyl disulphide (Wyllie and Leach 1992; Wyllie et al. 1995). In strawberry plant, alanine serves as a precursor for synthesis of volatile ethyl esters which is catalysed by the enzyme, strawberry alcohol acyl transferase, SAAT (Perez et al. 1992; Aharoni et al. 2000; Beekwilder et al. 2004). Alcohol acyl transferases are also involved in the formation of many volatile esters from alcohols and acyl-CoAs, which are derived from amino acids (Dudareva et al. 1998).

15.9 Herbivore-Induced Volatiles in Plant Defence

Plants, in response to attack by herbivory, synthesize and release a mixture volatile compounds from their damaged and also from health undamaged tissues which are specific to individual herbivore species. For example, terpenoids are synthesized *de nova* in cotton leaves in response to damage caused by the caterpillar, and similar chemicals were also reported from undamaged leaves. This clearly suggests that the chemical messenger is transported from the damage location to the undamaged leaves to trigger synthesis and release of plant volatiles. The mixtures of compounds collected from undamaged leaf differ from the volatiles from entire plant (Pare et al. 1998). The blend of volatile odour is very specific and can be useful in identifying the herbivore and its developmental stages (Takabayashi and Dicke 1996).

The attack of herbivore species triggers the release of different odour blends of volatiles, which can attract both parasitic and predatory insects to the host plant bearing prey species (De Moraes et al. 1998). Plant volatiles also induce defence responses in neighbouring plants and serve as a messenger between and among members of same species. These chemical signalling molecules are known as semiochemicals (Law and Regnier 1971). David Rhoades reported that willow plant (*Salix sitchensis*) showed increased resistance as they were warned by an airborne signal about the upcoming danger from the herbivore-infested neighbouring plants (Rhoades 1983). Altered levels of defence-related metabolites like terpenoids, proteinase inhibitors and phenolic compounds were also observed in plants exposed to HIPVs (War et al. 2011). In different plant species, the volatile chemicals released in response to herbivory also differ in their chemical composition both quantitatively and qualitatively. For example, in tobacco plant, higher amount of jasmonic acid (defence-signalling compound) is released during damage by hornworm caterpillars compared to mechanical damage (McCloud and Baldwin 1998). The corn plant infested with beet armyworm larvae produces linalool, (3E)-4,8-dimethyl1,3,7-nonatriene, (trans)-α-bergamotene and (E)-β-farnesene as major compounds; however, these volatiles are not released from soybean plants infested by the same herbivore (Turlings et al. 1993a).

15.10 Jasmonates (JA)

Jasmonic acid and its methyl ester (methyl jasmonate) are important volatile compounds released in plants in response to defence against herbivore (Fig. 15.3a). JA biosynthesized from linolenic acid by octadecanoid pathway plays a major role in plant defence against insects and biotrophic pathogens. In plants, JA accumulates in large amount in wounded tissues and activates genes coding for protease inhibitors that protect the plants from insect damage. Jasmonic acid also induces the expression of the genes coding for antifungal proteins, phytoalexin biosynthesis and phenolic compounds that are involved in plant defence mechanism (Creelman et al. 1992). The oxylipin pathway, another major source for JA, also leads to the synthesis of other volatile aldehydes and alcohols that function in wound healing and plant defence. For example, the growth of *Pseudomonas syringae* and *E. coli* was

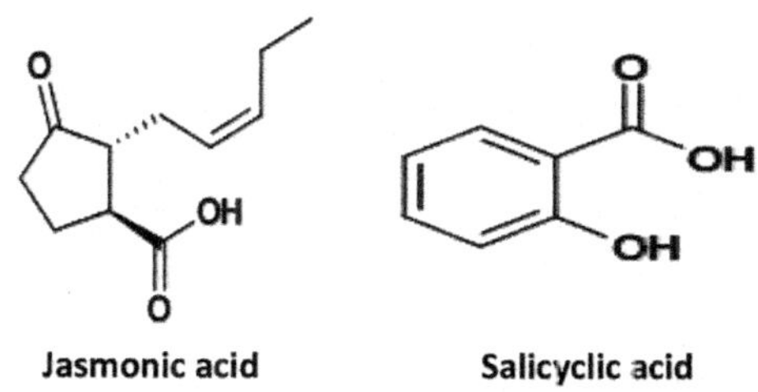

Fig. 15.3 Structure of Jasmonic acid and Salicylic acid

completely inhibited by the activity of C_6-aldehyde 2-hexenal (Deng et al. 1993). Similarly, in many plant species, the aphid fecundity is reduced by C6-aldehydes and alcohols (Hildebrand et al. 1993). These compounds are synthesized from 13-hydroperoxy linolenic acid by the catalytic action of hydroperoxy lyase enzyme. Presence of jasmonate and tissue wounding will elicit the expression of lipoxygenase gene, which in turn stimulates the hydroperoxy lyase enzyme. The increased activity of the enzyme enhances the ability of plants to produce the six-carbon compounds which contribute to plant protection (Creelman and Mullet 1997).

Exogenous application of JA in lima bean plants induces a specific mixture of volatile compounds, which is similar to the blend induced by the real attack of the herbivorous spider mite, *Tetranychus urticae*.. Further it was reported that JA-induced plants were visited by more carnivorous insects than non-induced plants suggesting that JA treatment may be used effectively in pest management (Radhika Venkateshan 2015).

15.11 Salicylic Acid (SA)

Salicylic acid is a phytohormone involved in signalling of plant defence against pathogens. The salicylic acid is synthesized in two pathways, one is from chorismate by the action of phenylalanine ammonia lyase (PAL), and the other pathway is from chorismate via isochorismate, catalysed by isochorismate synthase enzyme (ICS). SA induces resistance in plants against herbivores and other pathogens by production of pathogen-related proteins (Van Huijsduijnen et al. 1986) (Fig. 15.3b). Salicylic acid-induced plant communication is mediated by at least two mechanisms, one requiring the non-expressor of PR1 (NPR1) and second independent of NPR1 gene. It also induces systemic acquired resistance (SAR) in plants. SA helps in signalling nearby plants by converting into the volatile ester, methyl salicylate (MeSA). For example, MeSA is reported in the headspace volatiles of lima bean, *Arabidopsis*, tomato and soybean within hours of damage. MeSA also attracts insect predators singly and in combination with other HIPVs (Shah 2003). Treatment of lima bean plants with methyl salicylate and jasmonic acid produces a volatile blend closely resembling to that of herbivorous spider mite (*Tetranychus urticae*)-induced volatiles rather than JA treatment alone. Thus, JA and SA signalling pathways are synergistically involved in induction of herbivore-induced plant volatiles (Ozawa et al. 2000).

15.12 Herbivore-Induced Plant Volatile Emission

The amount of plant volatiles released in response to herbivore attack depends on various physical and environmental factors (Gouinguene and Turling 2002). Takabayashi et al. (1994) studied the importance of light in production of volatile on uninfested lima bean leaves (*Phaseolus lunata*). The study reports that more number of predatory mites was attracted in the leaves placed under high light intensity than the low light which is due to differential release of volatiles under two different light regimes. Similarly, the importance of light on emission of induced volatiles was reported by Maeda et al. (2000) in kidney bean (*Phaseolus vulgaris*) plants attacked by the spider mites (*Tetranychus urticae*), which are more active during light periods. Besides the intensity of light, temperature, relative humidity, atmospheric CO_2 concentration and nutrient status of the plant also influence the volatile emissions. Apart from the environmental factors, synthesis and release of various plant volatiles are regulated at the level of gene expression (Dudareva et al. 1996; Mc Conkey et al. 2000). The spatial and temporal release of herbivore-induced volatiles is under the control of transcriptional regulation. Not only that, the exposure of undamaged plants to herbivore-induced volatiles released by the damaged parts also plays a key role in regulation of gene expression involved in defence metabolism (Arimura et al. 2000).

The volatile emission and elicitation of other plant defence responses are induced by feeding behaviour of herbivore, which enables the plant to differentiate between general wounding and damage caused by chewing insects. In cotton, high amount of volatiles were released in response to wounding, created by caterpillar feeding rather than mechanical damage in the plants (Pare and Tumlinson 1997a). Similarly, in tobacco also, high concentrations of jasmonic acid are produced as result of damage by herbivore caterpillars than the mechanical injury to the plant tissues (McCloud and Baldwin 1998). This clearly indicates that various components like enzyme, elicitors and other factors associated with chewing and behaviour of insect herbivores trigger the emission of plant volatiles.

Different herbivore species produce elicitors with various structures to release plant volatiles. However, the biochemical mechanisms by which the elicitors trigger the synthesis and emission of volatiles in herbivore-infested plant are unknown. So far only one, volicitin, herbivore-specific volatile elicitor has been identified. But in some oral secretions of insect herbivores, other compounds analogous in structure to volicitin were also reported (Pare and Tumlinson 1999).

15.13 Fate of Volatiles After Release

The plant volatiles, after release into the atmosphere, are exposed to UV radiation and various reactive gases which affect their life span (Kroll and Seinfeld 2008). Hence, the function and activity of plant volatiles are only for limited time and mainly depend on its dispersal and reactivity with other compounds. The released HIPVs react with the gaseous, liquid and solid particulate matter present on the atmosphere and result in the formation of several reactive products such as secondary organic aerosols (SOA) (Joutsensaari et al. 2005; Virtanen et al. 2010). UV radiation triggers the oxidation of volatile compounds and results in the production of hydroxyl radicals (OH) and accumulation of ozone. The OH radicals lead to the conversion of NO to NO_2, and O_3 radicals react with other volatile organic compounds emitted in to the atmosphere (Pinto et al. 2010). In the presence of OH radicals and NO, the isoprene, monoterpenes and sesquiterpenes react with NO_3 radicals and result in the formation of organic nitrates ($RONO_2$). These organic nitrates are removed from the air during precipitation; thus, plant volatile organic compounds (VOCs) help to remove NO_x from the lower troposphere zone (Lerdau and Slobodkin 2002). It is also reported that VOCs released from plants also remove ozone from the troposphere and promote formation of secondary aerosol by ozonolysis (Lerdau and Slobodkin 2002; Virtanen et al. 2010). The organic nitrates of atmosphere act as reservoirs of NOx which help in formation of ozone and secondary organic aerosols (Pratt et al. 2012). The ability of plant volatiles to control and promote the formation of ozone in troposphere can benefit plants by eliciting many defence reactions against herbivores and many fungal pathogens (Sandermann et al. 1998; Cui et al. 2012). The tomato plants, when exposed to elevated O_3 levels, prolonged the developmental time and reduced the fecundity in whiteflies, *Bemisia tabaci* (Holopainen and Blande 2012). Plessl et al. (2005) reported that barley plants at high concentration of atmospheric O_3 reduced the aggressiveness of *Drechslera teres*, plant pathogenic fungi.

Herbivore-induced plant volatiles, with their chemical and physical properties, attract the natural enemies of herbivore from longer distances. However, the concentration of volatiles in the atmosphere will rapidly reduce if the distance is increased from the point of plant damage (McFrederick et al. 2009).

15.14 Conclusion

Plants produce a large number of volatile organic compounds. Some of these are specifically produced upon herbivore attack and are called as herbivore-induced plant volatiles (HIPVs). The HIPVs play a major role in plant-plant and plant-insect signalling pathways to combat herbivores. HIPVs are specific, nontoxic, biodegradable and can be used as alternative to synthetic pesticides.

In the recent years, there is a significant progress in studies of biochemical and molecular mechanisms involved in host plant-herbivore-insect interactions. Advanced technologies for qualitative and quantitative analyses of plant volatiles will contribute to the identification of key compound in herbivore-induced defences.

Developments in molecular biology have opened new perspectives in identifying the genes responsible for the induction of plant volatiles and developing crops with induced resistance to insect pests. Thus, research in the field of HIPVs will be helpful for employing volatile compounds as potential biocontrol agent in plant pest management.

References

Aharoni A, Keizer LCP, Bouwmeester HJ, Sun ZK, Alvarez-Huerta M, Verhoeven HA, Blaas J, van Houwelingen AMML, De Vos RCH, van der Voet H, Jansen RC, Guis M, Mol J, Davis RW, Schena M, van Tunen AJ, O'Connell AP (2000) Identification of the SAAT gene involved in strawberry flavor biogenesis by use of DNA microarrays. Plant Cell 12:647–661

Arimura G, Ozawa R, Shimoda T, Nishioka T, Boland W, Takabayashi J (2000) Herbivory-induced volatiles elicit defence genes in lima bean leaves. Nature 406:512–513

Arimura G, Ozawa R, Nishioka T, Boland W, Koch T, Kuhnemann F, Takabayashi J (2002) Herbivore-induced volatiles induce the emission of ethylene in neighboring lima bean plants. Plant J 29:87–98

Beekwilder J, Alvarez-Huerta M, Neef E, Verstappen FWA, Bouwmeester HJ, Aharoni A (2004) Substrate usage by recombinant alcohol acyltransferases from various fruit species. Plant Physiol 135:1865–1878

Bernasconi ML, Turlings TCJ, Ambrosetti L, Bassetti P, Dorn S (1998) Herbivore-induced emissions of maize volatiles repel the corn leaf aphid, *Rhopalosiphum maidis*. Entomol Exp Appl 87:133–142

Boatright J, Negre F, Chen XL, Kish CM, Wood B, Peel G, Orlova I, Gang D, Rhodes D, Dudareva N (2004) Understanding in vivo benzenoid metabolism in Petunia petal tissue. Plant Physiol 135:1993–2011

Bouwmeester HJ, Gershenzon J, Konings MCJM, Croteau R (1998) Biosynthesis of the monoterpenes limonene and carvone in the fruit of caraway—I. Demonstration of enzyme activities and their changes with development. Plant Physiol 117:901–912

Bruinsma M, Posthumus MA, Mumm R, Mueller MJ, van Loon JJA, Dicke M (2009) Jasmonic acid-induced volatiles of Brassica oleracea attract parasitoids: effects of time and dose, and comparison with induction by herbivores. J Exp Bot 60(9):2575–2587

Buchmann SL, Nabhan GP (1996) The forgotten pollinators. Island Press, Washington, DC

Chen F, D'Auria JC, Tholl D, Ross JR, Gershenzon J, Noel JP, Pichersky E (2003) An Arabidopsis thaliana gene for methyl salicylate biosynthesis, identified by a biochemical genomics approach, has a role in defense. Plant J 36:577–588

Chen F, Ro D-K, Petri J, Gershenzon J, Bohlmann J, Pichersky E, Tholl D (2004) Characterization of root-specific Arabidopsis terpene synthase responsible for the formation of the volatile monoterpene 1,8-cineole. Plant Physiol 135:1956–1966

Creelman RA, Mullet JE (1997) Biosynthesis and action of jasmonates in plants. Ann Rev Plant Physiol Plant Mol Biol 48:355–381

Creelman RA, Tierney ML, Mullet JE (1992) Jasmonic acid/methyl jasmonate accumulate in wounded soybean hypocotyls and modulate wound gene expression. Proc Natl Acad Sci U S A 89(11):4938–4941

Cui H, Sun Y, Su J, Ren Q, Li C, Ge F (2012) Elevated O_3 reduces the fitness of *Bemisia tabaci* via enhancement of the SA-dependent defense of the tomato plant. Arthropod-Plant Interact 6:425–437

De Moraes CM, Lewis WJ, Pare PW, Tumlinson JH (1998) Herbivore infested plants selectively attract parasitoids. Nature 393:570–574

Deng W, Hamilton-Kemp TR, Nielsen MT, Andersen RA, Collins GB, Hildebrand DF (1993) Effects of six-carbon aldehydes and alcohols on bacterial proliferation. J Agric Food Chem 41:506–510

Dudareva, N., Pichersky, E. (2006) Metabolic engineering of floral scent of ornamentals. J Crop Improv (in press)

Dudareva N, Cseke L, Blanc VM, Pichersky E (1996) Evolution of floral scent in Clarkia: novel patterns of S-linalool synthase gene expression in the *C. breweri* flower. Plant Cell 8:1137–1148

Dudareva N, Dauria JC, Nam KH, Raguso RA, Pichersky E (1998) Acetyl-coa-benzylalcohol acetyltransferase—an enzyme involved in floral scent production in Clarkia breweri. Plant J 14:297–304

Dudareva N, Murfitt LM, Mann CJ, Gorenstein N, Kolosova N, Kish CM, Bonham C, Wood K (2000) Developmental regulation of methyl benzoate biosynthesis and emission in snapdragon flowers. Plant Cell 12:949–961

Dudareva N, Pichersky E, Gershenzon J (2004) Biochemistry of plant volatiles. Plant Physiol 135:1893–1902

Eisenreich W, Schwarz M, Cartayrade A, Arigoni D, Zenk MH, Bacher A (1998) The deoxy xylulose phosphate pathway of terpenoid biosynthesis in plants and microorganisms. Chem Biol 5:R221–R233

Feussner I, Wasternack C (2002) The lipoxygenase pathway. Annu Rev Plant Biol 53:275–297
Friedman M, Henika PR, Mandrell RE (2002) Bactericidal activities of plant essential oils and some of their isolated constituents against *Campylobacter jejuni, Escherichia coli, Listeria monocytogenes, and Salmonella enterica*. J Food Protect 65:1545–1560
Gouinguene SP, Turlings TCJ (2002) The effect of abiotic factors on induced volatile emissions in corn plants. J Plant Physiol 129(3):1296–1307
Hammer KA, Carson CF, Riley TV (2003) Antifungal activity of the components of *Melaleuca alternifolia* (tea tree) oil. J Appl Microbiol 95:853–860
Hildebrand DF, Brown GC, Jackson DM, Hamilton-Kemp TR (1993) Effects of some leaf-emitted volatile compounds on aphid population increase. J Chem Ecol 19:1875–1887
Holopainen JK, Blande JD (2012) Molecular plant volatile communication, sensing in nature edited by Carlos Lopez-Larrea, 17–31
Humphreys JM, Chapple C (2002) Rewriting the lignin roadmap. Curr Opin Plant Biol 5:224–229
Joutsensaari J, Loivamaki M, Vuorine T, Miettinen P, Nerg AM, Holopainen JK, Laaksonen A (2005) Nanoparticle formation by ozonolysis of inducible plant volatiles. Atmos Chem Phys 5:11489–11495
Kessler A, Baldwin IT (2001) Defensive function of herbivore-induced plant volatile emissions in nature. Science 291:2142–2143
Kolosova N, Sherman D, Karlson D, Dudareva N (2001) Cellular and subcellular localization of S-adenosyl-L-methionine: benzoic acid carboxyl methyltransferase, the enzyme responsible for biosynthesis of the volatile ester methyl benzoate in snapdragon flowers. Plant Physiol 126:956–964
Kroll JH, Seinfeld JH (2008) Chemistry of secondary organic aerosol: formation and evolution of low-volatility organics in the atmosphere. Atmos Environ 42:3593–3624
Law JH, Regnier FE (1971) Pheromones. Ann Rev Biochem 40:533–548
Lerdau M, Slobodkin K (2002) Trace gas emissions and species-dependent ecosystem services. Trends Ecol Evol 17:309–312
Loreto F, Mannozzi M, Maris C, Nascetti P, Ferranti F, Pasqualini S (2001) Ozone quenching properties of isoprene and its antioxidant role in leaves. Plant Physiol 126:993–1000
Luft S, Curio E, Tacud B (2003) The use of olfaction in the foraging behaviour of the golden-mantled flying fox, *Pteropus pumilus*, and the greater musky fruit bat, *Ptenochirus jagori* (Megachiroptera : Pteropodidae). Naturwiss 90:84–87
Maeda T, Takabayashi J, Yano S et al (2000) Effects of light on the Tritrophic interaction between kidney bean plants, two-spotted spider mites and predatory mites, *Amblyseius womersleyi* (Acari: Phytoseiidae). Exp Appl Aracol 24(5):415–425
McCaskill D, Croteau R (1995) Monoterpene and sesquiterpene biosynthesis in glandular trichomes of peppermint (*Mentha piperita*) rely exclusively on plastid-derived isopentenyl diphosphate. Planta 197:49–56
McCloud ES, Baldwin IT (1998) Herbivore and caterpillar regurgitants amplify the wound-induced increases in jasmonic acid but not nicotine in *Nicotiana sylvestris*. Planta 203:430–435
McConkey ME, Gershenzon J, Croteau RB (2000) Developmental regulation of monoterpene biosynthesis in the glandular trichomes of peppermint. Plant Physiol 122:215–223
McFrederick Q. S., Fuentes J. D., Roulston T., Kathilankal J. C., Lerdau M. (2009). Effects of air pollution on biogenic volatiles and ecological interactions. Oecologia160: 411–420
Mercke P, Kappers IF, Verstappen FWA, Vorst O, Dicke M, Bouwmeester HJ (2004) Combined transcript and metabolite analysis reveals genes involved in spider mite induced volatile formation in cucumber plants. Plant Physiol 135:2012–2024
Mumm R, Dicke M (2010) Variation in natural plant products and the attraction of bodyguards involved in indirect plant defence. Can J Zool 88:628–667
Ozawa R, Arimura G, Takabayashi J, Shimoda T, Nishioka T (2000) Involvement of jasmonate-and salicylate related signalling pathways for the production of specific herbivore induced volatiles in plants. Plant Cell Physiol 41:391–398

Pare PW, Tumlinson JH (1997) De novo biosynthesis of volatiles induced by insect herbivory in cotton plants. Plant Physiol 114:1161–1167

Pare PW, Tumlinson JH (1999) Plant volatiles as defence against insect herbivores. Plant Physiol 121:325–331

Pare PW, Alborn HT, Tumlinson JH (1998) Concerted biosynthesis of an insect elicitor of plant volatiles. Proc Natl Acad Sci U S A 95:13971–13975

Perez AG, Rios JJ, Sanz C, Olias JM (1992) Aroma components and free amino acids in strawberry variety Chandler during ripening. J Agric Food Chem 40:2232–2235

Pichersky E, Gershenzon J (2002) The formation and function of plant volatiles: perfumes for pollinator attraction and defence. Curr Opin Plant Biol 5:237–243

Pinto DM, Blande JD, Souza SR, Nerg A, Holopainen JK (2010) Plant volatile organic compounds (VOCs) in ozone (O_3) polluted atmospheres: the ecological effects. J Chem Ecol 36:22–34

Plessl M, Heller W, Payer H, Elstner E, Habermeyer J, Heiser I (2005) Growth parameters and resistance against Drechslera teres of spring barley (*Hordeum vulgare* L. cv. Scarlett) grown at elevated ozone and carbon dioxide concentrations. Plant Biol 7:694–705

Pratt KA, Mielke LH, Shepson PB, Bryan AM, Steiner AL, Ortega J et al (2012) Contributions of individual reactive biogenic volatile organic compounds to organic nitrates above a mixed forest. Atmos Chem Phys 12:10125–10143

Price PW, Bouton CE, Gross P, McPheron BA, Thompson JN, Weis AE (1980) Interactions among 3trophic levels—influence of plants on interactions between insect herbivores and natural enemies. Ann Rev Ecol Syst 11:41–65

Raguso RA, Pichersky E (1995) Floral volatiles from *Clarkia breweri* and *C. concinna* (Onagraceae): recent evolution of floral scent and moth pollination. Plant Sys Evol 194:55–67

Reineccius G (2006) Flavor chemistry and technology. CRC Press, Boca Raton

Reinhard J, Srivivasan MV, Zhang S (2004) Scent-triggered navigation in honeybees. Nature 427:411

Rhoades DF (1983) Responses of alder and willow to attack by tent caterpillars and webworms: evidence for phenomenal sensitivity of willows. In: Hedin PA (ed) Plant resistance to insects, Symposium Series, vol 208. American Chemical Society, Washington, DC, pp 55–68

Ruther J, Kleier S (2005) Plant-plant signaling: ethylene synergizes volatile emission in *Zea mays* induced by exposure to(Z)-3-hexen-1-ol. J Chem Ecol 31:2217–2222

Sandermann H, Ernst D, Heller W, Langebartels C (1998) Ozone: an a biotic elicitor of plant defence reactions. Trends Plant Sci 3:47–50

Shah J (2003) The salicylic acid loop in plant defence. Curr Opin Plant Biol 6(4):365–371

Sharkey TD, Yeh S, Wiberley AE, Falbel TG, Gong DM, Fernandez DE (2005) Evolution of the isoprene biosynthetic pathway in Kudzu. Plant Physiol 137:700–712

Takabayashi J, Dicke M (1996) Plant–carnivore mutualism through herbivore-induced carnivore attractants. Trends Plant Sci 1:109–113

Takabayashi J, Dicke M, Posthumus MA (1994) Volatile herbivore-induced terpenoids in plant-mite interactions: variation caused by biotic and abiotic factors. J Chem Ecol 20:1329–1354

Turlings TCJ, McCall PJ, Alborn HT, Tumlinson JH (1993a) J Chem Ecol 19:411–425

Van Huijsduijnen RH, Alblas SW, De Rijk RH, Bol JF (1986) Induction by salicylic acid of pathogenesis-related proteins and resistance to alfalfa mosaic virus infection in various plant species. J Gen Virol 67:2135–2143

Vancanneyt G, Sanz C, Farmaki T, Paneque M, Ortego F, Castanera P, Sanchez-Serrano JJ (2001) Hydroperoxide lyase depletion in transgenic potato plants leads to an increase in aphid performance. Proc Natl Acad Sci USA 98:8139–8144

Venkateshan R (2015) Biosynthesis and regulation of herbivore induced plant volatile emission. J Indian Inst Sci 95(1):25–34

Vet LEM, Dicke M (1992) Ecology of infochemical use by natural enemies in a tritrophic context. Annu Rev Entomol 37:141–172

Vickers CE, Gershenzon J, Lerdau MT, Loreto F (2009) A unified mechanism of action for volatile isoprenoids in plant abiotic stress. Nat Chem Biol 5:283–291

Virtanen A, Joutsensaari J, Koop T, Kannosto J, Yli-Pirilä P, Leskinen J et al (2010) An amorphous solid state of biogenic secondary organic aerosol particles. Nature 467:824–827

War AR, Sharma HC, Paulraj MG, War MY, Ignacimuthu S (2011) Herbivore induced plant volatiles: their role in pant defense for pest management. J Plant Signal Behav 6(12):1973–1978

Wyllie SG, Leach DN (1992) Sulfur-containing compounds in the aroma volatiles of melons (*Cucumis melo*). J Agric Food Chem 40:253–256

Wyllie SG, Leach DN, Wang Y, Shewfelt RL (1995) Key aroma compounds in melons: their development and cultivar dependence. In: Rouseff RL, Leachy MN (eds) Fruit flavors: biogenesis, characterization, and authentication. American Chemical Society, Washington, pp 248–257

Bacterial Volatiles in Promotion of Plant Under Biotic Stress

16

Shekhar Jain, Ajit Varma, Narendra Tuteja, and Devendra Kumar Choudhary

Abstract

Confined to a narrow region but rich in nutrients, rhizosphere is always favorable to interactions between microorganisms and plants. While many soil microorganisms have no obvious effects on plants, some have negative as well as positive effects on plant growth. Plant growth-promoting rhizobacteria (PGPR) are beneficial microbes which have been fruitfully applied in agriculture to enhance seedling emergence, plant weight, crop yield, and disease resistance. Among these, some PGPR strains mediate plant growth promotion in direct and/or indirect manner by releasing volatile organic compounds (VOCs). Bacteria emitted a wide array of volatiles ranging from inorganic such as hydrogen cyanide (HCN) and nitric oxide (NO) to organic such as hydrocarbon, ketone, acids, terpenes, etc. Bacterial VOCs promote plant growth by eliciting different signaling pathway and show correlation with plant growth hormones also. In particular, 2,3-butanediol and acetoin were reported for the heightened level of plant growth promotion and triggering induced resistance against fungal pathogens. This chapter focuses on recent research study and role of bacterial volatiles in plant growth promotion and protection against pathogens.

Keywords

Plant growth-promoting bacteria • Pathogen • Volatile organic compounds • Induced systemic resistance

---

S. Jain
Division of Plant Pathology, ICAR-Indian Agricultural Research Institute, New Delhi, India

A. Varma • N. Tuteja • D.K. Choudhary (✉)
Amity Institute of Microbial Technology (AIMT), Block 'E-3', 4th Floor, Amity University Campus, Sector-125, Noida 201303, Gautam Buddha Nagar, Uttar Pradesh, India
e-mail: dkchoudhary1@amity.edu

D.K. Choudhary et al. (eds.), *Volatiles and Food Security*,
DOI 10.1007/978-981-10-5553-9_16

16.1 Introduction

For the good health, healthy food is the main concern which is directly related to the present agroecosystem. Agricultural products are the primary dietary source of proteins, carbohydrates, fat, vitamins, and other nutrients. With the increasing population, the need of dietary sources is increasing day by day drastically that ultimately leads to use of different chemicals in the form of phytohormones and pesticides to promote plants growth and amelioration of pathogens, respectively, to get quick result. Presently, the primary concern for the cultivator is to cultivate disease-free plant and to gain high yield in any adverse condition. Plants act in response to attack by pathogens or herbivores by inducing resistance phenomena through extensive changes in the gene expression, resulting in expression of various traits de novo or at much higher intensities, which diminish or put off further tissue damage (Karban and Baldwin 1997). Such plant responses expressed systemically because pathogens and herbivores can spread from the preliminary site of attack to other organs. This systemic response gets regulated by long-distance signaling of three plant hormones, namely, jasmonic acid (JA), ethylene, and salicylic acid (SA). SA and JA are transported either themselves or in the form of derivatives within the plant to elicit systemic responses against pathogen (Jain et al. 2017; Wasternack 2007; Heil and Ton 2008). Belowground, around the roots, rhizosphere is a narrow region where root-secreted exudates in the form of secondary metabolites affect the growth of indigenous microflora. This area can also be considered as a battlefield where beneficial bacteria and deleterious microorganisms compete for the resources. Among the diverse microbial population of the rhizosphere, the beneficial root-associated bacteria, also referred to as plant growth-promoting rhizobacteria (PGPR), have been used to control plant diseases and enhance plant growth and yield. Such a disease controlling strategy is referred to as biological control in contrast to chemical control. Several mechanisms have been explained till now to elucidate PGPR-mediated plant growth stimulation. These mechanisms are broadly categorized as direct and indirect where phosphate solubilization and production of different plant hormones, i.e., indole acetic acid (IAA), gibberellic acid (GA), and cytokinin, directly enhance the plant growth, while biocontrol of plant pathogens and deleterious rhizosphere microorganisms indirectly promotes growth by avoiding growth-limiting conditions (Glick 1995; Idriss et al. 2002; Kloepper 1992). This indirect mechanism of suppression of harmful microorganisms is known as induced systemic resistance (ISR) (Kloepper 1992; Dobbelaere et al. 2003).

Research studies have revealed that signaling molecules involved in the long-distance signaling through the vascular system can also be volatile compounds that move in the headspace outside the plant (Heil and Ton 2008). Volatile organic compounds (VOCs) released from bacteria act as signaling molecules for other bacteria, animals, insects, plants, and microorganisms (Farag et al. 2013). Zoller and Clark first of all reported microbial VOCs. These signaling molecules have been reported to have considerable role in plant growth, ameliorating abiotic stress in plants and as inhibitor of fungal pathogens (Baily and Weisskopf 2012; Davis et al. 2013). Plant-plant communication has been reported in several taxonomically unrelated plants due to freely moving nature of VOCs in the air (Tscharntke et al. 2001; Engelberth et al. 2004; Heil and Kost 2006; Paschold et al. 2006; Heil and Bueno 2007; Ton et al. 2007; Godard et al. 2008). These studies show the role of VOCs in herbivore resistance via plant-plant communication phenomenon, and similar volatile compounds from beneficial bacteria can also mediate the beneficial effects (Ryu et al. 2003, 2004b). Volatile derivatives of two major signaling molecules SA and JA play a major role in plant defense. Methyl salicylate (MeSA), volatile derivative of SA, has been projected as the most probable systemic signal in the plant (Park et al. 2007). In tobacco, active resistance-inducing compounds were formed by SA after being converted back from MeSA that shows resistance induction in tobacco plants by higher exposure of MeSA (Kumar and Klessig 2003; Forouhar et al. 2005; Shulaev et al. 1997).

In context to the plant defense, PGPB-produced VOCs elicited plant growth promotion, and induced systemic resistance provides a new insight in plant growth-promoting bacteria (PGPB)–plant interaction (Chung et al. 2016; Weisskopf et al. 2016). Bacteria produce a wide range of VOCs that have been reported to play a crucial role in plant defense. Some of the most common VOCs included dodecane, 2-undecanone, 2-tridecanone, 2-tridecanol, tetramethylpyrazine 2, 3-butanediol, 3-hydroxy-2-butanone (acetoin), etc. Volatile compounds, 2, 3-butanediol and 3-hydroxy-2-butanone, were reported to be the most important one and found to be consistently released by the bacterial strain *B. subtilis* GB03 and *B. amyloliquefaciens* IN937a. *A. thaliana* plants primed with these strains have shown that significant resistance upon challenge inoculation of *Erwinia carotovora* subsp. *carotovora* SCC1 concluded the role of VOCs in elicitation of ISR. Furthermore, genetically modified *Bacillus* strain, lacking VOC-producing ability, was also used to justify the priming activity of such VOCs to induce resistance against diseases (Ryu et al. 2003). Besides *Bacillus*, several strains of *P. fluorescens* were also reported for the production of VOCs and shown more effectiveness in controlling root and seedling diseases (Landa et al. 2002).

16.2 Bacterial VOCs: Diversified in Nature

Bacterial VOCs are diversified in nature as bacteria that produce a wide variety of volatiles ranging from inorganic to organic compound with different subcategories. In addition to the role in communication, several bacterial volatiles have been reported for the antifungal activity, and most probably ketones have been proposed to be responsible for this effect. Bacterial volatiles were found effective against a wide range of fungi such as VOCs from *Streptomyces* that inhibit growth of entomopathogenic fungi while from *Staphylococcus pasteuri* inhibit mycorrhizal fungi. Furthermore, VOCs were also reported for the plant growth-promoting activity and protection against bacterial phytopathogens (Schulz and Dickschat 2007). Bacteria produce a wide array of inorganic and organic volatile compounds.

16.2.1 Inorganic Compounds

Different types of inorganic VOCs such as nitric oxide (NO), hydrogen sulfide (H_2S), ammonia, or hydrogen cyanide (HCN) have been reported to release from bacteria. Biosynthesis of HCN is catalyzed by HCN synthase, encoded by *hcnABC* gene that converts glycine into HCN and CO_2. Few bacterial species including *Pseudomonas*, *Chromobacterium*, and *Rhizobium* have been reported for the production of HCN, and it is produced under low-oxygen concentration at the end of the exponential phase. Another volatile ammonia production is catalyzed by aspartate ammonia lyase via conversion of aspartate to fumarate resulting in release of ammonia (Bernier et al. 2011). In addition to these, NO is produced mostly from L-arginine by nitric oxide synthases, while most H_2S-producing bacteria generate this gas through degradation of cysteine (Mattila and Thomas 2014).

16.2.2 Organic Compounds

Organic VOCs released from bacteria include different types of chemical classes such as fatty acid derivatives, acids, sulfur, and nitrogen-containing compounds and terpenes.

16.2.2.1 Fatty Acid Derivatives

These types of volatiles are most likely formed by alteration of products of the fatty acid biosynthetic pathway, for example, hydrocarbons, aliphatic alcohols, ketones, and other components. Short-chain hydrocarbons such as decane to tetradecane are rarely found in microbes, while longer hydrocarbons such as hexadecane reported particularly in abundant manner in cyanobacteria (Ladygina et al. 2006). Hydrocarbons and acids occur fewer often than ketones and alcohols. Acetoin (3-hydroxy-2-butanone), one of the most notable VOCs, belongs to ketone class and is derived from pyruvate fermentation under anaerobic conditions (Ryu et al. 2003), while another important compound 2,3-butanediol falls under short-chain alcohols and is produced by *Proteobacteria* and *Firmicutes* under low-oxygen conditions (Farag et al. 2013; Whiteson et al. 2014). Comparative to ketones and alcohols, organic acids are found less abundant in bacterial volatiles which include several short-chain fatty acids such as acetic, propionic, or butyric acids (Schulz and Dickschat 2007).

16.2.2.2 Sulfur Compounds

Being one of the important components of proteins and primary metabolism in plants and animal, sulfur is the most important macronutrient after nitrogen, phosphorus, and potassium, and its deficiency affects plant and animal health. Due to sulfur deficiency, chlorophyll content and synthesis of ribulose-1,5-bisphosphate carboxylase/oxygenase are reduced that restricts CO_2 assimilation and leads to suppression of the photosynthetic machinery (Burke et al. 1986; Gilbert et al. 1997; Aziz et al. 2016). The biogenesis of S-containing compounds, such as methanethiol, dimethyl disulfide (DMDS), dimethyl sulfide, or dimethyl trisulfide and 1-(methyl thio)-3-pentanone, is often mediated by bacteria. Meldau et al. (2013) reported the role of VOC DMDS emitted by B55 in plant growth promotion by priming the *N. attenuata 35S-etr1*plant which is impaired in SO_4^{-2} uptake, with B55, and found significant promotion in seedling growth.

16.2.2.3 Terpenes

Terpenes are derived via mevalonate pathway or deoxyxylulose phosphate pathway from the terpene building units dimethylallyl pyrophosphate and isopentenyl pyrophosphate. Only monoterpenes (C_{10}), sesquiterpenes (C_{15}), and their derivatives or degradation products have been reported from bacterial volatile blends. Some of the notable terpene volatiles from bacteria are earthy odorant geosmin and antibiotic albaflavenone out of which the former is reported in several bacterial species, while the latter one is exclusively found in *Streptomyces* (Schulz and Dickschat 2007).

16.2.2.4 Nitrogen-Containing Compounds

Bacteria released several nitrogen-containing VOCs such as trimethylamine (TMA), 2-amino-acetophenone (2-AA), indole, etc. Among these, indole is reported to be produced by most of the PGPB and play an important role in plant growth. In *Escherichia coli*, tryptophanase enzyme catalyzed the biosynthesis of indole as it converts substrate tryptophan into indole, pyruvate, and ammonia (Lee and Lee 2010). Another VOC, 2-AA, can be used for the diagnosis of *P. aeruginosa* infections as it is aromatic in nature and responsible for the grape-like aroma of *P. aeruginosa* (Que et al. 2013; Schulz and Dickschat 2007).

16.3 Bacterial VOCs: Plant Growth Booster

Arabidopsis thaliana, the model of the plant world, has been used for the most of VOC-related studies using I-plate methods (divided plates), where compartments are separated by a plastic border. This facilitates exchange of volatile compounds without any physical diffusion of nonvolatile metabolites. Apart from that, research studies have also been done on some other plants. Initially Ryu et al. (2003) revealed that volatiles released from specific bacterial strains of PGPR have positive role on the growth of *Arabidopsis thaliana* seedlings. PGPR consist of a broad range of beneficial root-colonizing bacteria which can promote plant growth by means of enhancing different plant growth-promoting traits such as seed emergence, plant weight, and crop yields (Kloepper 1992; Ciancio et al. 2016). PGPR have been widely used for seed or seedling treatment to increase the growth of several crops and to restrain the growth of plant pathogens and harmful rhizosphere microorganisms (Glick 1995). Apart from enhancing plant growth, certain PGPR provide protection against pathogen also by means of triggering ISR in plant through VOCs. Several in vitro condition experiments have shown role of VOCs emitted by different genera of PGPR strains in the promotion of plant growth and elicitation of ISR against phytopathogens (Ryu et al. 2003, 2004a, b, c; Li et al. 2016). Two VOCs

2,3-butanediol and its precursor 3-hydroxy-2-butanone (acetoin) released from the strains GB03 and IN937a have been reported for the plant growth promotion through I-plate method. Furthermore, *A. thaliana* plants were reported with dose-dependent stimulation in growth upon exogenous application of commercial acetoin and 2,3-butanediol, which suggest the effects of the volatile blend produced by the two *Bacillus* sp. Later on, the role of these compounds as plant growth-promoting volatile determinants was confirmed by applying 2,3-butanediol mutant strains of *B. subtilis*, that results in no effect on plants' growth (Ryu et al. 2003, 2004a). VOCs released by bacteria also play an important role in signaling of plant growth hormones by regulating expression of some pathway genes. Correlation between some hormones and VOCs has been described here.

16.3.1 Ethylene

Ethylene is a gaseous plant growth hormone in nature and plays an important role in the VOC-mediated bacteria–plant interaction, and loss of the positive regulator of the ethylene pathway EIN2 led to different growth behaviors in response to bacterial strain IN937a and GB03 VOCs (Stepanova and Alonso 2009). VOCs from GB03 were found to be unable to trigger biomass increase in mutant *ein2 (ethylene insensitive2)* (Guzm'n and Ecker 1990; Ryu et al. 2003), while in another study VOCs of IN937a failed to show significant change in the growth promotion effect in the ethylene-insensitive mutants *etr1* (*ethylene response1*) (Bleecker et al. 1988), *ein2*, and *eir1* (*ethylene-insensitive root1*) (Luschnig et al. 1998). It implies that the plant amends its response by means of a strain-specific signal transduction pathway. Further in transcriptomic study, a subset of ethylene biosynthesis (*ACO2*, *ACS4*, *ACS12*, and *SAM-2*) and ethylene response (*CHIB*, *ERF1*, and *GST1*) genes have been shown to respond to bacterial volatiles (Kwon et al. 2010). Different expression of four enzymes, namely, aspartate aminotransferase, aspartate semialdehyde dehydrogenase precursor, methionine adenosyltransferase (MAT3), and S-adenosylmethionine synthetase 2 (SAM-2), has been reported upon exposure to bacterial volatiles. Conclusively, it shows significant role of ethylene pathway in the growth-promoting effect of bacterial volatiles.

16.3.2 Cytokinins

Plant growth hormone cytokinins play a major role in cell division processes and control the leaf size, root and shoot meristem maintenance, and root architecture (Argueso et al. 2010). Earlier, Arkhipova et al. (2005) have described the relationship between microbial production of cytokinins and plant development. It is also reported that *B. megaterium* failed to promote plant growth in plants devoid of the histidine kinase cytokinin receptors and cytokinin receptor-deficient (CRE1)/AHK4, AHK2, and AHK3. Ryu et al. (2003) have also proven cytokinin role in mediation of PGPR signals as *Arabidopsis* mutant's *cre1* and *ein2* were found insensitive to GB03 volatiles. Though no recognition of cytokinin-related genes was reported in the succeeding microarray or proteomics studies (Kwon et al. 2010). Taken together, it shows that cytokinins are crucial for the root development and physiology; hence the importance of cytokinin pathway with respect to change in plant growth mediated by bacterial volatiles appears to be worthy for deep investigation.

16.3.3 Abscisic Acid

The abscisic acid (ABA) originally known to be involved in abscission and bud dormancy also plays an important role in plant responses to environmental stresses and pathogens. Its signaling pathway overlaps widely with sugar sensing *in planta* (Rolland et al. 2006). Sugars, being a product of photosynthesis, play an important role in plant growth and development. In this context, increased level of sugar concentration via increase in photosynthetic activity and chlorophyll content was reported in the *Arabidopsis* seedlings treated with GB03 volatiles that conclude role of VOCs in photosynthesis (Zhang et al. 2008). In another study, although increased photosystem efficiency was observed in *Arabidopsis* lines *gin1* and *gin2*, impaired in hexokinase-dependent sugar sensing but found with no response to GB03 VOCs (Moore et al. 2003). These studies suggested that VOCs promote photosynthesis by means of triggering the repression of the hexokinase-dependent glucose signaling pathway. Further, Zhang et al. (2009) have also confirmed the VOCs' role in plant growth by showing reduced concentration of ABA contents in aerial parts of the plants exposed to bacterial VOCs in comparison to control plants. It was also observed that GB03 volatiles increase photosynthetic capacity in the plants via acidification of the rhizosphere and consequent increase in iron uptake.

16.3.4 Auxin

Auxin is one of the main plant growth hormones found in different natural forms. It conducts a complex signaling pathway that involves the spatial and temporal coordination of auxin synthesis, transport, and perception. Plant growth-promoting rhizobacteria have been reported to synthesize IAA to trigger plant growth promotion (Kim et al. 2011; Kochar et al. 2011). Zhang et al. (2007) have shown in a transcriptome study that genes associated with auxin synthesis and response showed differential regulation upon exposure of *Arabidopsis* to volatiles of GB03. Müller et al. (1998) have reported upregulation of enzymes, namely, tryptophan synthase, an anthranilate synthase, and three nitrilases (two of these expressed specifically in the aerial tissues of the plant), involved in the tryptophan-dependent IAA biosynthesis pathway. Zhang et al. (2007) further supported this finding by using auxin transport inhibitor naphthylphthalamic acid (NPA) that results in abolishment of this response.

Differential expression of expansin genes associated with cell wall remodeling and cell expansion was also reported in the *Arabidopsis* leaves exposed to bacterial

VOCs (Zhang et al. 2007). Similar upregulation has been reported for *Lactuca sativa EXPA5* and *Nicotiana tabacum EXP2* and *EXP6* (Wang et al. 2009; Minerdi et al. 2011). Conclusively, these studies suggested the role of IAA biosynthesis, its transportation and local tissue concentration in the enhancement of aerial part of the plants, and hence its superlative role for the auxin machinery in the VOC-mediated plant growth promotion.

16.4 Bacterial VOCs: Invisible Weapons Against Pathogens

Bacterial VOCs work as bacterial determinants that play an important role in plant protection against invading pathogens. Volatiles 2,3-butanediol and phenazines are known to possess antimicrobial activity and often been considered harmful to a wide variety of biological systems. Priming the defense pathways by external elicitors allows the potentiated induction of defense response while not forthwith activating the defense signal cascades, which might be accompanied by the outflow of energy for defense mobilization (Paré et al. 2005). PGPR priming is not directly associated with direct changes in gene expression in leaves, but it is assumed that elicitation ends in a rise within the quantity or activity of cellular elements that play vital roles in defense signaling (Lee et al. 2012). Transcriptional expression of the marker genes *PR1*, *ChiB*, and *VSP2* involved in salicylic acid, jasmonic acid, and ethylene signaling, respectively, was observed in the plants exposed to long-chain VOC tridecane, released from *P. polymyxa* against the biotrophic pathogen *P. syringae* pv. *tomato*. In another study, Han et al. (2006) reported that volatile 2,3-butanediol and acetoin from *B. subtilis* elicited a stronger ISR against *P. carotovorum* subsp. *carotovorum*, whereas direct application of 2,3-butanediol failed to trigger ISR against *P. syringae* pv. *tabaci*. In another study, Sharifi and Ryu (2016) found increased resistance in *Arabidopsis* plants against the necrotrophic fungus *B. cinerea* via ISR elicitation through bacterial VOCs exposure and suggested major role of volatiles in biocontrol against fungal phytopathogens. Recently, Tahir et al. (2017) have reported induced protection in tobacco against bacterial wilt-causing pathogen *Ralstonia solanacearum* upon priming with *Bacillus* whose volatiles adversely affect the physiology and ultrastructure of the pathogens. This study suggests that different defensive cascades are elicited in response to different pathogens.

Furthermore, transcriptomic and proteomic analysis of *A. thaliana* exposed to bacterial volatiles revealed that three major plant defense signaling pathways, including salicylic acid, jasmonic acid, and ethylene, mediate these effects (Kwon et al. 2010; Zhang et al. 2010). This study has shown that upon treatment with GB03, significant upregulation of VOCs was observed in the transcript of five ET biosynthesis-related genes, namely, *MAT3*, *SAM-2*, *ACS4*, *ACS12*, and *ACO2*, while strong induction was also found in the ET marker genes *ERF1*, *GST2*, and *CHIB*. Another proteome study has shown the independency of ISR elicited by *B. subtilis* FB17 to JA pathway but dependency on SA and ET signaling pathways against *P. syringae* pv. tomato DC3000.

Ethylene-insensitive plants treated with 2,3-butanediol did not elicit ISR, suggesting that ethylene is a major player in this interaction (Farag 2014). Different signaling pathways are responsible for the VOC-mediated induction of ISR. Mutants and transgenic plant lines of plants exposed to volatiles have been used to reveal the pathway (Han et al. 2006; Ramos et al. 2000; Ryu et al. 2003). Comparative study of the signal transduction pathways potentiated by individual VOCs in the presence or absence of exogenous priming agents offers more insight into the roles of elicitors and priming agents in eliciting plant defense responses and in increasing long-term protection of plants (Lee et al. 2012). Different VOCs produced by bacteria may help plants to change defense signaling pathways resulting in protection them against various types of pathogens, i.e., necrotrophs and biotrophs. Ryu et al. (2004a) reported independency of ISR triggered by GB03 emissions to the salicylic acid, NPR1, and jasmonic acid signaling pathways although this response seems to be mediated via ethylene. On the other hand, independency of ISR triggered by strain IN937a to all the signaling pathways tested shows the possibility that additional VOCs utilize alternative pathways *in planta.* In fact, the ethylene (ET)-dependent plant defense signaling pathway is more effective against a necrotrophic pathogen such as *P. carotovorum* subsp. *carotovorum* than against *P. syringae*, which requires an SA-dependent resistance response (Pieterse et al. 2009). Failure to elicit ISR by C4-alcohol biosynthetic pathway mutants stain of *Bacillus subtilis* confirms the role of acetoin and 2,3-butanediol as ISR volatile determinants. Significant upregulation in the transcript profile of *PDF1.2* gene was also observed upon priming with *B. subtilis* strain GB03 and strain FB17 (Ryu et al. 2004a). However, how plants modulate and recognize PGPR VOC-elicited defense straight away after pathogen infection remains to be fully elucidated.

16.5 Conclusion

With the diverse and rapidly evolving pathogens and global climate changes, adverse effects have been increasing in food security by reduction in the world crop yield. Although day-by-day increased application of pesticides and chemical fertilizers provides instant solutions for the plant disease and crop yield problems, respectively, it ultimately leads to affect human and environment health in negative manner. A wide array of research study on bacterial VOC–plant interactions has led to an ever more conceptual understanding of the nature of bacterial VOCs and their potential role in enhancing plant protection and productivity in a sustainable manner. Significant modulation of plant metabolomics, physiology, and transcriptional status upon exposing to bacterial VOCs shows the plants' ability to recognize and react to bacterial VOCs. These studies evidently revealed the necessity for implementation of bacterial VOCs in open-field conditions and stress their multiple roles to extend pathogen resistance, protection against herbivores, and in general as biocontrol agents. Bacterial VOC–plant interaction studies opened up a new era of bacterial VOC application for a sustainable development of agriculture with respect to crop protection and production strategies as a possible replacement for hazardous pesticides and chemical fertilizers. Although bacterial VOCs are equivalent to biopesticides or biofertilizers, effective use of bacterial VOCs still, however, remains a big challenge.

References

Argueso CT, Raines T, Kieber JJ (2010) Cytokinin signaling and transcriptional networks. Curr Opin Plant Biol 13:533–539

Arkhipova TN, Veselov SU, Melentiev AI, Martynenko EV, Kudoyarova GR (2005) Ability of bacterium *Bacillus subtilis* to produce cytokinins and to influence the growth and endogenous hormone content of lettuce plants. Plant Soil 272:201–209

Aziz M, Nadipalli RK, Xie X, Sun Y, Surowiec K, Zhang JL, Paré PW (2016) Augmenting sulfur metabolism and herbivore defense in Arabidopsis by bacterial volatile signaling. Front Plant Sci 7:458

Bailly A, Weisskopf L (2012) The modulating effect of bacterial volatiles on plant growth: current knowledge and future challenges. Plant Signal Behav 7:79–85

Bernier SP, Létoffé S, Delepierre M, Ghigo JM (2011) Biogenic ammonia modifies antibiotic resistance at a distance in physically separated bacteria. Mole Microbiol 81(3):705–716

Bleecker AB, Estelle MA, Somerville C, Kende H (1988) Insensitivity to ethylene conferred by a dominant mutation in *Arabidopsis thaliana*. Science 241:1086–1089

Burke JJ, Holloway P, Dalling MJ (1986) The effect of sulfur deficiency on the organization and photosynthetic capability of wheat leaves. J Plant Physiol 125:371–375

Chung JH, Song GC, Ryu CM (2016) Sweet scents from good bacteria: case studies on bacterial volatile compounds for plant growth and immunity. Plant Mole Biol 90(6):677–687

Ciancio A, Pieterse CM, Mercado-Blanco J (2016) Editorial: harnessing useful rhizosphere microorganisms for pathogen and pest biocontrol. Front Microbiol 7:2031

Davis TS, Crippen TL, Hofstetter RW et al (2013) Microbial volatile emissions as insect semiochemicals. J Chem Ecol 39:840–859

Dobbelaere S, Vanderleyden J, Okon Y (2003) Plant growth-promoting effects of diazotrophs in the rhizosphere. Crit Rev Plant Sci 22(2):107–149

Engelberth J, Alborn HT, Schmelz EA, Tumlinson JH (2004) Airborne signals prime plants against insect herbivore attack. Proceed Nat Acad Sci USA 101(6):1781–1785

Farag MA (2014) Comparative mass spectrometry and nuclear magnetic resonance metabolomic approaches for nutraceuticals quality control analysis: a brief review. Recent Pat Biotechnol 8(1):17–24

Farag MA, Zhang H, Ryu CM (2013) Dynamic chemical communication between plants and bacteria through airborne signals: induced resistance by bacterial volatiles. J Chem Ecol 39(7):1007–1018

Forouhar F, Yang Y, Kumar D, Chen Y, Fridman E, Park SW, Chiang Y, Acton TB, Montelione GT, Pichersky E, Klessig DF (2005) Structural and biochemical studies identify tobacco SABP2 as a methyl salicylate esterase and implicate it in plant innate immunity. Proceed Nat Acad Sci USA 102(5):1773–1778

Gilbert SM, Clarkson DT, Cambridge M, Lambers H, Hawkesford MJ (1997) SO_4^{2-} deprivation has an early effect on the content of ribulose-1,5-bisphosphate carboxylase/oxygenase and photosynthesis in young leaves of wheat. Plant Physiol 115:1231–1239

Glick BR (1995) The enhancement of plant growth by free-living bacteria. Can J Microiol 41:109–117

Godard KA, White R, Bohlmann J (2008) Monoterpene-induced molecular responses in *Arabidopsis thaliana*. Phytochemistry 69(9):1838–1849

Guzm'n P, Ecker JR (1990) Exploiting the triple response of *Arabidopsis* to identify ethylene-related mutants. Plant Cell 2:513–523

Han SH, Lee SJ, Moon JH, Park KH, Yang KY, Cho BH, Kim KY, Kim YW, Lee MC, Anderson AJ, Kim YC (2006) GacS-dependent production of 2R, 3R-butanediol by *Pseudomonas chlororaphis* O6 is a major determinant for eliciting systemic resistance against *Erwinia carotovora* but not against *Pseudomonas syringae* pv. tabaci in tobacco. Mol Plant-Microbe Interact 19:924–930

Heil M, Bueno JC (2007) Within-plant signaling by volatiles leads to induction and priming of an indirect plant defense in nature. Proceed Nat Acad Sci 104(13):5467–5472

Heil M, Kost C (2006) Priming of indirect defences. Ecol Lett 9(7):813–817

Heil M, Ton J (2008) Long-distance signalling in plant defence. Trends Plant Sci 13:264–227

Idriss EE, Makarewicz O, Farouk A, Rosner K, Greiner R, Bochow H, Richter T, Borriss R (2002) Extracellular phytase activity of *Bacillus amyloliquefaciens* FZB45 contributes to its plant-growth-promoting effect. Microbiology 148(7):2097–2109

Jain S, Varma A, Tuteja N, Choudhary DK (2017) Bacteria-mediated elicitation of induced resistance in plants upon fungal phytopathogen. In: Choudhary DK, Varma A, Tuteja N (eds) Plant-microbe interaction: an approach to sustainable agriculture. Springer, Singapore, pp 249–269

Karban R, Baldwin IT (1997) Induced responses to herbivory. University of Chicago Press, Chicago

Kim YC, Leveau J, McSpadden Gardener BB, Pierson EA, Pierson LS, Ryu CM (2011) The multifactorial basis for plant health promotion by plant-associated bacteria. Appl Environ Microbiol 77:1548–1555

Kloepper JW (1992) Plant growth-promoting rhizobacteria as biological control agents. In: Metting FB Jr (ed) Soil microbial ecology: applications in agricultural and environmental management. Marcel Dekker, New York, pp 255–274

Kochar M, Upadhyay A, Srivastava S (2011) Indole-3-acetic acid biosynthesis in the biocontrol strain *Pseudomonas fluorescens* Psd and plant growth regulation by hormone overexpression. Res Microbiol 162:426–435

Kumar D, Klessig DF (2003) High-affinity salicylic acid-binding protein 2 is required for plant innate immunity and has salicylic acid-stimulated lipase activity. Proceed Nat Acad Sci 100(26):16101–16106

Kwon YS, Ryu CM, Lee S, Park HB, Han KS, Lee JH et al (2010) Proteome analysis of *Arabidopsis* seedlings exposed to bacterial volatiles. Planta 232:1355–1370

Ladygina N, Dedyukhina EG, Vainshtein MB (2006) A review on microbial synthesis of hydrocarbons. Process Biochem 41(5):1001–1014

Landa BB, Mavrodi OV, Raaijmakers JM, Gardene BBM, Thomashow LS, Weller DM (2002) Differential ability of genotypes of 2,4-Diacetylphloroglucinol-producing *Pseudomonas fluorescens* strains to colonize the roots of pea plants. Appl Environ Microbiol 68:3226–3237

Lee JH, Lee J (2010) Indole as an intercellular signal in microbial communities. FEMS Microbiol Rev 34(4):426–444

Lee B, Farag MA, Park HB, Kloepper JW, Lee SH, Ryu CM (2012) Induced resistance by a long-chain bacterial volatile: elicitation of plant systemic defense by a C13 volatile produced by *Paenibacillus polymyxa*. PLoS One 7:e48744

Li N, Alfiky A, Vaughan MM, Kang S (2016) Stop and smell the fungi: fungal volatile metabolites are overlooked signals involved in fungal interaction with plants. Fungal Biol Rev 30:134–144

Luschnig C, Gaxiola RA, Grisafi P, Fink GR (1998) EIR1, a root-specific protein involved in auxin transport, is required for gravitropism in *Arabidopsis thaliana*. Genes Dev 12:2175–2187

Mattila JT, Thomas AC (2014) Nitric oxide synthase: non-canonical expression patterns. Front Immunol 5:478

Meldau DG, Meldau S, Hoang LH, Underberg S, Wünsche H, Baldwin IT (2013) Dimethyl disulfide produced by the naturally associated bacterium *Bacillus* sp B55 promotes *Nicotiana attenuata* growth by enhancing sulfur nutrition. Plant Cell 25(7):2731–2747

Minerdi D, Bossi S, Maffei ME, Gullino ML, Garibaldi A (2011) *Fusarium oxysporum* and its bacterial consortium promote lettuce growth and expansin A5 gene expression through microbial volatile organic compound (MVOC) emission. FEMS Microbiol Ecol 76:342–351

Moore B, Zhou L, Rolland F, Hall Q, Cheng W-H, Liu Y-X et al (2003) Role of the *Arabidopsis* glucose sensor HXK1 in nutrient, light, and hormonal signaling. Science 300:332–336

Müller A, Hillebrand H, Weiler EW (1998) Indole-3-acetic acid is synthesized from L-tryptophan in roots of *Arabidopsis thaliana*. Planta 206:362–369

Paré PW, Farag MA, Zhang H, Ryu CM, Kloepper JW (2005) Elicitors and priming agents initiate plant defense responses. Photosynth Res 85:149–159

Park S-W, Kaimoyo E, Kumar D, Mosher S, Klessig DF (2007) Methyl salicylate is a critical mobile signal for plant systemic acquired resistance. Science 318:113–116

Paschold A, Halitschke R, Baldwin IT (2006) Using 'mute'plants to translate volatile signals. Plant J 45(2):275–291

Pieterse CM, Leon-Reyes A, Van der Ent S, Van Wees SC (2009) Networking by small-molecule hormones in plant immunity. Nat Chem Biol 5:308–316

Que YA, Hazan R, Strobel B, Maura D, He J, Kesarwani M, Panopoulos P, Tsurumi A, Giddey M, Wilhelmy J, Mindrinos MN (2013) A quorum sensing small volatile molecule promotes antibiotic tolerance in bacteria. PLoS One 8(12):e80140

Ramos HC, Hoffmann T, Marino M, Nedjari H, PresecanSiedel E, Dreesen O, Glaser P, Jahn D (2000) Fermentative metabolism of *Bacillus subtilis*: physiology and regulation of gene expression. J Bacteriol 182:3072–3080

Rolland F, Baena-Gonzalez E, Sheen J (2006) Sugar sensing and signaling in plants: conserved and novel mechanisms. Annu Rev Plant Biol 57:675–709

Ryu CM, Farag MA, Hu CH, Reddy MS, Wei HX, Paré PW, Kloepper JW (2003) Bacterial volatiles promote growth in *Arabidopsis*. Proc Natl Acad Sci U S A 100:4927–4932

Ryu CM, Farag MA, Hu CH, Reddy MS, Kloepper JW, Paré PW (2004a) Bacterial volatiles induce systemic resistance in *Arabidopsis*. Plant Physiol 134:1017–1026

Ryu CM, Murphy JF, Mysore KS, Kloepper JW (2004b) Plant growth-promoting rhizobacteria protect systemically *Arabidopsis thaliana* against *Cucumber mosaic virus* by a salicylic acid and NPR1-independent and jasmonic acid dependent signaling pathway. Plant J 39:381–392

Ryu CM, Hu CH, Locy RD, Kloepper JW (2004c) Study of mechanisms for plant growth promotion elicited by rhizobacteria in *Arabidopsis thaliana*. Plant Soil 268:285–292

Schulz S, Dickschat JS (2007) Bacterial volatiles: the smell of small organisms. Nat Prod Rep 24(4):814–842

Sharifi R, Ryu CM (2016) Are bacterial volatile compounds poisonous odors to a fungal pathogen *Botrytis cinerea*, alarm signals to Arabidopsis seedlings for eliciting induced resistance, or both? Front Microbiol 7:196

Shulaev V, Silverman P, Raskin I (1997) Airborne signalling by methyl salicylate in plant pathogen resistance. Nature 385:718–721

Stepanova AN, Alonso JM (2009) Ethylene signaling and response: where different regulatory modules meet. Curr Opin Plant Biol 12:548–555

Tahir HA, Gu Q, Wu H, Niu Y, Huo R, Gao X (2017) *Bacillus* volatiles adversely affect the physiology and ultra-structure of *Ralstonia solanacearum* and induce systemic resistance in tobacco against bacterial wilt. Scie Rep 7:40481

Ton J, D'alessandro M, Jourdie V, Jakab G, Karlen D, Held M, Mauch-Mani B, Turlings TC (2007) Priming by airborne signals boosts direct and indirect resistance in maize. Plant J 49(1):16–26

Tscharntke T, Thiessen S, Dolch R, Boland W (2001) Herbivory, induced resistance, and interplant signal transfer in *Alnus glutinosa*. Biochem Systemat Ecol 29(10):1025–1047

Wang S, Wu H, Qiao J, Ma L, Liu J, Xia Y et al (2009) Molecular mechanism of plant growth promotion and induced systemic resistance to tobacco mosaic virus by *Bacillus* spp. J Microbiol Biotechnol 19:1250–1258

Wasternack C (2007) Jasmonates: an update on biosynthesis, signal transduction and action in plant stress response, growth and development. Ann Bot (Lond) 100:681–697

Weisskopf L, Ryu CM, Raaijmakers JM, Garbeva P (2016) Editorial: smelly fumes: volatile-mediated communication between bacteria and other organisms. Front Microbiol 7:2031

Whiteson KL, Meinardi S, Lim YW, Schmieder R, Maughan H, Quinn R, Blake DR, Conrad D, Rohwer F (2014) Breath gas metabolites and bacterial metagenomes from cystic fibrosis airways indicate active pH neutral 2, 3-butanedione fermentation. ISME J 8(6):1247–1245

Zhang H, Kim MS, Krishnamachari V, Payton P, Sun Y, Grimson M et al (2007) Rhizobacterial volatile emissions regulate auxin homeostasis and cell expansion in *Arabidopsis*. Planta 226:839–851

Zhang H, Xie X, Kim MS, Kornyeyev DA, Holaday S, Pare PW (2008) Soil bacteria augment *Arabidopsis* photosynthesis by decreasing glucose sensing and abscisic acid levels in planta. Plant J 56:264–273

Zhang H, Sun Y, Xie X, Kim MS, Dowd SE, Pare PW (2009) A soil bacterium regulates plant acquisition of iron via deficiency-inducible mechanisms. Plant J 58:568–577

Zhang H, Murzello C, Sun Y, Kim MS, Xie X, Jeter RM, Zak JC, Dowd SE, Paré PW (2010) Choline and osmotic-stress tolerance induced in *Arabidopsis* by the soil microbe *Bacillus subtilis* (GB03). Mol Plant-Microbe Interact 23:1097–1104

Fungal Volatile Organic Compounds: Emphasis on Their Plant Growth-Promoting

17

Shafiquzzaman Siddiquee

Abstract

Fungal volatile organic compounds (VOCs) commonly formed bioactive interface between plants and countless of microorganisms on the above- and below-ground plant–fungus interactions. Fungal–plant interactions symbolize intriguingly biochemical complex and challenging scenarios that are discovered by metabolomic approaches. Remarkably secondary metabolites (SMs) played a significant role in the virulence and existence with plant–fungal pathogen interaction; only 25% of the fungal gene clusters have been functionally identified, even though these numbers are too low as compared with plant secondary metabolites. The current insights on fungal VOCs are conducted under lab environments and to apply small numbers of microbes; its molecules have significant effects on growth, development, and defense system of plants. Many fungal VOCs supported dynamic processes, leading to countless interactions between plants, antagonists, and mutualistic symbionts. The fundamental role of fungal VOCs at field level is required for better understanding, so more studies will offer further constructive scientific evidences that can show the cost-effectiveness of ecofriendly and ecologically produced fungal VOCs for crop welfare.

Keywords

Fungal volatile organic compounds • Peptaibols • Diketopiperazine-like compounds • Polyketides • Pyrones • Terpenes • Nonribosomal peptides • Plant hormones

---

S. Siddiquee (✉)
Biotechnology Research Institute, Universiti Malaysia Sabah, Jln UMS, 88400 Kota Kinabalu, Sabah, Malaysia
e-mail: shafiqpab@ums.edu.my

D.K. Choudhary et al. (eds.), *Volatiles and Food Security*,
DOI 10.1007/978-981-10-5553-9_17

17.1 Introduction

Fungal volatile organic compounds (VOCs) mostly occurred as multifaceted concoction with low-molecular-mass compounds resulting from diverse biosynthetic pathways. VOCs are commonly carbon-based solids and liquids that enter the gas phase by vaporizing at 0.01 kPa at temperature of approximately 20 °C (Pagans et al. 2006). Fungi enable to produce carbon-based compounds from several blends of gas phase, so it is called volatile organic compounds. Fungi have countless number of VOCs, which comprise of low-molecular-mass compounds, but still now their actual roles have not yet been identified (Perez-Nadales et al. 2014; Scharf et al. 2014). An approximately 479 of VOCs have been identified from only *Trichoderma* fungi such as hydrocarbons; aldehydes; ketones; heterocycles; alcohols; thioalcohols; phenols; thioesters and their derivatives, including benzene derivatives; and cyclohexanes (Siddiquee 2014; Korpi et al. 2009; Ortiz-Castro et al. 2009). Fungal VOCs are found respectively from primary and secondary metabolism pathways (Korpi et al. 2009). VOCs can diffuse through the soil and atmosphere; they have model as "infochemicals" as shown in Table 17.1.

Fungi VOCs have different chemical classes such as peptaibols, gliotoxin, gliovirin, polyketides, pyrenes, terpenoids, terpenes, derived compounds of shikimic acid, and nonribosomal peptides. Considering the structures of *Trichoderma*, antibiotic molecules have been presented in nature mainly in two types (Sivasithamparam and Ghisalberti 1998; Szekeres et al. 2005; Reino et al. 2008): (1) low-mass volatile compounds (VCs) (nonpolar compounds) such as simple aromatic compounds, specific polyketides like pyrones and butenolides, terpenes, and isocyanate, and (2) high-mass compounds (polar compounds), like peptaibols and diketopiperazine such as gliovirin and gliotoxin compounds. Quantification of the exact number of VCs produced by fungi is not so easy task since they can produce more than 1000 compounds, depending on the effective strains, environment conditions, and the detection methods.

Additionally, hybrid compounds are composed of moieties from common classes as in the meroterpenoids, which synthesizes between terpenes and polyketides. In the analysis of fungal genomes, more secondary metabolism genes of ascomycetes are found compared to basidiomycetes, archeo-ascomycetes, and chytridiomycetes; none of the secondary metabolism genes have been found in hemi-ascomycetes and zygomycetes (Collemare et al. 2008). Genome code of ascomycete on the average was found to be 16 for polyketide synthases (PKS), 10 for nonribosomal protein synthases (NRPS), 2 for tryptophan synthetases (TS), and 2 for dimethylallyl tryptophan synthetases (DMATS). In ascomycetes only PKS–NRPSs was found, on the average of three genes per species. *Neurospora crassa*, *Coccidioides* spp., and *Histoplasma capsulatum* have been found, respectively, on 1–9 genes for PKSs, 3–6 genes for NRPS, and 0–2 genes for PKS–NRPSs than other ascomycetes. High numbers of fungal species have been found in more than 40 genes encoding NRPS, PKS, TS, hybrids, and DMATS, for example, *Magnaporthe grisea* was found on 45 genes (Collemare et al. 2008).

Table 17.1 Structure, function, and odors of selected volatile compounds produced from fungi

Molecule	Structure	Potential Function(s); odors
1-Octen-3-ol	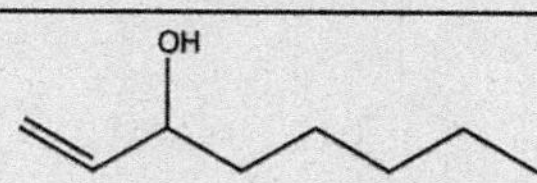	Semiochemical; earthy, "mushroomy" odor
1-Butanol-3-, methyl-, acetate	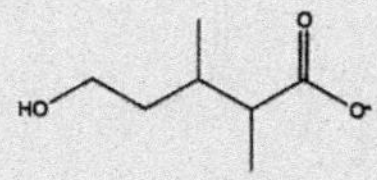	Antifungal; banana odor
Sabinene	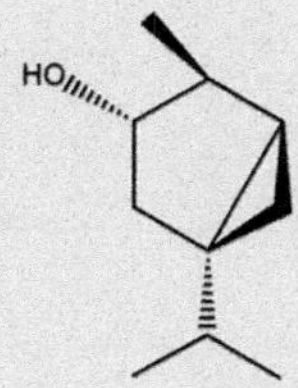	Unknown; peppery odor
6-Pentyl-α-pyrone	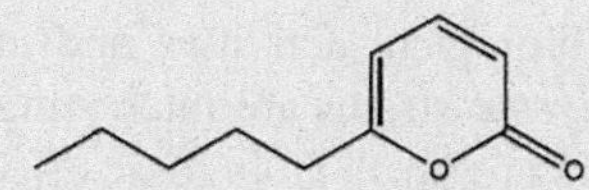	Antibiotic; coconut odor
β-Caryophyllene	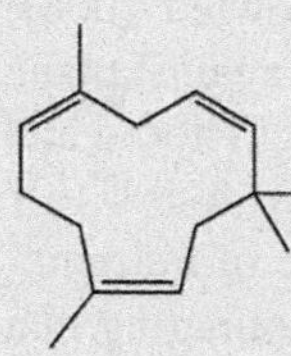	Plant-growth promoting; woody-spicy odor
Isobutyric acid	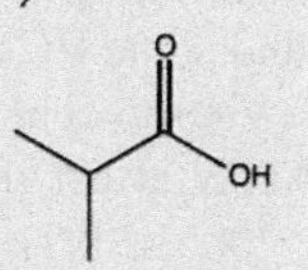	Antifungal; rancid cheese-like odor
Benzyl aldehyde	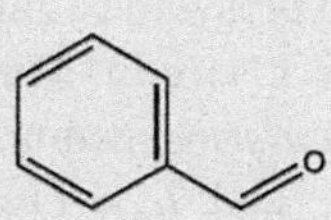	Anti-microbial; almond odor
1,8-Cineole	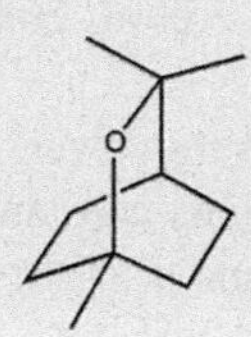	Antifungal; camphor-like odor

Table 1 – (continued)

Molecule	Structure	Potential Function(s); odors
2-Methyl-1-propanol	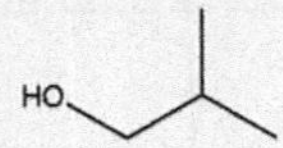	Fungivore attractant; mild alcohol odor
2-Heptanone	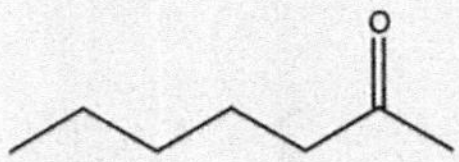	Unknown; cheese odor
3-Methyl-butanol	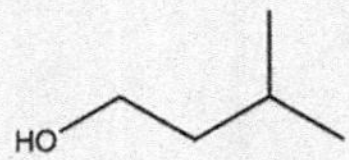	Unknown; component of truffle odor

Source: Morath et al. (2012)

Filamentous fungi of *Trichoderma* are prolific producers of VOCs; these are natural products that exhibited biological activities and also have a remarkable impact on society. Some *Trichoderma* strains are interesting to the pharmaceutical industry (antibiotics), while others are involved in toxic effects and disease interactions with plants or animals (Fox and Howlett 2008). Some strains have capability to decrease the severity of plant diseases against plant pathogens, through their high antagonistic and mycoparasitic activities (Viterbo and Horwitz 2010). Some strains of *Trichoderma* in the rhizosphere were found to have direct effects on plants, enhancing plant growth and high uptake of nutrient, fertilizer efficiency, and high percentage of seed germination and encouraging plant defenses against biotic and abiotic damage (Shoresh et al. 2010), and some VCs are produced by *Trichoderma* that have positive effects on the plants (Viterbo et al. 2007).

Heterotrimeric G-proteins and MAPKs affect the production of *Trichoderma* antifungal compounds (Reithner et al. 2005, 2007); the vel1 velvet gene has been implicated in the regulation of gliotoxin biosynthesis and other secondary metabolite (SM) genes, as well as conidiation in *T. virens* (Mukherjee and Kenerley 2010). Bioinformatics analysis of three *Trichoderma* genomes has distinguished that the mycoparasitic species of *T. atroviride* and *T. virens* are enriched in SM-related genes as compared to *T. reesei* (Martinez et al. 2008; Kubicek et al. 2011; Baker et al. 2012; Mukherjee et al. 2012b).

Some VOCs have distinguished odors; however, it is not astonishing that notice in VOCs began with the fungi that humans can smell, for example, the divergent odor of macrofungi such as mushrooms and truffles; different VOCs have high value in the culinary arts, of which aldehydes, terpenes, aromatics, alcohols, and thiols dominate (Tirillini et al. 2000; Splivallo et al. 2007; Cho et al. 2008; Fraatz and Zorn 2010). Moldy odors consisted eight carbon-based compounds like 1-octen-3-ol and 3-octanone (Morey et al. 1997). Chemical ecologists significantly characterized an important role of VOCs as semiochemicals that are applied as attractants and deterrents to insects and other invertebrates. Fungal VOCs have been applied as biocontrol strategies to inhibit plant pathogenic fungi in agriculture. In food industry, biocontrol approaches are also applied to inhibit postharvest fungal growth, termed as "myco-fumigation." Recently, fungal VOCs are referred to as "mycodiesel." This book chapter elaborated some important fungal VOCs with emphasis on their development of plant growth enhancement.

17.2 Peptaibols

Peptaibols are biosynthesized in the large family of antibiotic peptides by several fungi. These peptides comprised of approximately 7–20 amino acids that have high content of 2-aminoisobutyric acid (Aib), alkyl-N-terminal amino acid, and C-terminal hydroxyl-amino acid. The first compound of alamethicin F30 is isolated from *T. viride* (Brewer et al. 1987; Meyer and Reusser 1967). It is usually a mixture of at least 12 compounds. To date, the peptaibol database (http://www.cryst.bbk.ac.uk/peptaibol) found about 317 peptaibol structures and classified them into nine distinct subfamilies, and more than 190 peptaibols are found from the genus *Trichoderma/Hypocrea* (Neuhof et al. 2007; Degenkolb et al. 2008a).

Trichoderma peptaibols are classified into four subfamilies, namely, 1, 4, 5, and 9 (Fig. 17.1). Subfamily 1 is comprised mostly of well-known structures and included 18–20 amino acids in peptaibols. Eighteen-residue peptaibols of trichorzins HA, MA, and PA, trichovirins II, trichotoxin and trichokindins are isolated from

Subfamily / group	Sequence	Peptaibols
SUBFAMILY 1		
18 residues		
cs	AcUXXX(U,J)Q(J,U)XUXX UPLxUUQ(L,I,V)ol	
Group 18 (G/A pos 2)	AcUGXL U Q(J,U)XUXX UPL UUQ(L,V) ol	Trichotoxin, Trichovirins II
Group 19 (S pos 2)	AcUSAX U Q(J,U)XUXX UPL UUQ(I,V) ol	Hypomurocins, Trichokindins, Trichorzins MA
Group 21 (A pos 2)	AcUAUA UU Q U VUQ UPLxUUQ Lx ol	Trichostromaticins
Group 22 (W,F pos 18)	AcUSAU J Q U VUGL UPL UUQ W ol	Trichorzins PA
19 residues		
	AcUAUULxQUUUSLxUPVxUVxQQLxol	Trichostrigocin A
	AcUGFUU QUUUSLxUPVxUVxQQLxol	Tricholongin II
20 residues		
cs	AcU(P,A)UAU(A,U)QU(V,L)UG(L,U)UPVUU(G,Q)QFol	
Alamethicin type (P pos 2)	AcU P UAU A QU V UG L UPVUU G QFol	Alamethicin
Paracelsin type (A pos 2)	AcU A UAU A QU L UG U UPVUU Q QFol	Paracelsin B
SUBFAMILY 4		
11 residues		
cs	AcUN(Vx,Lx)(V,Lx)UPLxLxUPX ol	
	AcUN V V UPL L UPV ol	Trichovorins
14 residues		
cs	Ac(U,J,Vx)(N,Q)(Vx,Lx)UP(A,S,Vx,Lx)(Lx,Vx)UP(U,Vx,Lx)(Vx,Lx)UP(Vx,Lx)ol	
	Ac U (N,Q) L UP(A,S) (V,I) UP(J,U) L UP(L,V) ol	Harzianins
	Ac(U,J) (N,Q) L UP(A,S) V UP(J,U) L UP L ol	Trichovirins
SUBFAMILY 5 (lipopeptaibols)		
7 residues	Oc VxG LxUGLxLxol	Trichodecenins
10 residue	OcUGVxU VxLxUGLxLxol	Lipopubescins
11 residues	OcUGVxUGGVx UGLxLxol	Lipostrigocins
SUBFAMILY 9 (aminolipopeptides)		Trichopolyns

Fig. 17.1 Schematic diagram showed the potential *Trichoderma* peptaibols of subfamilies (SF) 1, 4, 5, and 9. Sequences are given in standard single-letter code (Ac, acetyl; U, Aib; J, Iva; Vx, Val/Iva; Lx, Leu/Ile; ol represents the C-terminal amino alcohol). Bold letters indicated those residues conserved within a given group. *cs* Consensus sequence (Source: These sequences have been summarized from Neuhof et al. (2007), Degenkolb et al. (2008a), and Mukherjee et al. (2011))

different potential strains of *T. harzianum*; respectively, trichotoxins from *T. asperellum*, hypomurocins B from *T. atroviridis*, and trichostromaticins from *T. stromaticum*. *T. strigosum* has produced 19-residue peptaibols like trikoningin type, trichostrigocin, and tricholongin (Degenkolb et al. 2006a; Neuhof et al. 2007). The paracelsin of 20-residue peptaibols was found from different *Trichoderma/Hypocrea* species in *T. longibrachiatum*, *T. citrinoviride*, *T. reesei/H. jecorina*, *T. strictipile*, and *T. pubescens*; longibrachins were found from *T. longibrachiatum* and *T. ghanense* and the alamethicin type from *H. atroviridis*, *T. hamatum*, and *T. brevicompactum*, respectively (Hermosa et al. 2014). Peptides of 11 or 14 amino acids consisted of subfamily 4 (SF4). Eleven-residue types of trichovorin peptaibols are diversely distributed in *T. viride*, harzianin HK VI isolated from *T. pseudokoningii*, trichorzins and harzianin HBI isolated from *T. harzianum/H. lixii*, hypomurocins A isolated from *H. atroviridis*, and trichofumin A and B isolated from *Trichoderma* species (Hermosa et al. 2014). The 14-residue types contain in trichovirins isolated from *T. viride* (Brückner and Koza 2003) and harzianins isolated from *T. harzianum/H. lixii* (Rebuffat et al. 1995). Neuhof et al. (2007) reported the low-molecular-weight SF4 type found from *T. croceum* and *T. hamatum*, an intermediate SF4 type found from *H. virens*, and high-mass SF4 found from *T. tomentosum*, *T. longipile*, *T. oblongisporum*, and *H. semiorbis*. Subfamily 5 (SF5) holds numerous lipopeptaibols, 7-, 10-, or 11-peptide residues, where the N-terminus is acylated by short fatty acid chain, i.e., octanoic acid. However trichogins, trikoningins, and trichodecenins are isolated from *T. longibrachiatum* (Auvin-Guette et al. 1992), *T. koningii* (Auvin-Guette et al. 1993), and *T. viride* (Fujita et al. 1994) respectively. Some lipostrigocins are isolated from *T. strigosum* and *T. pubescens* (Degenkolb et al. 2006b). Subfamily 9 obtains only a limited number of peptaibols such as trichopolyns cultured from *T. polysporum*. Peptaibols frequently exhibited antifungal, antibacterial, and anticancer properties (Degenkolb et al. 2008b; Sang and Blecha 2008; Pruksakorn et al. 2010).

Peptaibols usually synergistically allied with cell wall-degrading enzymes to prevent the growth of fungal pathogens (Schirmböck et al. 1994); the β-glucan isolated plasma membranes of *B. cinerea* that suppressed by the peptaibols trichorzianins TA and TB (Lorito et al. 1996), bioactivity of trichostromaticins A–E found in *T. stromaticum* strains toward the cacao disease (*Moniliophthora perniciosa*) (Samuels et al. 2000; Aime and Phillips-Mora 2005), and 20-residue peptaibols isolated from *T. citrinoviride* and applied antifungal activity toward forest tree pathogens (Maddau et al. 2009). Shi et al. (2012) reported that the trichokonin VI isolated from *T. pseudokoningii* cultures induces metacaspase-independent apoptosis in parallel with the accumulation of cytoplasmic vacuoles in the plant pathogen of *Fusarium oxysporum*. Peptaibols are involved in the biocontrol properties isolated from the effective *Trichoderma* strains.

Peptaibols elicited resistance to the plants pathogens. A 20-residue peptaibol of alamethicin from *T. viride* produced jasmonic acid (JA) and salicylic acid (SA) biosynthesis in lima bean (Engelberth et al. 2001) and induced native and long-distance electrical signals and defense in plants responses (Maischak et al. 2010; Rippa et al. 2010), while 18-residue peptaibols isolated from *T. virens* stimulated defenses in cucumber against *Pseudomonas syringae* pv. *lachrymans* (Viterbo et al. 2007). In 20-fold of alamethicin significantly applied indirect plant defense than JA in Brussels sprouts plants (Bruinsma et al. 2009).

The pbs1 gene of *T. atroviride* which encoded a long peptaibol synthetase has been cloned in silico and related to a 19-module enzyme able to produce 19-residue peptaibols (i.e., atroviridin) (Komon-Zelazowska et al. 2007). Two peptaibol synthetase genes such as Salps1 and Salps2 are partially cloned by *T. harzianum* CECT2413. Salps1 showed to be a pseudogene (Vizcaíno et al. 2005), and Salps2 contains a C-terminal end identical to Tex1 (Vizcaíno et al. 2006). Subsequently 11- and 18-residue peptaibols are isolated from *T. harzianum*.

17.3 Diketopiperazine-Like Compounds

Trichoderma species are nonribosomal peptide synthetases (NRPSs) that are involved in the secondary biosynthesis of compounds other than peptaibols. The first metabolite of gliotoxin has been found in *Trichoderma virens* strains (Brian 1944), and gliovirin is associated with epipolythiodioxopiperazine class of toxins characterized by disulfide bridged cyclic dipeptide. These classes of compounds have bind to protein thiols and generate ROS. *T. virens* strain produced gliotoxin, however, not yet gliovirin. It is more an effective antagonist against *Rhizoctonia solani*, while the strain "P" produced gliovirin, not able to antagonize toward *R. solani* (Howell et al. 1993).

Gliotoxin biosynthesis has been absorbed by a gli gene cluster and applied in the opportunistic human pathogenic fungus *A. fumigatus* (Mukherjee et al. (2012b). The gli cluster is composed of 12 genes from *A. fumigatus*, in the length of 28 kb (Gardiner and Howlett 2005). Dioxopiperazine synthase (gliP) is derived in the gliotoxin biosynthesis, producing the diketopiperazine ring. Gliotoxin can offer several advantages to *A. fumigatus* when infecting a host that harbors other opportunistic fungi (Coleman et al. 2011). Besides, the gli biosynthetic cluster, which encodes the reductase gliT, gives full safeguard to *A. fumigatus* against gliotoxin (Schrettl et al. 2010; Davis et al. 2011). Gliotoxin compound only was found in *T. virens* "Q" strains, even though not found either in *T. atroviride* or in *T. reesei*. In *T. virens*, the gliP gene clustered with another seven gli genes with high similarity to eight of the twelve described in *A. fumigatus* genes. Gli cluster is not produced from *T. atroviride* genome. Seven gli genes, including gliP, are identified in *T. reesei* genome; however this species is not gliotoxin producer (Mukherjee et al. 2012b).

17.4 Polyketides

Polyketides (PKs) have more diverse structural SMs, including antibiotic activity such as macrolides, tetracyclines, and polyenes or aflatoxins; it can be produced by eukaryotic cells and bacteria. PKs are synthesized by highly complex polyketide synthase (PKS) enzymes. PKSs catalyzed the natural products from simple precursors such as propionyl-CoA and methylmalonyl-CoA that are closely related to fatty acid biosynthesis. Intermediate fatty acid biosynthesis is entirely reduced into unfunctionalized alkyl chain form, and the intermediate PK biosynthesis might be partially managed, providing the complex patterns of functional groups. However, many PKs have several importance to pharmacology due to their antimicrobial, anticancer, and immunosuppressive properties (Khosla 2009).

Several genes are observed in the *T. virens*, *T. reesei*, and *T. atroviride* that encode polyketide synthases. Divergent difference between mycoparasitic and saprophytic species is found: encodes 11 predicted PKSs for *T. reesei* (Martinez et al. 2008) and encodes 18 predicted PKSs from *T. virens* and *T. atroviride* (Kubicek et al. 2011). Several genes of the genera *Fusarium* and *Aspergillus* have been obtained in the biosynthetic pathway of PKs (Chiang et al. 2011; Frandsen et al. 2011).

NRPS–PKS may have calculated the specificity of adenylation and acyltransferase domains with high accuracy. These characteristics of NRPS–PKS created valuable resource for the identification of natural products (Yadav et al. 2009) and comparatively common practices in fungi and bacteria (Keller et al. 2005). Plant–fungus interactions discovered the PKS/NRPS hybrid enzyme Ace1 of the rice blast fungus *Magnaporthe grisea* (Böhnert et al. 2004; Collemare et al. 2008).

Bioinformatics and comparative genome analysis showed some genes encoding PKS/NRPS hybrid enzymes in the *Trichoderma* genomes (Mukherjee et al. 2012a,b), for example, four in *T. virens*, one in *T. atroviride*, and two in *T. reesei*. Mukherjee et al. (2012a) screened the expression of genes in the encoding of 22 for NRPSs and four for PK/NRPS hybrid enzymes when *T. virens* mycelia contacted with maize roots. The results showed among all mutants retained have the ability to colonize maize roots. PKS/NRPS mutant is reduced in the stimulation of the defense response pal gene, encoding phenylalanine ammonia lyase, but retained its capability to influence the defense response aox gene, which encodes allene oxide synthase, providing evidence that PKS/NRPS enzyme is involved in *Trichoderma*–plant interactions.

17.5 Pyrones

The 6-pentyl-alpha-pyrone (6-PP) is one of the most important volatile compounds used as biocontrol agent in agriculture and food industry. 6-PP has a yellow pigmentation and coconut odor. This volatile compound was firstly cultured from *T. viride* (Collins and Halim 1972), which subsequently isolated from different *Trichoderma* species (Keszler et al. 2000; Vinale et al. 2008; Rubio et al. 2009). 6-PP has showed good inhibitory properties against several phytopathogenic fungi such as *R. solani* (Claydon et al. 1987), *F. oxysporum* (Scarselletti and Faull 1994), *Ganoderma boninense* (Siddiquee et al. 2009), and *Botrytis* spp. (Poole et al. 1998).

Interference of the Thctf1 gene in *T. harzianum*, which codes for a cutinase transcription factor 1 alpha, produced a yellow pigmentation, which associated with the absence of two secondary metabolites derived from 6-PP. The Δthctf1 strain decreased antimicrobial activities against *Fusarium* spp. (Rubio et al. 2009).

The 6-substituted 2H-pyran-2-one compound was isolated from *T. virens* and showed antifungal activity against *S. rolfsii* (Evidente et al. 2003). Hill et al. (1995) found two volatile antifungal compounds of massoilactone and γ-decanolactone, used as biocontrol agents against several plant pathogenic diseases. Viridepyronone has showed antagonistic activity against *S. rolfsii* (Evidente et al. 2003). *T. cerinum* cultures produced a new metabolite compound of cerinolactone and showed good antagonistic activity against *B. cinerea*, *R. solani*, and *Pythium ultimum* (Vinale et al. 2012).

The biosynthetic pathway observed that the G protein subunit Tga1 in *Trichoderma* is involved in the production of 6-PP (Reithner et al. 2005). The deletion of the tag1 gene in *T. atroviride* resulted in mutants with decreased production of 6-PP and other metabolites with a sesquiterpene structure, although other low-molecular-weight antifungal metabolites are produced in these mutants. Rubio et al. (2009) reported the transcription factor THCTF1 involved in 6-PP production and applied antifungal activity of *T. harzianum*.

Disruption of the Thctf1 gene resulted in mutants that never produced two 6-PP by-products that showed lower antimicrobial effect than the wild-type strain. In addition, a comparative study of the SMs of two wild-type *Trichoderma* species with those of the Thctf1 gene null mutant strain has been carried out by Daoubi et al. (2009).

17.6 Terpenes

Terpenes constitute one of the largest groups of natural products comprising numerous SMs with important pharmacological activities such as antibacterial, antiviral, antimalarial, anti-inflammatory actions, the inhibition of cholesterol synthesis, and anticancer activity. Terpene compounds are mostly involved as biocontrol activities due to their antifungal properties (i.e., viridins and ergokonins), even though their structural function in cell membranes (i.e., ergosterol) is necessary. *Trichoderma* produced a large series of these compounds (Degenkolb et al. 2008a; Reino et al. 2008; Cardoza et al. 2011).

The biosynthesis of terpenes is split out into four steps (Fig. 17.2): (1) the synthesis of precursors, isopentenyl diphosphate (IPP) and dimethylallyl diphosphate (DMAPP), through two different pathways; (2) repetitive addition of the precursors to form a series of homologues of prenyl diphosphate (PPP), which are the different classes of terpenes; (3) the formation of terpene backbones through the activity of specific synthases; and (4) secondary enzymatic modification of these backbones, which provides functional properties and family diversity.

The isomerization of IPP produced the allylic isomer DMAPP, which is considered the first PPP. IPP and DMAPP are combined to form geranyl diphosphate (GPP), a key factor of monoterpene biosynthesis. This molecule is summarized again with IPP, giving (1) farnesyl diphosphate (FPP), a common intermediate in the production of geranylgeranyl diphosphate and its diterpene derivatives; (2) trichodiene and the sesquiterpene compounds, including trichothecene toxins; and

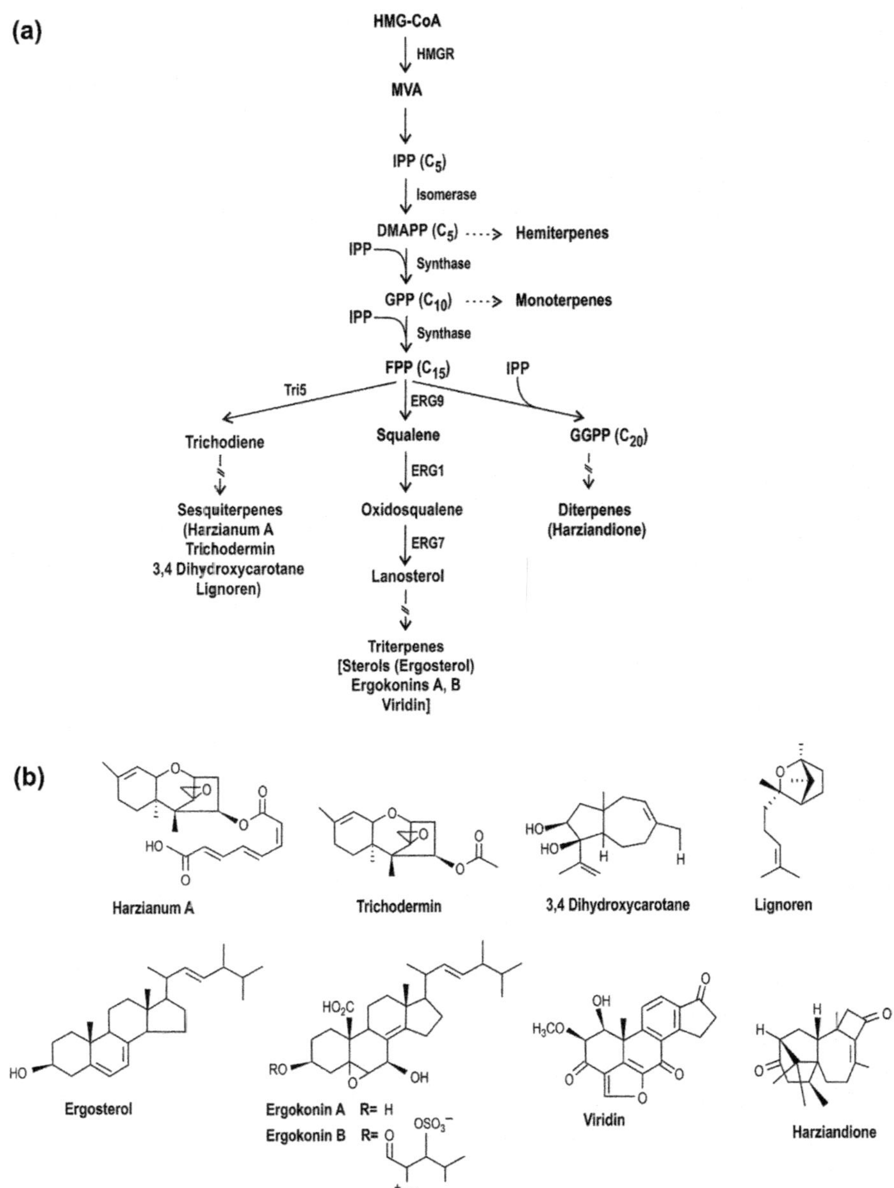

Fig. 17.2 (**a**) Terpene biosynthesis from hydroxymethylglutaryl-coenzyme A (HMG-CoA) to farnesyl pyrophosphate (FPP) through the so-called mevalonate pathway, followed by the biosynthetic schematic pathway of several sesquiterpenes, triterpenes, and diterpenes that have been identified in *Trichoderma* species. (**b**) Chemical structures of the most representative *Trichoderma* terpenes

(3) squalene, as a precursor of the triterpene pathway to produce prenylated proteins, sterols, ubiquinones, dolichols, and other SMs of pharmacological relevance (Chang and Keasling 2006), as well as plant carotenoids, gibberellins, tocopherols, and chlorophylls (Kirby and Keasling 2009).

No individual monoterpenes are isolated from *Trichoderma*, although the formations of volatile terpenes, lactones, and alcohols are separately reported (Zeppa et al. 1990). Harziandione is called to be the first diterpene cultured from *Trichoderma* species (Ghisalberti et al. 1992). Sesquiterpenes in *Trichoderma* have showed antibacterial, antifungal, and neuroleptic activities (Bennett and Klich 2003; Berg et al. 2003). Some examples of sesquiterpenes with antifungal activity are daucane sesquiterpene, 3,4-dihydroxycarotane, and lignoren, which displayed antimicrobial activity against *Mycobacterium smegmatis*, *B. subtilis*, and *P. aeruginosa* and a moderate antiyeast activity against *Sporobolomyces salmonicolor* and *Rhodotorula rubra* (Berg et al. 2004). One particular group of sesquiterpenes includes the fungal toxins known as trichothecenes. Only some *Trichoderma* species produced the trichothecenes trichodermin and harzianum A (HA) (Cardoza et al. 2011). Trichodermin displays antiyeast activity and phytotoxicity (Tijerino et al. 2011a), while HA showed beneficial effects in plants and growth inhibition of phytopathogenic fungi (Malmierca et al. 2012). Ergokonins A and B are examples of *Trichoderma* steroid antibiotics with antifungal properties against species of *Candida* and *Aspergillus* (Vicente et al. 2001). *T. virens* produced the fungistatic and anticancer triterpene/steroid viridin, which reduced to viridiol, a compound with herbicidal properties (Howell et al. 1993; Druzhinina et al. 2011). Mukherjee et al. (2006) described a gene cluster for viridin biosynthesis and indicated that a gene cluster is not involved in viridin production but in the synthesis of volatile sesquiterpenes (Mukherjee et al. 2012b).

The terpene formation in *Trichoderma* produced mevalonic acid (MVA) (Fig. 17.2). The MVA biosynthetic pathway is the first enzyme produced of glycoprotein hydroxymethylglutaryl-coenzyme A reductase (HMGR). In *T. harzianum*, partial gene silencing of hmgR always provided transformants with higher level of sensitivity to lovastatin. However, these hmgR-silenced transformants produced lower levels of ergosterol than the wild type and showed a reduction in their antifungal activity against *R. solani* and *F. oxysporum* (Cardoza et al. 2007).

The biosynthetic pathway of triterpenes is one of the key steps of the stereospecific epoxidation of squalene to 2,3(S)-oxidosqualene, produced by squalene epoxidase (Liu et al. 2004). Squalene epoxidase is encoded by the erg1 gene, isolated from *T. harzianum* (Cardoza et al. 2006). The trichothecene biosynthetic pathway of *Fusarium* species has been studied by Brown et al. (2004). The cyclization of FPP to trichodiene is the first committed step in the biosynthesis of closely 100 different trichothecene toxins and antibiotics produced by many fungi, for example, mostly T-2 toxin uses (Rynkiewicz et al. 2001). The first step is catalyzed by trichodiene synthase, which is encoded by the tri5 gene. Trichothecene compounds are produced by different *Trichoderma* species and classified within the clade brevicompactum (Degenkolb et al. 2008a). This clade has a common pathway from FPP to trichodiene, to trichodiol, and to trichodermol; the HA production in *T.*

arundinaceum, *T. turrialbense*, and *T. protrudens*; and the biosynthesis of the other nonmacrocyclic trichothecene-type toxin trichodermin in *T. brevicompactum*. Cardoza et al. (2011) described trichothecene biosynthesis and a tri-cluster in *T. arundinaceum* and *T. brevicompactum*. The tri5 gene regulates the FPP–farnesol–squalene–trichodiene balance during the different growth stages of cell; thus its overexpression results in an increase in the level of transcription of three trichothecene genes, namely, tri4, tri6, and tri10 (Tijerino et al. 2011b).

17.7 Nonribosomal Peptides

Peptaibols are synthesized by nonribosomal peptide synthetases (NRPSs). NRPS compounds contain amino acid monomers (Strieker et al. 2010). The nonribosomal peptides (NRPs) or peptaibols showed different group of molecules that play fundamental role in fungal physiology. These enzymes are synthesized from several varieties of compounds from monomers to molecules with over 20 proteinogenic residues, even though some are not proteinogenic or amino acid. Peptaibols are commonly formed by the α-aminoisobutyric acid and C-terminal 1,2-amino alcohol. NRPS accessible database found more than 700 entries (http://bioinfo.lifl.fr/norine/). The NRPS repertoire encoded in the three genome sequencing of *Trichoderma* species; *T. atroviride* and *T. virens* are enriched in, respectively, 16 NRPS genes and 28 NRPS genes as compared to only 10 NRPS genes of *T. reesei* (Kubicek et al. 2011). The above three species comprised of two peptaibol synthetases, one for short (10–14 residues) and one for long (18–25 residues) peptaibols. The long peptaibol synthetase-encoding genes have lacked introns and produced about 60–80 kb long of mRNA that encodes about 25,000 amino acids (Druzhinina et al. 2011). Long peptaibol synthetase gene (Tex1) (18-residue peptaibols) was isolated from *T. virens* Gv29–8 strain and latterly sequenced by Wiest et al. (2002). A short peptaibol synthetase gene (Tex2) encoding a 14-module enzyme is assisted to accumulate both 11-residue and 14-residue peptaibols found from *T. virens* (Mukherjee et al. 2011).

Phosphopantetheinyl transferases (PPTases) belong to a superfamily enzymes and found in prokaryotes and eukaryotes which are syntheses varied range of compounds, such as amino acids, fatty acids, polyketides, and nonribosomal peptides. PPT1 enzyme was isolated from *T. virens*, and it played an important role in secondary metabolism. A mutant in PPT1 gene in *T. virens* is auxotrophic for lysine. It has formed nonpigmented conidia and impotent to synthesize NRPs. Furthermore, the Δppt1 mutants are incapable to inhibit the growth of phytopathogenic fungi (Velázquez-Robledo et al. 2011).

In *T. virens*, the VELVET protein always played a significant role in the morphogenesis and secondary metabolism. The mutant is incapable to produce gliotoxin and associated with low levels of gliP gene transcription, which encoded the NRPS of glyotoxin synthesis in *A. fumigatus*. A low transcription level of secondary metabolism-related genes was found in the Δvel1 strain, and it is applied in mycoparasitism and biocontrol efficacy. However, the Δvel1 mutant is impotent to safeguard cotton plants in *P. ultimum*- and *R. solani*-infested soil (Mukherjee and Kenerley 2010).

Gliotoxin belongs to epipolythiodioxopiperazine (ETP) peptides which is responsible of such lethal effect. It is important that all of the *Trichoderma* species or even strains do not produce similar antibiotics. For example, *T. virens* strain "Q" is capable to produce gliotoxin, whereas strains "P" cannot. However, the strains "P" produced the ETP class peptide, gliovirin, which has antimicrobial properties, although a restricted spectrum activity against actinomycetes, oomycetes, and bacteria, whereas gliotoxin showed more potential against a wide range of microorganisms (Howell 2006).

17.8 Production of Plant Hormones

Several numbers of researchers have concentrated on the main role of phytohormones in microbially induced plant growth promotion, in cases either the microorganisms/fungi release the phytohormones or instead stimulate the plant to produce phytohormones (Sofo et al. 2011). Vinale et al. (2008) found major secondary metabolites in liquid culture from the effective *Trichoderma* biocontrol strains such as *T. atroviride* P1 and *T. harzianum* T22, T39, and A6. These effective strains produced several compounds such as butenolide, harzianopyridone, azaphilone, 1-hydroxy-3-methylanthraquinone, harzianolide, 6-n-pentyl-6H-pyran-2-one (6PP), and 1,8-dihydroxy-3-methyl-anthraquinone. Six-day old wheat seedlings are grown in liquid suspensions with different concentrations in the range of 1–10 mgl^{-1}. Compounds of T39 butenolide (1 mgl^{-1}), anthraquinone (10 mgl^{-1}), and harzianolide (1 mgl^{-1}) are significantly increased in the range of 23–24 cm for stem and leaf length as compared to the control of 18.6 cm, and harzianolide (1 mgl^{-1}) also potentially increased about 43 mg of dry weight as compared to 34 mg of the control. *T. harzianum* T22 enhanced hormone levels in cherry rootstocks. T22 inoculation of cherry resulted in significant increases in the order of 76% for root and 61% for shoot growth, respectively.

High-performance liquid chromatography/mass spectrometry analysis found gibberellic acid (GA3) and indole acetic acid (IAA) levels that are significantly enhanced in both roots (40% and 143%, respectively) and leaves (49% and 71%, respectively) with no variation in abscisic acid levels and decline in trans-zeatin riboside levels (Sofo et al. 2011). Subsequently, auxins are stimulated by root induction and growth, and axial buds and leaves are inhibited by cytokinins and GA3 promoters (Blake et al. 2000). T22-treated plants observed the hormone changes that could describe the higher root growth and shoot elongation recorded in these plants. It is the upregulation of genes required for hormone biosynthesis or the downregulation of genes that is involved in hormone catabolism induced by elicitors secreted into the medium by T22 or via direct transfer from the fungus to the root cells as suggested by Harman et al. (2004).

Vinale et al. (2008) found that harzionalide and 6-PP two of the major secondary compounds produced by *T. atroviride* and *T. harzianum* provoked an auxin-like effect on etiolated tomato (*Lycopersicum esculentum*), pea (*Pisum sativum*), and canola (*Brassica napus*) seedlings. Benítez et al. (1998) found that coculture of tomato seeds with *Trichoderma*, separated by a cellophane membrane, allowed promotion of seed germination. These results suggested that *Trichoderma* produces and secretes plant growth-promoting and seed-germinating factors. Recently, *T. atroviride* and *T. virens* produced 3-indoleacetic acid-related indoles. In *Trichoderma* liquid cultures, IAA-related indoles are produced that are stimulated by the addition of l-tryptophan (Gravel et al. 2007; Salas-Marina et al. 2011). Inoculation of *A. thaliana* or tomato plants with *Trichoderma* showed the stimulation of plant growth and biomass production (Gravel et al. 2007; Salas-Marina et al. 2011). The production of IAA-related indoles could be one of the mechanisms employed by *Trichoderma* to enhance the plant growth and rise up the number of secondary roots, leading to higher biomass accumulation in *Arabidopsis* (Salas-Marina et al. 2011). Gravel et al. (2007) proposed that the degradation of IAA by *Trichoderma* allowed a reduction on the negative effect of IAA on root elongation, which could decrease ethylene (ET) production, resultant from a reduced in its precursor 1-aminocyclopr opane-1-carboxylic acid (ACC) and/or ACC deaminase activity existence in *Trichoderma*. It demonstrated that *T. asperellum* T203 has high 1-amino cyclopropane-1-carboxylic acid deaminase (ACCD) activity when growing in ACC as the solitary nitrogen source. The ACCD encoding gene Tas-acdS is upregulated when ACC is added to the culture medium. Silencing of Tas-acdS showed the reduced ability of silenced transformants to stimulate root elongation of canola plants (Viterbo and Horwitz 2010). These mechanisms could be the main responsible for plant growth-promoting activity (Gravel et al. 2007; Viterbo and Harwitz 2010; Salas-Marina et al. 2011).

Martinez-Medina et al. (2011) analyzed five major hormones in the endogenous levels (zeatin (Ze), IAA, jasmonic acid, salicylic acid, and the ethylene precursor 1 aminocyclopropane-1-carboxylic acid (ACC)) in the plants of melon inoculated with *T. harzianum* CECT 20714. In shoot and root growth, recorded are the above concentrations of 30% for Ze, 40% of IAA, and standard ACC concentrations. IAA is a common natural auxin in plants and positively affects root morphology and growth (Vessey 2003). Roots are mainly sensitive at varying levels of IAA, and root enhancement could be significantly affected by exterior bases of plant growth regulator including those produced by soil microbes (Tanimoto 2005). IAA has l-tryptophan metabolism produced by several groups of microbes (Arshad and Frankenberger 1992). For example, *Azotobacter paspali* releases IAA into culture medium and enables to increase the dry weight of leaves and roots with different plant species (Barea and Brown 1974). In an in vitro study, inoculation of canola seeds with *Pseudomonas putida* GR12–2 produced IAA and later increased the length of seedling roots (Caron et al. 1995).

Trichoderma virens spores are applied to *Arabidopsis* seedlings for increasing shoot and root growth. There is a four- to sixfold stimulation in lateral root formation. *T. virens* inoculation increased in auxin-regulated gene expression within the plant (Contreras-Cornejo et al. 2009). It also produced auxin compounds for promoting root development and plant defense mechanisms. Mutation genes are involved in auxin transport or signaling such as BIG, AUX1, EIR1, and AXR1 which declined the growth-promoting effects of *T. virens* inoculum.

In liquid culture of *T. virens*, IAA and IAA-related compounds are isolated. Production of IAA is 17 times higher in medium supplemented with tryptophan. Fungus induces shoot and root growth including lateral root formation, which provides more root surface area for colonization. The branched roots have increased water and nutrient uptake capacity (Contreras-Cornejo et al. 2009). Plant growth-promoting rhizobacteria, bacteria growing in the rhizosphere of plants, are used with tryptophan to synthesize IAA. Bacteria produced ACC deaminase that stimulated plant growth. The ACC enzyme cleaves the plant hormone ethylene to produce α-ketobutyrate and ammonia (Todorovic and Glick 2008). Ethylene is an important signaling molecule in plants when under attacked by pathogens or exposed to abiotic stresses that affect plant growth (Abels et al. 1992; Pierik et al. 2006). Inoculation of plants with bacteria that produced ACC deaminase has lower ethylene levels, and the effects are longer roots and reduced plant growth inhibition (Glick et al. 1998). Subsequently, IAA can stimulate the activity of the enzyme ACC deaminase; IAA synthesized by bacteria cannot only direct stimulation of cell elongation and division; however, it also promoted plant growth by inhibiting ethylene synthesis.

Trichoderma atroviride promoted plant growth and subsequently increased the yield of tomato in hydroponic conditions. Production is increased based on the addition of l-tryptophan, tryptamine, and tryptophol in cultural medium. Growth promotion of tomato seedlings is stimulated in the presence of cumulative concentrations of l-tryptophan; as a result the fungus could synthesize IAA through tryptophan-dependent pathway (Gravel et al. 2007). Tanimoto (2005) widely reported that high levels of exogenous IAA can be deleterious to root elongation. This effect has been linked with an increase in ethylene levels in the plant (Glick et al.1997, 1998), because of IAA-induced stimulation of ACC synthase, which catalyzes the transformation of S-adenosyl methionine to ACC, the precursor of ethylene in plants (Kende 1993). When exogenous IAA is applied to the roots of tomato seedlings, root length has decreased; however when seedlings are inoculated with *T. atroviride* with the presence of exogenous IAA, root length is significantly increased. Results suggested that *T. atroviride* could partially degrade IAA in vitro, so the fungus is able to produce the enzyme ACC deaminase. Based on these observations, the authors hypothesized two ways in which the fungus could reduce the detrimental effect of IAA on root elongation: (1) *T. atroviride* could degrade IAA close to the roots at the presence level that will not inhibit root elongation, and (2) *T. atroviride* could control the ethylene concentration through the reduction of its precursor ACC by the action of ACC deaminase. *T. asperellum* has been grown in a medium with ACC, and only nitrogen source produced high levels of ACC deaminase. Brotman et al. (2010) also reported plant growth promotion mediated by the synthesis of auxin by *Trichoderma* spp. and the activity of the enzyme ACC deaminase.

References

Abels FB, Morgan PW, Saltveit ME (1992) Ethylene in plant Biology. Academic, New York

Aime MC, Phillips-Mora W (2005) The causal agents of witches' broom and frosty pod rot of cacao (chocolate, Theobroma cacao) form a new lineage of Marasmiaceae. Mycologia 97:1012–1022

Arshad M, Frankenberger WTJ (1992) Microbial production of plant growth regulators. In: Metting FBJ (ed) Soil microbial ecology. Marcel Dekker, New York, pp 307–347

Auvin-Guette C, Rebuffat S, Prigent Y, Bodo B (1992) Trichogin A IV, an 11-residue lipopeptaibol from Trichoderma longibrachiatum. J Am Chem Soc 114:2170–2174

Auvin-Guette C, Rebuffat S, Vuidepot I, Massias M, Bodo B (1993) Structural elucidation of trikoningins KA and KB, peptaibols from Trichoderma koningii. J Chem Soc Perkin Trans 1:249–255

Baker SE, Perrone G, Richardson NM, Gallo A, Kubicek CP (2012) Phylogenetic analysis of polyketide synthase-encoding genes in Trichoderma. Microbiology 158:35–45

Barea JM, Brown ME (1974) Effects on plant growth produced by Azotobacter paspali related to synthesis of plant growth regulating substances. J Appl Microbiol 37(4):583–593

Benítez T, Delgado-Jarana J, Rey M, Limón J (1998) Biofungicides: Trichoderma as biocontrol agent against phytopathogenic fungi. In: Pandalia SG (ed) Recent research developments in microbiology, vol 2. Research Signpost, Trivandrum, pp 129–150

Bennett JW, Klich M (2003) Mycotoxins Clin Microbiol Rev 16:497–516

Berg A, Grigoriev PA, Degenkolb T, Neuhof A, Haertl A, Schlegel B, Graefe U (2003) Isolation, structure elucidation and biological activities of trichofumins A, B, C and D, new 11 and 13mer peptaibols from Trichoderma sp. HKI 0276. J Pept Sci 9:810–816

Berg A, Wangun HVK, Nkengfack AE, Schlegel B (2004) Lignoren, a new sesquiterpenoid metabolite from Trichoderma lignorum HKI 0257. J Basic Microbiol 44:317–319

Blake PS, Browning G, Benjamin LJ, Mander LN (2000) Gibberellins in seedlings and flowering trees of Prunus avium L. Phytochemistry 53(4):519–528

Böhnert HU, Fudal I, Dioh W, Tharreau D, Notteghem JL, Lebrun MH (2004) A putative polyketide synthase/peptide synthase from Magnaporthe grisea signals pathogen attack to resistance rice. Plant Cell 16:2499–2513

Brewer D, Mason FG, Taylor A (1987) The production of alamethicins by. Trichoderma spp Can J Microbiol 33:619–625

Brian PW (1944) Production of gliotoxin by Trichoderma viride. Nature 154:667–668

Brotman Y, Kapuganti JG, Viterbo A (2010) Trichoderma Curr Biol 20(9):R390–R391

Brown DW, Dyer RB, McCormick SP, Kendra DF, Plattner RD (2004) Functional demarcation of the Fusarium core trichothecene gene cluster. Fungal Genet Biol 41:454–462

Brückner H, Koza A (2003) Solution phase synthesis of the 14-residue peptaibol antibiotic trichovirin I. Amino Acids 24:311–323

Bruinsma M, Pang B, Mumm R, van Loon JJA, Dicke M (2009) Comparing induction at an early and late step in signal transduction mediating indirect defence in Brassica oleracea. J Exp Bot 60:2589–2599

Cardoza RE, Vizcaíno JA, Hermosa MR, Sousa S, González FJ, Llobell A, Monte E, Gutiérrez S (2006) Cloning and characterization of the erg1 gene of Trichoderma harzianum effect of the erg1silencing on ergosterol biosynthesis and resistance to terbinafine. Fungal Genet Biol 43:164–178

Cardoza RE, Hermosa MR, Vizcaíno JA, González FJ, Llobell A, Monte E, Gutiérrez S (2007) Partial silencing of a hydroxy-methylglutaryl-CoA reductase encoding gene in Trichoderma harzianum CECT 2413 results in a lower level of resistance to lovastatin and a lower antifungal activity. Fungal Genet Biol 44:269–283

Cardoza RE, Malmierca MG, Hermosa R, Alexander NJ, McCormick SP, Proctor RH, Tijerino AM, Rumbero A, Monte E, Gutiérrez S (2011) Identification of loci and functional characterization of trichothecene biosynthetic genes in the filamentous fungus Trichoderma. Appl Environ Microbiol 77:4867–4877

Caron M, Patten CL, Ghosh S (1995) Effects of plant growth promoting rhizobacteria Psuedomonas putida GR-122 on the physiology of canolla roots. Proc Plant Growth Reg Soc Am 7:18–20

Chang MC, Keasling JD (2006) Production of isoprenoid pharmaceuticals by engineering isoprenoids. Nat Chem Biol 2:674–681

Chiang YM, Meyer KM, Praseuth M, Baker SE, Bruno KS, Wang CC (2011) Characterization of a polyketide synthase in Aspergillus niger whose product is a precursor for both dihydroxynaphthalene (DHN) melanin and naphtha-γ-pyrone. Fungal Genet Biol 48:430–437

Cho IH, Namgung H-J, Choi H-K, Kim YS (2008) Volatiles and key odorants in the pileus and stipe of pine-mushroom (Tricholoma matsutake Sing). Food Chem 106:71e76

Claydon N, Allan M, Hanson JR, Avent AG (1987) Antifungal alkyl pyrones of Trichoderma harzianum. Trans Br Mycol Soc 88:503–513

Coleman JJ, Ghosh S, Okoli I, Mylonakis E (2011) Antifungal activity of microbial secondary metabolites. PLoS One 6:e25321

Collemare J, Pianfetti M, Houlle AE, Morin D, Camborde L, Gagey MJ, Barbisan C, Fudal I, Lebrun MH, Böhnert HU (2008) Magnaporthe grisea avirulence gene ACE1 belongs to an infection-specific gene cluster involved in secondary metabolism. New Phytol 179:196–208

Collins RP, Halim AF (1972) Characterization of the major aroma constituent of the fungus Trichoderma viride. J Agric Food Chem 20:437–438

Contreras-Cornejo HA, Macias-Rodriguez L, Cortes-Penagos C, Lopez-Bucio J (2009) Trichoderma virens, a plant beneficial fungus, enhances biomass production and promotes lateral root growth through an auxin-dependent mechanism in Arabidopsis. Plant Physiol 149(3):1579–1592

Daoubi M, Pinedo-Rivilla C, Rubio MB, Hermosa R, Monte E, Aleu J, Collado IG (2009) Hemisynthesis and absolute configuration of novel 6-pentyl-2H-pyran-2-one derivatives from Trichoderma spp. Tetrahedron 65:4834–4840

Davis C, Carberry S, Schrettl M, Singh I, Stephens JC, Barry SM, Kavanagh K, Challis GL, Brougham D, Doyle S (2011) The role of S-transferase GliG in gliotoxin biosynthesis in Aspergillus fumigatus. Chem Biol 18:542–552

Degenkolb T, Gräfenhan T, Berg A, Nirenberg HI, Gams W, Brückner H (2006a) Peptaibiomics: screening for polypeptide antibiotics (peptaibiotics) from plant-protective Trichoderma species. Chem Biodivers 3:593–610

Degenkolb T, Gräfenhan T, Nirenberg HI, Gams W, Brückner H (2006b) Trichoderma brevicompactum complex: rich source of novel and recurrent plant-protective polypeptide antibiotics (peptaibiotics). J Agric Food Chem 54:7047–7061

Degenkolb T, Dieckmann R, Nielsen KF, Gräfenhan T, Theis C, Zafari H, Chaverri P, Ismaiel A, Brückner H, von Döhren H, Thrane U, Petrini O, Samuels GJ (2008a) The Trichoderma brevicompactum clade: a separate lineage with new species, new peptaibiotics, and mycotoxins. Mycol Prog 7:177–219

Degenkolb T, von Döhren H, Nielsen KF, Samuels GJ, Brückner H (2008b) Recent advances and future prospects in peptaibiotics, hydrophobin, and mycotoxin research, and their importance for chemotaxonomy of Trichoderma and Hypocrea. Chem Biodivers 5:671–680

Druzhinina IS, Seidl-Seiboth V, Herrera-Estrella A, Horwitz BA, Kenerley CM, Monte E, Mukherjee PK, Zeilinger S, Grigoriev IV, Kubicek CP (2011) Trichoderma: the genomics of opportunistic success. Nat Rev Microbiol 9:749–759

Engelberth J, Koch T, Schüler G, Bachmann N, Rechtenbach J, Boland W (2001) Ion channel-forming alamethicin is a potent elicitor of volatile biosynthesis and tendril coiling. Cross talk between jasmonate and salicylate signaling in lima bean. Plant Physiol 125:369–377

Evidente A, Cabras A, Maddau L, Serra S, Andolfi A, Motta A (2003) Viridepyronone, a new antifungal 6-substituted 2H-pyran-2-one produced by Trichoderma viride. J Agric Food Chem 51:6957–6960

Fox EM, Howlett B (2008) Secondary metabolism: regulation and role in fungal biology. Curr Opin Microbiol 11:481–487

Fraatz MA, Zorn H (2010) Fungal flavours. In: Hofrichter M (ed) The Mycota X: industrial applications, 2nd edn. Springer-Verlag, Berlin, p 249e264

Frandsen RJ, Schütt C, Lund BW, Staerk D, Nielsen L, Olsson S, Giese H (2011) Two novel classes of enzymes are required for the biosynthesis of aurofusarin in Fusarium graminearum. J Biol Chem 286:10419–10428

Fujita T, Wada S, Iida A, Nishimura T, Kanai M, Toyama N (1994) Fungal metabolites. XIII. Isolation and structural elucidation of new peptaibols, trichodecenins-I and -II, from Trichoderma viride. Chem Pharm Bull 42:489–494

Gardiner DM, Howlett BJ (2005) Bioinformatic and expression analysis of the putative gliotoxin biosynthetic gene cluster of Aspergillus fumigatus. FEMS Microbiol Lett 248:241–248

Ghisalberti EL, Hockless DCR, Rowland C, White AH (1992) Harziandione, a new class of diterpene from Trichoderma harzianum. J Nat Prod 55:1690–1694

Glick BR, Liu CP, Ghosh S, Dumbroff EB (1997) Early development of canola seedlings in the presence of the plant growth promoting rhizobacterium Pseudomonas putida GR12-2. Soil Biol Biochem 29(8):1233–1239

Glick BR, Penrose DM, Li JP (1998) A model for the lowering of plant ethylene concentrations by plant growth-promoting bacteria. J Theor Biol 190(1):63–68

Gravel V, Antoun H, Tweddell RJ (2007) Growth stimulation and fruit yield improvement of greenhouse tomato plants by inoculation with Pseudomonas putida or Trichoderma atroviride: possible role of indole acetic acid (IAA). Soil Biol Biochem 39(8):1968–1977

Harman GE, Howell CR, Viterbo A, Chet I, Lorito M (2004) Trichoderma species opportunistic, avirulent plant symbionts. Nat Rev Microbiol 84:377–393

Hermosal R, Cardoza RE, Rubiol MB, Gutiérrez S, Monte E (2014) Secondary metabolism and antimicrobial metabolites of trichoderma. In: Gupta VK, Schmoll M, Herrera-Estrella A, Upadhyay RS, Druzhinina I, Tuohy MG (eds) Biotechnology and biology of trichoderma, pp 125–137

Hill RA, Cutler HG, Parker SR, (1995) Trichoderma and metabolites as control agents for microbial plant diseases. PCT Patent Appl 9520879

Howell CR (2006) Understanding the mechanisms employed by Trichoderma virens to affect biological control of cotton diseases. Phytopathology 96:178–180

Howell CR, Stipanovic RD, Lumsden R (1993) Antibiotic production by strains of Gliocladium virens and its relation to biocontrol of cotton seedling diseases. Biocontrol Sci Tech 3:435–441

Keller NP, Turner G, Bennett JW (2005) Fungal secondary metabolism—from biochemistry to genomics. Nat Rev Microbiol 3:937–947

Kende H (1993) Ethylene biosynthesis. Annu Rev Plant Physiol Plant Mol Biol 44:283–307

Keszler A, Forgacs E, Kotai L, Vizcaíno JA, Monte E, García-Acha I (2000) Separation and identification of volatile components in the fermentation broth of Trichoderma atroviride by solid-phase extraction and gas chromatography-mass spectrometry. J Chromatogr Sci 38:421–424

Khosla C (2009) Structures and mechanisms of polyketide synthases. J Org Chem 74:6416–6420

Kirby J, Keasling JD (2009) Biosynthesis of plant isoprenoids: perspectives for microbial engineering. Annu Rev Plant Biol 60:335–355

Komon-Zelazowska M, Neuhof T, Dieckmann R, von Döhren H, Herrera-Estrella A, Kubicek CP (2007) Formation of atroviridin by Hypocrea atroviridis is conidiation associated and positively regulated by blue light and the G protein GNA3. Eukaryot Cell 6:2332–2342

Korpi A, Jarnberg J, Pasanen A-L (2009) Microbial volatile organic compounds. Crit Rev Toxicol 39:139–193

Kubicek CP, Herrera-Estrella A, Seidl-Seiboth V, Martínez DA, Druzhinina IS, Thon M, Zeilinger S, Casas-Flores S, Horwitz BA et al (2011) Comparative genome sequence analysis underscores mycoparasitism as the ancestral life style of Trichoderma. Genome Biol 12:R40

Liu W, May GS, Lionakis MS, Lewis RE, Kontoyiannis DP (2004) Extra copies of the Aspergillus fumigatus squalene epoxidase gene confer resistance to terbinafine: genetic approach to studying gene dose-dependent resistance to antifungals in A. fumigatus. Antimicrob. Agents Chemother 48:2490–2496

Lorito M, Farkas V, Rebuffat S, Bodo B, Kubicek CP (1996) Cell wall synthesis is a major target of mycoparasitic antagonism by Trichoderma harzianum. J Bacteriol 178:6382–6385

Maddau L, Cabras A, Franceschini A, Linaldeddu BT, Crobu S, Roggio T, Pagnozzi D (2009) Occurrence and characterization of peptaibols from Trichoderma citrinoviride, an endophytic fungus of cork oak, using electrospray ionization quadrupole time-of-flight mass spectrometry. Microbiology 155:3371–3381

Maischak H, Zimmermann MR, Felle HH, Boland W, Mithöfer A (2010) Alamethicin-induced electrical long distance signaling in plants. Plant Signal Behav 5:988–990

Malmierca MG, Cardoza RE, Alexander NJ, McCormick SP, Hermosa R, Monte E, Gutiérrez S (2012) Involvement of Trichoderma Trichothecenes in the biocontrol activity and induction of plant defense-related genes. Appl Environ Microbiol 78:4856–4868

Martinez D, Berka RM, Henrissat B, Saloheimo M, Arvas M, Baker SE, Chapman J, Chertkov O, Coutinho PM et al (2008) Genome sequencing and analysis of the biomass-degrading fungus Trichoderma reesei (syn. Hypocrea jecorina). Nat Biotechnol 26:553–560

Martinez-Medina A, Roldan A, Albacete A, Perez-Alfocea F, Pascual JA, 2011 Hormonal signalling of the Trichoderma harzianum-induced resistance to Fusarium oxysporum and growth promotion effect in melon plants. In: Pascual JA, Perez Alfocea F (eds) Acta Horticulturae, pp 61–67

Meyer CE, Reusser F (1967) A polypeptide antibacterial agent from Trichoderma viride. Experientia 23:85–86

Morath SU, Hung R, Bennett JW (2012) Fungal volatile organic compounds: a review with emphasis on their biotechnological potential. Fungal Biol Rev 26(2):73–83

Morey P, Wortham A, Weber A, Horner E, Black M, Muller W (1997) Microbial VOCs in moisture damaged buildings. Healthy Build 1:245–250

Mukherjee PK, Kenerley C (2010) Regulation of morphogenesis and biocontrol properties in Trichoderma virens by a VELVET protein Vel1. Appl Environ Microbiol 76:2345–2352

Mukherjee M, Horwitz BA, Sherkhane PD, Hadar R, Mukherjee PK (2006) A secondary metabolite biosynthesis cluster in Trichoderma virens: evidence from analysis of genes underexpressed in a mutant defective in morphogenesis and antibiotic production. Curr Genet 50:193–202

Mukherjee PK, Wiest A, Ruíz N, Keightley A, Morán-Diez ME, McCluskey K, Pouchus YF, Kenerley CM (2011) Two classes of new peptaibols are synthesized by a single non-ribosomal peptide synthetase of *Trichoderma virens*. J Biol Chem 286:4544–4554

Mukherjee PK, Horwitz BA, Kenerley CM (2012a) Secondary metabolism in Trichoderma—a genomic perspective. Microbiology 158:35–45

Mukherjee PK, Buensanteai N, Morán-Diez ME, Druzhinina I, Kenerley CM (2012b) Functional analysis of non-ribosomal peptide synthetases (NRPSs) in Trichoderma virens reveals a polyketide synthase (PKS)/NRPS hybrid enzyme involved in the induced systemic resistance response in maize. Microbiology 158:155–165

Neuhof T, Dieckmann R, Druzhinina I, Kubicek CP, von Döhren H (2007) Intact-cell MALDI-TOF mass spectrometry analysis of peptaibol formation by the genus Trichoderma: can molecular phylogenic knowledge predict peptaibol structures? Microbiology 153:3417–3437

Ortiz-Castro R, Contreras-Cornejo H, Macias-Rodriguez L, Lopez-Bucio J (2009) The role of microbial signals in plant growth and development. Plant Signal Behav 4:701–712

Pagans E, Font X, Sanchez A (2006) Emission of volatile organic compounds from composting of different solid wastes: abatement by biofiltration. J Hazard Mater 131:179e186

Perez-Nadales E, Almeida Nogueira MF, Baldin C, Castanheira S, El Ghalid M, Grund E et al (2014) Fungal model systems and the elucidation of pathogenicity determinants. Fungal Genet Biol 70:42–67

Pierik R, Tholen D, Poorter H, Visser EJW, Voesenek LACJ (2006) The Janus face of ethylene: growth inhibition and stimulation. Trends Plant Sci 11(4):176–183

Poole PR, Ward BG, Whitaker G (1998) The effects of topical treatments with 6-pentyl-2-pyrone and structural analogs on stem end post-harvest rots in kiwi fruit due to Botrytis Cinerea. J Agric Food Chem 77:81–86

Pruksakorn P, Arai M, Kotoku N, Vilcheze C, Baughn AD, Moodley P, Jacobs WR Jr, Kobayashi M (2010) Trichoderins, novel aminolipopeptides from a marine sponge-derived Trichoderma sp., are active against dormant mycobacteria. Bioorg Med Chem Lett 20:3658–3663

Rebuffat S, Goulard C, Bodo B (1995) Antibiotic peptides from Trichoderma harzianum: harzianins HC, proline-rich 14-residue peptaibols. J Chem Soc Perkin Trans 1:1849–1855

Reino JL, Guerrero RF, Hernández-Galán R, Collado IG (2008) Secondary metabolites from species of the biocontrol agent Trichoderma. Phytochem Rev 7:89–123

Reithner B, Brunner K, Schuhmacher R, Peissl I, Seidl V, Krska R, Zeilinger S (2005) The G protein alpha subunit Tga1 of Trichoderma atroviride is involved in chitinase formation and differential production of antifungal metabolites. Fungal Genet Biol 42:749–760

Reithner B, Schuhmacher R, Stoppacher N, Pucher M, Brunner K, Zeilinger S (2007) Signaling via the Trichoderma atroviride mitogen-activated protein kinase Tmk1 differentially affects mycoparasitism and plant protection. Fungal Genet Biol 44:1123–1133

Rippa S, Eid M, Formaggio F, Toniolo C, Beven L (2010) Hypersensitive-like response to the pore-former peptaibol alamethicin in Arabidopsis thaliana. Chembiochem 11:2042–2049

Rubio MB, Hermosa R, Reino JL, Collado IG, Monte E (2009) Thctf1 transcription factor of Trichoderma harzianum is involved in 6-pentyl-2H-pyran-2-one production and antifungal activity. Fungal Genet Biol 46:17–27

Rynkiewicz MJ, Cane DE, Christianson DW (2001) Structure of trichodiene synthase from Fusarium sporotrichioides provides mechanistic inferences on the terpene cyclization cascade. Proc Natl Acad Sci U S A 98:13543–13548

Salas-Marina MA, Silva-Flores MA, Uresti-Rivera EE, Castro-Longoria E, Herrera-Estrella A, Casas-Flores S (2011) Colonization of Arabidopsis roots by Trichoderma atroviride promotes growth and enhances systemic disease resistance through jasmonic acid/ethylene and salicylic acid pathways. Eur J Plant Pathol 131:15–26

Samuels GJ, Pardo-Schultheiss R, Hebbar KP, Lumsden RD, Bastos CN, Costa JCB, Bezerra JL (2000) Trichoderma stromaticum sp. nov. A parasite of the cacao witches broom pathogen. Mycol Res 104:760–764

Sang Y, Blecha F (2008) Antimicrobial peptides and bacteriocins: alternatives to traditional antibiotics. Anim Health Res Rev 9:227–235

Scarselletti R, Faull JL (1994) In vitro activity of 6-pentyl-α-pyrone, a metabolite of Trichoderma harzianum in the inhibition of Rhizoctonia solani and Fusarium oxysporum f. sp. lycopersici. Mycol Res 98:1207–1209

Scharf DH, Heinekamp T, Brakhage AA (2014) Human and plant fungal pathogens: the role of secondary metabolites. PLoS Pathog 10:e1003859. doi:10.1371/journal.ppat.1003859

Schirmböck M, Lorito M, Wang YL, Hayes CK, Arisan-Atac I, Scala F, Harman GE, Kubicek CP (1994) Parallel formation and synergism of hydrolytic enzymes and peptaibol antibiotics, molecular mechanisms involved in the antagonistic action of Trichoderma harzianum against phytopathogenic fungi. Appl Environ Microbiol 60:4364–4370

Schrettl M, Carberry S, Kavanagh K, Haas H, Jones GW, O'Brien J, Nolan A, Stephens J, Fenelon O, Doyle S (2010) Self-protection against gliotoxin—a component of the gliotoxin biosynthetic cluster, GliT, completely protects Aspergillus fumigatus against exogenous gliotoxin. PLoS Pathog 6:e1000952

Shi M, Chen L, Wang XW, Zhang T, Zhao PB, Song XY, Sun CY, Chen XL, Zhou BC, Zhang YZ (2012) Antimicrobial peptaibols from Trichoderma pseudokoningii induce programmed cell death in plant fungal pathogens. Microbiology 158:166–175

Shoresh M, Harman GE, Mastoury F (2010) Induced systemic resistance and plant responses to fungal biocontrol agents. Annu Rev Phytopathol 48:21–43

Siddiquee S, Yusuf UK, Hossain K, Jahan S (2009) In vitro studies on the potential Trichoderma harzianum for antagonistic properties against Ganoderma boninense. J Food Agric Environ 7(3&4):970–976

Siddiquee S (2014) Recent advancements on the role and analysis of Volatile Compounds (VOCs) from Trichoderma. In: Gupta VK, Schmoll M, Herrera-Estrella A, Upadhyay RS, Druzhinina I, Tuohy MG (eds) Biotechnology and biology of trichoderma, pp 138–175

Sivasithamparam K, Ghisalberti E (1998) Secondary metabolism in Trichoderma and Gliocladium. In: Kubicek CP, Harman GE (eds) Trichoderma and Gliocladium basis Biology, taxonomy and genetics, vol 1. Taylor and Francis, London, pp 139–191

Sofo A, Scopa A, Manfra M, De Nisco M, Tenore G, Troisi J et al (2011) Trichoderma harzianum strain T-22 induces changes in phytohormone levels in cherry rootstocks (Prunus cerasus × P-canescens). Plant Growth Regul 65(2):421–425

Splivallo R, Novero M, Bertea CM, Bossi S, Bonfante P (2007) Truffle volatiles inhibit growth and induce an oxidative burst in Arabidopsis thaliana. New Phytol 175:417–424
Strieker M, Tanovic A, Marahiel MA (2010) Nonribosomal peptide synthetases: structures and dynamics. Curr Opin Struct Biol 20:234–240
Szekeres A, Leitgeb B, Kredics L, Antal Z, Hatvani L, Manczinger L, Vagvölgyi C (2005) Peptaibols and related peptaibiotics of Trichoderma. Acta Microbiol Immunol Hung 52:137–168
Tanimoto E (2005) Regulation of root growth by plant hormones: roles for auxin and gibberellin. Crit Rev Plant Sci 24(4):249–265
Tijerino A, Cardoza RE, Moraga J, Malmierca MG, Vicente F, Aleu J, Collado IG, Gutiérrez S, Monte E, Hermosa R (2011a) Overexpression of the trichodiene synthase gene tri5 increases trichodermin production and antimicrobial activity in Trichoderma brevicompactum. Fungal Genet Biol 48:285–296
Tijerino A, Cardoza RE, Moraga J, Malmierca MG, Aleu J, Collado IG, Gutiérrez S, Monte E, Hermosa R (2011b) Overexpression of the trichodiene synthase gene tri5: effect on the expression of the trichodermin biosynthetic genes and on tomato seedlings. Toxins 3:1220–1232
Tirillini B, Verdelli G, Paolocci F, Ciccioli P, Frattoni M (2000) The volatile organic compounds from the mycelium of tuber borchii Vitt. Phytochemistry 55:983e985
Todorovic B, Glick BR (2008) The interconversion of ACC deaminase and D-cysteine desulfhydrase by directed mutagenesis. Planta 229(1):193–205
Velázquez-Robledo R, Contreras-Cornejo HA, Macías-Rodríguez L, Hernández-Morales A, Aguirre J, Casas-Flores S, López-Bucio J, Herrera-Estrella A (2011) Role of the 4-phosphopantetheinyl transferase of Trichoderma virens in secondary metabolism and induction of plant defense responses. Molecular plant-microbe interactions 24(12):1459–1471
Vessey JK (2003) Plant growth promoting rhizobacteria as biofertilizers. Plant Soil 255(2):571–586
Vicente MF, Cabello A, Platas G, Basilio A, Diez MT, Dreikorn S, Giacobbe RA, Onishi JC, Meinz M, Kurtz MB, Rosenbach M, Thompson J, Abruzzo G, Flattery A, Kong L, Tsipouras A, Wilson KE, Pelaez F (2001) Antimicrobial activity of ergokonin a from Trichoderma longibrachiatum. J Appl Microbiol 91:806–813
Vinale F, Sivasithamparam K, Ghisalberti EL, Marra R, Barbetti MJ, Li H, Woo SL, Lorito M (2008) A novel role for the Trichoderma–plant interaction Trichoderma secondary metabolites in the interactions with plants. Physiol Mol Plant Pathol 72:80–86
Vinale F, Girona IA, Nigro M, Mazzei P, Piccolo A, Ruocco M, Woo SL, Rosa DR, Herrera CL, Lorito M (2012) Cerinolactone, a hydroxyl-lactone derivative from Trichoderma cerinum. J Nat Prod 27:103–106
Viterbo A, Horwitz BA (2010) Mycoparasitism. In: Borkovich KA, Ebbole DJ (eds) Cellular and molecular biology of filamentous fungi, vol 42. American Society for Microbiology, Washington, DC, pp 676–693
Viterbo A, Wiest A, Brotman Y, Chet I, Kenerley C (2007) The 18mer peptaibols from Trichoderma virens elicit plant defense responses. Mol Plant Pathol 8:737–746
Vizcaíno JA, Sanz L, Cardoza RE, Monte E, Gutiérrez S (2005) Detection of putative peptide synthetase genes in Trichoderma species: application of this method to the cloning of a gene from T. harzianum CECT2413. FEMS Microbiol Lett 244:139–148
Vizcaíno JA, González FJ, Suárez MB, Redondo J, Heinrich J, Delgado-Jarana J, Hermosa R, Gutiérrez S, Monte E, Llobell A, Rey M (2006) Generation, annotation and analysis of ESTs from Trichoderma harzianum CECT2413. BMC Genomics 7:193–206
Wiest A, Grzegorski D, Xu BW, Goulard C, Rebuffat S, Ebbole DJ, Bodo B, Kenerley C (2002) Identification of peptaibols from Trichoderma virens and cloning of a peptaibol synthetase. J Chem Biol 277:20862–20868
Yadav G, Gokhale R, Mohanty D (2009) Towards prediction of metabolic products of polyketide synthases: an in silico analysis. PLoS Comput Biol 5:1–14
Zeppa G, Allegrone G, Barbeni M, Guarda PA (1990) Variability in the production of volatile metabolites by Trichoderma viride. Ann Microbiol Enzyme 40:171–176

Bacterial Volatiles for Plant Growth

18

Sreedevi Sarsan

Abstract

The rhizosphere is defined as the zone of soil surrounding the plant roots. Plant roots secrete a variety of plant exudates rich in nutrients resulting in accumulation of more bacteria in the rhizosphere region, generally 10–100 times higher than in the bulk soil. The bacteria colonizing this rhizosphere region are called as rhizobacteria, and those which help in promoting the growth of plants are called as plant growth-promoting rhizobacteria (PGPR). Currently, many biological approaches have gained importance for improving the crop production. One of the approaches includes using microbes as bioinoculants to promote growth and development of plants. Many rhizobacteria are presently being used as bioinoculants. They possess different mechanisms to enhance the plant growth such as nitrogen fixation, phosphate solubilization, production of siderophores, production of 1-aminocyclopropane-1-carboxylate deaminase (ACC), phytohormone production exhibiting antifungal activity, quorum sensing (QS) signal interference, induction of systemic resistance, interference with pathogen toxin production, and production of volatile organic compounds (VOCs). The production of VOCs by microorganisms can be considered as a novel characteristic property of PGPR in promoting the plant growth. The chemicals produced by microorganisms like bacteria and fungi as a part of their metabolism are called as microbial volatile organic compounds (MVOCs). These can modulate the physiology of plants and microorganisms and thus can provide an alternative method to use of chemicals in protecting plants from pathogens and increasing crop yield. MVOCs can be considered as ecofriendly and cost-effective strategy for sustainable agriculture.

S. Sarsan (✉)
Department of Microbiology, St. Pious X Degree & PG College for Women,
Snehapuri Colony, Nacharam, Hyderabad 500076, Telangana, India
e-mail: sreedevi163@yahoo.com

D.K. Choudhary et al. (eds.), *Volatiles and Food Security*,
DOI 10.1007/978-981-10-5553-9_18

Keywords
Rhizosphere • Plant growth promoting Rhiobacteria (PGPR) • Bioinoculants • Volatile organic compounds

18.1 Introduction

The soil has the most dynamic biological component with diverse types of living organisms – microorganisms as well as larger soil fauna such as nematodes, earthworms, ants, insects, rodents, etc. A wide range of microorganisms inhabit soil, but the most important ones are bacteria, actinomycetes, fungi, algae, protozoa, and viruses. These microbes vary in numbers and types owing to the vast differences in the physical and chemical characteristics of soils and also differences in agricultural practices. Microorganisms in soil are important in affecting soil structure and maintaining soil fertility. Microbes play an important role in cycling of nutrient elements (C, N, P, S, Fe) and are sources of industrial products such as enzymes, antibiotics, organic acids, vitamins, etc.

The bacteria are the most dominant of all the soil microorganisms, and they range from 10^5 to 10^8 per gram of soil. They vary in both numbers and types depending upon the physical, chemical, and biological conditions of the soil. Bacteria are the vital components of soils involved in various activities in the soil improving the soil fertility and enhancing the crop production (Ahmad et al. 2008; Chandler et al. 2008; Ahemad and Khan 2009). They promote plant growth by solubilizing or providing nutrients, producing plant growth hormones, controlling or inhibiting plant pathogens, sequestering toxic heavy metals from soils, etc. (Braud et al. 2009; Hayat et al. 2010; Rajkumar et al. 2010; Ahemad and Malik 2011; Ahemad 2012).

18.1.1 Rhizosphere and Rhizobacteria

The narrow zone of soil directly surrounding the root system (plant roots and root hairs) is referred to as rhizosphere. There are three separate but interacting components in the rhizosphere region: the rhizosphere (soil), the rhizoplane, and the root itself (Fig. 18.1). Many microorganisms are especially abundant in this rhizosphere. The zone of soil surrounding the roots is called rhizosphere and is influenced by substrates released by roots which affect microbial activity. The root surface including the strongly adhering soil particles is called rhizoplane also harbors microbes. The root itself is also a component of the system and many microorganisms (like endophytes) colonize the root tissues (Barea et al. 2005).

Plant roots majorly play a role by providing the mechanical support and facilitating water and nutrient uptake. Apart from this, plant roots also synthesize and secrete a variety of compounds which act as chemical attractants for diverse kinds of soil microbes (Walker et al. 2003). The bacteria lodging around the plant roots are called as rhizobacteria (Kloepper et al. 1991; Dakora and Phillips 2002). The

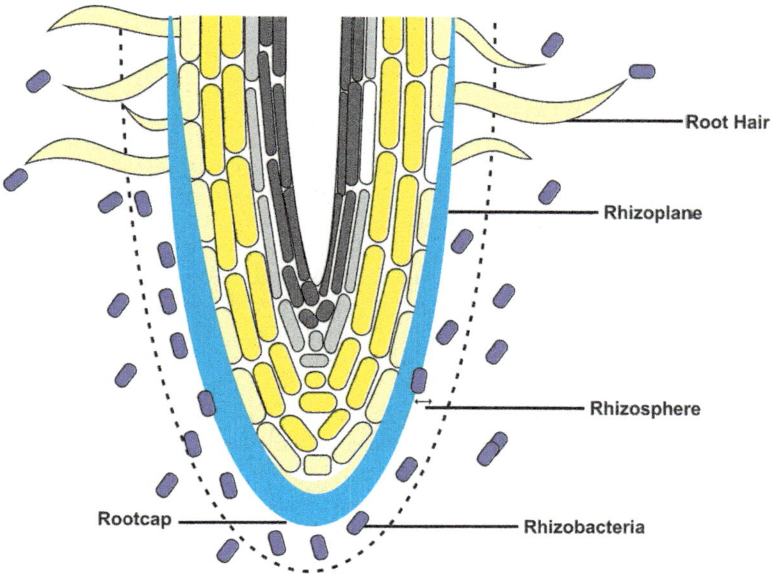

Fig. 18.1 Longitudinal cross section of a root with the surrounding rhizosphere and rhizobacteria

chemicals which are secreted by roots of different plant species into the soils are called as root exudates. These include a wide range of chemical compounds such as amino acids, organic acids, sugars, vitamins, enzymes, inorganic ions, and gaseous molecules (Table 18.1). These exudates regulate the structure of soil microbial community in the immediate vicinity of root surface by modifying the chemical and physical properties of the soil.

18.1.2 Plant Growth-Promoting Rhizobacteria

The bacteria colonizing the rhizosphere region around plant roots are called as rhizobacteria. Rhizobacteria are more versatile and efficient in transforming, mobilizing, and solubilizing the nutrients when compared to bacteria from bulk soils. Therefore, the rhizobacteria are considered as important in recycling of soil nutrients and thus improving the soil fertility (Hayat et al. 2010; Glick 2012). All these attributes of rhizobacteria help in promoting growth of plants, and thus these bacteria are termed as plant growth-promoting bacteria (PGPR). PGPR should have the following inherent characteristics: (1) they must colonize the root surface or the rhizosphere region; (2) they must be able to survive, multiply, and compete with other microorganisms and establish themselves; and (3) they must express the plant growth promotion activities and result in plant growth (Ahmed and Kibret 2014).

Table 18.1 Various compounds in root exudates of different plant species

S. No.	Type of root exudate compounds	Examples
1	Amino acids	α-Alanine, β-alanine, c-aminobutyric acid, a-aminoadipic acid, arginine, asparagines, aspartate, cysteine, cystine, glutamate, glycine, histidine homoserine, isoleucine, leucine, lysine, methionine, ornithine, phenylalanine, proline, serine, threonine tryptophan, and valine
2	Organic acids	Acetic acid, aconitic acid, aldonic acid, butyric acid, citric acid, formic acid, fumaric acid, glutaric acid, glycolic acid, lactic acid, malic acid, malonic acid, oxalic acid, pyruvic acid, succinic acid, tetronic acid, and valeric acid
3	Sugars	Hexoses and pentoses such as glucose, fructose, galactose, ribose, xylose, rhamnose, arabinose, oligosaccharides, raffinose, and maltose
4	Vitamins	Biotin, niacin, pantothenic acid, riboflavin, and thiamine
5	Purines	Adenine and guanine
6	Enzymes	Amylase, invertase, protease, and phosphatase
7	Inorganic ions and gaseous molecules	H^+, OH^-, HCO^{-3}, CO_2, and H_2

Source: Dakora and Phillips (2002)

PGPR can be divided on the basis of their location into two types:

1. Extracellular (ePGPR) – those which exist in the rhizosphere or on the rhizoplane or in the spaces between root cortical cells. Examples include *Bacillus*, *Azotobacter*, *Azospirillum*, *Pseudomonas*, *Agrobacterium*, *Arthrobacter*, *Burkholderia*, *Caulobacter*, *Chromobacterium*, *Erwinia*, *Flavobacterium*, *Micrococcus*, *Serratia*, etc.
2. Intracellular (iPGPR) – those which exist inside root cells, generally in specialized nodular structures. Most of rhizobacteria belonging to this group are Gram-negative rods, and examples include genera such as *Allorhizobium*, *Azorhizobium*, *Bradyrhizobium*, *Mesorhizobium*, and *Rhizobium* (Figueiredo et al. 2011; Bhattacharya and Jha 2012).

PGPR can also be classified based on their functional activities as (1) biofertilizers increasing the availability of nutrients to plants, (2) phytostimulators causing plant growth promotion through phytohormone production, (3) rhizoremediators degrading organic pollutants, and (4) biopesticides controlling diseases, mainly by the production of antibiotics and antifungal metabolites (Somers et al. 2004; Antoun and Prevost 2005).

18.1.3 Mechanism of Plant Growth Promotion by PGPR

Plant growth promotion mediated by PGPR occurs through the production of various substances and mechanisms resulting in the alteration of the whole microbial community in rhizosphere region. Large quantities of growth-promoting substances

Table 18.2 Growth-promoting substances released by PGPR

S. No.	Plant growth-promoting substances	PGPR
1.	IAA	*Pseudomonas putida, Pseudomonas aeruginosa, Klebsiella* sp., *Rhizobium* sp., *Enterobacter* sp., *Bacillus* sp., *Bradyrhizobium* sp., *Mesorhizobium* sp., *Paenibacillus polymyxa, Acinetobacter* spp., *Azospirillum* sp., *Rahnella aquatilis, Serratia marcescens, Stenotrophomonas maltophilia, Burkholderia, Azotobacter* sp., *Brevibacillus* spp., *Xanthomonas* sp., *Sphingomonas* sp., *Mycobacterium* sp., *Rhodococcus* sp., and *Cellulomonas* sp.
2.	Gibberellin and kinetin	*Azotobacter chroococcum*
3.	HCN and ammonia	*Pseudomonas* sp., *Rhizobium* sp., *Bradyrhizobium* sp., *Enterobacter* sp., *Bacillus* sp., *Klebsiella* sp., *Mesorhizobium* sp., and *Serratia marcescens*
4.	Cytokinins	*Rhizobium leguminosarum*
5.	ACC deaminase	*Kluyvera ascorbata, Rahnella aquatilis, Stenotrophomonas maltophilia, Acinetobacter* sp., *Pseudomonas* sp., and *Burkholderia*
6.	Siderophores	*Brevibacterium* sp., *Bacillus* sp., *Azotobacter* sp., *Bradyrhizobium, Rhizobium, Mesorhizobium, Kluyvera ascorbata, Variovorax paradoxus, Rhodococcus* sp., *Pseudomonas* sp., *Burkholderia, Enterobacter* sp., *Serratia marcescens, Paenibacillus polymyxa, Proteus vulgaris, Ralstonia metallidurans, Acinetobacter* spp., and *Klebsiella* sp.
7.	Exopolysaccharides	*Pseudomonas* sp., *Rhizobium* sp., *Bradyrhizobium* sp., *Mesorhizobium* sp., *Enterobacter* sp., *Bacillus* sp., and *Klebsiella* sp.

Adapted from Ahemad and Kibret (2014)

are produced by these rhizosphere microorganisms that influence the overall morphology and growth of the plants (Table 18.2). The beneficial effects of these rhizobacteria on plant growth can be direct or indirect. PGPR promote plant growth directly by either facilitating or increasing the availability of nutrients such as nitrogen, phosphorus, and other essential minerals or modulating plant hormone levels, stimulating of root growth, or degrading organic pollutants and plant stress control. Rhizobacteria can also promote plant growth indirectly by mechanisms of biological control by reducing the disease incidence by inhibiting various plant pathogens and development in the form of biocontrol agents which include antibiosis, induction of systemic resistance, and competition for nutrients and niches (Kloepper and Schroth 1981; Glick 2012).

A wide range of plant growth-promoting traits have been studied in various microbes (Fig. 18.2). They include phytohormone production (Joo et al. 2005; Tank and Saraf 2010; Ahemad and Khan 2012c); production of siderophores (Tian et al. 2009; Jahanian et al. 2012); production of compounds such as 1-aminocyclopropane-1-carboxylate (ACC), hydrogen cyanate (HCN), and ammonia; nitrogenase

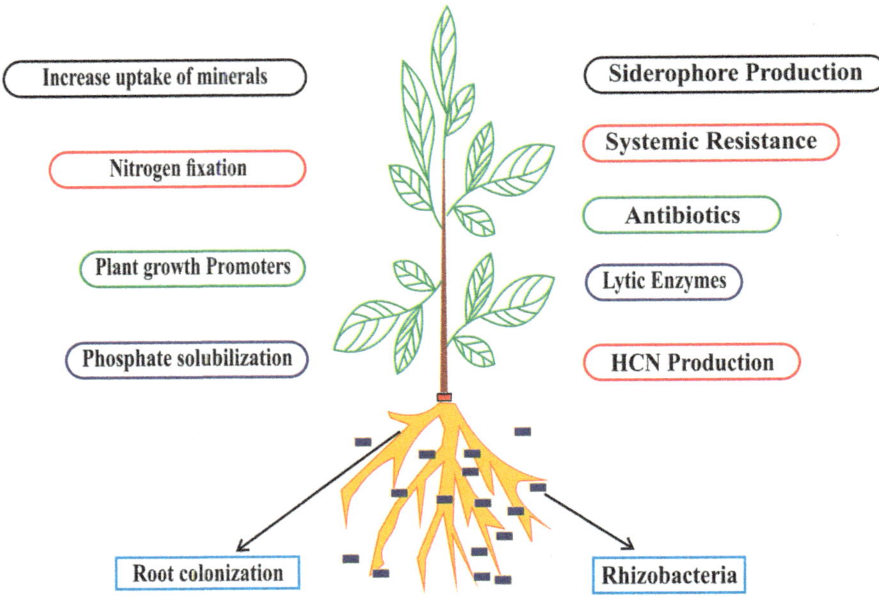

Fig. 18.2 Mechanisms of plant growth promotion by rhizobacteria

activity (Glick 2005); phosphate solubilization (Ahemad and Khan 2012c); detoxification of heavy metals (Khan 2005; Wani and Khan 2010; Ma et al. 2011a); pesticide degradation or tolerance (Ahemad and Khan 2012a, b); salinity tolerance (Mayak et al. 2004; Tank and Saraf 2010); and biological control of phytopathogens and insects (Murphy et al. 2000; Hynes et al. 2008; Russo et al. 2008). These traits of microorganisms are responsible for the plant growth promotion and increased yield through the action of multiple mechanisms (Bashan and Holguin 1997). Thus PGPR offers an attractive way to replace the use of chemical fertilizers, pesticides, and other supplements, increasing their potentiality in agriculture.

18.2 Novel Attributes of PGPR in Plant Growth Promotion

Agricultural crop yield and food security are alarmingly scaring due to rapidly evolving plant pathogens and climate changes all over the world. The increased use of chemical fertilizers and pesticides provides immediate solutions for the plant disease control and increase crop yield. But their excessive use negatively effects human health and environment. Therefore biological approaches have become important for enhancing/ increasing the crop production especially among agronomists and environmentalists. Diverse rhizobacteria possessing different mechanisms are now being used as bioinoculants all over the world to promote growth and development in plants. Bioinoculants are easy to deliver and cause an increase in biomass production and crop yield. Although hazardous synthetic fertilizers and pesticides are replaced by biofertilizers, biopesticides, and biocontrol agents derived from living microbes, alternatives are searched for, owing to their high costs, their reduced efficiency, and inconsistent field performance (Glare et al. 2012). Research is being carried out to explore new attributes of microbes in promoting plant growth and crop protection. Emission of volatile organic compounds (VOCs) is one of the most widespread mechanisms by which microorganisms modulate growth and development of plants. Microbial volatile organic compounds (MVOCs) can become an alternative to chemicals in providing disease resistance against plant pathogens and can be exploited as a cost-effective strategy for enhancing plant growth and productivity.

18.2.1 Volatile Organic Compounds (VOCs)

Volatiles organic compounds (VOCs) are low-molecular-weight compounds with high vapor pressure. They exist in the gaseous state at room temperature and are characterized by low to medium water solubility. All these properties of VOCs allow them to easily evaporate into air. VOCs can be produced through industrial processes, usually formed as by-products during the manufacture of paints, petroleum fuels, pharmaceuticals, refrigerants, household cleaners, and other products. VOCs can also be produced by microorganisms as a part of their metabolism called as microbial volatile organic compounds (MVOCs).

18.2.2 Microbial Volatile Organic Compounds (MVOCs)

Microbial volatile organic compounds (MVOCs) are a type of volatile compounds produced by microorganisms during their metabolism. Microbial volatile organic compounds (MVOCs) are produced by different groups of microbes especially bacteria and fungi. Very few, around 1,000 MVOCs released by 400 bacteria and fungi, have been described so far in the literature (Effmert et al. 2012; Lemfack et al. 2014). A high proportion of unknown compounds are yet unexplored as revealed by GC-MS analyses, suggesting a great potential for the discovery of new compounds.

MVOCs are a complex mixture of low-molecular-weight lipophilic compounds and include low-molecular-weight alcohols, aldehydes, ketones, amines, terpenes, aromatic and chlorinated hydrocarbons, and sulfur-based carbon compounds. Furfural, butanoic acid, propanoic acid, 5-hydroxy-methyl-furfural, β-caryophyllene, geosmin, 2-methyl-isoborneol, 1-octen-3-ol, α-pinene, camphene, camphor, methanol, and acetaldehyde are among the most frequently emitted compounds. MVOCs are derived from different biosynthetic pathways during microbial metabolism – primary and secondary metabolism. In primary metabolism, MVOCs are produced as by-products, while the organism breaks down food in the environment to extract nutrients needed for the maintenance of cell structures. Examples include ethanol, 1-octen-3-ol, 2-octen-1-ol, and benzyl cyanide produced by some fungi such as

Aspergillus niger, *Aspergillus flavus*, and *Penicillium roqueforti* and *Botrytis cinerea*. In secondary metabolism, the production of MVOCs is driven by the competition for resources in a nutrient-poor environment. Examples include 2-methyl-isoborneol, geosmin (1-10-dimethyl-*trans*-9-decalol), and terpenes produced by fungi such as *Chaetomium* sp., *Penicillium aurantiogriseum* and *Penicillium vulpinum*, and *Aspergillus*. Few bacteria such as *Streptomyces griseus* and *Streptomyces odorifer* produce MVOCs such as geosmin, 3-methyl-butanol, and 2-methyl-isoborneol. These emitted volatile compounds vary quantitatively and qualitatively depending on the age and genetic profiles of the producing species as well as also on external variables such as substrate, temperature, moisture level, and pH of microbial growth (Sunesson et al. 1995; Claeson et al. 2002; Matysik et al. 2008).

A number of microorganisms interact with different ecological components via the volatiles they release. MVOCs are produced both below- and aboveground and interfere with the rhizosphere and the atmosphere in different ways (Tirranen and Gitelson 2006; Wenke et al. 2010; Kanchiswamy et al. 2015). VOCs play an important role in signaling between species that are present in a common ecological niche as they can diffuse through the atmosphere and the soil. Different organisms respond variedly to different MVOCs produced, and thus complex interactions can result at trophic levels. Moreover, species-specific MVOCs can also serve as marker compounds for the selective detection of microorganisms in the environment (Fiedler et al. 2001).

Microbial volatile organic compounds can travel through the atmosphere, porous soils, and liquids and serve as chemical windows through which the information about the molecular basis of microbial activities is released. They function as semiochemicals and help in mediating both short- and long-distance interactions at intercellular and organisms level (Liang et al. 2008; Korpi et al. 2009). Thus MVOCs are considered as ideal info-chemicals and are responsible for inter- and intraorganismic communication and interactions between plants, antagonists, and symbionts both below- and aboveground (Beattie and Torrey 1986; Maffei 2010; Maffei et al. 2011; Morath et al. 2012; Kanchiswamy et al. 2015).

18.3 Bacterial VOCs

Bacteria produce a wide variety of volatile compounds depending on the specific metabolism or metabolic pathway(s) active in the bacteria. Bacterial VOCs comprise of hydrocarbons, alcohols, ketones, alkanes, alkenes, esters, sulfur compounds, and terpenoids (Fig. 18.3). They occur over a range of concentrations and can act over long distances (Wheatley 2002; Schulz and Dickschat 2007; Kai et al. 2009). These volatiles emitted by bacteria trigger many physiological changes in a broad range of organisms and influence interactions among various populations and communities. Bacterial volatiles play an important role in bacterial–plant, bacterial–bacterial, and bacterial–fungal interactions and affect either positively or negatively.

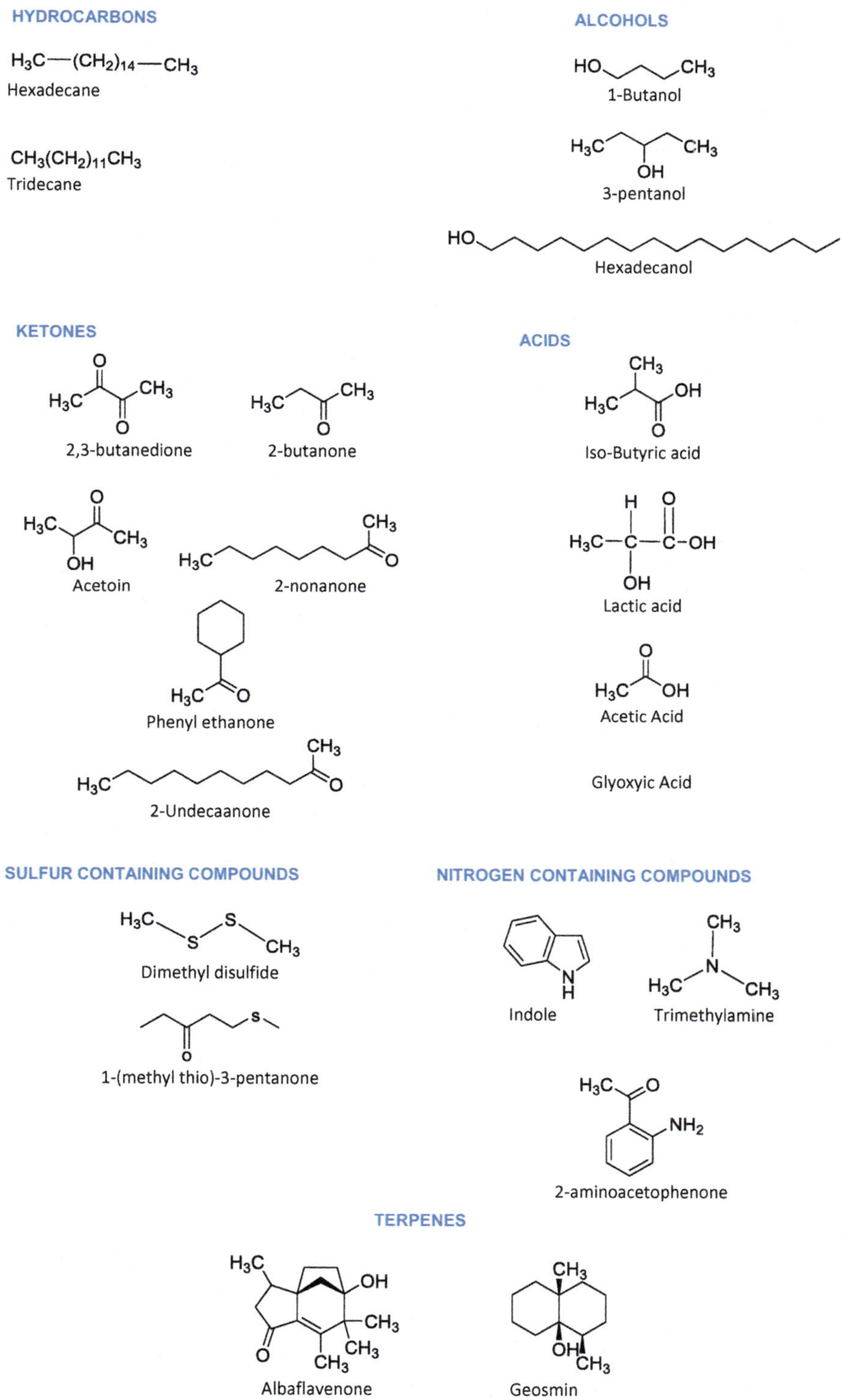

Fig. 18.3 Bacterial volatile organic compounds (VOCs)

They cause plant growth promotion, induce systemic resistance in plants, are effective against a wide range of plant-pathogenic bacteria and fungi, and act as biocontrol agents (Chen et al. 2008; Kai et al. 2008; Wan et al. 2008; Lee et al. 2010; Leroy et al. 2011; Davis et al. 2013; D'Alessandro et al. 2014; Kanchiswamy et al. 2015). These progressive studies on MVOCs made by various scientists demonstrate critical roles in multitrophic interactions affecting agriculture and entire ecosystem.

18.4 MVOCs in Bacterial–Plant Interactions

Many bacteria colonize around the host plant and establish themselves in the rhizosphere region. Extensive communication occurs between soil microorganisms and plants during the different stages of plant development by signal molecules produced from the two partners. Bacteria play an important role by producing plant growth-regulating substance such as auxins and/or cytokinins (Ortiz-Castro et al. 2009). Apart from these, the MVOCs produced are known to involve in various inter- and intraspecific interactions, above- and belowground, resulting in genetic, phenotypic, and morphologic alteration of both the interacting organisms (Effmert et al. 2012; Penuelas et al. 2014; Piechulla and Degenhardt 2014).

VOC-mediated interactions between bacteria and plants are widespread. Bacterial volatile organic compounds (VOCs) are highly complex in nature and act as signaling molecules causing an interaction between bacteria and plants in the ecosystem. They are found to have varied effects causing positive or negative impact on plant growth. These VOCs released by bacteria can promote plant growth. Many studies have shown bacteria to produce volatiles that cause plant growth promotion such as *Bacillus* sp. (Ryu et al. 2003, 2004, 2005; Farag et al. 2006; Zhang et al. 2007; Yang et al. 2009) and *Pseudomonas chlororaphis* (Han et al. 2006). Deleterious effects of bacterial volatiles on plants such as chlorosis, inhibition of plant growth, and plant death have also been observed (Vespermann et al. 2007; Blom et al. 2011b). The negative effects observed on plant growth include both inorganic and organic compounds. Inorganic volatiles include hydrogen cyanide (HCN) and NH3, and organic volatile compounds include dimethyl disulfide and 3-phenylpropionic acid (Rudrappa et al. 2008; Chung et al. 2010; Kai et al. 2010; Blom et al. 2011b; Weise et al. 2013).

18.4.1 MVOC Role in Plant Growth Promotion

Rhizosphere bacteria are the bacteria which preferentially live in the soil closely associated with the plant roots. A characteristic property of these rhizobacteria is production of plant growth-modulating volatiles. These MVOCs usually have molecular mass below 300 daltons with relatively low boiling points. They are lipophilic in nature. Rhizosphere bacteria emit different VOCs that can modulate plant growth promotion. VOCs produced by rhizobacteria show antimicrobial property against plant-pathogenic microorganisms and help in plant growth promotion. A number of research works have been published reporting the effect of volatiles on plant growth promotion. Meldau et al. (2013) found that *Bacillus* B55 strain isolated from *Nicotiana attenuata* (coyote tobacco) roots released VOCs which has strong plant growth promotion effects on wild-type *N. attenuata*. A study by Blom et al. (2011a) showed that many soil and rhizosphere strains produced volatiles which had significant effects on plant growth of *Arabidopsis thaliana*. Studies by Kai et al. (2010) have shown that *Bacillus subtilis* (GB03) and *Bacillus amyloliquefaciens* (IN937a) emitted two volatiles, namely, 2,3-butanediol and acetoin, which had a positive effect on the growth of *Arabidopsis thaliana*.

18.4.2 MVOC Role in Inducing Phenotypic Plant Responses

N-Acyl homoserine lactones (AHLs) belong to a class of amino compound-containing lipids produced by many plant-associated bacteria, especially PGPR. They mediate communication between bacterial cells and produce MVOCs inside the plant and play a role in morphogenetic processes of plants. AHLs alter gene expression in roots and shoots and modulate defense and cell growth responses in plants. Studies by Ortiz-Castro et al. (2008) and Von Rad et al. (2008) have shown that AHL compounds such as *N*-hexanoyl-homoserine lactone and *N*-3-oxo-hexanoyl-homoserine lactone, *N*-octanoyl-homoserine lactone, etc., showed an effect on root architecture and altered primary root growth, lateral root formation, and root hair development of *Arabidopsis*. Volatiles released from different microbial species are also found to have an effect on leaf starch metabolism and promoted starch accumulation in leaves of both mono-and dicotyledonous plants (Ezquer et al. 2010).

18.4.3 MVOC Role in Induced Systemic Response

MVOCs also act as initiators of defense responses in plants mediated through induced systemic response (ISR) (Ryu et al. 2003, 2004; Cortes-Barco et al. 2010a, b; D'Alessandro et al. 2014). Application of 2,3-BD to the soil protected against fungal pathogens *Microdochium nivale*, *R. solani*, or *Sclerotinia homoeocarpa* and had shown to reduce the diseased leaf area of *Agrostis stolonifera* by 20–40 %. Application of 3-pentanol and 2-butanone on cucumber seedlings triggered plant systemic defense responses against *P. syringae* pv. *lachrymans*. These compounds play a role in an indirect defense strategy that protects plants from herbivores by inducing gene expression of plant green leaf volatile signaling pathway to attract natural enemies of pests (Scala et al. 2013). Studies by Song and Ryu (2013) also showed that these compounds result in a significant increase in the number of ladybird beetles, *Coccinella septempunctata*, a natural enemy of aphids.

18.5 MVOCs in Bacterial–Bacterial Interactions

MVOCs play key roles in interspecific interactions among bacteria. VOCs of bacteria can influence the metabolism in certain bacterial species and stimulate their growth. For example, the growth of *Pseudomonas fluorescens* is stimulated by volatiles produced by *Collimonas pratensis* and *Serratia plymuthica.* A number of unique volatile compounds emitted by *C. pratensis* and *S. plymuthica* include benzonitrile, methyl thiocyanate, *S*-methyl thioacetate, and DMDS. Specific MVOCs are also by these bacteria such as 2-methyl propanal, ethenyl acetate, 3-methyl-2-pentanone, methyl 2-methylbutanoate, 3-hexanone, myrcene, terpinene, methyl salicylate, etc., produced by *C. pratensis* and chlorobenzene, dimethylsulfone, ethyl butanoate, 2-pentadecanone 1H-pyrrole, 2-octanone, and 5-dodecanone produced by *S. plymuthica* (Garbeva et al. 2014a).

MVOCs are also known to inhibit the growth of certain bacteria. Examples include 1-undecene as produced by some *P. fluorescens* and DMDS as produced by *S. plymuthica* strains which inhibited the growth of *Agrobacterium tumefaciens* and *A. vitis* strains in vitro. It has also been demonstrated that the growth of *Burkholderia cepacia* complex (Bcc) strains is inhibited through the synthesis of VOCs by bacteria. The main VOCs emitted by the *P. chlororaphis* strain 449 such as 1-undecene, 2-nonanone, 2-heptanone, and 2-undecanone DMDS were effective against *A. tumefaciens* strain C58 and the cyanobacterium *Synechococcus* sp. (Papaleo et al. 2012, 2013; Orlandini et al. 2014; Popova et al. 2014).

The AHLs produced by certain bacteria have the ability to disrupt quorum sensing networks in Gram-negative bacteria which usually regulates characteristics such as bioluminescence, biofilm formation, and production of virulence factors, antibiotics, and pigments. This phenomenon is termed as quorum quenching (QQ) and can be considered as a new approach in controlling plant pathogens (Rasmussen and Givskov 2006; Chernin et al. 2011).

18.6 MVOCs in Bacterial–Fungal Interactions

MVOCs have both positive and antagonistic interactions between bacteria and fungi in the rhizosphere (Effmert et al. 2012). Many bacterial volatiles have suppressive effects on soil fungi that might be harmful to agricultural crops. Rhizobacterial isolates like *Serratia* sp. (*S. plymuthica, S. odorifera*), *Pseudomonas* sp. (*P. fluorescens, P. trivialis*), and *Stenotrophomonas* sp. (*S. maltophilia* and *S. rhizophila*) synthesize and emit complex blends of MVOCs that inhibit growth of fungi – phytopathogens and nonpathogens (Vespermann et al. 2007; Zou et al. 2007; Kai et al. 2010; Verginer et al. 2010; Garbeva et al. 2014a, b). *Staphylococcus pasteuri* strain inhibited the mycelia growth of fungi *T. borchii* due to the production of MVOCs such as γ-patchoulene (antifungal), 3-methyl butanal, and 1-octen 3-ol. Typical metabolites of the *Staphylococcus* sp. were 2-undecanone, 3-methylbutanoate, 2-nonanone, ethanethioic acid, 2-methylbutan-1-ol, 4-methyl-2-heptanone, and dimethyl trisulfide (Barbieri et al. 2005).

Studies have shown the negative effects of VOCs produced by bacteria on growth of *R. solani*, a soilborne plant pathogen. Kai et al. (2007) reported that MVOCs such as β-phenylethanol and 2-(benzyloxy)benzonitrile emitted by *Stenotrophomonas* sp., *Serratia* sp., *Pseudomonas* sp., and *Bacillus subtilis* showed strong negative influence on the mycelial growth of *R. solani*. Weise et al. (2012) have reported that *Xanthomonas campestris* pv. *vesicatoria 85–10* emit more than more than 50 volatile compounds (mostly ketones and methylketones) to cause either promotion or inhibition effects on the fungus *R. solani*. Elshafie et al. (2012) reported that *Burkholderia gladioli* pv. *agaricola* strain produced MVOCs, the most effective being limonene compound, which inhibited the growth of fungi *F. oxysporum and R. solani.*

Zhang et al.'s (2013) studies showed that *Bacillus atrophaeus CAB-1* produced many volatile compounds, the most abundant being the O-anisaldehyde, hexadecane, and 2,3-dimethoxybenzamide, and were found to inhibit the growth of the fungal pathogen *Botrytis cinerea.* Also MVOCs such as DMDS, dimethyl trisulfide, 2-undecanone, S-methyl methanethiosulfonate 4-octanone, and 1-phenylpropan-1-one emitted by *Burkholderia ambifaria* caused a significant growth inhibition of two phytopathogenic fungi, *R. solani* and *Alternaria alternata* (Groenhagen et al. 2013). *B. tropica* strain MTo431 emitted a range of VOCs such as α-pinene, DMDS, ocimene, limonene, and fenchone which are known to have antagonistic effect on fungi and significantly inhibited the mycelial growth of four plant-pathogenic fungi – *Colletotrichum gloeosporioides*, *Fusarium culmorum*, *F. oxysporum*, and *Sclerotum rolffsi* (Tenorio-Salgado et al. 2013).

18.7 Conclusion

Volatile organic compounds emitted by microorganisms (MVOCs) are involved in various kinds of interactions between plants and microbes – antagonistic effects, mutualistic effects, and regulation of cellular and developmental processes at intra- and interspecies level as well as modification of their surrounding environments. MVOCs can modulate the physiology of plants and microorganisms especially impacting plant health. Bacterial VOCs are chemically heterogeneous molecules and produced in low concentrations. They are fully biodegradable and have no hazardous effects as in case of use of chemical pesticides or fertilizers. Thus MVOCs can be exploited as an ecofriendly and cost-effective approach for sustainable agriculture. The vast diversity of microbial communities in nature and their importance in modulating ecology, health, and growth of plants necessitate the need to systematically explore and characterize the biological functions and ecological roles of plethora of microbial VOCs. This will not only help in discovering novel mechanisms for controlling diverse biological processes critical to plant health but also provide practical solutions to various agricultural and environmental problems.

- Microbes produce a wide range of MVOCs and exert an immense effect on plant growth and development. Till now, only a few VOCs emitted by microbes (bacteria and fungi) have been identified and characterized. There is a need to explore new microbial volatile compounds and characterize their biological functions and ecological roles and study the underlying mechanisms so that they can be employed for controlling plant health thus reducing agricultural and environmental problems.

- Most of the studies conducted on MVOCs are in vitro studies (laboratory). In situ studies have to be performed, i.e., field trials are to be done to demonstrate their effects on growth and development of plants. Moreover physiological and molecular studies should be conducted for a better understanding of the role of MVOCs at field level. New understanding of the importance of MVOCs for crop plants both at the lab and open field conditions will provide further scientific evidence to adopt them for sustainable crop protection. All these studies can be used to develop production strategies and assess the cost effectiveness of naturally produced MVOCs for crop welfare and sustainable agriculture.
- MVOCs may modulate growth or defense of plants in a species-dependent manner. The effect of these volatile compounds varies depending on crop species, soil types, and environmental conditions. Thus there is difference in results obtained between lab scale studies and field trials. This implies the necessity to evaluate single MVOCs as well as blends of different MVOCs in modulating growth and defense of different crop species both at lab and field conditions.
- MVOCs should be characterized for their bioactive molecules, their proper bioactive dosage, and their role on plant growth. The requirement of energy and resources for the synthesis of MVOCs should also be considered. MVOCs may have side effects, many of them exert inhibitory effects, and some of them are also toxic. Hence the dose-response effect on specific crops is assessed before their use and then can be suggested for crop protection and productivity.
- Another challenging aspect is the manner of application of MVOCs since most of them have rapid evaporation rates, thus making them difficult to use in open field conditions. An appropriate method of MVOCs delivery in the field is still lacking, and therefore future studies are needed to understand and devise a better, cost-effective, durable, and efficient delivery of MVOCs.

The future research works on MVOCs should focus on expanding the knowledge on the MVOC biodiversity, exploring the physiological and ecological roles of single as well as blends of MVOC mixtures, establishing plant response analyses to MVOCs, and devising a more efficient delivery system to crop fields. Thus the potential MVOCs will provide a plethora of applications for protection, growth, and development of plants paving a way for sustainable agriculture.

References

Ahemad M (2012) Implications of bacterial resistance against heavy metals in bioremediation: a review. IIOAB J 3:39–46

Ahemad M, Khan MS (2009) Effect of insecticide-tolerant and plant growth promoting *Mesorhizobium* on the performance of chickpea grown in insecticide stressed alluvial soils. J Crop Sci Biotechnol 12:213–222

Ahemad M, Khan MS (2012a) Effect of fungicides on plant growth promoting activities of phosphate solubilizing *Pseudomonas putida* isolated from mustard (Brassica compestris) rhizosphere. Chemosphere 86:945–950

Ahemad M, Khan MS (2012b) Ecological assessment of biotoxicity of pesticides towards plant growth promoting activities of pea (*Pisum sativum*)-specific *Rhizobium* sp. strain MRP1. Emirates J Food Agric 24:334–343

Ahemad M, Khan MS (2012c) Evaluation of plant growth promoting activities of rhizobacterium *Pseudomonas putida* under herbicide-stress. Ann Microbiol 62:1531–1540

Ahemad M, Kibret M (2014) Mechanisms and applications of plant growth promoting rhizobacteria: current perspective. J King Saud Univ Sci 26:1–20

Ahemad M, Malik A (2011) Bioaccumulation of heavy metals by zinc resistant bacteria isolated from agricultural soils irrigated with wastewater. Bacteriol J 2:12–21

Ahmad F, Ahmad I, Khan MS (2008) Screening of free-living rhizospheric bacteria for their multiple plant growth promoting activities. Microbiol Res 163:173–181

Antoun H, Prevost D (2005) Ecology of plant growth promoting rhizobacteria. In: Siddiqui ZA (ed) PGPR: biocontrol and biofertilization. Springer, Dordrecht, pp 1–38

Barbieri E, Gioacchini AM, Zambonelli A, Bertini L, Stocchi V (2005) Determination of microbial volatile organic compounds from *Staphylococcus pasteuri* against tuber borchii using solid-phase microextraction and gas chromatography/ion trap mass spectrometry. Rap Comm Mass Spectrom 19:3411–3415

Barea JM, Pozo MJ, Azcon R, Aguilar CA (2005) Microbial co-operation in the rhizosphere. J Exp Bot 56:1761–1778

Bashan Y, Holguin G (1997) Azospirillum-plant relationships: environmental and physiological advances (1990–1996). Can J Microbiol 43:103–121

Beattie SE, Torrey GS (1986) Toxicity of methanethiol produced by Brevibacterium linens toward *Penicillium expansum*. J Agr Food Sci 34:102–104

Bhattacharyya PN, Jha DK (2012) Plant growth-promoting rhizobacteria (PGPR): emergence in agriculture. World J Microbiol Biotechnol 28:1327–1350

Blom D, Fabbri C, Connor EC, Schiestl FP, Klauser DR et al (2011a) Production of plant growth modulating volatiles is widespread among rhizosphere bacteria and strongly depends on culture conditions. Environ Microbiol 13:3047–3058

Blom D, Fabbri C, Eberl L, Weisskopf L (2011b) Volatile-mediated killing of *Arabidopsis thaliana* by bacteria is mainly due to hydrogen cyanide. Appl Environ Microb 77:1000–1008

Braud A, Jezequel K, Bazot S, Lebeau T (2009) Enhanced phytoextraction of an agricultural Cr-, Hg- and Pb-contaminated soil by bioaugmentation with siderophore producing bacteria. Chemosphere 74:280–286

Chandler D, Davidson G, Grant WP, Greaves J, Tatchell GM (2008) Microbial biopesticides for integrated crop management: an assessment of environmental and regulatory sustainability. Trends Food Sci Tech 19:275–283

Chen H, Xiao X, Wang J, Wu LJ, Zheng ZM et al (2008) Antagonistic effects of volatiles generated by *Bacillus subtilis* on spore germination and hyphal growth of the plant pathogen, *Botrytis cinerea*. Biotechnol Lett 30:919–923

Chernin L, Toklikishvili N, Ovadis M, Kim S, Ben-Ari J et al (2011) Quorum-sensing quenching by rhizobacterial volatiles. Environ Microbiol Rep 3:698–704

Chung EJ, Park JH, Park TS, Ahn JW, Chung YR (2010) Production of a phytotoxic compound, 3-phenylpropionic acid by a bacterial endophyte, *Arthrobacter humicola* YC6002 isolated from the root of *Zoysia japonica*. Plant Pathol J 26:245–252

Claeson AS, Levin JO, Blomquist G, Sunesson AL (2002) Volatile metabolites from microorganisms grown on humid building materials and synthetic media. J Environ Monit 4:667–672

Cortes-Barco AM, Goodwin PH, Hsiang T (2010a) Comparison of induced resistance activated by benzothiadiazole, (2R,3R)-butanediol and an isoparaffin mixture against anthracnose of *Nicotiana benthamiana*. Plant Pathol 59:643–653

Cortes-Barco AM, Hsiang T, Goodwin PH (2010b) Induced systemic resistance against three foliar diseases of *Agrostis stolonifera* by (2R,3R)-butanediol or an isoparaffin mixture. Ann Appl Biol 157:179–189

D'Alessandro M, Erb M, Ton J, Brandenburg A, Karlen D et al (2014) Volatiles produced by soil-borne endophytic bacteria increase plant pathogen resistance and affect tritrophic interactions. Plant Cell Environ 37:813–826

Dakora FD, Phillips DA (2002) Root exudates as mediators of mineral acquisition in low-nutrient environments. Plant Soil 245:35–47

Davis TS, Crippen TL, Hofstetter RW, Tomberlin JK (2013) Microbial volatile emissions as insect semiochemicals. J Chem Ecol 39:840–859

Effmert U, Kalderas J, Warnke R, Piechulla B (2012) Volatile mediated interactions between bacteria and fungi in the soil. J Chem Ecol 38:665–703

Elshafie HS, Camele I, Racioppi R, Scrano L, Iacobellis NS et al (2012) In vitro antifungal activity of Burkholderia gladioli pv. agaricicola against some phytopathogenic fungi. Int J Mol Sci 13:16291–16302

Ezquer I, Li J, Ovecka M, Baroja-Fernandez E, Jose Munoz F et al (2010) Microbial volatile emissions promote accumulation of exceptionally high levels of starch in leaves in mono- and dicotyledonous plants. Plant Cell Physiol 51:1674–1693

Farag MA, Ryu CM, Sumner LW, Pare PW (2006) GC-MS SPME profiling of rhizobacterial volatiles reveals prospective inducers of growth promotion and induced systemic resistance in plants. Phytochemistry 67:2262–2268

Fiedler K, Schutz E, Geh S (2001) Detection of microbial volatile organic compounds (MVOCs) produced by moulds on various materials. Int J Hyg Environ Health 204:111–121

Figueiredo MVB, Seldin L, Araujo FF, Mariano RLR (2011) Plant growth promoting rhizobacteria:fundamentals and applications. In: Maheshwari DK (ed) Plant growth and health promoting Bacteria. Springer-Verlag, Berlin, pp 21–42

Garbeva P, Hordijk C, Gerards S, De Boer W (2014a) Volatile-mediated interactions between phylogenetically different soil bacteria. Front Microbiol 5:289

Garbeva P, Hordijk C, Gerards S, De Boer W (2014b) Volatiles produced by the mycophagous soil bacterium Collimonas. FEMS Microbiol Ecol 87:639–649

GlareT CJ, Gelernter W, Jackson T, Keyhani N et al (2012) Have biopesticides come of age? Trends Biotechnol 30:250–258

Glick BR (2005) Modulation of plant ethylene levels by the bacterial enzyme ACC deaminase. FEMS Microbiol Lett 251:1–7

Glick BR (2012) Plant growth-promoting Bacteria: mechanisms and applications. Hindawi Publ Corp Sci 2012:1–15

Groenhagen U, Baumgartner R, Bailly A, Gardiner A, Eberl L et al (2013) Production of bioactive volatiles by different *Burkholderia ambifaria* strains. J Chem Ecol 39:892–906

Han SH, Lee SJ, Moon JH, Park KH, Yang KY et al (2006) GacS-dependent production of 2R, 3R-butanediol by *Pseudomonas chlororaphis* O6 is a major determinant for eliciting systemic resistance against *Erwinia carotovora* but not against *Pseudomonas syringae pv. tabaci* in tobacco. Mol Plant Mol Plant Microb Inter 19:924–930

Hayat R, Ali S, Amara U, Khalid R, Ahmed I (2010) Soil beneficial bacteria and their role in plant growth promotion: a review. Ann Microbiol 60:579–598

Hynes RK, Leung GC, Hirkala DL, Nelson LM (2008) Isolation, selection, and characterization of beneficial rhizobacteria from pea, lentil and chickpea grown in Western Canada. Can J Microbiol 54:248–258

Jahanian A, Chaichi MR, Rezaei K, Rezayazdi K, Khavazi K (2012) The effect of plant growth promoting rhizobacteria (pgpr) on germination and primary growth of artichoke (Cynara scolymus). Int J Agric Crop Sci 4:923–929

Joo GJ, Kin YM, Kim JT, Rhee IK, Kim JH et al (2005) Gibberellins-producing rhizobacteria increase endogenous gibberellins content and promote growth of red peppers. J Microbiol 43:510–515

Kai M, Effmert U, Berg G, Piechulla B (2007) Volatiles of bacterial antagonists inhibit mycelial growth of the plant pathogen Rhizoctonia solani. Arch Microbiol 187:351–360

Kai M, Vespermann A, Piechulla B (2008) The growth of fungi and *Arabidopsis thaliana* is influenced by bacterial volatiles. Plant Signal Behav 3:482–484

Kai M, Haustein M, Molina F, Petri A, Scholz B et al (2009) Bacterial volatiles and their action potential. Appl Microbiol Biotechnol 81:1001–1012

Kai M, Crespo E, Cristescu S, Harren F, Francke W et al (2010) Serratia odorifera: analysis of volatile emission and biological impact of volatile compounds on *Arabidopsis thaliana*. Appl Microbiol Biotechnol 88:965–976

Kanchiswamy CN, Malnoy M, Maffei ME (2015) Chemical diversity of microbial volatiles and their potential for plant growth and productivity. Front Plant Sci 6(151):1–23

Khan AG (2005) Role of soil microbes in the rhizospheres of plants growing on trace metal contaminated soils in phytoremediation. J Trace Elem Med Biol 18:355–364

Kloepper JW, Schroth MN (1981) Relationship of in vitro antibiosis of plant growth promoting rhizobacteria to plant growth and the displacement of root microflora. Phytopathology 71:1020–1024

Kloepper JW, Zablotowick RM, Tipping EM, Lifshitz R (1991) Plant growth promotion mediated by bacterial rhizosphere colonizers. In: Keister DL, Cregan PB (eds) The rhizosphere and plant growth. Kluwer Academic Publishers, Dordrecht, pp 315–326

Korpi A, Jarnberg J, Pasanen AL (2009) Microbial volatile organic compounds. Crit Rev Toxicol 39:139–193

Lee HH, Molla MN, Cantor CR, Collins JJ (2010) Bacterial charity work leads to population-wide resistance. Nature 467:82–85

Lemfack MC, Nickel J, Dunkel M, Preissner R, Piechulla B (2014) mVOC: a database of microbial volatiles. Nucleic Acids Res 42:D744–D748

Leroy PD, Sabri A, Heuskin S, Thonart P, Lognay G et al (2011) Microorganisms from aphid honeydew attract and enhance the efficacy of natural enemies. Nat Commun 2:348

Liang H, Zhang X, Jun RAO, Huanwen CHEN (2008) Microbial volatile organic compounds: generation pathways and mass spectrometric detection. J Chin Biotechnol 28:124–133

Maffei M (2010) Sites of synthesis, biochemistry and functional role of plant volatiles. S Afr J Bot 76:612–631

Maffei ME, Gertsch J, Appendino G (2011) Plant volatiles: production, function and pharmacology. Nat Prod Rep 28:1359–1380

Matysik S, Herbarth O, Mueller A (2008) Determination of volatile metabolites originating from mould growth on wall paper and synthetic media. J Microbiol Methods 75:182–187

MaY RM, Luo Y, Freitas H (2011) Inoculation of endophytic bacteria on host and non-host plants-effects on plant growth and Ni uptake. J Hazard Mater 195:230–237

Mayak S, Tirosh T, Glick BR (2004) Plant growth-promoting bacteria confer resistance in tomato plants to salt stress. Plant Physiol Biochem 42:565–572

Meldau DG, Meldau S, Hoang LH, Underberg S, Wuensche H et al (2013) Dimethyl disulfide produced by the naturally associated bacterium *Bacillus* sp. B55 promotes nicotiana attenuata growth by enhancing sulfur nutrition. Plant Cell 25:2731–2747

Morath SU, Hung R, Bennett JW (2012) Fungal volatile organic compounds: a review with emphasis on their biotechnological potential. Fungal Biol Rev 26:73–83

Murphy JF, Zehnder GW, Schuster DJ, Sikora EJ, Polston JE et al (2000) Plant growth-promoting rhizobacterial mediated protection in tomato against tomato mottle virus. Plant Dis 84:779–784

Orlandini V, Maida I, Fondi M, Perrin E, Papaleo MC et al (2014) Genomic analysis of three sponge-associatedArthrobacter Antarctic strains, inhibiting the growth of *Burkholderia cepacia* complex bacteria by synthesizing volatile organic compounds. Microbiol Res 169:593–601

Ortiz-Castro R, Martinez-Trujillo M, Lopez-Bucio J (2008) N-acyl-L-homoserine lactones: a class of bacterial quorum-sensing signals alter post-embryonic root development in *Arabidopsis thaliana*. Plant Cell Environ 31:1497–1509

Ortiz-Castro R, Contreras-Cornejo HA, Ias-Rodriguez L, Lopez-Bucio J (2009) The role of microbial signals in plant growth and development. Plant Sign Behav 4:701–712

Papaleo MC, Fondi M, Maida I, Perrin E, Lo Giudice A et al (2012) Sponge-associated microbial Antarctic communities exhibiting antimicrobial activity against *Burkholderia cepacia* complex bacteria. Biotechnol Adv 30:272–293

Papaleo MC, Romoli R, Bartolucci G, Maida I, Perrin E et al (2013) Bioactive volatile organic compounds from Antarctic (sponges) bacteria. New Biotechnol 30:824–838

Penuelas J, Asensio D, Tholl D, Wenke K, Rosenkranz M et al (2014) Biogenic volatile emissions from the soil. Plant Cell Environ 37:1866–1891

Piechulla B, Degenhardt J (2014) The emerging importance of microbial volatile organic compounds. Plant Cell Environ 37:811–812

Popova AA, Koksharova OA, Lipasova VA, Zaitseva JV, Katkova-Zhukotskaya OA et al (2014) Inhibitory and toxic effects of volatiles emitted by strains of Pseudomonas and Serratia on growth and survival of selected microorganisms,Caenorhabditis elegans, and Drosophila melanogaster. Biomed Res Int 2014:11

Rajkumar M, Ae N, Prasad MNV, Freitas H (2010) Potential of siderophore-producing bacteria for improving heavy metal phytoextraction. Trends Biotechnol 28:142–149

Rasmussen TB, Givskov M (2006) Quorum-sensing inhibitors as anti-pathogenic drugs. Int J Med Microbiol 296:149–161

Rudrappa T, Splaine RE, Biedrzycki ML, Bais HP (2008) Cyanogenic pseudomonads influence multitrophic interactions in the rhizosphere. PLoS One 3(4):e2073

Russo A, Vettori L, Felici C, Fiaschi G, Morini S et al (2008) Enhanced micropropagation response and biocontrol effect of *Azospirillum brasilense* Sp245 on *Prunus cerasifera* L. clone Mr.S 2/5 plants. J Biotechnol 134:312–319

Ryu CM, Farag MA, Hu CH, Reddy MS, Wei HX et al (2003) Bacterial volatiles promote growth in *Arabidopsis*. Proc Natl Acad Sci U S A 100:4927–4932

Ryu CM, Farag MA, Hu CH, Reddy MS, Kloepper JW et al (2004) Bacterial volatiles induce systemic resistance in *Arabidopsis*. Plant Physiol 134:1017–1026

Ryu CM, Hu CH, Locy RD, Kloepper JW (2005) Study of mechanisms for plant growth promotion elicited by rhizobacteria in *Arabidopsis thaliana*. Plant Soil 268:285–292

Scala A, Allmann S, Mirabella R, Haring MA, Schuurink RC (2013) Green leaf volatiles: a plant's multifunctional weapon against herbivores and pathogens. Int J Mol Sci 14:17781–17811

Schulz S, Dickschat JS (2007) Bacterial volatiles: the smell of small organisms. Nat Prod Rep 24:814–842

Somers E, Vanderleyden J, Srinivasan M (2004) Rhizosphere bacterial signalling: a love parade beneath our feet. Crit Rev Microbiol 30:205–240

Song GC, Ryu CM (2013) Two volatile organic compounds trigger plant self-defense against a bacterial pathogen and a sucking insect in cucumber under open field conditions. Int J Mol Sci 14:9803–9819

Sunesson AL, Vaes WHJ, Nilsson CA, Blomquist GR, Andersson B et al (1995) Identification of volatile metabolites from five fungal species cultivated on two media. Appl Environ Microbiol 61:2911–2918

Tank N, Saraf M (2010) Salinity-resistant plant growth promoting rhizobacteria ameliorates sodium chloride stress on tomato plants. J Plant Interact 5:51–58

Tenorio-Salgado S, Tinoco R, Vazquez-Duhalt R, Caballero-Mellado J, Perez-Rueda E (2013) Identification of volatile compounds produced by the bacterium *Burkholderia tropica* that inhibit the growth of fungal pathogens. Bioengineered 4:236–243

Tian F, Ding Y, Zhu H, Yao L, Du B (2009) Genetic diversity of siderophore-producing bacteria of tobacco rhizosphere. Braz J Microbiol 40:276–284

Tirranen LS, Gitelson II (2006) The role of volatile metabolites in microbial communities of the LSS higher plant link. Adv Space Res 38:1227–1232

Verginer M, Leitner E, Berg G (2010) Production of volatile metabolites by grape-associated microorganisms. J Agric Food Chem 58:8344–8350

Vespermann A, Kai M, Piechulla B (2007) Rhizobacterial volatiles affect the growth of fungi and *Arabidopsis thaliana*. Appl Environ Microb 73:5639–5641

Von Rad U, Klein I, Dobrev PI, Kottova J, Zazimalova E et al (2008) Response of *Arabidopsis thaliana* to N-hexanoyl-DL-homoserine-lactone, a bacterial quorum sensing molecule produced in the rhizosphere. Planta 229:73–85

Walker TS, Bais HP, Grotewold E, Vivanco JM (2003) Root exudation and rhizosphere biology. Plant Physiol 132:44–51

Wan MG, Li GQ, Zhang JB, Jiang D, Huang HC (2008) Effect of volatile substances of *Streptomyces platensis* F-1 on control of plant fungal diseases. Biol Control 46:552–559

Wani PA, Khan MS (2010) Bacillus species enhance growth parameters of chickpea (Cicer arietinum L.) in chromium stressed soils. Food Chem Toxicol 48:3262–3267

Weise T, Kai M, Gummesson A, Troeger A, von Reuss S et al (2012) Volatile organic compounds produced by the phytopathogenic bacterium *Xanthomonas campestris* pv. vesicatoria 85–10. Beilstein J Org Chem 8:579–596

Weise T, Kai M, Piechulla B (2013) Bacterial ammonia causes significant plant growth inhibition. PLoS One 8:e63538

Wenke K, Kai M, Piechulla B (2010) Belowground volatiles facilitate interactions between plant roots and soil organisms. Planta 231:499–506

Wheatley RE (2002) The consequences of volatile organic compound mediated bacterial and fungal interactions. Antonie Van Leeuwenhoek Int J Gen Mol Microbiol 81:357–364

Yang J, Kloepper JW, Ryu CM (2009) Rhizosphere bacteria help plants tolerate abiotic stress. Trends Plant Sci 14:1–4

Zhang H, Kim MS, Krishnamachari V, Payton P, Sun Y et al (2007) Rhizobacterial volatile emissions regulate auxin homeostasis and cell expansion in *Arabidopsis*. Planta 226:839–851

Zhang X, Li B, Wang Y, Guo Q, Lu X et al (2013) Lipopeptides, a novel protein, and volatile compounds contribute to the antifungal activity of the biocontrol agent *Bacillus atrophaeus* CAB-1. Appl Microbiol Biotechnol 97:9525–9534

Zou CS, Mo MH, Gu YQ, Zhou JP, Zhang KQ (2007) Possible contributions of volatile-producing bacteria to soil fungistasis. Soil Biol Biochem 39:2371–2379

Microbial Volatiles and Plant Defense

19

Kalaivani K. Nadarajah

Abstract

Microbes, insects, and other organisms within the plant environment emit volatiles responsible for various biological and physiological processes that affect the well-being and growth of plants. These volatile organic compounds (VOCs) that are produced by organisms can travel a distance from the locality of emission into the atmosphere, soils, and liquids making them an efficient means of regulating short as well as long-distance interactions between organisms. In this chapter we look into the plethora of VOCs that are emitted in bacteria-bacteria, bacteria-fungi, and bacteria-plant interactions. These VOCs are involved in mutualistic, antagonistic, or symbiotic relationships that regulate cellular, developmental, and biological processes which include growth and plant defenses. Due to the large abundance of microorganisms, the function of these VOCs on other organisms remains poorly understood. Here we have reviewed the role of MVOCs with special emphasis on plant defense and health, alluding to the potential of these compounds to control various processes that result in the plants' elevated health and defense.

Keywords

Microbial volatiles • Plant defense and growth • Rhizopheric • Phyllospheric

K.K. Nadarajah (✉)
School of Environmental and Natural Resource Sciences, Faculty of Science and Technology, National University of Malaysia, Bangi, Selangor, Malaysia
e-mail: vani@ukm.edu.my

D.K. Choudhary et al. (eds.), *Volatiles and Food Security*,
DOI 10.1007/978-981-10-5553-9_19

19.1 Introduction

The plant root system emits essential and auxiliary metabolites, macromolecules, and even cells into the rhizosphere that bolsters nutrient acquisition and formation of microbial communities (Driouich et al. 2013). The chemical constituents of the soil encourages rhizobacteria (PGPR) to synthesize substances such as antimicrobials and phytohormones which increases nutrient uptake and volatile production while remaining in proximity with the root system (Farag et al. 2006; Ryu et al. 2003, 2004). Over a decade's worth of work has eluded that besides the antimicrobial compounds secreted by this group of organisms, various VOCs are also produced (Audrain et al. 2015; Garbeva et al. 2014a, b; Hunziker et al. 2015; Kanchiswamy et al. 2015; Schmidt et al. 2015). Microbial volatile organic compounds (MVOCs) are produced as part of the microbe's metabolism. These volatile compounds are found both in the phyllosphere and rhizosphere (Kai et al. 2009; Rasmann et al. 2005). PGPRs have been credited in transmitting VOCs that improve plant growth in addition to enhancing the plant's immune system (Bailly and Weisskopf 2012; Zou et al. 2010). There seems, by all accounts, to be a multipartite reaction to MVOCs and a complex trophic association attributed towards the production of MVOCs. In addition, species-specific MVOCs may be utilized in determining the presence of specific bacteria or fungi within a given environment (Fiedler et al. 2001). Bacterial VOCs serve as signals in inter- and intraspecies cell-to-cell communication that contributes towards stimulation or inhibition of plant growth, arresting phytopathogens, and the activation of induced systemic resistance within a plant system (Marmulla and Harder 2014; Wenke et al. 2010; Zou et al. 2010).

Hundreds of MVOCs are produced by bacteria and fungi (Effmert et al. 2012). Years of study have shown that volatile metabolites exhibit better contribution towards microbial-microbial or microbial and environmental interactions compared to nonvolatile compounds. Volatiles are discharged frequently, and these compounds regulate various microorganisms and microbial-related activities (Tirranen and Gitelson 2006). VOCs discharged by microorganisms are inhibitory and are sometimes bactericidal in their effect on microorganisms. The stimulative effects of VOCs however happens at a lower frequency (Tirranen and Gitelson 2006). Some examples of VOCs that are frequently emitted by organisms are furfural, butanoic acid, propanoic acid, geosmin, β-caryophyllene, 2-methylisoborneol, 1-octen-3ol, αpinene, 5-hydroxy-methyl-furfural, camphene, camphor, methanol, and acetaldehyde (Gray et al. 2010; Insam and Seewald 2010; Juenger et al. 2012; Leff and Fierer 2008; Li et al. 2004; Muller et al. 2004a, b; Perl et al. 2011; Ramirez et al. 2010; Stahl and Parkin 1996; Sundberg et al. 2013; Wenke et al. 2010). Bacterial VOCs can have a wide range of structures where mixes, for example, amines, benzaldehyde, benzothiazole, decanal, dimethyl trisulfide, cyclohexanol 2,3-butanediol, nonanal, and 2-ethyl-1-hexanol, have been classified as fungicidals that act as plant defense triggers (Cortes-Barco et al. 2010a, b; Kai et al. 2009; Ryu et al. 2004).

Although the influence of BVOCs in disease resistance and growth development has been widely recorded (Bailly and Weisskopf 2012), few have alluded to their roles on insects and other predators. Based on the reviews concerning volatiles, we

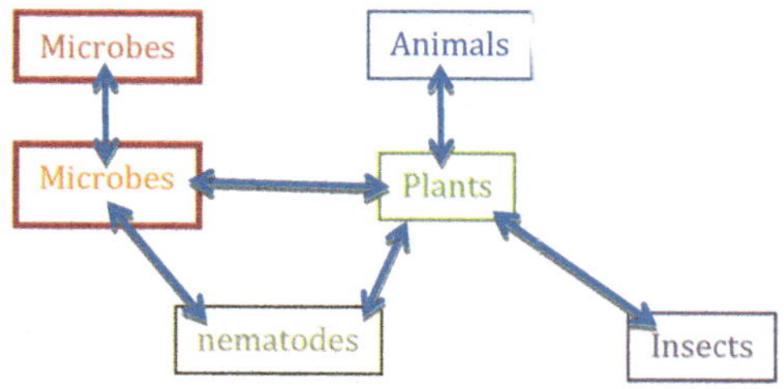

Fig. 19.1 The role of volatiles in inter and intraspecies communications between organisms in the environment

may conclude that their potential may further be harnessed to provide protection ranging from diseases to insect predation (Pineda et al. 2010, 2013; Neal and Ton 2013). In nature, VOCs have been implicated in processes ranging from cell-to-cell communication, quorum sensing, development, protection, to original differentiation (Bailly and Weisskopf 2012; Bos et al. 2013; Kanchiswamy et al. 2015; Vespermann et al. 2007). Plants themselves appear to respond to microbial volatiles, bringing about immediate or indirect contribution towards the plant's well-being and development (Bailly et al. 2014; Bitas et al. 2013; Farag et al. 2013; Vespermann et al. 2007; Weisskopf and Bailly 2013). In both under- or above ground plant structures, bacterial populaces are likely to change chemical constituents found in the environment into chemical effectors that are viable especially in proximity with the pathogen's point of entry.

With the advent of the omics platforms, bioinformatic softwares have been programmed to study metabolomics with the intention of providing a better picture of the role of volatiles in the below and above ground interactions (Aponte et al. 2013; Bunge et al. 2008; Dolch et al. 2012; Juenger et al. 2012; Kunze et al. 2013; Kuske et al. 2005; LeBouf et al. 2010; Liang et al. 2008; Mcneal and Herbert 2009; O'Hara and Mayhew 2009; Ramirez et al. 2010; Rasanen et al. 2010; Scotter et al. 2005; Thorn and Greenman 2012; Wihlborg et al. 2008). Various studies have been conducted on MVOCs, and several reviews have dealt with its biosynthesis. In this chapter, we address the microbial volatiles and their role in plant defense mechanisms, organismal interactions and possible application and involvement in sustainable agriculture. Figure 19.1 displays all the possible interactions that involve volatiles.

19.2 The Active Nature of Microbial Volatiles

Volatiles form a small percentage of metabolites exuded by microorganisms involved in the communication and induction of several biological processes. Due to their small molecular size, they are easily diffused in water and air (Insam and Seewald 2010; Wheatley 2002; Ryan and Dow 2008). In the past decade, the emanation of a significantly large number of bacterial volatiles has been detected and investigated hence demonstrating that bacterial volatiles exert their influence on other organisms within an environment or community (Romoli et al. 2014). The information garnered has increased our understanding of the concepts and complexities involved in the dynamic processes between MVOCs and plants which contribute towards the blueprints of sustainable crop protection. As numerous chemicals are exuded by bacteria and are known to impact other organisms in bacteria-plant, bacteria-fungi, and bacterial interactions, evidence shows that the initial interaction between plants and certain microbes are able to induce an immune response. These organisms, specifically *Bacillus* spp. and *Pseudomonas* spp., have been implicated as the main players in the induction of ISR. Factors such as volatiles, antibiotics, phytohormones, lipopolysaccharides, salicylic acid, and siderophore have been reported as compounds that are associated with induction of ISR in hosts (Bordiec et al. 2010; Nadarajah 2016).

19.3 Microbial Volatiles in Bacteria-Plant Interactions

Microbial volatiles are involved in metabolic pathways and are produced in varying concentrations over distances within the biospheres (Wheatley 2002). MVOCs such as alcohols, alkenes, alkanes, ketones, esters, terpenoids, and sulfur compounds have been identified in various interactions between microbes and of other organisms. PGPR especially pseudomonads are known to produce a large number of volatiles (Bhattacharyya and Jha 2012; Goswami et al. 2013; Mendes et al. 2013). These organisms, in specific the fluorescent *Pseudomonas* strains, maintain rhizospheric health and protects the plants against pathogens and predators (Hol et al. 2013). Rhizospheric bacterial strains produce differential VOC emissions. Through the recent developments in analytical techniques, the profiles of these volatiles have been expounded and in brief have been classified as low-molecular-weight, low-boiling-point, and lipophilic compounds (Gutierrez-Luna et al. 2010).

Blom et al. (2011) had analyzed the effect of BVOCs against *Arabidopsis thaliana* and identified compounds such as 1-hexanol, indole, and pentadecane as able to promote growth. In another study by Vespermann et al. (2007), ammonia and dimethyl disulfide (DMDS) were isolated from *Serratia* and *Bacillus sp.*. *Bacillus cereus*-derived DMDS significantly inhibited infections by *Botrytis cinerea* and *Cochliobolus heterostrophus* on tobacco and corn, respectively (Huang et al. 2012; Kai et al. 2010). It was also reported by Meldau et al. (2013) that the DMDS and VOCs from *Bacillus* strain B55 rescued many of *Nicotiana attenuata* wild type and ET mutant phenotypes. He also observed that the presence of DMDS inhibited the expression of sulfur assimilating genes as well as those involved in methionine biosynthesis and recycling. This is beneficial for plants growing in sulfur-deficient environments. Besides these compounds, volatiles such as butanediol (2,3-BD) and acetoin produced by *Bacillus subtilis* GB03 and *B. amyloliquefaciens* IN937a enhanced systemic resistance (ISR) in *A. thaliana* to *Erwinia carotovora* subsp. *carotovora* (Ryu et al. 2003). Hence isomers of 2,3-BD have been used to trigger ISR. Mutant of *B. subtilis* GB03 with lower levels of 2,3-BD exhibited reduced resistance to pathogens. Via transgenics in *Arabidopsis*, D'Alessandro et al. (2014) were able to evidence GB03 was dependent on ET signaling and independent of SA/JA triggered defense response in host. MVOC like 2,3-BD produced by

Enterobacter aerogenes, induced resistance of corn against *Setosphaeria turcica* (Rudrappa et al. 2010).

Rhizobacteria that colonize roots of lemon plants have been reported to produce emission of volatiles such as acetophenone, tetradecanal, and benzaldehyde. The effects of these volatiles are dependent on the concentration and the type of volatiles emitted (Gutierrez-Luna et al. 2010). *Burkholderia ambifaria* produced acetophenone, 3-hexanone, and DMDS that contribute towards increase in tissue biomass of *Arabidopsis* (Groenhagen et al. 2013). Maffei (2002) reported that a wide spectrum of proteobacteria, Gram-positive bacteria, and *Fibrobacteres* are able to live in vascular cells of *Vetiveria zizanioides*. These organisms live within the cells and utilize oil sesquiterpenes as a nutrient source and in return emit a large number of chemicals that are typical of Vetiver oils (Alifano et al. 2010; Maffei et al. 2011). Del Giudice et al. (2008) observed that when Vetiver plants were fed with cuparene, the root-associated bacteria produced a number of sesquiterpenes such as ledene, β-bourbonene, β-humulene, β-copaene, δ-cadinene, spathulenol, α-muurolene, β-caryophyllene, and viridiflorol. Pseudomonad volatiles exhibited mycelial inhibition and sporicidal activity against *Phytopthora infestans.* For these volatiles to exert any inhibitory effect on soil pathogens, sufficient accumulation of volatiles surrounding the root cap is required to enable bacteria-pathogen interactions. Mutant and wild-type *Pseudomonas fluorescens* B16 were studied for their role in plant protection and growth promotion, where it was identified that the sequestered pyrroloquinoline quinone (PQQ) was responsible for the increased growth observed in tomato and cucumber. Choi et al. (2008) reported that the application of PQQ on cucumber leaf disks brought about the foraging of oxygen radicals indicating PQQ as an antioxidant in plants.

19.4 Microbial Volatiles in Bacteria-Bacteria Interactions

Although very little is known about volatile interaction between bacteria, volatiles are known to play a role in bacterial metabolism (Garbeva et al. 2014a). While rhizobacteria may produce MVOCs that affect plants positively, their effects on other bacteria may either be positive or negative. This can be seen from the positive effects of volatiles from *Collimonas pratensis* and *Serratia plymuthica* on *Pseudomonas fluorescens'* growth, while most reports on *Pseudomonas* strains have been inhibitory on other strains (Romoli et al. 2011). One important component of bacteria cell-to-cell communication is *N*-acyl homoserine lactone (AHL), which controls quorum sensing (Chernin et al. 2011; Ng and Bassler 2009; Rasmussen and Givskov 2006; Waters and Bassler 2005). Chemical components that inhibit the quorum sensing networks form quorum quenching and are a potential approach in controlling pathogens (Ortiz-Castro et al. 2008). Chemicals such as DMDS (produced by *S. plymuthica* and *S. proteamaculans* 94), 1-undecene, 2-nonanone, and 2-undecanone (produced by *P. fluorescens* and *P. chlororaphis* strain 449) are strong suppressors of oncogenicity in plants (Dandurishvili et al. 2011; Popova et al. 2014). These

MVOCs differentially inhibited *A. tumefaciens* C58 and *Synechococcus* PCC7942 (Popova et al. 2014; Rasmussen and Givskov 2006).

Phylogenetically different bacteria (*Burkholderia* sp., *Dyella* sp., *Janthinobacterium* sp., *Pseudomonas* sp., and *Paenibacillus* sp.) were used in a study on the role of VOCs in communication and competition. Multi- and monocultures of these organisms were inoculated into soils containing artificial root exudates before analyzing the volatiles produced. In addition the effects of these volatiles were also monitored on nutrient-deprived organisms in the rhizosphere. The results revealed that any changes to the microbial interactions and the communities influenced the volatiles emitted within the environment. This study disclosed that the slower-growing *Paenibacillus* strain significantly affected the volatile production by other strains. Volatiles released by organisms that feed on root exudates in certain cases induced the development of other nutrient-deprived organisms in the soil. The present work proposes that volatiles created amid microbial interactions in the rhizosphere have profound effects on microorganisms in nutrient-depleted soils (Nadarajah 2016).

19.5 Microbial Volatiles in Bacteria-Fungi Interactions

Fungi are a major problem in agricultural areas resulting in research being conducted on the inhibition potential of these organisms by beneficial microbes such as PGPRs (Ryu et al. 2003). Organisms such as *Rhizoctonia solani*, *Fusarium* sp., *Phytopthora infestans*, *Puccinia graminis*, *Ustilago maydis*, and *Uromyces* spp. are major threats especially to countries that are dependent on their cereal production, hence threatening food security worldwide (Xu et al. 2004). Oomycetes and fungi such as *Botrytis*, *Colletotrichum*, *Alternaria*, *Fusarium*, *Pythium*, *Phytophthora*, *Verticillium*, *Rhizoctonia*, and *Sclerotinia* are responsible for crop diseases and have been tested for their response toward BVOCs. From the research conducted on the above pathogens, *Bacillus* spp. (*B. cereus* and *B. subtilis*), *Pseudomonas* spp. (*P. fluorescens* and *P. trivialis)*, and *Streptomyces* spp. (*S. plymuthica* and *S. odorifera*) have exhibited the best effects with sporadic results observed in other genera such as *Serratia* and *Burkholderia* spp. (Xu et al. 2004; Yuan et al. 2011, 2012). Kai et al. (2007) observed that both *Pseudomonas* spp. and *Serratia* spp showed overlapping similarities in the VOCs produced. However this observation may be bias due to a larger number and the ease in which the former genera are isolated and cultured hence recording more positive outcomes. A systematic analysis comparing the expanse of VOCs and their effects on fungi would be required for conclusively indicating *Pseudomonas* and *Bacillus* spp. as main players in bacteria-fungi volatile interactions (Garbeva et al. 2014a, b; Verginer et al. 2010; Yuan et al. 2012).

One key component produced by these organisms in the soil is pyrrolnitrin (PRN). PRN produced by *Burkholderia, Pseudomonas*, *Enterobacter*, *Myxococcus*, and *Serratia* exhibit broad-spectrum reactivity towards fungi (Garbeva et al. 2014a), including commercially significant crop pathogens. Hwang et al. (2002) reported the inhibition of *R. solani* infections in poinsettia when treated with PRN from

Burkholderia cepacia strain 5.5B highlighting its significance in development of biofungicides (Ligon et al. 2000). Further, volatiles emitted by *Burkholderia cepacia, B. subtilis, P. fluorescens, Pseudomonas trivialis, Serratia odorifera, S. plymuthica, Stenotrophomonas maltophilia*, and *Stenotrophomonas rhizophila* hinder mycelial growth of phytopathogens (Kai et al. 2007, 2009; Vespermann et al. 2007; Zou et al. 2007). While organisms like *Xanthomonas campestris* pv. *vesicatoria* may produce a plethora of VOCs, the main players are ketones, methyl ketones, and acetones. *Bacillus atrophaeus* CAB-1 on the other hand is reported to produce volatiles like alcohols, phenols, amines, and alkane amides. From the various studies conducted on the VOCs released by rhizospheric organisms, conflicting results were obtained on their effects on fungi and other organisms (Weise et al. 2012). Collectively compounds like DMDS, acetones, ketones, methyl ketones, amines, and amides inhibit mycelial growth of *Botrytis cinerea* (Zhang et al. 2013), *R. solani, Alternaria alternata, and F. oxysporum* (Elshafie et al. 2012; Groenhagen et al. 2013). Similarly the volatiles emitted by *Burkholderia tropica* significantly inhibited *Colletotrichum gloeosporioides, Fusarium culmorum, F. oxysporum*, and *Sclerotum rolffsi* (Minerdi et al. 2009; Tenorio-Salgado et al. 2013). In certain cases, these fungi live in symbiotic associations with bacteria and are kept in check and balance with no negative effect to the host as seen in the association between *Serratia* and *Achromobacter* spp. against the pathogenic fungi *F. oxysporum* (Minerdi et al. 2011). The main VOCs in these interactions were also reported to be DMDS, alcohols, acetones, and ketones (Minerdi et al. 2011). One largely disregarded group of fungi in the antimicrobial studies are the oomycetes. This group has gained some significance since *P. infestans* caused the great potato blight. Over these past decades, cyanogenic organisms such as *Pseudomonas* strains have been identified with potential against *P. infestans*. The chemical volatile that has been shown to inhibit *P. infestans* through headspace studies is 1-undecene. This substance reduced mycelial growth, sporangium formation, germination, and zoospore release (Hunziker et al. 2015).

19.6 The Role of MVOCs in Plant Defense

Various research teams have addressed the role of MVOCs in defense response, organismal interactions (plant, bacteria, fungi, nematode, insects), chemical attractants, and as biocontrols for application in sustainable agriculture (D'Alessandro et al. 2014; Davis et al. 2013; Leroy et al. 2011; Weise et al. 2013). Their involvement is clearly exhibited in the early stages of plant-environment interactions where the multitrophic nature of this interaction is demonstrated (Nadarajah 2016). Though not clearly understood, MVOCs have been implicated in inter- and intra-changes at the genetic, phenotypic, and morphological levels within the interacting organisms. (Effmert et al. 2012; Piechulla and Degenhardt 2014). Studies have shown the involvements of MVOCs in the regulation of metabolic, genomic, and proteomic functions within crops (Cho et al. 2008; Kwon et al. 2010; Park et al. 2013; Song and Ryu 2013). Several reviews have covered extensively the contribution of this group of chemicals in the function of hormones, induced systemic resistance, resistance against pathogens and pests, and enhanced physiological changes to crops that relate to improvement in biomass, growth, and yield (Bailly and Weisskopf 2012; Bitas et al. 2013; Effmert et al. 2012; Ortiz-Castro et al. 2009; Penuelas et al. 2014; Piechulla and Degenhardt 2014; Shannon et al. 2012). In this section, emphasis is given to the contribution of MVOCs on the above-mentioned.

The microorganisms within the soil produce an arsenal of volatiles which are either distinct or overlapping in nature and components (Vespermann et al. 2007). One major MVOC that has been implicated in the interactions between *Bacillus* spp. and *A. thaliana* and *P. chlororaphis* on *Nicotiana tabacum* is 2,3-BD (Farag et al. 2006; Ryu et al. 2003). This chemical is activated through the GacS kinase-dependent pathway (Han et al. 2006) that is modulated by ethylene and auxin homeostasis. However 2,3-BD is activated by either jasmonic acid (JA) and salicylic acid (SA) or other phytohormones resulting in cross talking between the biotic and abiotic pathways (Cho et al. 2008; Han et al. 2006; Ryu et al. 2004; Zhang et al. 2007, 2008). The same pathway is also responsible for AHL synthesis indicating that this pathway and chemical classes are crucial in microbial interactions (Han et al. 2006). AHLs trigger alteration of gene expression in various plant organs bringing about the induction of defenses and the enhancement of growth responses (Ortiz-Castro et al. 2008, 2009). In addition, medium-chain AHLs (e.g., *N*-hexanoyl-homoserine lactone, *N*-octanoyl-homoserine lactone) affect organ architecture in a dose-dependent manner (Ortiz-Castro et al. 2008, 2009).

A quick review shows that a large number of microbes in the soil trigger ISR in hosts via SA-independent pathways, which are triggered by JA, methyl JA, and ethylene (ETH). Ryu et al. (2004) observed that VOC produced by *B amyloliquefaciens* IN937a triggers ISR via ETH-independent signaling pathway, while *B. subtilis* GB03 used an ETH-dependent pathway. In this study too, in addition to triggering ISR, these microbes were shown to facilitate growth promotion (Ryu et al. 2003, 2004). As in most other reports, the major volatile that plays a crucial role in defense and growth induction is 2,3-BD and its precursor acetoin (Ryu et al. 2003, 2004). However in general the volatile blends emitted by PGPRs qualitatively and quantitatively differ in their composition within the soil and effects on hosts (Ryu et al. 2003; Nadarajah 2016).

Most reports on PGPRs and PGPFs have shown that volatiles exuded by these organisms can induce growth and at the same time enhance resistance within hosts. One such example is seen in *Phoma* sp. GS8-3 that enhances growth in tobacco while inducing resistance to *Colletotrichum higginsianum* (Naznin et al. 2014; Yamagiwa et al. 2011). In a study by Pare et al. (2005), two compounds that simulate microbial volatiles from *Bacillus* spp. (acetoin and 2,3-BD), when applied exogenously, resulted in both growth and defense induction in hosts. The defense induction through exogenous stimulant bypassed the activation of defense cascade that requires energy. This saw an increase in the activity at the cell level which consequently primed plant defense with no clear changes in gene expression in leaves (D'Alessandro et al. 2014; Lee et al. 2012). Cortes-Barco et al. (2010b) reported that when 2,3-BD was applied against soilborne pathogens such as *Microdochium nivale*, *R. solani*, or *Sclerotinia homoeocarpa*, there was a reduced impact of disease on the hosts. However when this observation was compared against that of Rudrappa et al. (2010), there was a variation in the level of protection afforded by 2,3-BD on *Pseudomonas syringae* pv. *tabaci* and *P. carotovora* subsp. *carotovora*, i.e., ISR was induced in the latter but not the former indicating that the defense pathways activated are pathogen dependent (Nadarajah 2016). In addition Rudrappa et al. (2010) also reported that *P. syringae's* ISR response was triggered by acetoin. Some of these volatiles such as 3-pentanol and 2-butanone have induced ISR against *P. syringae* pv. *lachrymans* which also indirectly caused the synthesis of other chemicals which served as an attractant for natural predators and provided protection against insect predation (Scala et al. 2013; Song and Ryu 2013). Besides growth induction, some microorganisms such as *Serratia*, *Chromobacterium*, and *Stenotrophomans* produced volatiles that inhibit plant growth and development (Weise et al. 2013; Ortiz-Castro et al. 2009). The expressional analysis of inhibitory volatiles exuded by *S. plymuthica* and *S. maltophilia* shows the involvement of W-box motif-related genes (Kanchiswamy et al. 2015). Analysis of volatile responsive genes showed some overlap in genes that were triggered by abiotic stresses. A closer examination of these genes indicated the presence of W-box motifs in the promoter regions and transcription factors (MYB73, ERF2, WRKY18) (Wenke et al. 2012). Mirmajlessi et al. (2016) reported that volatile from *Trichoderma viride* was able to improve plant root morphology and flowering phenotype. Volatiles such as 1-Octen-3-ol produced by some species of *Trichoderma*, *Penicillium*, and *Alternaria* spp. are able to enhance plant resistance to necrotrophic organisms such as *Botrytis cinerea* (Contreras-Cornejo et al. 2014; Kishimoto et al. 2007), promote growth and increase starch accumulation in plants (Ezquer et al. 2010). A nonpathogenic strain (Minerdi et al. 2009, 2011) of *Fusarium oxysporum*, MSA35, was able to produce a major volatile sequiterpene (β-caryophyllene) which is responsible for growth enhancement (Minerdi et al. 2011). While collectively the volatile compounds from bacteria and fungi are able to promote plant growth and well-being, most of this work has only been tested at the laboratory or greenhouse levels. However as the information on MVOCs continue to build on its efficiency in growth and defense promotion, recent studies have been directed to field testing of these microbes as to enhance yield as well as to reduce crop loss to disease in a sustainable manner (Song and Ryu 2013). Through the utilization of MVOCs we hope to substitute the unsustainable chemicals that are damaging to environment and human with safer methods via the utilization of beneficial microbes. Table 19.1 presents the data on some of the microbial volatiles and their effect on pathogens and hosts.

19.7 Application of MVOCs in the Field

Although the utilization of pesticides and fertilizers have provided a solution to control disease and yield enhancement, it still remains crucial for farmers and scientists to look into solutions that are geared towards sustainable agriculture and not one based on conventional norms (Nadarajah 2016). The potential of biopesticides, biocontrols, and biofertilizers in providing solution to agricultural problems (Glare et al. 2012) requires continuous isolation and identification of potent candidates. As mentioned above, MVOCs are able to interact with host and affect the metabolome, proteome, transcriptome, and physiology of the plants thus providing sustainable crop protection.

Most of the volatiles listed in Table 19.1 have been compiled from works utilizing laboratory techniques. Fiddaman and Rossall (1994) were the pioneers to study the effects of bacterial volatiles from *B. subtilis* in protecting rapeseed plants from

Table 19.1 List of MVOCs and their effects on pathogens and hosts

Beneficial organisms	Volatiles	Effect on hosts
B. amyloliquefaciens IN937a	2,3-BD and acetoin	Enhance growth and induced systemic resistance (ISR)
B. subtilis GBO3	2,3-BD and acetoin	Enhanced growth and ISR
Muscodor albus	Isoamyl acetate, 2-methyl butanol, and isobutyric acid	Inhibits a broad range of pathogens
Muscodor crispans	Wide variety of volatiles	Acts on *Pythium* sp. and *Phytophthora cinnamomi*
Phoma sp.	Mixture of VOCs, e.g., sesquiterpenoids	*Verticillium*, *Ceratocystis*, *Cercospora*, and *Sclerotinia*
Phomopsis sp.	Sabinene, 2-methyl propanol, isoamyl alcohol, and 2-propanone	Effective against a host of phytopathogens
		Proven effective in inhibition of *Colletotrichum*, *Fusarium*, and *Rhizoctonia*
P. aeruginosa PAO1, PAO14	HCN	Growth inhibition
P. chlororaphis	2,3-BD	Enhanced growth, biotic, abiotic stresses, and ISR
P. fluorescens A112	Unknown	Growth inhibition
P. trivialis 3Re2–7	Unknown	Growth inhibition
S. marcescens MG-1	Unknown	Growth inhibition
Stenotrophomonas rhizophila P69	Unknown	Growth inhibition
S. plymuthica 3Re4–18	Unknown	Growth inhibition
Stenotrophomonas maltophilia R3089	Unknown	Growth inhibition
Bacillus subtilis	Unknown	Growth inhibition and distortions in *Alternaria alternate*
Pseudomonas corrugata	Unknown	Mycelial inhibition
Burkholderia ambifaria	Dimethyl trisulfide, 2-nonanone, undecanone	Mycelial inhibition
Paenibacillus polymyxa	1-octen-3-ol, benzothiazole	Inhibits *Alternaria brassica*, *Botrytis cinerea*, and *Fusarium oxysporum*

(continued)

Table 19.1 (continued)

Beneficial organisms	Volatiles	Effect on hosts
Streptomyces globisporus JK-1	Not determined	Inhibits *Alternaria solani* and *Colletotrichum*
Bacillus cereus	Dimethyl disulfide benzenes, ketones, and aldehydes	Inhibits *Botrytis cinerea*
Bacillus amyloliquefaciens		Inhibit mycelial growth of *Fusarium oxysporum*
Pseudomonas sp.	Dimethyl trisulfide, 2-ethyl 1-hexanol, cyclohexanol, nonanal, benzothiazole	Inhibiting growth of *Sclerotinia sclerotiorum*
Xanthomonas campestris pv. *vesicatoria* 85–10	Decan-2-one	Inhibits mycelial growth of *Rhizoctonia solani*
Burkholderia ambifaria	DMDS, 2-Undecanone, dimethyl trisulfide, 4-octanone, phenylpropan-1-one, 1-phenyl-1	Inhibition of mycelial growth in *R. solani*
Paenibacillus polymyxa BMP-11	1-octen-3-ol, benzothiazole, citronellol	*Phytophthora capsici*

R. solani in soil and sand. From their studies, it was reported that nutrients play a critical role in the emission of volatiles by microbes. When void of carbon sources, there was a significant decrease in volatiles emitted as opposed to when soil was supplemented with nutrients. This suggests that in niche communities such as in the rhizosphere, the exudates from roots of plants and other microorganisms may very well play a role in providing the carboxylates required for antimicrobial volatile emission to antagonize phytopathogenic fungi. Fernando et al. (2005) also observed similar results in the inhibition of *Sclerotinia sclerotiorum* by *Pseudomonas* strains which produced sufficient and strong levels of antimicrobial agents. While there were reports that organisms showing good antagonism in vitro were also successful in providing similar results in soil, there were also reports to the contrary. This perhaps is attributed to the differences in agar nutrients versus the soil constituents (Berrada et al. 2012). However, what was clear is that the level of inhibition shown by the organisms varied strain to strain and with the hosts (Berrada et al. 2012). Volatiles inhibit the effect of pathogens such as *B. cinerea* and *Fusarium graminearum* by either direct application of the antagonistic organism or drenching the soil with volatile compounds (Alimi et al. 2012; Huang et al. 2012). Other than the reports on volatiles being used efficiently through direct application of organisms or soil drenching, only two reports have shown the efficiency of antifungal volatiles in dealing with pathogens that grow on and in plants. The first report articulates the efficiency in which *Streptomyces platensis* was able to inhibit three phytopathogens: *S. sclerotiorum*, *R. solani*, and *B. cinerea* (Wan et al. 2008). The second report was on the effect of volatiles from *Streptomyces globisporus* on *B. cinerea* infections in tomatoes.

Following the proof of concept on the applicability of volatiles as growth promoters and defense inducers, crop studies have ensued to identify the best practice in taking these volatiles to the field. In all reports concerning the use of volatiles, soil drenching or fumigation (2,3-BD, acetoin, dimethyl disulfide) with volatiles and the direct inoculation of soils with beneficial microbes were applied. The biofumigation technique utilizing chemical volatiles such as DMDS has already been utilized in suppression of pathogens in agriculture through the enrichment with volatile-producing antagonistic microbes for specific pathogens (Morales-Rodriguez et al. 2012; Meldau et al. 2013). However, the direct inoculation method would be preferred as it benefits the plants and microbes through the volatile-mediated activities generated as a consequence of the interaction between plant, microbe, insect, and nematode in any given environment. Hence continuous emphasis has to be given to the isolation, identification, and utilization of native plant-associated microbes where studies should also incorporate the identification of active compounds from the volatilome that may prove useful in sustainable agriculture specifically in disease suppression and defense activation. Jointly these applications will contribute significantly to the above- and below-ground protection of the host from potential pathogens and pests.

19.8 Conclusion and Future Prospects

Currently other than improving the plant varieties through breeding or genetic modification, and good management practices, agriculture is heavily dependent on pesticides and fertilizers to ensure yield and productivity. Farmers require an alternative to the current chemical control techniques with negative effects to the environments and human health (Weisskopf 2013). MVOCs form an interface between the host and the organisms. Although MVOCs are a good targets to adopt in providing a cheaper, efficient, sustainable, and eco-friendly solution to farmers, the breadth of knowledge in this area is still in its infancy (Blom et al. 2011; Bailly and Weisskopf 2012). Although there are thousands of microbial species, only approximately 1000 volatiles have been studied from dominant bacteria and fungi inhabitants of the soil (Lemfack et al. 2014). A fraction of these volatiles have been studied extensively at the laboratory and greenhouse level, while many more require further study. All information derived from lab and greenhouse needs to be translated to field application before it can be of any worth to the farmers (Cortes-Barco et al. 2010a, b; Han et al. 2006; Song and Ryu 2013; Weisskopf 2013).

As seen in Table 19.1, microbes produce a whole plethora of MVOCs that modulate various processes in plants such as growth enhancement and defense. Hence MVOCs may modulate multiple or very specific functions in plants. While the multifunctional property of MVOCs makes it rather appealing for use in agriculture, challenges lie in fine-tuning its application and making sure that these chemicals do not evaporate and lose their effect on plants when used in the field. Preliminary test conducted in the field with a few volatile compounds such as 2,3-BD, 3-pentanol, and 2-butanone provided inconsistent results on crops indicating the need to look into suitable delivery methods for MVOCs in the field (Cortes-Barco et al. 2010a, b; Song and Ryu 2013). In addition to the delivery mechanism of MVOCs in field, there also needs to be some form of evaluation on the "allocation of fitness costs" or "trade-off" in form of required energy/resources for its synthesis (Heil 2001; Heil and Baldwin 2002). There is much work ahead in determining the characteristic of these bioactive compounds, dosage, and specific roles on hosts (Piechulla and Degenhardt 2014). The delivery mechanism/mode of the most potent MVOCs would require study into best carrier or molecules that may be used to ensure stability, safety in storage, and ease of application in the field (Song and Ryu 2013).

Although MVOCs are perfect candidates for defense and plant well-being, many questions still remain unanswered. Therefore it isn't sufficient to just identify the microbes and the MVOCs, but it is essential to know the mechanism of action of these volatiles and to know what would be the best method of application of these chemicals in the field. Following are some questions that require answering and elucidation before MVOCs translate into agents of plant defense in sustainable agriculture.

1. How are these MVOCs perceived by plant or other microbes?
2. What are the benefits derived from these perceptions?
3. How will the perception of MVOCs benefit the plants?
4. What are the pathways involved in the perception of these MOVCs?
5. Are there multiple pathways involved in each host perception of MVOCs?
6. Are there distinct classes of MVOCs that are responsible for plant defense and plant growth?
7. Are the plants able to distinguish between MVOCs emitted by the beneficial microbes and that of pathogens or pests?
8. What makes the effect of MVOCs on host specific?
9. What is the scope of utilization for MVOCs?

References

Alifano P, Del Giudice L et al (2010) Microbes at work in perfumery: the microbial community of vetiver root and its involvement in essential oil biogenesis. Flavour Fragr J 25:121–122. doi:10.1002/ffj.1978

Alimi M, Soleimani MJ, Darzi MT (2012) Characterization and application of microbial antagonists for control of fusarium head blight of wheat caused by *Fusarium graminearum* using single and mixture strain of antagonistic bacteria on resistance and susceptible cultivars. Afr J Microbiol Res 6:326–334

Aponte M, Boscaino F et al (2013) Volatile compounds and bacterial community dynamics of chestnut-flour-based sourdoughs. Food Chem 141:2394–2404. doi:10.1016/j.foodchem.2013.05.052

Audrain B, Farag MA, Ryu CM et al (2015) Role of bacterial volatile compounds in bacterial biology. FEMS Microbiol Rev 39:222–233. doi:10.1093/femsre/fuu013

Bailly A, Weisskopf L (2012) The modulating effect of bacterial volatiles on plant growth: current knowledge and future challenges. Plant Signal Behav 7:79–85

Bailly A, Groenhagen U, Schulz S et al (2014) The inter-kingdom volatile signal indole promotes root development by interfering with auxin signalling. Plant J 80:758–771. doi:10.1111/tpj.12666

Berrada I, Benkhemmar O, Swings J et al (2012) Selection of halophilic bacteria for biological control of tomato gray mould caused by *Botrytis cinerea*. Phytopathol Mediterr 51:625–630

Bhattacharyya P, Jha D (2012) Plant growth-promoting rhizobacteria (PGPR): emergence in agriculture. World J Microbiol Biotechnol 28:1327–1350. doi:10.1007/s11274-011-0979-9

Bitas V, Kim HS et al (2013) Sniffing on microbes: diverse roles of microbial volatile organic compounds in plant health. Mol Plant-Microbe Interact 26:835–843. doi:10.1094/MPMI-10-12-0249-CR

Blom D, Fabbri C et al (2011) Production of plant growth modulating volatiles is widespread among rhizo-sphere bacteria and strongly depends on culture conditions. Environ Microbiol 3:3047–3058. doi:10.1111/j.1462-2920.2011.02582.x

Bordiec S, Paquis S, Lacroix H et al (2010) Comparative analysis of defence responses induced by the endophytic plant growth-promoting rhizobacterium Burkholderia phytofirmans strain PsJN and the non-host bacterium Pseudomonas syringae pv. pisi in grapevine cell suspensions. J Exp Bot 62:595–603

Bos LD, Sterk PJ, Schultz MJ (2013) Volatile metabolites of pathogens: a systematic review. PLoS Pathog 9:e1003311. doi:10.1371/journal.ppat.1003311

Bunge M, Araghipour N et al (2008) On-line monitoring of microbial volatile metabolites by proton transfer reaction-mass spectrometry. Appl Environ Microbiol 74:2179–2186. doi:10.1128/AEM.02069-07

Chernin L, Toklikishvili N et al (2011) Quorum-sensing quenching by rhizobacterial volatiles. Environ Microbiol Rep 3:698–704. doi:10.1111/j.1758-2229.2011.00284.x

Cho SM, Kang BR et al (2008) 2R, 3R-butanediol, a bacterial volatile produced by *Pseudomonas chlororaphis* O6, is involved in induction of systemic tolerance to drought in *Arabidopsis thaliana*. Mol Plant Microbe Interact 21:1067–1075. doi:10.1094/MPMI-21-8-1067

Choi GJ, Choi YH, Kim JC et al (2008) Nematicidal activity of malabaricones isolated from Myristica malabarica fruit rinds against *Bursaphelenchus xylophilus*. Nematol 10:801–807

Contreras-Cornejo H, Macías-Rodríguez L, Herrera-Estrella A et al (2014) The 4-phosphopantetheinyl transferase of Trichoderma virens plays a role in plant protection against Botrytis cinerea through volatile organic compound emission. Plant Soil 379:261–274. doi: 10.1007/s11104-014- 2069-x

Cortes-Barco A, Goodwin P, Hsiang T (2010a) Comparison of induced resistance activated by benzothiadiazole, (2R,3R)-butanediol and anisoparaffin mixture against anthracnose of *Nicotiana benthamiana*. Plant Pathol 59:643–653. doi:10.1111/j.1365-3059.2010.02283.x

Cortes-Barco A, Hsiang T, Goodwin P (2010b) Induced systemic resistance against three foliar diseases of *Agrostis stolonifera* by (2R,3R)-butanediol or an isoparaffin mixture. Ann Appl Biol 157:179–189. doi:10.1111/j.1744-7348.2010.00417.x

D'Alessandro M, Erb M, Ton J et al (2014) Volatiles produced by soil-borne endophytic bacteria increase plant pathogen resistance and affect tritrophic interactions. Plant Cell Environ 37:813–826. doi:10.1111/pce.12220

Dandurishvili N, Toklikishvili N et al (2011) Broad-range antagonistic rhizobacteria *Pseudomonas fluorescens* and *Serratia plymuthica* suppress agrobacterium crown gall tumours on tomato plants. J Appl Microbiol 110:341–352. doi:10.1111/j.1365-2672.2010.04891.x

Davis TS, Crippen TL et al (2013) Microbial volatile emissions as insect semiochemicals. J Chem Ecol 39:840–859. doi:10.1007/s10886-013-0306-z

Del Giudice L, Massardo DR et al (2008) The microbial community of vetiver root and its involvement into essential oil biogenesis. Environ Microbiol 10:2824–2841. doi:10.1111/j.1462-2920.2008.01703.x

Dolch M, Hornuss C et al (2012) Volatile compound profiling for the identification of gram-negative bacteria by ion-molecule reaction-mass spectrometry. J Appl Microbiol 113:1097–1105. doi:10.1111/j.1365-2672.2012.05414.x

Driouich A, Follet-Gueye M-L, Vicré-Gibouin M, Hawes M (2013) Root border cells and secretions as critical elements in plant host defense. Curr Opin Plant Biol 16:489–495. doi:10.1016/j.pbi.2013.06.010

Effmert U, Kalderas J et al (2012) Volatile mediated interactions between bacteria and fungi in the soil. J Chem Ecol 38:665–703. doi:10.1007/s10886-012-0135-5

Elshafie HS, Camele I, Racioppi R et al (2012) In vitro antifungal activity of *Burkholderia gladioli pv. agaricicola* against some phytopathogenic fungi. Int J Mol Sci 13:16291–16302

Ezquer I, Li J, Ovecka M et al (2010) Microbial volatile emissions promote accumulation of exceptionally high levels of starch in leaves in mono-and dicotyledonous plants. Plant Cell Physiol 51:1674–1693. doi:10.1093/pcp/pcq126

Farag MA, Ryu C-M, Sumner LW, Pare PW (2006) GC-MS SPME profiling of rhizobacterial volatiles reveals prospective inducers of growth promotion and induced systemic resistance in plants. Phytochemistry 67:2262–2268

Farag M, Zhang H, Ryu CM (2013) Dynamic chemical communication between plants and bacteria through airborne signals: induced resistance by bacterial volatiles. J Chem Ecol 39:1007–1018. doi:10.1007/s10886-013-0317-9

Fernando WGD, Ramarathnam R, Krishnamoorthy AS et al (2005) Identification and use of potential bacterial organic antifungal volatiles in biocontrol. Soil Biol Biochem 37:955–964

Fiddaman PJ, Rossall S (1994) Effect of substrate on the production of antifungal volatiles from *Bacillus subtilis*. J Appl Bacteriol 76:395–405

Fiedler K, Schutz E, Geh S (2001) Detection of microbial volatile organic compounds (MVOCs) produced by moulds on various materials. Int J Hyg Environ Health 204:111–121. doi:10.1078/1438-4639-00094

Garbeva P, Hordijk C et al (2014a) Volatile-mediated interactions between phylogenetically different soil bacteria. Front Microbiol 5:289. doi:10.3389/fmicb.2014.00289

Garbeva P, Hordijk C et al (2014b) Volatiles produced by the mycophagous soil bacterium Collimonas. FEMS Microbiol Ecol 87:639–649. doi:10.1111/1574-6941.12252

Glare T, Caradus J, Gelernter W et al (2012) Have biopesticides come of age? Trends Biotechnol 30:250–258. doi:10.1016/j.tibtech.2012.01.003

Goswami D, Vaghela H, Parmar S et al (2013) Plant growth promoting potentials of *Pseudomonas* spp. strain OG isolated from marine water. J Plant Interact 8:281–290. doi:10.1080/17429145.2013.768360

Gray CM, Monson RK, Fierer N (2010) Emissions of volatile organic compounds during the decomposition of plant litter. J Geophys Res Biogeosci 115:G03015

Groenhagen U, Baumgartner R, Bailly A et al (2013) Production of bioactive volatiles by different *Burkholderia ambifaria* strains. J Chem Ecol 39:892–906. doi:10.1007/s10886-013-0315-y

Gutierrez-Luna FM, Lopez-Bucio J, Tamirano-Hernandez J et al (2010) Plant growth-promoting rhizobacteria modulate root-system architecture in *Arabidopsis thaliana* through volatile organic compound emission. Symbiosis 51:75–83. doi:10.1007/s13199-010-0066-2

Han SH, Lee SJ, Moon JH et al (2006) GacS-dependent production of 2R,3R-butanediol by *Pseudomonas chlororaphis* O6 is a major determinant for eliciting systemic resistance against *Erwinia carotovora* but not against *Pseudomonas syringae pv. tabaci* in tobacco. Mol Plant-Microbe Interact 19:924–930. doi:10.1094/MPMI-19-0924

Heil M (2001) The ecological concept of costs of induced systemic resistance (ISR). Eur J Plant Pathol 107:137–146. doi:10.1023/A:1008793009517

Heil M, Baldwin IT (2002) Fitness costs of induced resistance: emerging experimental support for a slippery concept. Trends Plant Sci 7:61–67. doi:10.1016/S1360-1385(01)02186-0

Hol W, Bezemer T, Biere A (2013) Getting the ecology into interactions between plants and the plant growth-promoting bacterium *Pseudomonas fluorescens*. Front Plant Sci 4:81. doi:10.3389/fpls.2013.00081

Huang CJ, Tsay JF, Chang SY (2012) Dimethyl disulfide is an induced systemic resistance elicitor produced by Bacillus cereus C1L. Pest Manag Sci 68:1306–1310. doi: 10.1002/ps.3301

Hunziker L, Bonisch D, Groenhagen U et al (2015) Pseudomonas strains naturally associated with potato plants produce volatiles with high potential for inhibition of *Phytophthora infestans*. Appl Environ Microbiol 81:821–830. doi:10.1128/AEM.02999-14

Hwang J, Chilton WS, Benson DM (2002) Pyrrolnitrin production by Burkholderia cepacia and biocontrol of Rhizoctonia stem rot of poinsettia. BiolControl 25:56–63

Insam H, Seewald MS (2010) Volatile organic compounds (VOCs) in soils. Biol Fertil Soils 46:199–213. doi:10.1007/s00374-010-0442-3

Juenger M, Vautz W, Kuhns M et al (2012) Ion mobility spectrometry for microbial volatile organic compounds: a new identification tool for human pathogenic bacteria. Appl Microbiol Biotechnol 93:2603–2614. doi:10.1007/s00253-012-3924-4

Kai M, Effmert U, Berg G et al (2007) Volatiles of bacterial antagonists inhibit mycelial growth of the plant pathogen *Rhizoctonia solani*. Arch Microbiol 187:351–360. doi:10.1007/s00203-006-0199-0

Kai M, Haustein M, Molina F et al (2009) Bacterial volatiles and their action potential. Appl Microbiol Biotechnol 81:1001–1012. doi:10.1007/s00253-008-1760-3

Kai M, Crespo E, Cristescu SM et al (2010) *Serratia odorifera*: analysis of volatile emission and biological impact of volatile compounds on *Arabidopsis thaliana*. Appl Microbiol Biotechnol 88:965–976. doi:10.1007/s00253-010-2810-1

Kanchiswamy CN, Malnoy M, Maffei ME (2015) Bioprospecting bacterial and fungal volatiles for sustainable agriculture. Trends Plant Sci. doi:10.1016/j.tplants.2015.01.004

Kishimoto K, Matsui K, Ozawa R et al (2007) Volatile1-octen-3-ol induces a defensive response in *Arabidopsis thaliana*. J Gen Plant Pathol 73:35–37. doi:10.1007/s10327-006-0314-8

Kunze N, Goepel J, Kuhns M et al (2013) Detection and validation of volatile metabolic patterns over different strains of two human pathogenic bacteria during their growth in a complex medium using multi-capillary column-ion mobility spectrometry (MCC-IMS). Appl Microbiol Biotechnol 97:3665–3676

Kuske M, Romain AC, Nicolas J (2005) Microbial volatile organic compound as indicator of fungi. Can an electronic nose detect fungi in indoor environments? Build Environ 40:824–831. doi:10.1016/j.buildenv.2004.08.012

Kwon YS, Ryu CM, Lee S et al (2010) Proteome analysis of *Arabidopsis* seedlings exposed to bacterial volatiles. Planta 232:1355–1370

LeBouf RF, Schuckers SA, Rossner A (2010) Preliminary assessment of a model top mold contamination based on microbial volatile organic compound profiles. Sci Total Environ 408:3648–3653. doi:10.1016/j.scitotenv.2010.04.054

Lee B, Farag MA, Park HB (2012) Induced resistance by along-chain bacterial volatile: elicitation of plant systemic defense by a C13 volatile produced by *Paenibacillus polymyxa*. PLoS ONE 7:e48744. doi:10.1371/journal.pone.0048744

Leff JW, Fierer N (2008) Volatile organic compound (VOC) emissions from soil and litter samples. Soil Biol Biochem 40:1629–1636. doi:10.1016/j.soilbio.2008.01.018

Lemfack MC, Nickel J, Dunkel M et al (2014) mVOC: a database of microbial volatiles. Nucleic Acids Res 42:D744–D748. doi:10.1093/nar/gkt1250

Leroy PD, Sabri A, Heuskin S et al (2011) Microorganisms from aphid honeydew attracts and enhance the efficacy of natural enemies. Nat Commun 2:348. doi:10.1038/ncomms1347

Li HF, Imai T, Ukita M et al (2004) Compost stability assessment using a secondary metabolite: geosmin. Environ Technol 25:1305–1312. doi:10.1080/09593332508618374

Liang H, Zhang X, Jun RAO et al (2008) Microbial volatile organic compounds: generation pathways and mass spectrometric detection. J Chin Biotechnol 28:124–133

Ligon JM, Hill DS, Hammer PE et al (2000) Natural products with antifungal activity from *Pseudomonas biocontrol* bacteria. Pest Manag Sci 56:688–695. doi:10.1002/1526 4998(200008). 56:8<688::AID-PS186>3.0.CO;2-V

Maffei M (2002) Vetiveria, the genus Vetiveria. Taylor & Francis, London. doi:10.4324/9780203218730

Maffei ME, Gertsch J, Appendino G (2011) Plant volatiles: production, function and pharmacology. Nat Prod Rep 28:1359–1380. doi:10.1039/c1np00021g

Marmulla R, Harder J (2014) Microbial mono terpene transformations – a review. Front Microbiol 5:346. doi:10.3389/fmicb.2014.00346

Mcneal KS, Herbert BE (2009) Volatile organic metabolites as indicators of soil microbial activity and community composition shifts. Soil Sci Soc Am J 73:579–588. doi:10.2136/sssaj2007.0245

Meldau DG, Meldau S, Hoang LH et al (2013) Dimethyl disulfide produced by the naturally associated bacterium *Bacillus* sp. B55 promotes *Nicotiana attenuate* growth by enhancing sulfur nutrition. Plant Cell 25:2731–2747. doi:10.1105/tpc.113.114744

Mendes R, Garbeva P, Raaijmakers JM (2013) The rhizosphere microbiome: significance of plant beneficial, plant pathogenic, and human pathogenic microorganisms. FEMS Microbiol Rev 37:634–663. doi:10.1111/1574-6976.12028

Minerdi D, Bossi S, Gullino ML et al (2009) Volatile organic compounds: a potential direct long-distance mechanism for antagonistic action of *Fusarium oxysporum* strain MSA35. Environ Microbiol 11:844–854. doi:10.1111/j.1462-2920.2008.01805.x

Minerdi D, Bossi S, Maffei ME et al (2011) *Fusarium oxysporum* and its bacterial consortium promote lettuce growth and expansin A5 gene expression through microbial volatile organic compound (MVOC) emission. FEMS Microbiol Ecol 76:342–351. doi:10.1111/j.1574-6941.2011.01051.x

Mirmajlessi SM, Mänd M, Najdabbasi N et al (2016) Screening of native Trichoderma harzianum isolates for their ability to control Verticillium wilt of strawberry. Zemdirbyste-Agricul 103:397–404. doi:10.13080/z-a.2016.103.051

Morales-Rodriguez C, Picon-Toro J, Palo C, Palo EJ et al (2012) In vitro inhibition of mycelial growth of *Phytophthora nicotianae* from different hosts by Brassicaceae species. Effect of the developmental stage of the biofumigant plants. Pest Manag Sci 68:1317–1322. doi:10.1002/ps.3310

Muller T, Thissen R, Braun S et al (2004a) MVOC and composting facilities-part 2: MVOC dispersal in the environment. Environ Sci Pollut Res 11:152–157. doi:10.1007/BF02979669

Muller T, Thissen R, Braun S et al (2004b) MVOC and composting facilities. Part 1: MVOC emissions from municipal biowaste and plant refuse. Environ Sci Pollut Res 11:91–97. doi:10.1007/BF02979708

Nadarajah K (2016) Rhizosphere interactions: life below ground. In: Choudhary KD, Varma A (eds) Plant microbe interactions: an approach to sustainable agriculture. Springer, Singapore, pp 3–24. doi:10.1007/978-981-10-2854-0

Naznin HA, Kiyohara D, Kimura M et al (2014) Systemic resistance induced by volatile organic compounds emitted by plant growth-promoting fungi in *Arabidopsis thaliana*. PLoS ONE 9:e86882. doi:10.1371/journal.pone.0086882

Neal AL, Ton J (2013) Systemic defense priming by *Pseudomonas putida* KT2440 in maize depends on benzoxazinoid exudation from the roots. Plant Signal Behav 8(1):e22655

Ng WL, Bassler BL (2009) Bacterial quorum-sensing network architectures. Annu Rev Genet 43(197):222. doi:10.1146/annurev-genet-102108-134304

O'Hara M, Mayhew C (2009) A preliminary comparison of volatile organic compounds in the headspace of cultures of *Staphylococcus aureus* grown in nutrient, dextrose and brain heart bovine broths measured using a proton transferre action mass spectrometer. J Breath Res 3:027001

Ortiz-Castro R, Contreras-Cornejo HA et al (2009) The role of microbial signals in plant growth and development. Plant Signal Behav 4:701–712. doi:10.4161/psb.4.8.9047

Ortiz-Castro R, Martinez-Trujillo M, Lopez-Bucio J (2008) N-acyl-L homoserine lactones: a class of bacterial quorum-sensing signals alter post-embryonic root development in *Arabidopsis thaliana*. Plant Cell Environ 31:1497–1509. doi:10.1111/j.1365-3040.2008.01863.x

Paré PW, Farag MA, Krishnamachari V et al (2005) Elicitors and priming agents initiate plant defense responses. Photosyn Res 85:149–159

Park HB, Lee B, Kloepper JW et al (2013) One shot-two pathogens blocked: exposure of *Arabidopsis* to hexadecane, along chain volatile organic compound, confers induced resistance against both *Pectobacterium carotovorum* and *Pseudomonas syringae*. Plant Signal Behav 8:e24619. doi:10.4161/psb.24619

Penuelas J, Asensio D, Tholl D et al (2014) Biogenic volatile emissions from the soil. Plant Cell Environ 37:1866–1891. doi:10.1111/pce.12340

Perl T, Juenger M, Vautz W et al (2011) Detection of characteristic metabolites of *Aspergillus fumigatus* and *Candida species* using ion mobility spectrometry-metabolic profiling by volatile organic compounds. Mycoses 54:E828–E837. doi:10.1111/j.1439-0507.2011.02037.x

Piechulla B, Degenhardt J (2014) The emerging importance of microbial volatile organic compounds. Plant Cell Environ 37:811–812

Pineda A, Zheng SJ, van Loon JJA et al (2010) Helping plants to deal with insects: the role of beneficial soil-borne microbes. Trends in Plant Sci 15:507–514

Pineda ANA, Soler R, Weldegergis BT et al (2013) Non-pathogenic rhizobacteria interfere with the attraction of parasitoids to aphid-induced plant volatiles via jasmonic acid signalling. Plant Cell Environ 36:393–404. doi:10.1111/j.1365-3040.2012.02581.x

Popova AA, Koksharova OA, Lipasova VA et al (2014) Inhibitory and toxic effects of volatiles emitted by strains of *Pseudomonas* and *Serratia* on growth and survival of selected microorganisms, *Caenorhabditis elegans*, and *Drosophila melanogaster*. Biomed Res Int 2014:11. doi:10.1155/2014/125704

Ramirez KS, Lauber CL, Fierer N (2010) Microbial consumption and production of volatile organic compounds at the soil-litter interface. Biogeochemistry 99:97–107. doi:10.1007/s10533-009-9393-x

Rasmann S, Köllner TG, Degenhardt J et al (2005) Recruitment of entomopathogenic nematodes by insect-damaged maize roots. Nature 434:732–737

Rasanen RM, Hakansson M, Viljanen M (2010) Differentiation of air samples with and without microbial volatile organic compounds by aspiration ion mobility spectrometry and semiconductor sensors. Build Environ 45:2184–2191. doi:10.1016/j.buildenv.2010.03.018

Rasmussen TB, Givskov M (2006) Quorum-sensing inhibitors as anti-pathogenic drugs. Int J Med Microbiol 296:149–161. doi:10.1016/j.ijmm.2006.02.005

Romoli R, Papaleo MC, de Pascale D et al (2011) Characterization of the volatile profile of Antarctic bacteria by using solid-phase microextraction-gas chromatography-mass spectrometry. J. Mass Spectrom 46:1051–1059. doi:10.1002/jms.1987

Romoli R, Papaleo M, De Pascale D et al (2014) GC MS volatolomic approach to study the antimicrobial activity of the Antarctic bacterium *Pseudoalteromonas* sp TB41. Metabolomics 10:42–51. doi:10.1007/s11306-013-0549-2

Rudrappa T, Biedrzycki ML, Kunjeti SG, Donofrio NM et al (2010) The rhizobacterial elicitor acetoin induces systemic resistance in *Arabidopsis thaliana*. Commun Integ Biol 3:130–138. doi:10.4161/cib.3.2.10584

Ryan RP, Dow J (2008) Diffusible signals and interspecies communication in bacteria. Microbiol Sgm 154:1845–1858. doi:10.1099/mic.0.2008/017871-0

Ryu CM, Farag MA, Hu CH et al (2003) Bacterial volatiles promote growth in *Arabidopsis*. Proc Natl Acad Sci U S A 100:4927–4932. doi:10.1073/pnas.0730845100

Ryu CM, Farag MA, Hu CH et al (2004) Bacterial volatiles induce systemic resistance in *Arabidopsis*. Plant Physiol 134:1017–1026. doi:10.1104/pp.103.026583

Scala A, Allmann S, Mirabella R et al (2013) Green leaf volatiles: a Plant's multifunctional weapon against herbivores and pathogens. Int J Mol Sci 14:17781–17811. doi:10.3390/ijms140917781

Schmidt R, Cordovez V, DeBoer W et al (2015) Volatile affairs in microbial interactions. ISMEJ 9:2329–2335

Scotter JM, Langford VS, Wilson PF et al (2005) Real-time detection of common microbial volatile organic compounds from medically important fungi by selected ion flowtube-mass spectrometry (SIFT-MS). J Microbiol Methods 63:127–134. doi:10.1016/j.mimet.2005.02.022

Shannon U et al (2012) Fungal volatile organic compounds: a review with emphasis on their biotechnological potential. Fungal Biol Rev 26:73–83

Song GC, Ryu CM (2013) Two volatile organic compounds trigger plant self defense against a bacterial pathogen and a sucking insect in cucumber under open field conditions. Int J Mol Sci 14:9803–9819. doi:10.3390/ijms14059803

Stahl PD, Parkin TB (1996) Microbial production of volatile organic compounds in soil microcosms. Soil Sci Soc Am J 60:821–828. doi:10.2136/sssaj1996.03615995006000030020x

Sundberg C, Yu D, Franke-Whittle I et al (2013) Effects of pH and microbial composition on odour in food waste composting. Waste Manag 33:204–211. doi:10.1016/j.wasman.2012.09.017

Tenorio-Salgado S, Tinoco R, Vazquez-Duhalt R et al (2013) Identification of volatile compounds produced by the bacterium *Burkholderia tropica* that inhibit the growth of fungal pathogens. Bioengineered 4:236–243. doi:10.4161/bioe.23808

Thorn RMS, Greenman J (2012) Microbial volatile compounds in health and disease conditions. J Breath Res 6:024001. doi:10.1088/1752-7155/6/2/024001

Tirranen LS, Gitelson I (2006) The role of volatile metabolites in microbial communities of the LSS higher plant link. Adv Space Res 38:1227–1232. doi:10.1016/j.asr.2006.02.038

Verginer M, Leitner E, Berg G (2010) Production of volatile metabolites by grape-associated microorganisms. J Agric Food Chem 58:8344–8350. doi:10.1021/jf100393w

Vespermann A, Kai M, Piechulla B (2007) Rhizobacterial volatiles affect the growth of fungi and *Arabidopsis thaliana*. Appl Environ Microbiol 73:5639–5641. doi:10.1128/AEM.01078-07

Wan M, Li G, Zhang J et al (2008) Effect of volatile substances of Streptomyces platensis F-1 on control of plant fungal diseases. Biol Control 46:552–559

Waters CM, Bassler BL (2005) Quorum sensing: cell-to-cell communication in bacteria. Annu Rev Cell Dev Biol 21:319–346. doi:10.1146/annurev.cellbio.21.012704.131001

Weise T, Kai M, Gummesson A et al (2012) Volatile organic compounds produced by the phytopathogenic bacterium *Xanthomonas campestris* pv. *vesicatoria* 85–10. Beilstein J Org Chem 8:579–596. doi:10.3762/bjoc.8.65

Weise T, Kai M, Piechulla B (2013) Bacterial ammonia causes significant plant growth inhibition. PLoS ONE 8:e63538. doi:10.1371/jour-nal.pone.0063538

Weisskopf L, Bailly A (2013) Plant growth modulation by bacterial volatiles—a focus on Burkholderia species. In de Bruijn FJ (ed) Molecular microbial ecology of the rhizosphere. Wiley, New York, pp 665–675

Weisskopf L (2013) The potential of bacterial volatiles for crop protection against phytopathogenic fungi. In: Méndez-Vilas A (ed) Microbial pathogens and strategies for combating them: science, technology and education, pp 1352–1363

Wenke K, Kai M, Piechulla B (2010) Below ground volatiles facilitate interactions between plant roots and soil organisms. Planta 231:499–506. doi:10.1007/s00425-009-1076-2

Wenke K, Wanke D, Kilian J et al (2012) Volatiles of two growth-inhibiting rhizobacteria commonly engage AtWRKY18 function. Plant J 70:445–459. doi:10.1111/j.1365-313X.2011.04891.x

Wheatley RE (2002) The consequences of volatile organic compound mediated bacterial and fungal interactions. Anton Van Leeuwenhoek Int J Gen Mol Microbiol 81:357–364. doi:10.1023/A:1020592802234

Wihlborg R, Pippitt D, Marsili R (2008) Headspace sorptive extraction and GC-TOFMS for the identification of volatile fungal metabolites. J Microbiol Methods 75:244–250. doi:10.1016/j.mimet.2008.06.011

Xu CK, Mo MH, Zhang LM et al (2004) Soil volatile fungistasis and volatile fungistatic compounds. Soil Biol Biochem 36:1997–2004

Yamagiwa Y, Inagaki Y, Ichinose Y et al (2011) *Talaromyces wortmannii* FS2 emits beta-caryophyllene, which promotes plant growth and induces resistance. J Gen Plant Pathol 77:336–341. doi:10.1007/s10327-011-0340-z

Yuan Z, Chen Y, Xu B, Zhang C (2012) Current perspectives on the volatile-producing fungal endophytes. Crit Rev Biotechnol 32:363–373. doi:10.3109/07388551.2011.651429

Yuan J, Raza W, Huang Q, Shen Q (2011) Quantification of the antifungal lipopeptide iturin A by high performance liquid chromatography coupled with aqueous two-phase extraction. J Chromatogr B 879:2746–2750

Zhang H, Kim MS, Krishnamachari V et al (2007) Rhizobacterial volatile emissions regulate auxin homeostasis and cell expansion in Arabidopsis. Planta 226:839–851

Zhang H, Kim MS, Sun Y et al (2008) Soil bacteria confer plant salt tolerance by tissue-specific regulation of the sodium transporter HKT1. Mol Plant Microbe Interact 21:737–744

Zhang X, Li B, Wang Y et al (2013) Lipopeptides, a novel protein, and volatile compounds contribute to the antifungal activity of the biocontrolagent *Bacillus atrophaeus* CAB-1. Appl Microbiol Biotechnol 97:9525–9534. doi:10.1007/s00253-013-5198-x

Zou C, Li Z, Yu D (2010) *Bacillus megaterium* strain XTBG34 promotes plant growth by producing 2-pentylfuran. J Microbiol 48:460–466. doi:10.1007/s12275-010-0068-z

Zou CS, Mo MH, Gu YQ et al (2007) Possible contributions of volatile-producing bacteria to soil fungistasis. Soil Biol Biochem 39:2371–2379. doi:10.1016/j.soilbio.2007.04.009